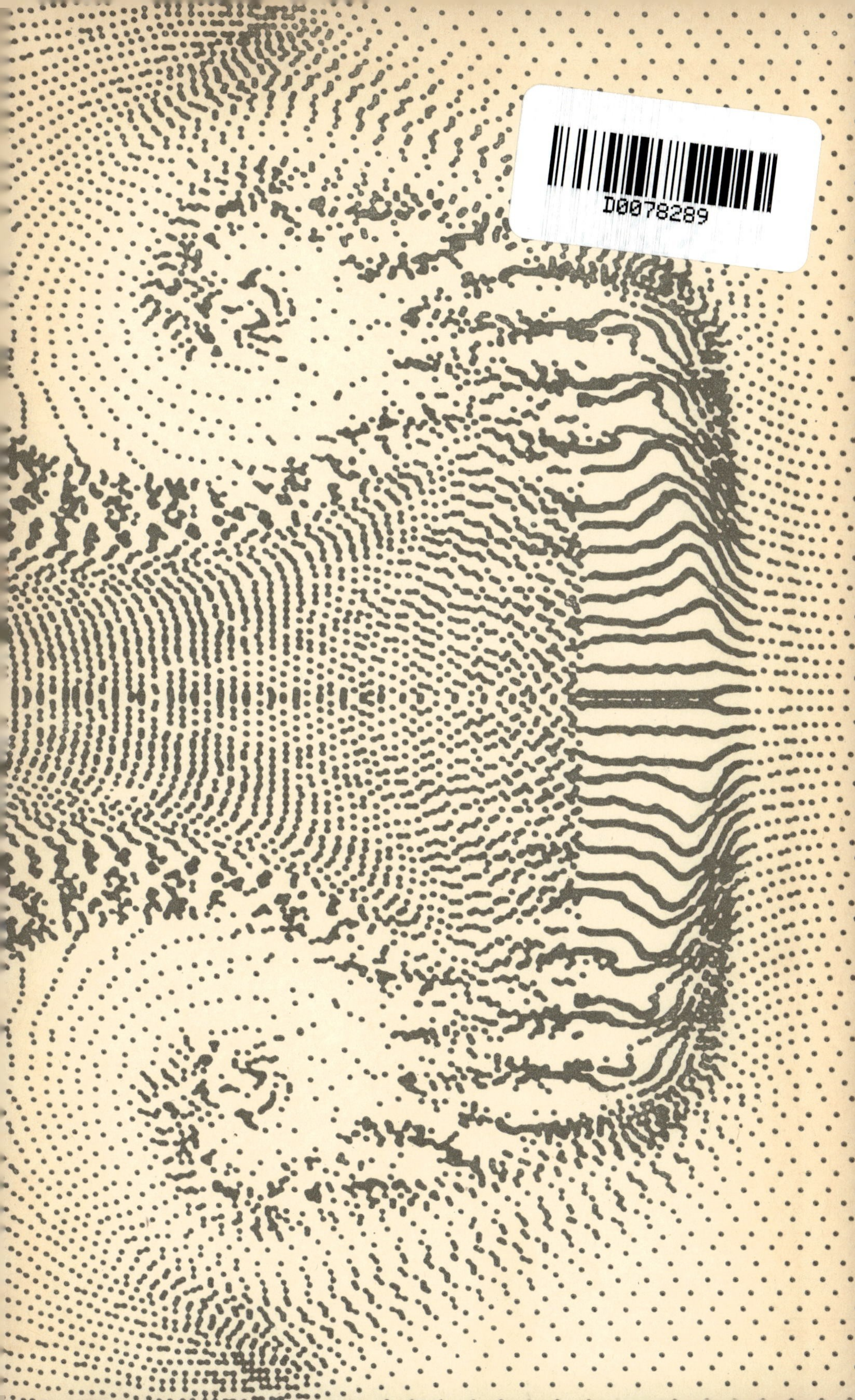
D0078289

COMPUTERS AND THEIR ROLE IN THE PHYSICAL SCIENCES

COMPUTERS
AND THEIR ROLE IN THE
PHYSICAL SCIENCES

Edited by

S. Fernbach and **A. Taub**

University of California

GORDON AND BREACH SCIENCE PUBLISHERS

New York **London** **Paris**

Copyright © 1970 by
Gordon and Breach, Science Publishers, Inc.
150 Fifth Avenue
New York, N.Y. 10011

Editorial office for the United Kingdom
Gordon and Breach, Science Publishers Ltd.
12 Bloomsbury Way
London W.C. 1

Editorial office for France
Gordon & Breach
7–9 rue Emile Dubois
Paris 14[e]

Library of Congress catalog card number 70–111943. ISBN 0 677 14030 4. Printed in East Germany.

Preface

MODERN DIGITAL COMPUTERS have been developed and used extensively for less than two decades. They have had a very important impact on mathematics and the physical sciences in that they have allowed the practitioners of these subjects to carry out computations many orders of magnitude more complex than those possible twenty years ago. Even more important they have provided scientists with new experimental tools in three ways: (1) Computers have been incorporated in the experimental data gathering and control mechanisms; (2) They have been used in novel data reduction procedures; and (3) They have provided the theorist a tool for testing various hypotheses more easily and economically than is possible through laboratory experiments.

In addition, and perhaps more important than the above, because computers and their use involves mathematical and logical techniques quite different from those based on mathematical continuum theory, scientists have been stimulated to give new mathematical formulations of their problems. These new techniques do not use equations which in turn have to be approximated for the computers but rather are logically set up to match the logic of the machine.

One of our purposes in presenting the collection of articles given below was to illustrate to the non-expert reader the various ways, mentioned above, that the computer has affected mathematics and the physical sciences. We were encouraged to do so by the Division of Research of the U.S. Atomic Energy Commission and in particular its Mathematics and Computers Branch. The Commission's research and development program has made extensive use of computers and in turn has provided great stimulation to the research and development of computers. It is another of our purposes to illustrate for the non-expert the interdependence of these two research and development programs.

These two purposes, and the willingness of authors to write expository articles have dictated the selection of articles presented herewith. The contribution by Hans Bethe which also serves as an introduction to this book provides a picture of the early use of computers in the Manhattan project. The next two articles are historical, "Pre-Electronic Aids to Digital Computation" by W. J. Worlton and "Early Electronic Computers" by Herman H. Goldstine, which are intended to give the reader an indication of the nature of aids to computation that were available before the stored program computer was invented and the story behind the invention of this device as seen by one of the participants.

The present day computer is referred to as the "third generation" computer whereas those that emerged from the period reviewed by Goldstine are referred to as "first generation" computers. This nomenclature attempts to describe an evolutionary process in the development of computers during which significant changes were made primarily in the components used in the logical organization. The second generation of computers differed from the first mainly in that the logical circuits were built from transistors instead of vacuum tubes, and memories were built from magnetic cores instead of oscilliscope tubes, mercury tanks, and other devices. During this phase of development, some new, interesting, and very useful devices such as "B boxes" or index registers were added. These made it easily possible to have the machine automatically change its sequence of operations depending on the results obtained during the course of the computation.

The first generation computers could do this, through straight forward programming, but the introduction of index registers made it so much easier that computer users reached a new level of sophistication in their development of computer programs. However, both first and second generation computers were otherwise quite similar in logical organization and had one feature in common: namely, they carried out one instruction at a time, even though it need not have arisen from a completely predetermined sequence.

Computer designers recognized that the existence and use of index registers made it possible for machines to have their parts so interconnected that various instructions could be carried out concurrently or overlapped instead of one at a time. The third generation computers differ from the second in that this feature is exploited. Primarily however, the third generation introduces the integrated circuits as basic building blocks.

The articles, "Computer Architecture" by J. E. Thornton and "Computer Memories" by Jan Rajchman describe present day computers, the problems involved in their design, and one of their most crucial components—their memories. They are followed by two articles describing the programs (software) needed for their use. Namely, the articles, "Computer Software" by Max Goldstein and Peter Capek, and "Software for Interaction with Computers" by J.I.Schwartz. The first of these deals with the organization of a problem for computer solution, the instructions the computer must have and the development of procedures by which the computer manages to process a stream of problems once they are formulated in a language it can understand. If it is do to this efficiently, a fairly rigid system must be devised and the user's interaction with the computer tends to become so formal that its usefulness as a scientific tool may be compromised. Thus there has arisen a great interest in the methods discussed in Schwartz's article for systems which allow for immediate interaction between the computer and the user. Since there is such a great discrepancy between the reaction time of the user and that of the computer, the software developers attempt to achieve an economical use of computers by allowing many users to share the computer during this reaction time period.

The article "Errors in Computing" by N.Metropolis discusses the errors involved in using a computer and stresses those that come about when the computer is used as an "arithmetic engine", namely a device for performing arithmetic at a prodigious rate. The dangers of "round-off", the absolutely necessary procedure of doing approximate arithmetic are pointed out and the necessity of devising schemes for recognizing and minimizing the effects of round-off are discussed.

John Pasta's article, "The Role of the University in the Computer Age", provides a transition from the discussion of computers, their history, description, software systems, and properties, to their impact on education and research as carried out in universities. The contribution "The Impact of Computers on Mathematics" by Peter D.Lax discusses the interaction between this subject, which for many years has been one of the primary tools of the theorist and the computer, the novel and powerful new tool whose potentiality is still not fully determined.

The four papers, "Monte Carlo Methods" by Malvin H.Kalos, "Impact on Statistical Mechanics" by B.J.Alder, "Plasma Physics" by Charles H.Birdsall and John M.Dawson, and "Computer Solutions in Continuum Mechanics" by Francis H.Harlow illustrate both the experimental use of the computer as a replacement and refinement of laboratory experiments and the results that can be achieved through the novel mathematical formulation of problems—a formulation more readily adaptable for computers than that which uses the conventional approach involving mathematical continuum theory.

The articles by Erwin Bareiss, "Computers and Reactor Design", by Charles L. Mader, "Numerical Calculation of Explosive Phenomena", and "Computers and the Optical Model for Nuclear Scattering" by T. Tamura and W. R. Coker illustrate the role that computers played in obtaining the numerical solution of the complex problems which lie at the core of the research and development programs in atomic energy.

D. Secrest in his article, "On-line Use of Computers in Chemistry", describes the ways in which computers have begun to aid chemists in controlling and conducting experiments, as well as data gathering, data reduction, and interpretation.

The role computers have played in the prodigious task of dealing with the photographs collected in experiments in high energy physics is discussed in the papers by James N. Snyder and L. Kowarski entitled "The Role of Computers in the Analysis of Bubble Chamber Film" and "Recognition of Visual Data in High Energy Physics", respectively. They point out the virtual impossibility of dealing with the vast amount of data produced without computer aids, and in addition the difficulty of designing completely automatic recognition systems.

The book closes with four survey articles which describe the impact that computers have had on the fields of molecular chemistry, geophysics, meterorology and astrophysics. These are the articles by Enrico Clementi, "Molecular Structures"; Bruce A. Bolt, "The Use of Computers in Studies of the Earth"; Akira Kasahara, "Computer Simulations of the Global Calculation of the Earth's Atmosphere"; and Icko Iben, Jr., "The Role of the Computer in Astrophysics". In each of these articles, there is to be found illustrations of the points made in the opening paragraphs of this preface concerning the ways in which computers have extended the scientists' capabilities.

It is evident from a reading of the contributions given in this book that the computer, which was originally developed as a sophisticated arithmetic engine, continues to be used as such, but at the same time plays a larger and larger role as a "logical engine"—i.e., a data analyzer and decision maker. As the theorist learns how to use the arithmetic engine, his demands for a more powerful one increase. Indeed ever since the first computers became available, the cry has arisen for a machine 100 times faster and with a memory capacity 100 times greater than the existing ones. Today's need for such more powerful computers is a real one, and their existence will aid in the solution of important technical problems.

At the same time, it is evident that in dealing with extremely large and complicated problems both the computer and the user are aided if they can interrupt each other and interact during the cause of a problem solution, instead of at preplanned intervals. Even when a computer is to be used primarily as an arithmetic engine, its logical engine capabilities must be comparable to those needed by a computer if it is to be used as a critical device.

Thus one can foresee a real continuing need for larger and faster computers with more sophisticated input–output devices—that is, devices which allow for ready communication between users and the machines without slowing down the latter. The development of such computers will continue to stimulate progress in the physical sciences and in addition will influence the development of almost all other fields.

Serious problems arise when one attempts to fulfill this need for there are limitations on the amount by which the speed of the elementary circuit elements can be increased and the speed with which various parts of a computer can communicate with each other. It should be remembered that a nanosecond (10^{-9} second) is the time it takes light to travel a distance of 30 centimeters, about a foot; and many present day computer circuits have operation time of the order of a few nanoseconds. Hence the time it takes an electrical signal to travel from one end of a computer which is a few feet long is comparable to the time it takes to generate the signal. There are, however, circuit design techniques for ameliorating this difficulty.

The computer architect may be able to satisfy the needs of computer users in spite of these physical limitations by exploiting novel logical organizations of computer parts. Computers are now under development which extend the principle of concurrent operation from relatively small portions of a system to larger ones. Such computers are referred to as parallel processors. They consist of a number of subcomputers, each one of which may communicate with its own as well as a common memory and carry out one portion of a single problem at the same time each other subcomputer is doing a similar task. Programmers are studying methods of formulating problems for solution by such machines. These developments may produce the computers that will answer the needs described below and those which arise in other areas.

Authors

Berni Alder	Lawrence Radiation Laboratory, Livermore, California
Erwin Bareiss	Argonne National Laboratory, Argonne, Illinois
Hans A. Bethe	Cornell University, Laboratory of Nuclear Studies, Ithaca, New York
Charles K. Birdsall	Electrical Engineering Department, University of California, Berkeley, California
Bruce A. Bolt	Department of Geology, University of California, Berkeley, California
Enrico Clementi	IBM Corporation, Research Division Laboratory, San Jose, California
William R. Coker	The University of Texas, Center for Nuclear Studies, Austin, Texas
Peter Capek	New York University, Courant Institute of Mathematical Sciences, New York, N.Y.
John M. Dawson	Plasma Physics Laboratory, Princeton University, James Forrestal Campus, Princeton, New Jersey
Max Goldstein	New York University, Courant Institute of Mathematical Sciences, New York, N.Y.
Herman H. Goldstine	I.B.M. Yorktown Heights, N.Y.
Francis H. Harlow	University of California, Los Alamos Scientific Laboratory, Los Alamos, New Mexico
Icko Iben, Jr.	Department of Physics, Massachusetts Institute of Technology, Cambridge, Massachusetts

Malvin H. Kalos	New York University, Courant Institute of Mathematical Sciences, New York, N.Y.
Akira Kasahara	National Center for Atmospheric Research, Boulder, Colorado
L. Kowarski	CERN, 1211 Geneva 23, Switzerland
Peter D. Lax	New York University, Courant Institute of Applied Mathematical Sciences, New York, N.Y.
Charles L. Mader	University of California, Los Alamos Scientific Laboratory, Los Alamos, New Mexico
Nicholas Metropolis	University of California, Los Alamos Scientific Laboratory, Los Alamos, New Mexico
John R. Pasta	Department of Computer Science, University of Illinois, Urbana, Illinois
Jan A. Rajchman	Staff Vice President, Data Processing Research, RCA Laboratories, Princeton, New Jersey
Jules I. Schwartz	King Resources Company, Computer Systems Division, Los Angeles, California
Donald Secrest	University of Illinois, Department of Chemistry & Chemical Engineering, The William Albert Noyes Laboratory, Urbana, Illinois
James N. Snyder	Department of Computer Science, University of Illinois at Urbana–Champaign, Urbana, Illinois
T. Tamura	The University of Texas, Center for Nuclear Studies, Austin, Texas
James E. Thornton	Control Data Corporation, Arden Hills, Minn.
William J. Worlton	University of California, Los Alamos Scientific Laboratory, Los Alamos, New Mexico

Table of Contents

CHAPTER 1

Introduction

HANS A. BETHE

BACK IN THE 1930's, I frequently visited the Physics Department of Columbia University. On one of these visits, I was taken up high in the building, to the Astronomy Department. They had a beautiful view of the Hudson River, but even more fascinating was the work of Professor Eckert: He was using an IBM computer to calculate the orbits of planets—asteroids, the earth itself, the moon. Gone were the frustrating attempts to solve the classical 3-body problem which had kept mathematicians and astronomers busy for over 150 years, gone were the attempts to put perturbation upon perturbation in finding the erratic motion of the moon. Instead of this, it was now possible to solve Newton's equations step by step, progressing by as small steps in time as necessary, and taking into account the attraction of as distant planets as desired. It was just a question of skill in writing the program, of time and of money to run the computer.

A few years later, at Los Alamos, we were faced with many numerical problems. The first was neutron diffusion: It was clear that the differential diffusion equation was not adequate for our purpose: The fast neutrons on which the atomic bomb was to operate, had a fission cross section σ_f almost as large as the scattering (transport) cross section σ_s, whereas the differential theory was based on the assumption $\sigma_f \ll \sigma_s$. So integral theory had to be used. This was rather easy for a bare sphere of fissionable material, not so easy but still soluble for a sphere surrounded by a spherical shell of inert, scattering material (reflector or "tamper").

But what about more complicated shapes? It was planned, in 1943, to assemble the bomb by a gun, shooting a cylinder of fissile material into a hollow cylinder of fissile material surrounded by reflector. In the process of this assembly, very complicated shapes would result. We wanted to know how neutrons would diffuse in such a complicated assembly, in order to assess the probability that the chain reaction might start prematurely, and the bomb explode with less than the full yield. Even in the final assembly, we might have a cylinder of fissile material rather than a sphere, because this would be much easier to fabricate: we wanted to know how much of the energy yield of the bomb we would lose by this. All these problems were insoluble by analytical means, and while we could set up integral equations describing the process, they were far too complicated to be solved by desk computing machines.

Fortunately, one of the members of the laboratory was Professor Dana Mitchell of Columbia University. He was in charge of procurement, and in this capacity, was a member of our Governing Board, which consisted of about 8 leaders of the various divisions of the laboratory, and Dr. Oppenheimer. In the weekly meetings of this Board, we reported on our problems and needs. When I mentioned the neutron diffusion computations, Dr. Mitchell immediately remembered the work of Dr. Eckert. After some discussion in the same meeting, we decided to get an IBM machine.

In those days, the machines could only be rented, and had to be constantly serviced and often repaired. Being at a very remote location, and behind many security fences, a Los Alamos machine would be difficult to service. Fortuitously, there was a solution: a repair man of the IBM company had been drafted into the Army, and Los Alamos was entitled to have army personnel assigned to it for technical work. So, on our request he came, about 3 days after the machines themselves.

The arrival of the machines was a great day. Richard Feynman, Eldred Nelson and Stanley Frankel had studied the manuals carefully. So when the machines arrived, in big boxes and disassembled into a thousand pieces, they set to work immediately to put them together. They were electromechanical machines in those days; the computation was done mechanically, much as in a desk calculator, the sensing was by an electrical contact through the holes in a punch card. The mechanical design was quite complicated, but Feynman, the mechanical wizard who had been repairing all the desk calculators in the laboratory whenever they went wrong, and the skillful Nelson and Frankel managed. When the official repairman arrived, he was most complimentary to our 3 physicist–mechanics. Nobody outside IBM had ever assembled a machine successfully. The repairman became a most valued and loyal member of our IBM computing group ever after.

In the meantime, the orientation of the laboratory had changed: the favored method of assembling the bomb was now the implosion. Von Neumann had given a strong impulse to this project, and had pointed out that the solid fissile material might be compressed by the implosive force thus reducing the amount of material needed—a most important point when plutonium was just beginning to be produced by the milligram, and when the critical mass was 6 kilograms!

How much compression could we expect? Metropolis, Teller and Feynman devised an equation of state for uranium and plutonium, interpolating between the Thomas–Fermi theory at high pressure, and experimental data at low pressure. (The solutions to the Thomas–Fermi differential equation at various densities were themselves obtained on the IBM machine.) But what about the hydrodynamics of the reflector and the fissile material when it was struck by a spherically converging detonation wave from the explosive surrounding the reflector? How, in fact, would the detonation wave itself behave? No analytical solution to these problems seemed possible, and in fact, none has been obtained to date, as far as I know.

For the computer, the problem was simple, at least in principle. Each element of mass moves according to the hydrodynamic equation

$$\frac{d^2 R}{dt^2} = -\frac{1}{\varrho}\frac{\partial p}{\partial R} \tag{1}$$

The pressure p is given by the equation of state,

$$p = p(\varrho, S) \tag{2}$$

where ϱ is the density and S the entropy. In the absence of shocks, S remains constant for a given element of mass. Then (1) is a simple second order differential equation whose solution gives $\varrho(R, t)$.

The first problem was to choose suitable intervals in space and time. In all calculations, we used Lagrange coordinates, i.e. we followed each element of mass in time. To make the calculation manageable, different intervals of mass had to be chosen in the interior and exterior parts of the problem, small intervals near the center in order to give sufficient detail, large intervals in the outer parts, especially in the reflector and the explosive, in order to keep down the total number of mass elements.

The time steps were at first chosen by intuition, but soon we noticed strange instabilities. Richard Courant, who was a consultant to Los Alamos, found the answer: the solution to the differential equation is stable if and only if

$$c\,\Delta t < \Delta r \tag{3}$$

where c is the local velocity of sound, Δt and Δr the steps in time and radius. This was the origin of the famous Courant criterion which is now observed in all numerical work on similar equations. It gave added reason to keep the mass intervals as large as compatible with sufficient detail in the description.

Another problem was the actual set-up of the calculation. The "program" consisted, in those days, of making electrical connections between conducting holes in a large, rectangular, insulating plate, the "plugboard". Especially Frankel became extremely expert at this. A good plugboard could save a factor two, five, ten in calculation time—quite important when the time for one problem might run into many weeks.

Finally, there was the problem of shock waves. The detonation of the explosive was clearly a shock, and this would send a shock wave into the reflector, then into the fissile material. In our wartime calculations, the Hugoniot conditions at the shock front were fitted by hand, at each time step in the calculation. The position of the shock front was also calculated by hand, using the shock velocity deduced from the Hugoniot equation at the previous time step. This was a somewhat laborious procedure; but each time step on the computer took about one hour, so that the hand calculation could easily keep step with the machine. The computer itself consisted of many separate machines: the multiplying machine itself, the card sorter, the key punch, the duplicator, the printer and two or three others. The cards had to be carried from one to the other to perform the operations needed for one step in time, and carrying them to the hand computer for shock fitting was just one more operation. After the war, Richtmyer and von Neumann developed their method of simulating a shock by an artificial viscosity: this made calculations involving shocks fully automatic, a necessity for modern computers.

Finally, about 3 months after it was started, the first implosion calculation was finished. The result was very satisfactory to us: the fissile material was strongly compressed. This showed that we could get a good energy yield by assembling a relatively small amount of fissile material. Essentially it proved that our project would be successful, provided we could produce a spherically symmetrical implosion. The Trinity Test at Alamogordo, on July 16, 1945, proved (among other things) that our calculation of compression by implosion was correct.

About a dozen implosion calculations were done before the end of the War, and the time per calculation gradually diminished. Nowadays, such calculations are routine, and take a minute or less.

Already during the War, the physicists in charge of the computer were very much tempted to "play" with it, to find new ways of solving a problem, and of saving space on a punch card so that more information could be stored on it. One of the chief aims in the latter connection was to have all the information for one mass point at one time step on a single card; clearly, this cut the computing time about in half, compared with two cards. Plugboards, of course, were also constantly improved and simplified.

Stanley Frankel was particularly good at this game, and became so enchanted that he sometimes forgot that the real aim of the project was to solve the implosion problem. In the end, in order to accelerate the implosion calculations, Frankel had to be transferred to another theoretical physics group. He was replaced by Nicholas Metropolis, and the whole IBM group was placed under the direction of Richard Feynman. But both Frankel and Metropolis retained their love for computers, and worked for many years after the War on their improvement, both in hardware and in logic. Under the direction of Feynman, Metropolis and Nelson, the IBM group, consisting of two dozen faithful coders and machine operators, accomplished all the work required by the Los Alamos Project.

John von Neumann was fascinated by the power of the computer. He foresaw that many problems in mathematical physics, which were too difficult for an analytical solution, would soon be solved by computers. One of his problems was the Mach reflection of an oblique shock wave from a solid surface which was of interest to Los Alamos because it is important for the effects of nuclear weapons.

Von Neumann also was convinced that computers could be greatly speeded up from the wartime models. Why use clumsy, mechanical devices to do the addition and multiplication? Clearly, electronics could be a thousand times faster or more. The big companies showed at first some reluctance to change their design so completely. So von Neumann master-minded an amateur project, at the Institute of Advanced Studies, with Government support, to build an electronic computer. A second machine, the Maniac (mechanical and numerical integrator and computer), was constructed simultaneously by Metropolis at Los Alamos, and was actually finished a little earlier than the Princeton machine. It got into operation in 1952, in time to be of great help in calculations for the world's first H-bomb, the "Mike" test in November

1952. Already in 1951, several electronic computers, of slightly smaller capacity, became available, at the National Bureau of Standards, The Rand Corporation and others. There was constant travel, in 1951 and 1952, between Los Alamos and these various computing facilities.

Soon afterwards, the IBM company released the first fully engineered electronic computer, the 701, followed by the 704, 709, 7090 and many more. Sperry Rand developed the Univac about the same time. From then on, private groups—universities, laboratories—no longer needed to engagein hardware development, nor could they possibly compete with the giants, except perhaps for small, special purpose computers.

Nobody in 1945, not even von Neumann, could foresee the incredible development of the computer industry. That computers were important for scientific calculations was clear, that they would become more important, both for science and business applications, if they could be speeded up, also. But nobody dreamed that computers would become a multi-billion dollar industry, that the market value of IBM stock would be the highest of any company in the U.S., topping that of giants like American Telephone and Telegraph or General Motors which have also grown at a healthy pace, and being more than four times that of General Electric! (Of course, stock market analysts are among the most dedicated users of computers).

Von Neumann also predicted at an early time that computers would be useful to direct other activities. We now have many industrial manufacturing processes which are computer-controlled. Railways control the flow of traffic by computers. Inventories obviously lend themselves well to computer control, but so does also the manifacture of precision components for machinery. Von Neumann's (and I believe, Norbert Wiener's) favorite project was to program a computer to manufacture another computer just like itself, given appropriate mechanical and electronic parts.

Many of the scientific problems of wartime Los Alamos have since been solved by computers, among them neutron diffusion which was the problem for which we first got computers. The first step here was the treatment of neutrons of many different velocities: This has been done by introducing neutron groups, each representing some velocity interval. Henry Hurwitz, at the Knolls Atomic Power Laboratory, was one of the pioneers of this method. Now multi-group calculations are indispensable for any reactor design. If the reactor can be approximated by a spherical (or-one-dimensional) arrangement, a calculation with many groups takes little computer time; two-dimensional calculations, in cylindrical geometry or in the *xy*-

plane, still require great skill of the programmer and substantial computing time. True three-dimensional problems with a sufficiently fine mesh are still beyond the capacity of even the most modern, high speed computers. Therefore it is still impossible to make a completely realistic calculation of a reactor with all its control rods, fuel rods, cooling tubes etc., and possibly a complicated core shape in addition: The solution must be pieced together by treating small regions (like the control rods) separately, then joining to an overall calculation of the reactor as a whole.

For the neutron diffusion in small regions, whose size (in one or more dimensions) is smaller than the (transport) mean free path, integral diffusion theory must be used. This may be approximated by the S_n method, invented by Bengt Carlson of Los Alamos and further developed by others. In some cases of neutron diffusion in regions of very complicated shape, the Monte Carlo method is useful, first proposed by Ulam and von Neumann, and this can also be used for problems in more than 3 dimensions, e.g. evaluating integrals over many independent variables. Of course, the accuracy of the result is limited, by statistics, to the order of one percent.

The number of applied physics problems to which computers can be applied, is legion. Among those with which I have been associated, are many hydro- and aerodynamics problems. Computers are especially useful when chemical reactions accompany the aerodynamics, as is quite common in high-speed flow. The aerodynamics changes temperature and density of the gas, which influences the rate of the chemical reactions, and conversely the chemical composition influences the aerodynamics. Even in pure air, 10 or more reactions may go on, involving atoms, ions, and molecules, and when other substances are added, the number of reactions increases rapidly.

Computers have been equally revolutionary in pure theoretical physics. The non-relativistic energy of the He atom has been calculated to 1 part in 10^{12}, which of course is far more accurate than our knowledge of the relativistic correction or of the experimental binding energy. Methods have been developed to calculate energy levels of atoms with 3 to 10 electrons to an accuracy of a few milli-electron-volts, or better than 1 part in 1000; at this accuracy, experimentally unknown levels can be predicted with confidence. Calculation of diatomic molecules is only slightly behind, and a general, quantitative understanding of the chemical bond in more complicated molecules is developing on this basis.

In my own field, low energy nuclear physics, the progress due to computers has been tremendous. All my Ph.D. candidates, of course, have to learn to write computer programs, and become quite expert at it. (I must confess that I have never learned this important skill myself).

Equally important as calculations in theoretical physics, has become the use of computers in experimental physics, to evaluate complicated experiments, e.g. bubble chamber photographs, for the physical quantities involved, and for the answer. Especially powerful is the use of computers "on line" for immediate evaluation of experiments.

With all the tremendous contribution of computers to pure and applied science, we must remember that they do not, by themselves, give us understanding. Far too often scientists (especially engineers, but also pure scientists) put their raw problem on the computer, without giving it any prior thought. Very often, a large part of the problem could be solved by analytical methods, and this would give the scientist a far greater insight into the solution than the computer answer. The analytical answer is usually much more general, and could be reproduced only by a large number of computer calculations. Not only would this be a great waste of money, but the result would be extremely bulky, very difficult to store and almost impossible to grasp intellectually.

Thus it seems to me that any problem should first be thoroughly understood, then solved analytically as far as possible, and only after this, the insoluble parts should be put on a computer. When the computer results are available, they must again be thoroughly analyzed, their significance understood, and explored for the possibility of further analytical approximations.

Another important limitation is that computers can only execute the orders they are given. Their answers cannot be more intelligent than the questions they are asked. Many people seem to believe that a result obtained on a computer must be right. If the underlying theory or assumption was wrong, or incomplete, so is the answer. The computer has merely obtained faithfully and accurately the result of a wrong formula.

At the other extreme, some people say that computers cannot obtain more than is put into them, implying that they are essentially useless. This, of course, is equally false: The main activity of theoretical science (and of much of mathematics) is to start from premises which are believed (or assumed) to be correct, or are to be tested, and to derive consequences therefrom. This is what a computer is superbly fitted to do.

These arguments apply quite particularly to Operational Analysis. In this area, even more than in science, the mystique of computer infallibility prevails. But often, especially in the earlier work, the assumptions put into the computer program were quite incomplete. The early computers just did not have enough capacity, and the scientists using them quite consciously left out some important inputs. In recent years, this situation has greatly improved, and I have seen excellent operational analysis work, especially some carried out by the Department of Defense. However, one must still remember that the computer answer is no better than the analyst's input. And if the answer appears to contradict common sense, the analyst had better stop and think what is wrong with his input.

The question has often been asked whether this situation will always remain so. Or will computers in the end become more intelligent than their operators? Can computers learn? This question was asked already by Wiener and von Neumann, and has been investigated over the last several years by psychologists and computer scientists. So far, I believe, the answer is quite inconclusive. Computers have been "trained" to recognize simple patterns, circles, triangles, etc. But where do we go from there?

A most intriguing question is whether we can learn anything about the functioning of the human brain from that of computers. Presumably, the (0, 1) or (no, yes) response of a single computer element is similar to that of certain elements in the brain. We might hope to learn about the connection between elements, the "program" and other features of brain activity, and perhaps the computer analogy will help. It has helped, I believe, in the solution of the genetic code, just by making us familiar with the concept of a code which may direct other processes.

Coming back to the learning of computers, I am not optimistic that this will lead to useful results soon, even on the question how humans learn, let alone to making computers independent of the operator. And I doubt whether we should desire such independence. The computer is a wonderful servant, reliable, obedient, and incredibly fast. Let us keep it in that role, and always remember that it is our servant, meant to make our life easier, and that it must never become our master.

This foreword has been written in a very personal vein, by a person who has been associated for many years with people who use computers, and who has made use of computer results, but who is in no way a computer expert. The names I mention are just the people I happen to know, and I must apologize to the hundreds I have not mentioned, who have contributed greatly to the development of the fascinating field of the application of computers to science.

CHAPTER 2

Pre-Electronic Aids to Digital Computation

WILLIAM J. WORLTON

1 INTRODUCTION

THIS CHAPTER describes the development of aids to digital computation prior to the advent of electronic computers. A definitive history of digital computing has yet to be written, and this presentation is certainly too limited in scope to correct that lack. Its purpose is to broaden the reader's understanding and appreciation of the contributions made in the past to a science that is usually thought to be limited to the last twenty years but which, in fact, extends back in human history more than twenty centuries.

The computer sciences are commonly described as being among the youngest of the scientific disciplines, and this point of view can be supported by restricting the definition of the term "digital computer" to include only electronic digital computers, the first of which was completed in 1946. A broader definition, however, would include all digital devices that have been used as aids to computation, and in this sense the history of digital computing includes electromechanical devices used since the late nineteenth century, mechanical devices used since the seventeenth century, physical but nonmechanical devices used since the eleventh century B.C., and that most ancient of aids to computation—the human hand—used since prehistoric times.

Man's earliest efforts to develop means of computation are, of course, lost in the mists of prehistory. A sense of number is shared by man and some of the lower animals, but the ability to communicate this sense is presumably limited to man. Every known language contains words which convey a sense of number, although some are limited to words for "one" and "two". As human societies became increasingly complex, the need also increased for more sophisticated ways of expressing and manipulating numerical information. Figure 1 is a classification of the various media which were, and in some cases still are, used to record and communicate numerals. It should not be inferred that oral numerals necessarily preceded finger numerals, or that finger numerals necessarily preceded recorded numerals, since it is entirely possible, for example, that marks made in the dust, or quantities expressed on the fingers, contributed to the development of oral numerals.

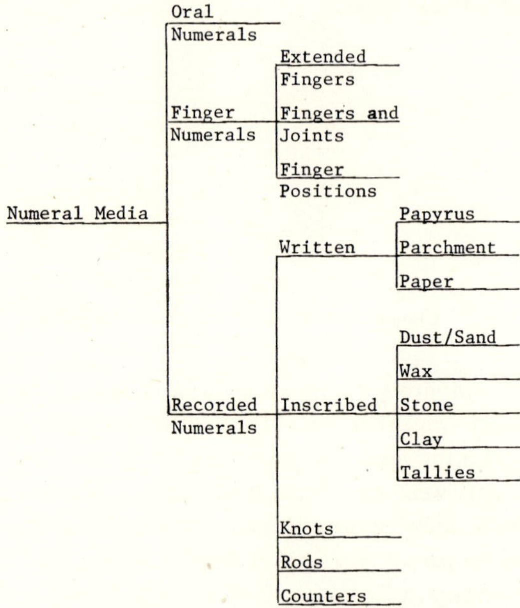

Fig. 1 Classification of early numeral media

Not all of the media used for the expression of numerals are amenable to manipulation for the purposes of computation; numerals inscribed in stone or recorded in the knots of the quipu, for example, are particularly inflexible. Other numeral media such as counters and rods can be readily manipulated, and they therefore can be used for computation with the numerals they express. Figure 2 provides a classification of the numeral media used for computation in ancient times. Not all of these are obsolete: finger reckoning, the abacus, and written numerals continue to serve mankind even in modern times.

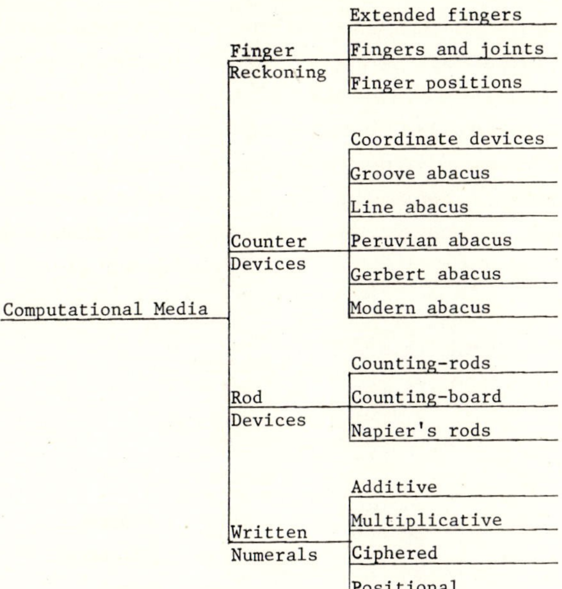

Fig. 2 Classification of early computational media

The history of computing suffers, as do other histories, from the lack of adequate materials to resolve many of the questions for which we would like to have answers. Our ignorance of these early developments comes not only because many events were never recorded, but also because many ancient libraries were deliberately destroyed, including the burning of the ancient Chinese books by Emperor Chin Chih Huang in 213 B.C., the destruction of the Alexandrian library in Egypt, the burning of the Octavian and Palatine libraries in Rome, and the burning of Mayan works in Mexico. The fragile nature of recording materials has undoubtedly caused the loss of other ancient copies. Questions such as the origin of the abacus will probably never be resolved because of the inadequacy of historical information.

Note that this discussion is limited to the history of *digital* computing devices; analog computing devices (such as the slide rule) are not covered.

2 FINGER RECKONING

The use of the fingers and hands for counting and computing was common prior to the widespread availability of inexpensive writing materials. Indeed, finger reckoning is the oldest and most generally used method of computing known to man; its use continues to the present time among primitive peoples and in the markets of the Far East.[1,2] Finger reckoning probably had no single location for its origin, but was developed independently in many cultures. Its ancient use can be documented among the Romans, Greeks, Egyptians, Babylonians, Persians, Hindus, and the Chinese.[3,4,5]

The three most common ways of expressing the finger numerals were

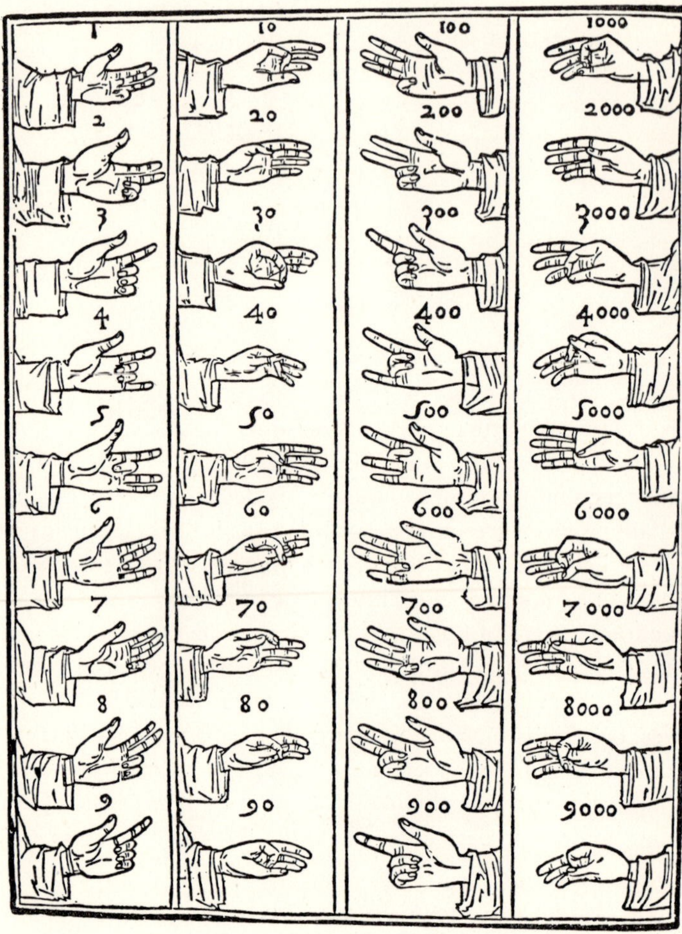

Fig. 3 Finger numerals. (From Luca di Burgo Pacioli, Summa de arithmetica geometria, second edition, Venice, 1523. Courtesy McGraw–Hill Book Company)

(a) using the extended fingers, (b) using the fingers and the joints, and (c) using finger positions. The finger position method, illustrated in Fig. 3, is the most sophisticated; the lower three fingers of the left hand designate the units digit, the upper two designate the tens digit; the lower three fingers of the right hand designate the hundreds digit, the upper two designate the thousands digit. Although certain of these figures appear ambiguous, the fingers touched the palm in some, not in others. In this method, therefore, numbers as high as 9999 could be expressed. When Juvenal wrote in his tenth satire, "Happy, no doubt! was he who for so many years put off his hour of death; and now begins to count his years on his right hand,"[6] he was referring to a finger numeral method which expressed numbers greater than or equal to one hundred on the right hand. Needham mentions a Chinese finger numeral method in which the fingers represented powers of ten.[5]

Finger reckoning included not only the simple counting used surreptitiously by modern school children, but also rather complex methods for addition, subtraction, multiplication, and division.[4] Figure 4 illustrates a finger-reckoning method for multiplication; it is assumed that the user knows the products for factors from one to five, and this method then allows the higher products to be derived. Suppose the product 7×9 is desired; 7 is expressed by extending two fingers and closing three on the left hand, 9 is expressed by extending four fingers and closing one on the right hand. The sum of the extended fingers (six) determines the number of tens; the product of the closed fingers (three) determines the units. Larger products and those which cross ranges are decomposed into simpler problems but are nevertheless extremely complicated tasks to perform. Finger reckoning played an important role in Europe until about the sixteenth century, when it was replaced by written positional numerals.

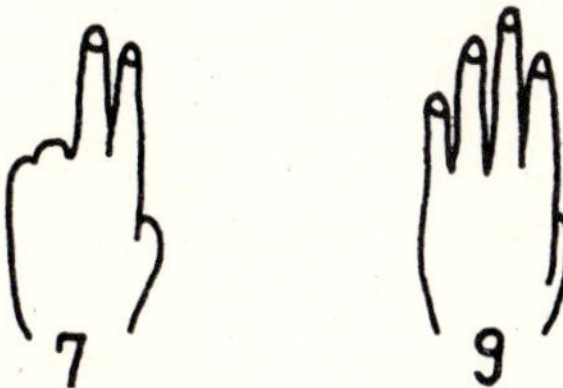

Fig. 4 Finger multiplication

3 PHYSICAL AIDS TO COMPUTATION

3.1 Counting-rods

Physical, but nonmechanical, aids to computation have been used since about the eleventh century B.C. when the Chinese began using "counting-rods."[7] Although the counting-rods were made in various sizes and materials, they were typically made of bamboo, about six inches long and one-tenth of an inch in diameter; other counting-rod materials mentioned in the literature include ivory, iron, bone, and horn, and they were sometimes made as long

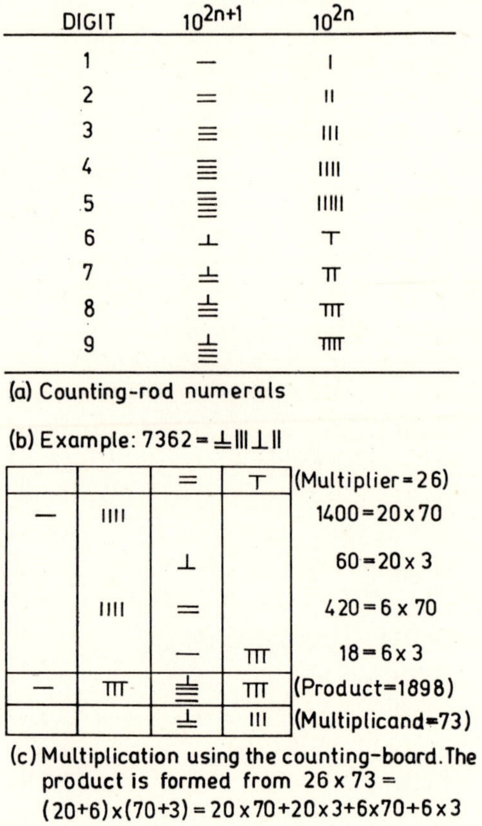

DIGIT	10^{2n+1}	10^{2n}
1	𝍩	𝍠
2	𝍪	𝍡
3	𝍫	𝍢
4	𝍬	𝍣
5	𝍭	𝍤
6	𝍮	𝍥
7	𝍯	𝍦
8	𝍰	𝍧
9	𝍱	𝍨

(a) Counting-rod numerals

(b) Example: 7362 = 𝍯𝍢𝍮𝍡

		𝍪	𝍥	(Multiplier = 26)
𝍩	𝍣			1400 = 20 x 70
		𝍮		60 = 20 x 3
	𝍣	𝍪		420 = 6 x 70
		𝍩	𝍧	18 = 6 x 3
𝍩	𝍧	𝍱	𝍧	(Product = 1898)
		𝍯	𝍢	(Multiplicand = 73)

(c) Multiplication using the counting-board. The product is formed from 26 x 73 = (20+6) x (70+3) = 20 x 70 + 20 x 3 + 6 x 70 + 6 x 3

Fig. 5 Counting-rod numerals and multiplication

as twelve inches. The counting-rods were originally used merely to form the "counting-rod numerals," but were used later in conjunction with a "counting-board" for computation.[5,8] Figure 5 illustrates the counting-rod numerals and the use of the counting-board for multiplication. The counting-board is merely an aid to positional notation, and as such is functionally the same as the Gerbert abacus (see Sec. 3.5). Neither on the Chinese counting-board nor the Gerbert abacus was there an explicit numeral for "zero", but neither needed one, since this was indicated by the absence of a numeral in the columns.

Counting-rods and the counting-board were in use in China until about the thirteenth century. Their use had passed to Japan by no later than the sixth century A.D.; the Japanese made the rods in a rectangular form so they would not roll, and these they called *sangi*. They referred to the counting-board as the "sangi-board", and references to sangi-board usage occur in Japanese texts as late as the eighteenth century. In both China and Japan, it appears that the counting-board was used at the same time as the modern form of the abacus, even though the abacus was easier and faster to use.

3.2 Coordinate devices

The counting-board has the significant disadvantage that it is necessary to form a counting-rod numeral in the columns in order to specify the desired digit. A simpler and faster method was employed in devices which had columns similar to the counting-board but a coordinate method of designating the numerals, as illustrated in Figs. 6 and 7.[5,7,9] Although these devices are described in the ancient Chinese literature, no drawings exist; the use of the counting-rod numerals in these figures is based on the fact that the devices and the numerals come from the Chou dynasty (sixth to third centuries B.C.). The term *Thai I* means "greater unity", and refers to the fact that there was a single ball in each column which had nine horizontal divisions to designate the digits. Addition with this device was accomplished by moving the balls higher, subtraction by moving the balls lower. A carry out of any column was accomplished by advancing the ball in the next higher column one position. This device is sometimes referred to as the "Chu Pan," from the Chinese *chu suan phan*, meaning "ball arithmetic plate"; the term is equally applicable to this device and the usual form of the abacus, to which it is also applied by various writers.

The device shown in Fig. 7 was used in the *Liang I* method; it was essen-

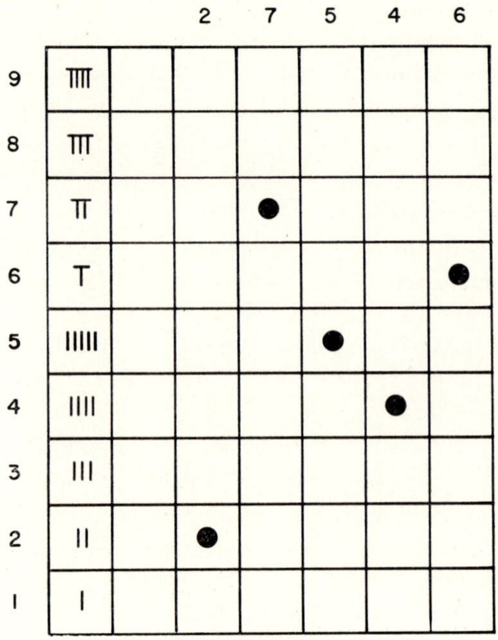

			2	7	5	4	6
9	𝍨						
8	𝍧						
7	𝍦			●			
6	𝍥						●
5	𝍤				●		
4	𝍣					●	
3	𝍢						
2	𝍡		●				
1	𝍠						

Fig. 6 The "Thai I" device

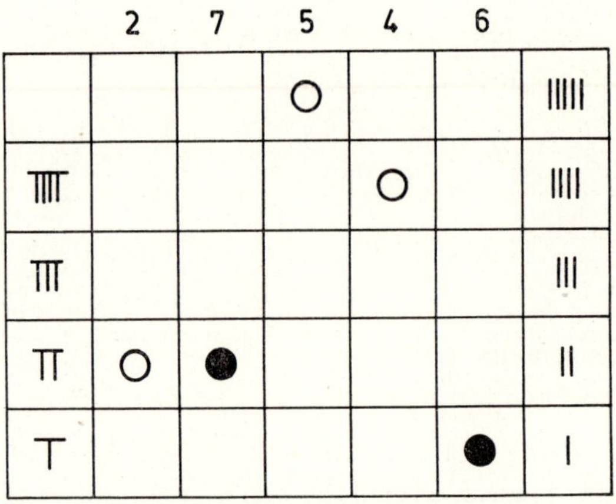

	2	7	5	4	6	
			○			𝍤
𝍨				○		𝍣
𝍧						𝍢
𝍦	○	●				𝍡
𝍥					●	𝍠

Fig. 7 The "Liang I" device. The open counters refer to the right-hand scale; the black counters refer to the left-hand scale

tially a half-sized version of the Thai I device. It used two colors of balls, yellow and blue, which referred to two different scales of value.

The relationship of these devices to the counting-board and the modern form of the abacus cannot be determined from the scanty sources available. It is possible that the order of development was from the counting-rods to the counting board, to the Thai I device, to the Liang I device, to the abacus with unattached counters, and finally to the modern form of the abacus with attached counters. In any event, various forms of coordinate ball arithmetic were used in China several centuries prior to the Christian era.

3.3 The groove abacus

The device most nearly like, and probably immediately antecedent to, our modern abacus is the groove abacus; a Roman specimen of uncertain date is illustrated in Fig. 8.[8] The groove abacus was used in both Rome and China and there are claims and denials of mutual influence. Our knowledge of the ancient use of the groove abacus in China comes from a document called the *Shu Shu Chi I* (Memoir on some Traditions of Mathematical Art), attributed to Hsu Yo at the end of the Later Han dynasty (early third century A.D.) and preserved through a commentary by Chen Luan in the sixth century. A portion of Chen Luans' commentary reads as follows.

Needham's translation
A board is carved with three horizontal divisions, the upper one and the lower one for suspending the travelling balls, and the middle one for fixing the digit. Each digit (column) has five balls. The color of the ball in the upper division is different from the color of the four in the lower ones. The upper one corresponds to five units, and each of the four lower balls corresponds to one unit.[5]

Kojima's translation
The abacus is divided into three sections. In the uppermost and lowest section, idle counters are kept. In the middle section designating the places of numbers, calculation is performed. Each column in the middle section may have five counters, one uppermost five-unit counter and four differently colored one-unit counters.[9]

Although these translations do not entirely agree, and their authors (and others) draw figures of devices which are quite different, it is nonetheless clear that an abacus-like device is being described.

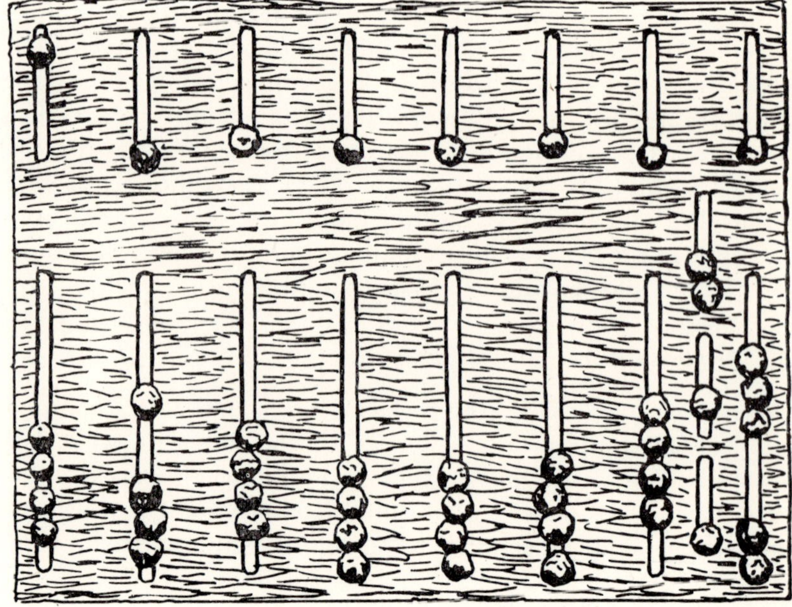

Fig. 8 Roman groove abacus

It is generally agreed that the modern form of the abacus, in which the counters are carried by rods or wires, evolved from the groove abacus, but there is little agreement on where or when this occurred.

3.4 The line abacus or counter-board

The line abacus was probably used in both ancient Greece and Rome, although the evidence is rather sketchy. A large line abacus made of white marble, about 30 by 60 inches, was found on the island of Salamis, but its discovery was not documented in sufficient detail to allow the device to be dated.[10] Several Roman authors mention tables and loose counters, but no description of the line abacus has been found in Roman literature. The line abacus was widely used in Europe, especially the countries north of the Alps where it provided almost the only means of commercial computation during the fifteenth century.[8] By the sixteenth century, it was still widely used in Germany, Holland, Poland, and Austria, but was falling into disuse in England and France, and had been abandoned in Italy. Texts on the line abacus appeared as late as the seventeenth century in Germany.

The device is extremely simple to make and use; some lines and loose counters are all that are necessary, as illustrated in Fig. 9. The lines are assigned ascending powers of ten, and the spaces between the lines are assigned the value of five times the units immediately below them. Four counters on the lines or one counter between the lines are allowed; if more occur, then a counter is carried to the next higher position. Addition and subtraction can be performed very rapidly, but multiplication and division require repeated addition and subtraction. Many operational short-cuts are possible, e.g., to add 99 to the value shown on the line abacus in Fig. 9, a single counter is moved from the 100 line to the 1000 line, thereby subtracting 100 and adding 1000 in one motion.

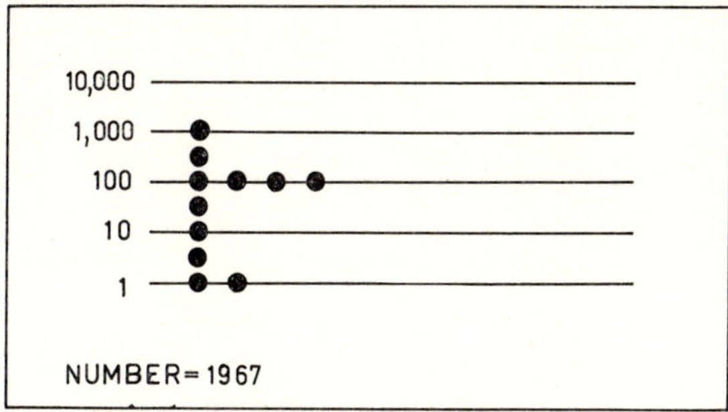

Fig. 9 Line abacus

Although the above description covers the most common form of the line abacus, several other forms were also used. Barnard, whose book *The Casting-Counter and the Counting Board* (1916) is still the most comprehensive study of the line abacus, classifies the varieties as follows.

"A Methods with lines.

- a Process in which lines only were used.
- b Process in which spaces only were used.
- c Process in which both lines and spaces were used.
- d Process in which both lines and spaces were used, and also 'Lyers', or stationary jettons, i.e., the 'Tree of Numeration'.

B Methods without lines.

- e Process in which 'Lyers' were used, forming the 'Tree of Numeration',
- f Process in which no guides to help the eye were used."[11]

The "Tree of Numeration" was formed by a column of counters with increasing values marked on them; as noted above, they were sometimes used with the lines and sometimes used alone. A Tree of Numeration could be made with United States coins using the penny, nickel, dime, half-dollar, and the dollar to mark values of 1, 5, 10, 50, and 100.

The counters used with the line abacus were referred to by such terms as *casting-counters* in English, *jetton* in French, *rechenpfennig* in German, *worpghelt* in Dutch, *jactus* in Latin and *contador* in Spanish. They were cast in gold, silver, latten (60% copper, 30% zinc, 10% lead), brass, copper, bronze, and lead. Whereas coinage is required to be uniform, no such requirement applied to these counters, and they were made with a considerable variety of religious, political, personal, portrait, explanatory, moral, quotation, motto, advertising, and even humorous themes imprinted on them. Dated counters appeared in the fifteenth century and were made as recently as the late eighteenth century. The number of counters needed for most computations was modest, and counter sets usually did not exceed 100.

As indicated by the name, the counting board itself was usually made of woods such as walnut and oak, although ebony and ivory are also throught to have been used. Specimens are very rare. Plain tables were often used; the lines were marked on them when computations were performed and then erased. Reckoning cloths were made with the appropriate lines sewn on them, to provide a portable aid to computation. The modern term "counter", used in reference to a merchandising table in a store, is a carry-over from the use of the counter board by merchants of several centuries ago.

In contrast to the Chinese abacus, which is still widely used, the line abacus is obsolete and almost unknown.

3.5 The Gerbert abacus

Just before 1000 A.D., a computing device was introduced into Europe that had functional similarities to the Chinese counting-board, i.e., it served as an aid to positional notation. The device is attributed to Gerbert (d. 1003 A.D.), who was elected Pope Sylvester II in 999. The Gerbert abacus, together with a problem in division illustrating its use, is shown in Fig. 10.[12] The circled numbers represent counters that had numerals written on them; they were moved to form values just as positional notation is used, but there was no counter for zero, which was shown by a blank column. The development of a symbol for zero made the Gerbert abacus unnecessary.

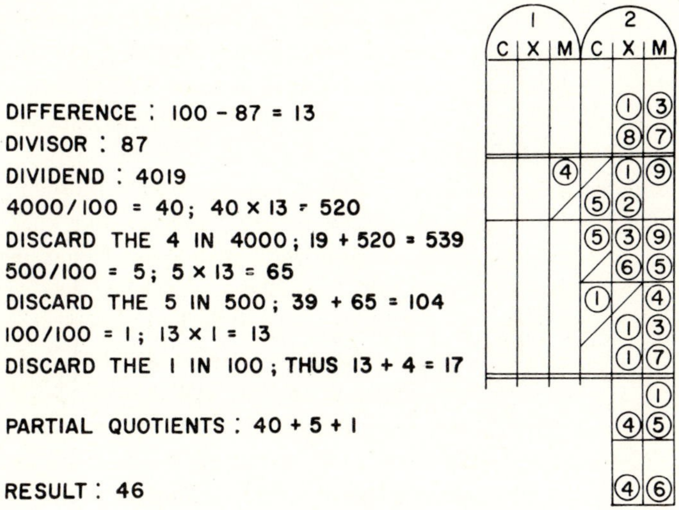

DIFFERENCE : 100 - 87 = 13
DIVISOR : 87
DIVIDEND : 4019
4000/100 = 40; 40 × 13 = 520
DISCARD THE 4 IN 4000; 19 + 520 = 539
500/100 = 5; 5 × 13 = 65
DISCARD THE 5 IN 500; 39 + 65 = 104
100/100 = 1; 13 × 1 = 13
DISCARD THE 1 IN 100; THUS 13 + 4 = 17

PARTIAL QUOTIENTS : 40 + 5 + 1

RESULT : 46

DIVISION ON THE GERBERT ABACUS

Fig. 10 Gerbert abacus

3.6 The Peruvian abacus

About 1600 a manuscript written in Spanish by a half-caste Peruvian Indian named Felipe Huaman Poma de Ayala described in text and sketches the people and customs of pre-Columbian Peru.[13] Included in the manuscript is the sketch shown in Fig. 11, portraying someone who is the "chief counter and treasurer". The device he is holding is a *quipu*, used for recording numerical information by knots tied in the dangling strings. To the man's right is the device now referred to as the "Peruvian abacus".

In an article published in *Comparative Ethnographical Studies*[14] in 1931, Henry Wassen described this device and presented a theory (attributed to a collaborator, K. G. Tengstrand) of how the device might have been used. The circles in each square represent holes, some of which are filled with counters such as pebbles or kernels of corn. According to this theory, the five holes in the square in the lower left-hand corner were used to count one each; the three holes in the next square to the right were used to count five each; the two holes in the next square to the right were used to count fifteen each; and the last square on the right counted thirty. Each successive higher row in the

Fig. 11 Peruvian abacus

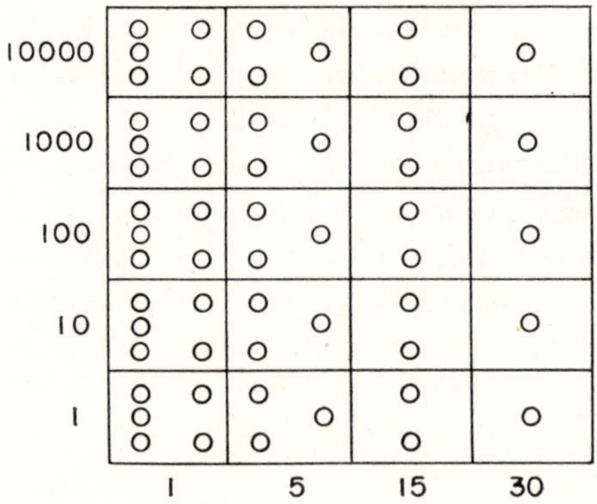

10000	○ ○ ○ ○ ○	○ ○ ○	○ ○	○
1000	○ ○ ○ ○ ○	○ ○ ○	○ ○	○
100	○ ○ ○ ○ ○	○ ○ ○	○ ○	○
10	○ ○ ○ ○ ○	○ ○ ○	○ ○	○
1	○ ○ ○ ○ ○	○ ○ ○	○ ○	○
	1	5	15	30

Fig. 12 Wassen's theory of the Peruvian abacus

device increased its units by a factor of ten, as illustrated in Fig. 12. This last feature is suggested by the manuscript which says, "... they number from one hundred thousand and from ten thousand and from one hundred and from ten..."

The assignment of values in the horizontal direction is not described in the manuscript, and Wassen's proposal of units of (1, 5, 15, 30) is most unlikely. First, since it does not provide an explicit value for 20, this theory ignores the fact that the vigesimal number system was used widely in the Americas. Second, the device, as described, is horizontally redundant: it could be cut off to the right of the two counters in the second column from the left, and the remaining portion would be exactly equivalent to a Chinese abacus. Third, the sum of the values expressible in any row equals 80 times the units of the row, and this is an unlikely sum.

If the horizontal values are assigned as (1, 5, 20, 40), then the first and third difficulties mentioned above are eliminated, i.e., the device provides an explicit representation for 1, 5, and 20 (as in the Mayan vigesimal system, for example), and the sum of the values expressed in each row is 100 times the units of that row. The problem of the horizontal redundancy remains, however, unless (a) the higher rows increased their units by two orders of magnitude instead of one, or (b) the device actually was horizontally redundant.

Wassen adduced other sources for the Peruvian abacus, including a dictionary of the Aymara language in which Aymara terms were given which meant "to count with pebbles", "to put two pebbles in the count when there is no more than one", and "to put five in the count when it is done with pebbles".

There is no other computing device known which is similar to the Peruvian abacus, and the principles of its use can only be speculated upon until further evidence is found.

3.7 The modern abacus

The term "abacus" is thought to have been derived from the Greek *abax*, in turn taken from the Hebrew *abak*, which means "dust"; the name was probably applied to the primitive dust-board which was used as a writing surface and then passed over to the computing device.[8,15] The abacus on which the counters are carried by rods or wires is usually called "the Chinese abacus", but there are actually several forms, as illustrated in Fig. 13. Note that the Chinese abacus and the Japanese *soroban* differ in the number of counters both above and below the dividing rod, and that the Russian *s'choty* has no dividing rod but uses different colored counters in the center to aid the user. The form of this device used by the Russians probably developed first in the Near East; this same device is known in Turkey as the *coulba* and in Armenia as the *choreb*.

Fig. 13 (d) shows a compound abacus with two sets of counters; the extra set is used for computing intermediate results without disturbing the previously obtained results. For example, an inner product ($\sum_i a_i b_i$) could be readily computed on the compound abacus. This device illustrates a fact that is often overlooked: the abacus is not just an interesting antique or a toy, but is a computing device which continues to be used and improved by a significant portion of the world's population. National examinations are given for the Abacus Operator's License by the Japan Chamber of Commerce and Industry; these are graded by difficulty from the relatively easy eighth-grade operator to the extremely difficult first-grade operator.[9] The first-grade test, for example, requires the entry of about 1200 digits in ten minutes, or an average of two digits per second, in the process of accumulating ten sums

Fig. 13 Several forms of the modern abacus (a) Chinese abacus

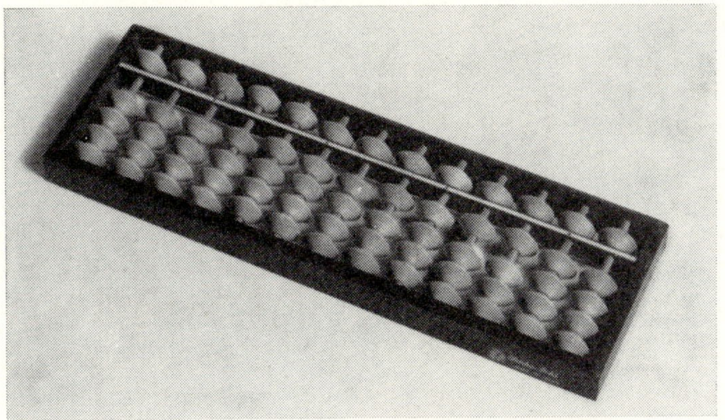

Fig. 13 Several forms of the modern abacus (b) Japanese *soroban*

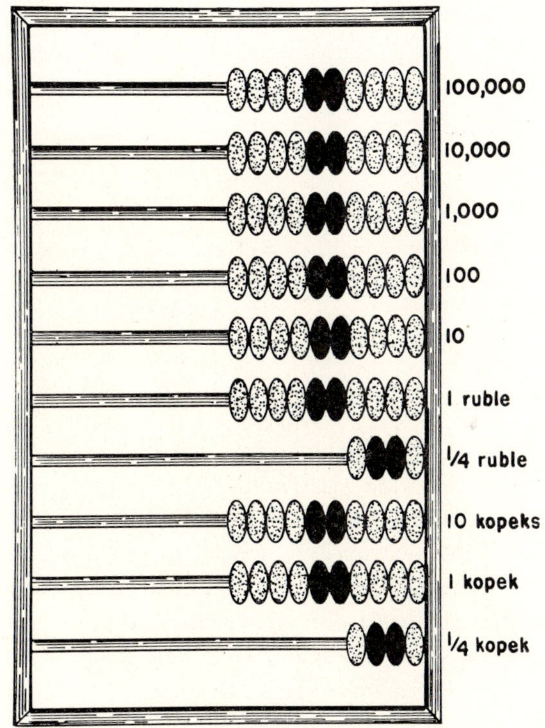

Fig. 13 Several forms of the modern abacus (c) Russian *s'choty*

Fig. 13 Several forms of the modern abacus (d) Compound abacus

using positive and negative numbers. Organizations in Japan concerned with various aspects of abacus usage include the Abacus Research Institute, the Federation of Abacus Workers, and the Japan Association of Abacus Calculation.

3.8 Napier's rods

In 1617 John Napier invented an aid to multiplication consisting of rods with integers inscribed at their top and the multiples of the integers inscribed down their length, as illustrated in Fig. 14.[5,8,16] By selecting the rods corresponding to the digits of the multiplicand, it was possible to record the product immediately, since the digit products had already been formed. The rods were sometimes made in the flat form, but repeated digits in a factor required many sets of the rods, so rectangular rods were made which had several integers and their multiples inscribed on them. A cylindrical form with the integers 0 through 9 and their multiples on them is shown in Fig. 15. Napier was as famous in the seventeenth century for this invention as for the invention of logarithms. His rods, sometimes called "Napier's bones", were used in many countries, including China and Japan; the Chinese set of rods included rods for the square and the cube of the integers.

Fig. 14 Napier's rods. (British Crown Copyright. Science Museum, London)

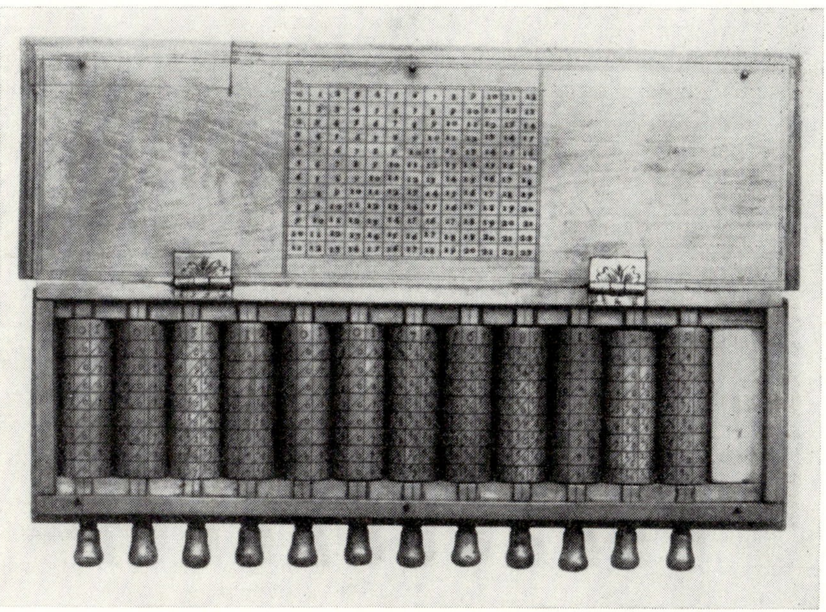

Fig. 15 Napier's rods in cylindrical form (British Crown Copyright. Science Museum, London)

4 MECHANICAL AIDS TO COMPUTATION

Four inventions should be noted in the history of mechanical aids to computation: the mechanical desk calculator, the punched card, the difference engine, and the analytical engine. The order stated is the general historical order of the development of these inventions.

4.1 The mechanical desk calculator

The invention of the first mechanical desk calculator is often attributed to Pascal, but this now appears to be an error.[17,18] In 1624 Wilhelm Schickhardt (1592–1635), a professor of mathematics and astronomy at Tübingen, Germany, wrote Kepler that he had invented a calculating machine, and a sketch by the mechanic who built the device for Schickhardt has been found in the State Library in Stuttgart. A model of this device has been reconstructed, as illustrated in Fig. 16; the upper register consisted of a cylindrical form of Napier's rods, and the lower portion was a stylus-operated adding mechanism providing for six digits with automatic carry.

In 1642 Pascal invented a small adding machine illustrated in Fig. 17. It also consisted of a stylus-operated adding mechanism, in which the wheels moved only in the forward direction. As any wheel was rotated, a number corresponding to the position of the wheel appeared at a window, and as the zero position was reached, a ratchet caused the next higher wheel to advance one position, thus implementing automatic carry. By using complements for subtraction, and repeated addition and subtraction for multiplication and division, the operator could perform all four arithmetic operations, even though the device itself was capable of only the single operation of addition. Pascal was evidently unaware of Schickhardt's earlier work; his initial advertisement of his machine was offered in competition with "the counters or with the pen".[19]

In 1671 Leibniz invented a calculator, shown in Fig. 18. A stepped cylinder was used to control repeated addition and subtraction of rotary mechanisms similar to Pascal's, thus implementing automatic multiplication and division.[20] The Leibniz method was not altered in principle for almost two hundred years.

These machines all suffered from the lack of adequate skills in the contemporaneous technology to make the devices reliable, and various attempts

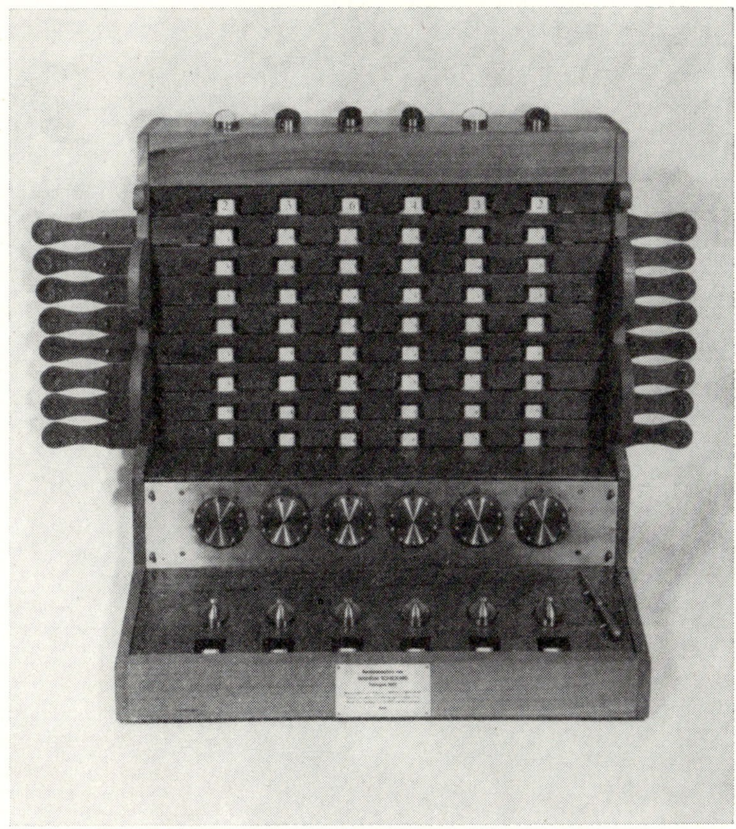

Fig. 16 Schickhardt's calculator (Courtesy IBM Corp. Photo by Geoffrey Clements)

were made to improve upon them during the next century, some of the more notable being the work of Giovanni Poleni in Italy in 1709,[21] the work of Philipp Hahn in Germany in 1770,[22] and the work of Stanhope in England in 1775. It was not until the early nineteenth century, however, that a mechanical calculator sufficiently reliable for commercial production was invented; this was the "Arithmometer" of Charles Xavier Thomas of Colmar, Alsace.[23] Thomas' first machine was produced in 1820; a Thomas machine of 1870 is illustrated in Fig. 19. The Thomas machines provided the point of departure for the development of later mechanical and electromechanical desk calculators. The first patent for a calculator which multiplied directly

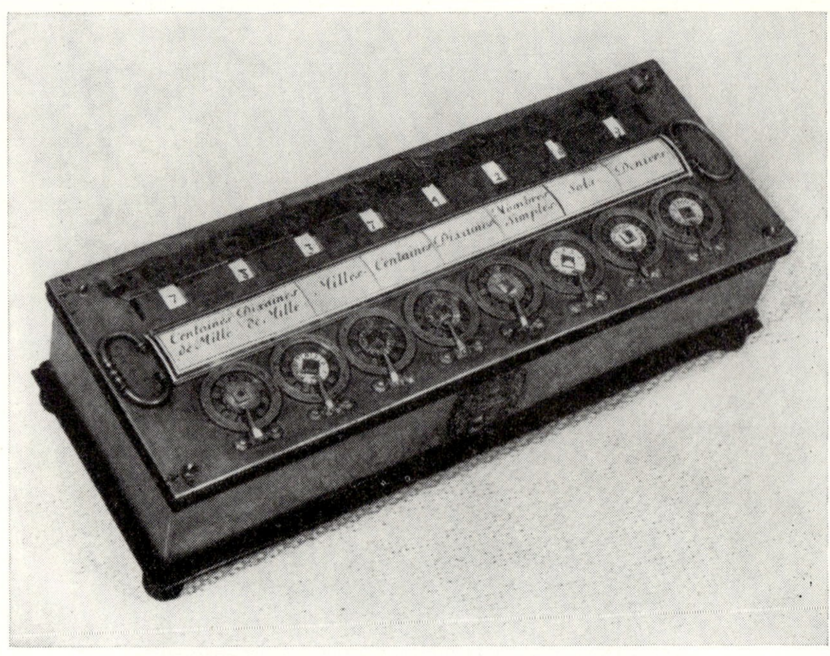

Fig. 17 Pascal's calculator (Courtesy IBM Corp.)

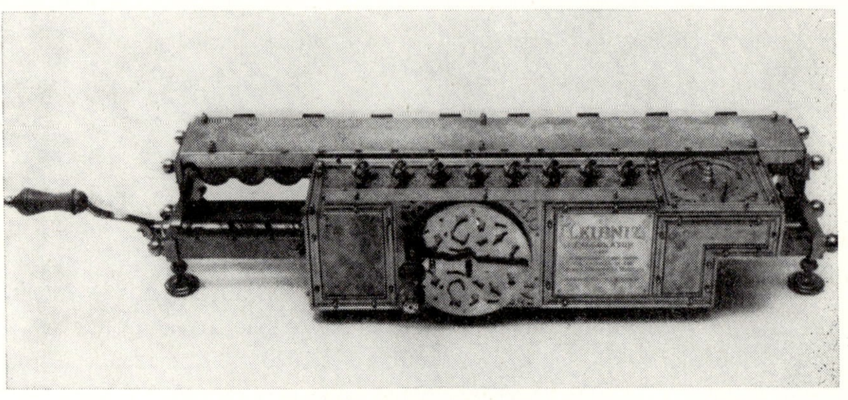

Fig. 18 Leibniz's calculator (Courtesy IBM Corp. Photo by Geoffrey Clements)

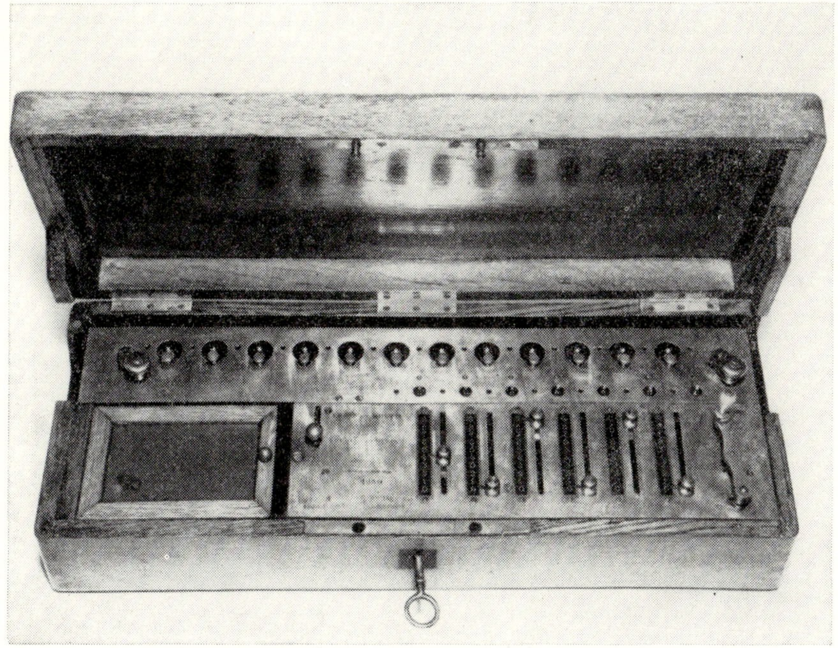

Fig. 19 Thomas calculator of 1870 (Courtesy IBM Corp.)

rather than by repeated addition was issued to Edmund D. Barbour of Boston in 1872. Barbour's ideas were not exploited commercially, and the development of direct multiplication machines followed the work of Ramon Verea of New York in 1878, Leon Bollee of France in 1889, and Otto Steiger of Munich in 1897. The first fully automatic electromechanical desk calculator is usually credited to Baldwin and Monroe in 1920.

4.2 Punched cards

The use of punched cards to control mechanical equipment was an invention of the early eighteenth-century French weaving industry.[24] In 1725 Basile Bouchon developed a loom that used an endless band of punched paper to control the lifting of warp threads of the loom. Three years later Falcon constructed a loom (see Fig. 20) controlled by punched cards rather than a band of paper, and this machine embodied the essential elements of the machine since called the Jacquard loom. Improvements were made by Reynier in 1740, Vaucanson in 1745, and by Jacquard in 1801 and 1804. The Jacquard loom of 1804 was so successful that by 1812 there were 11,000 of them in use in France alone.

Fig. 20 Falcon's loom (Courtesy IBM Corp.)

The Jacquard punched cards and loom are illustrated in Fig. 21. The cards were connected with loops of string to form a continuous chain that passed over a rectangular cylinder having pegs in each end of its four faces to hold and properly register the cards. The cylinder was then pushed against a set of horizontal needles, which contacted vertical hooks. Where there was a hole in the card, the needle passed through; where there was no hole the card pushed the needle and its associated hook back a short distance. A lifting device called the *griff* then lifted all of the hooks of the needles that had passed

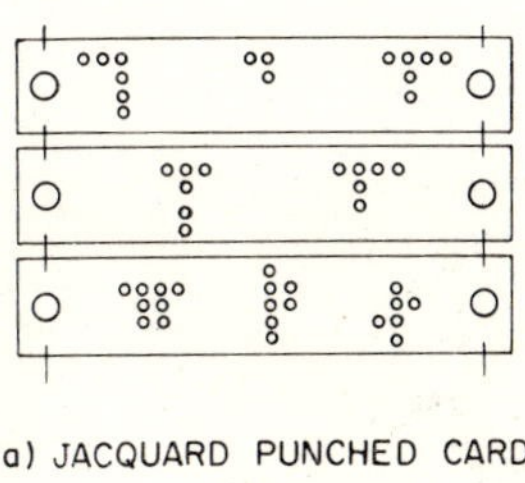

(a) JACQUARD PUNCHED CARDS

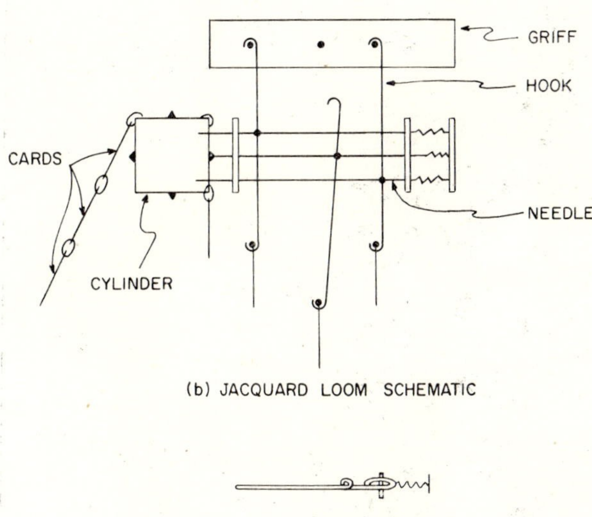

(b) JACQUARD LOOM SCHEMATIC

(c) JACQUARD NEEDLE

Fig. 21 Details of the Jacquard loom

through the holes; these hooks were attached to the warp threads and thereby lifted them to allow the passage of the weft thread across the warp.[25] Fig. 22 shows a portrait of Jacquard woven on a modern loom by the Rüti Machine Works in Switzerland.

Charles Babbage employed the Jacquard punched cards to control the sequence of operations and the transfer of variables in his Analytical Engine, and thereby became the first man to use punched cards to control mechanical computing equipment (see Sec. 4.4). Herman Hollerith designed electromechanical equipment for processing punched cards in the late nineteenth century (see Sec. 5.1); this is usually described as the beginning of modern information processing using punched cards, but it was a "beginning" already more than one hundred and fifty years old.

Fig. 22 Woven portrait of Jacquard (Courtesy Rüti Machine Works)

4.3 The difference engine

A third invention which advanced the development of mechanical computing machinery was the difference engine. A difference engine is a special-purpose calculator designed to tabulate the values of polynomials for equidistant values of the independent variable, using the fact that the Nth-order differences of a polynomial of order N are constant. A difference engine requires several registers, one more than the order of the polynomial to be tabulated, and the means to add the contents of adjacent registers. Table I illustrates

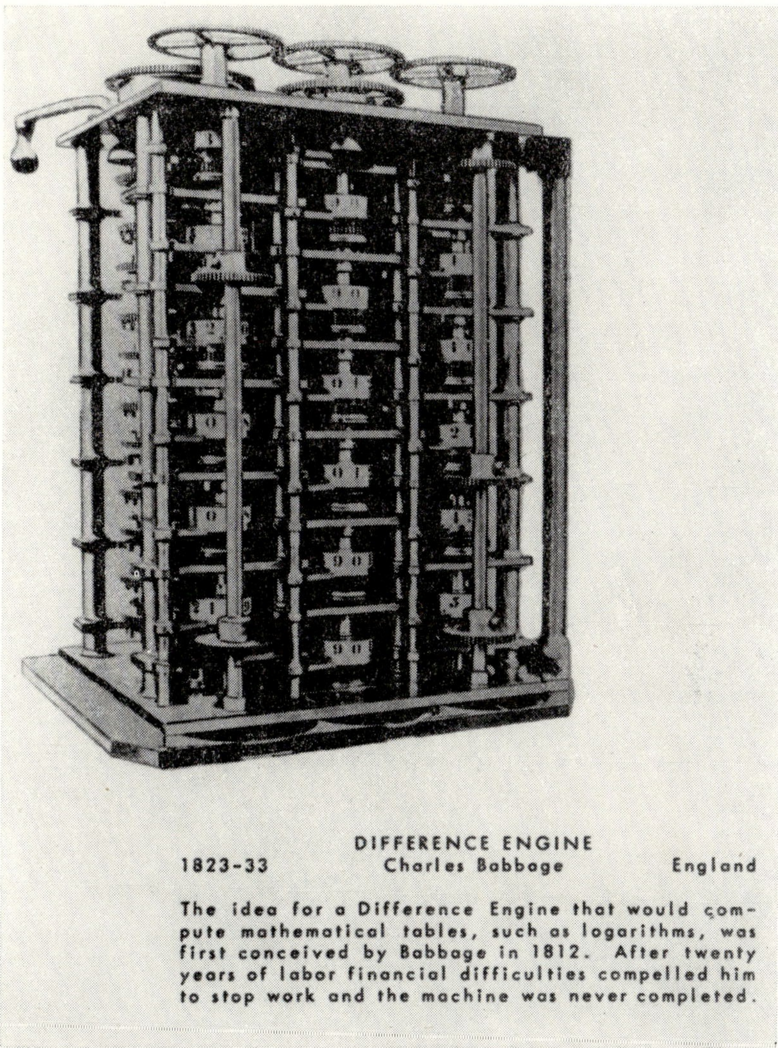

Fig. 23 Babbage's difference engine (Courtesy IBM Corp.)

how a difference engine would tabulate the simple polynomial $y = x^2$, given appropriate initial values of the function and first- and second-order differences. R_3 is first added to R_2, which is in turn added to R_1. This process is repeated, and each new value of R_1 is a new value of the polynomial.

The difference engine was first conceived by Müller in 1786,[17,23] but the

Table I Example of difference engine operation

	R_1	R_2	R_3
Initial values	1	1	2
First Cycle	1	3	2
	4	3	2
Second Cycle	4	5	2
	9	5	2
Third Cycle	9	7	2
	16	7	2
	:	:	:

first successful attempt at construction was the work of Charles Babbage who completed a difference engine capable of tabulating second-order polynomials to six decimal places in 1822.[25] Babbage then embarked on a much more ambitious project—the construction of a difference engine to tabulate sixth-order polynomials to twenty decimal places. The project was never completed, largely because the mechanical technology of the day was incapable of meeting Babbage's requirements. A portion of this difference engine is shown in Fig. 23. Babbage's design was described in an article in the *Edinburgh Review* in 1834, leading Georg Scheutz of Sweden to design and construct a working model of a difference engine.[26] This was demonstrated

Fig. 24 Scheutz's difference engine (Courtesy IBM Corp.)

to the Royal Swedish Academy of Sciences in 1843; a commercial model, illustrated in Fig. 24 capable of tabulating fourth-order polynomials to fourteen decimal places, was completed in 1853. Difference engines were also constructed by Deacon of London, Wiberg of Sweden, and Grant of the United States. The Grant difference engine was exhibited in 1876 at the Philadelphia Centennial Exposition.

4.4 The analytical engine

In 1833 Babbage conceived a generalization of his Difference Engine, which he called "the Analytical Engine",[27] whose basic elements are shown in the form of a block diagram in Fig. 25. This design is remarkably similar to the design of a modern digital computer. The "store" is to contain 1000 words of 50 decimals each (equivalent to about 167,000 binary bits); modern computers did not attain this capacity until the early 1950's. The "mill" is equivalent to the modern arithmetic unit; it was to be capable of adding (or subtracting) in one second and multiplying (or dividing) in one minute. Control of the analytical engine was implemented through two strings of Jacquard punched cards. A string of "operation" cards controlled the sequence of operations to be performed; a string of "variable" cards controlled the transfer of data between the store and the mill. This type of programming is similar to that used in card-programmed calculators of modern time. Output was to be punched in cards, printed directly onto paper, or impressed into a stereographic mold which could be used later for printing. In this last ap-

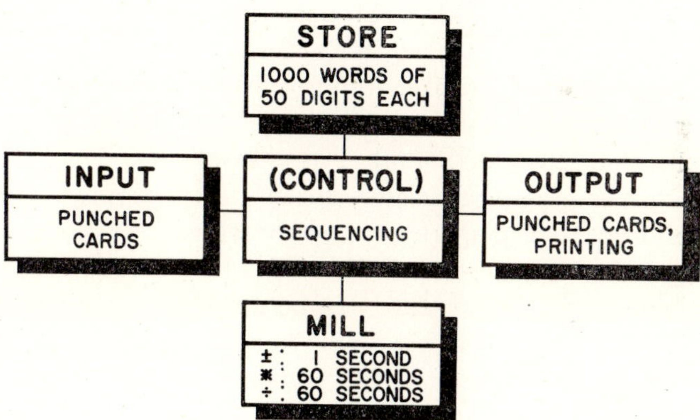

Fig. 25 Babbage's analytical engine

proach to output, Babbage anticipated modern computer-controlled typesetting by more than a century. He also anticipated several programming features of modern computers which give them their great versatility: the conditional branch, dynamic data modification, and the use of an index to control looping. The fact that Babbage's design was never completed due to the lack of an adequate technology in his day, does not detract from his greatness in designing an automatic general-purpose computer over 100 years before actual construction was accomplished. If we were to name a "Father of Computing" it would surely be Charles Babbage.

5 ELECTROMECHANICAL AIDS TO COMPUTATION

5.1 The pioneering work of Hollerith

The invention of electromechanical equipment for information processing is due to Herman Hollerith, a statistician from Buffalo, New York.[28,29] The 1880 census of the United States had taken seven years to tabulate and, with the growth in population, it was expected that the 1890 census might take as many as ten years to complete, with the result that the information would be obsolete before it became available. Hollerith designed a system of processing punched cards which was successfully used by the City of Baltimore to compile mortality statistics in 1887; this was the first application of electromechanical punched-card methods. The Hollerith equipment was later used by the Bureau of Vital Statistics of the State of New Jersey and the Board of Health of New York City with equal success. The Hollerith methods and equipment were used in the United States census of 1890 and completed the task in one-third of the time taken for the census of 1880, even though the population had grown by twelve million people in the interim.

The Hollerith machines used in the census of 1890 are shown in Fig. 27: a card punch, a nonprinting tabulator, and a sorting box. A card was placed on the lever-operated reading tray; the lever was then lowered and telescoping pins descended on the card. In the positions where there were no holes the pins merely telescoped; where there were holes, however, the pins dropped through the holes into cups of mercury and caused an electrical impulse to be sent to the tabulator and the sorter. The tabulator dial receiving the impulse would then advance one position; the impulse received by the sorter would unlatch the appropriate cover and the operator would then remove the card from the reading tray, place it in the uncovered bin, and relatch the cover. A comparison of the 1890 census card with a modern card is shown in Fig. 26. Crude as these devices seem now, they were the marvel of the times and Hollerith was internationally honored for these inventions. He received the Franklin Institute's Elliott Cresson Medal, the Paris Exposition's Medaille d'Or, an honorary Ph.D. from his alma mater, Columbia, and the Bronze Medal from the World's Fair of 1893.

Fig. 27 Hollerith's machines used in the census of 1890 (Courtesy IBM Corp.)

5.2 The development of commercial punched-card equipment

In 1896 Hollerith formed the Tabulating Machine Company to market his products. The census tabulator was a simple counting device not well adapted to commercial needs, and it was modified in several ways.[30] First, an adding mechanism was provided and the machine was then called the "Integrating Tabulator". The Integrating Tabulator still used the lever-operated reading tray, however, permitting only about fifty to eighty cards per minute to be tabulated, depending on operator speed. An automatic card feed was developed which could feed the tabulator at the rate of 150 cards per minute. A plugboard was designed to allow easy adaptation of the tabulator to tasks with varying definitions.

In 1911 the Tabulating Machine Company merged with two other companies to form the Computing Tabulating Recording Company (CTR).[31,32] In 1914 the product line of CTR included just four punched-card machines, illustrated in Fig. 28: the Type 001 mechanical keypunch, the Type 009 non-printing tabulator, the Type 070 vertical sorter, and a lever-set gang-punch. Improvements in this equipment line in the next few years included both numeric and alphabetic printing tabulators and horizontal one-deck sorters. In 1924 the name of the company was changed to International Business Machines.

Shortly before the census of 1910 Hollerith lost his contract with the Bureau of the Census, which began developing its own punched-card equipment through the efforts of James Powers.[33] Hollerith filed suit against the government, but his suit failed when the Supreme Court declined to render a judgment and President Theodore Roosevelt upheld the government against Hollerith. Powers designed the equipment used in the census of 1910, and in 1911 he formed the Powers Accounting Machine Company to market an improved punching machine and a two-deck sorter. Important improvements in punched-card equipment by this company in this early period included the first numerical printing tabulator, the first single-deck sorter, and the first alphabetical printing tabulator. In 1927 Power's company and seven others joined to form the Remington Rand Corporation.

Although this early punched-card equipment was intended primarily for commercial applications, L. J. Comrie in England[34] and W. J. Eckert in the United States[35] pioneered its use in scientific calculations. In 1928, for example, Comrie used commercial equipment to calculate the positions of the moon for the years 1935 through 2000, a literally staggering job of manipulating more than half a million punched cards.

The basic trend in the development of commercial punched-card equipment in the 1930's and 1940's was the combination into a single machine of functions previously performed by separate machines. In 1931, for example, the Model 600 Multiplying Punched was marketed; it would read cards,

Fig. 28 CTR punched-card machines of 1914 (Courtesy IBM Corp.)
(a) Type 001 mechanical keypunch

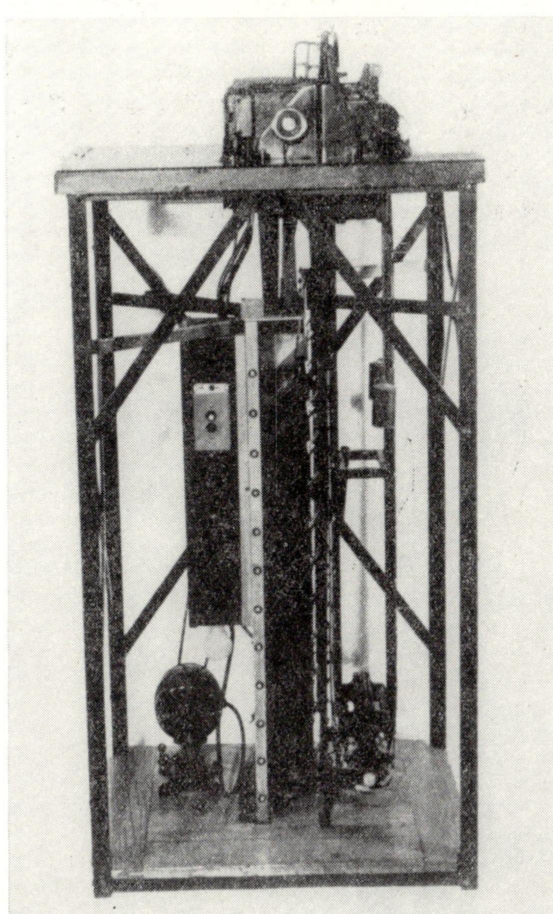

Fig. 28 CTR punched-card machines of 1914 (Courtesy IBM Corp.)
(b) Type 070 vertical sorter

Fig. 28 CTR punched-card machines of 1914 (Courtesy IBM Corp.)
(c) Type 009 nonprinting tabulator

Fig. 28 CTR punched-card machines of 1914 (Courtesy IBM Corp.)
(d) Lever-set gang-punch

multiply two numbers together, and punch the results. Later developments included calculating punches that could add, subtract, multiply, and divide under punched-card control, culminating in the "Card-Programmed-Calculator", or CPC.[36]

5.3 The development of scientific electromechanical computers

In parallel with the development of commercial punched-card equipment, there were several independent efforts in the late 1930's to produce electromechanical computers for scientific applications: the work of Stibitz and Williams at the Bell Telephone Laboratories, the work of Konrad Zuse in Germany, and the work of Howard Aiken and IBM at the Harvard Computational Laboratory. As noted by Aiken,[37] the commercial punched-card machines were inadequate for several reasons:

1 Accounting machines were designed to handle positive numbers, whereas both positive and negative numbers are needed for mathematical purposes.

2 In mathematical formulae, transcendental functions are often required.

3 The prevalence of iterative processes in scientific calculations requires automatic control of the same sequence of operations (looping) rather than a linear flow of different operations.

4 The existing machinery could handle calculations which allowed partial results to be obtained for all values of the independent variable, but many scientific calculations require that complete results be obtained before proceeding to the next, as in recursion relations.

Basically, the available commercial equipment was designed to handle relatively large amounts of input data with a small amount of calculation, whereas scientific calculations required relatively small amounts of input data with a large amount of calculation, thus inverting the emphasis on input-output and arithmetic capability.

The first Bell Labs computer, called the "Complex Calculator", was placed in operation in 1940 to handle calculations with complex numbers.[38] It was demonstrated to the American Mathematical Society in 1940 using a keyboard-printer at Dartmouth linked over telephone lines to the computer in New York City. It used binary circuitry implemented with electromechanical telephone relays. Succeeding machines built by Bell Labs included the Model II (Relay Interpolator) in 1943, the Model III (Ballistic Computer) in

1944, the Model IV (Error Detector) in 1945, the Model V in 1946, and the Model VI in 1950.

The pioneering work of Konrad Zuse in Germany in the 1930's and early 1940's in the development of computing equipment has often been ignored in historical reviews.[39,40] In 1941 Zuse completed the Zuse Z–3, a machine with some remarkably advanced features. Program control, using 8-bit one-address instructions, was punched on 32 mm cinefilm. The Arithmetic Unit operated on 22-bit binary floating-point data, with one bit for sign, 7 bits for exponent, and 14 bits for mantissa. The operations available were: add, subtract, multiply, divide, square-root, and decimal-to-binary and binary-to-decimal conversion. A relay memory of 64 words was provided to hold the data which was entered by a numerical keyboard. Four-decimal output was displayed on a light panel, together with a decimal point indicator. The speed of operation was about the same as the Harvard Mark I, described below. The original machine was destroyed in a bombing raid on Berlin in 1944, but a model, illustrated in Fig. 29, has been reconstructed from the original plans.

Fig. 29 Zuse Z–3 (Courtesy K. Zuse)

After World War II, Zuse continued his work in computer development and today his company is an important supplier of computing equipment.[41]

In 1944 the Automatic Sequence Controlled Calculator, or Harvard Mark I, illustrated in Fig. 30, was completed at Endicott, N.Y., by IBM, and presented to Harvard University. Professor Howard Aiken of Harvard originally proposed this machine in 1937,[37] and design and construction took place between 1939 and 1944. The machine was built largely with standard parts of IBM tabulating machinery through the efforts of Aiken's coinventors from IBM, B. M. Durfee, F. E. Hamilton, and C. D. Lake.[42]

Fig. 30 Harvard Mark–I (Courtesy IBM Corp.)

The essential parts of the machine consisted of:

1 A set of 60 registers for signed constants of 23 decimals each;

2 A set of 72 adding-storage registers;

3 A multiplier-divider unit;

4 Electromechanical tables of the functions $\log_{10} x$, 10^x, and $\sin x$;

5 Three tape-driven interpolators; and

6 A tape-driven sequencing unit for controlling the whole machine.[43]

Data input was made on punched cards; output was either punched on cards or printed on two electric typewriters. This machine was large not only in the logical sense but in the physical sense as well: it was 51 feet long, 8 feet high, and weighed about 5 tons. The operations of which the machine was capable and their associated excution times are shown in Table II. The Mark I remained in operation for more than 15 years, and produced results used throughout the world.

Table II Mark-I operations and their execution times

Operation	Execution time (seconds)
Addition	0.3
Subtraction	0.3
Multiplication	6.0
Division	11.4
$\text{Log}_{10}\, x$	72.6
10^x	61.2
Sin x	60.0

6 CONCLUSION

The modern electronic computer is rightly considered to be one of the most remarkable creations of the human mind; it has influenced modern society in such fields as science, engineering, education, business, law, government, warfare—even the arts. But in the sense that it produces a fixed response to a given input, it is not different in kind from its historical predecessors whose history extends back in human experience several thousand years. It is always difficult to determine why an invention occurred at a particular time—whether it was due to the brilliance of the inventor, the need for the invention, the availability of the necessary technology, or perhaps that it was the next logical step in an evolutionary chain. All these elements were an inherent part of the development of the electronic computer in the late 1940's, and certainly the long historical ancestry of electromechanical, mechanical, and physical aids to digital computation played an important role in the birth of the modern computer sciences.

Acknowledgment

The author is especially grateful to Jeanette Tegtmeier, both for the preparation of the manuscript and for many suggestions for the improvement of the paper.

References

1. Ore, O., *Number Theory and Its History*, New York: McGraw–Hill Book Company, Inc. (1948), pp. 5–7.
2. Bergamini, D., *Mathematics*, New York: Time, Inc. (1963), pp. 18, 19.
3. Smith, D.E., and Ginsburg, J., "From Numbers to Numerals and from Numerals to Computation", *The World of Mathematics* (James R. Newman, Ed.), Vol. 1, New York: Simon and Shuster (1956), pp. 442–464.
4. Richardson, L.J., "Digital Reckoning Among the Ancients", *American Mathematical Monthly*, Vol. 23, No. 1, January 1916, pp. 7–13.
5. Needham, J., *Science and Civilization in China*, Vol. 3, Cambridge: The University Press (1959), pp. 68–80.
6. Guinagh, K., and Dorjahn, A., *Latin Literature in Translation*, New York: Longmans, Green and Co. (1960), pp. 741, 742.
7. Shu-t'ien Li, "Origin and Development of the Chinese Abacus", *Journal of the Association for Computing Machinery*, Vol. 6, No. 1, January 1959, pp. 102–110.
8. Smith, D.E., *History of Mathematics*, Vol. II, New York: Dover Publications (1952), pp. 156–207.
9. Kojima, T., *Advanced Abacus*, Rutland, Vermont: Charles E. Tuttle Co. (1963).
10. Heath, T., *A History of Greek Mathematics*, Vol. 1, Oxford: The Clarendon Press (1921), pp. 46–52.
11. Barnard, F.P., *The Casting-Counter and the Counting Board*, Oxford: The University Press (1916).
12. Taton, R., *History of Science: Ancient and Medieval Science*, New York: Basic Books (1963), pp. 473, 474.
13. Poma de Ayala, Felipe Guamán, *Nueva coronica y buen gobierno (Codex perúvian illustre* . Institute d'Ethnologie, Travaux et Mémoires, Vol. 23, Paris. (Cited by Mason, J.A., *The Ancient Civilizations of Peru*, Baltimore, Md.: Penguin Books (1957)).
14. Wassen, H., "The Ancient Peruvian Abacus", *Comparative Ethnographical Studies*, Vol. 9 (1931), pp. 191–205.
15. Knott, C.G., "The Abacus, In Its Historic and Scientific Aspects", *Transactions of the Asiatic Society of Japan*, Vol. 14 (1886), pp. 18–71.
16. Wolf, A., *History of Science, Technology, and Philosophy in the 16th and 17th Centuries*, Vol. II, New York: Harper and Brothers (1959), pp. 556–563.
17. Serrell, R., *et al.*, "The Evolution of Computing Machines and Systems", *Proceedings of the IRE*, May 1962, pp. 1039–1058.
18. Czapla, V.P., "The Inventor of the First Desk Calculator", *Computers and Automation*, Vol. 10, No. 9, September 1961, p. 6.
19. Smith, D.E., "Pascal: On His Calculating Machine", *A Source Book in Mathematics*, New York: McGraw–Hill Book Co. (1929), pp. 165–172.
20. Smith, D.E., "Leibniz: On His Calculating Machine", *ibid.* pp. 173–181.
21. IBM Italia, *Il Calcolo Automatico Nella Storia* (1959).
22. Chapin, N., *An Introduction to Automatic Computers*, Princeton: D. Van Nostrand (1955), pp. 179–218.
23. Chase, G.C., "History of Mechanical Computing Machinery", *Proc. of the National Conference of the Association for Computing Machinery*, May 1952, pp. 1–28.

24. Fox, T.W., and Hanton, W.A., "Weaving", *Encyclopedia Brittanica*, Chicago: Encyclopedia Brittanica, Inc. (1961), p.861.
25. Wilkes, M.V., *Automatic Digital Computers*, London: Methuen and Co., Ltd. (1956).
26. Archibald, P.G., "P.G.Sheutz—Biography and Bibliography", *Math Tables and Other Aids to Computation*, Vol. II, No.18, April 1947, pp.238–245.
27. Morrison, P., and Morrison, E., *Charles Babbage and His Calculating Engines*, New York: Dover Publications (1961).
28. (The Editors), "Herman Hollerith", *Systems and Procedures Journal*, Vol. 14, Nov.–Dec. 1963, pp.18–24.
29. Rex, F.J., Jr., "Herman Hollerith, the First 'Statistical Engineer'", *Computers and Automation*, August 1961, pp.10–13.
30. (International Business Machines Corporation), *Electric Tabulating Machines*, No.AM–15 (1936).
31. (International Business Machines Corporation), *Highlights of IBM History*, August 20, 1966.
32. (International Business Machines Corporation), *Development of International Business Machines Corporation*, No.AM–1–1 (1936).
33. Jordan, G., *A Survey of Punched Card Development*, Unpublished Master's Thesis, MIT School of Industrial Management, May 21, 1956.
34. Comrie, L.J., "The Application of Commercial Calculating Machinery to Scientific Computing", *Math Tables and Other Aids to Computation*, Vol. II, No.16, October 1946, pp.149–159.
35. Eckert, W.J., *Punched Card Methods in Scientific Computation*, The Thomas J.Watson Astronomical Computing Bureau, Columbia University (1940).
36. Hurd, C.C., "The IBM Card-Programmed Electronic Calculator", *Proc. Seminar on Scientific Computation, November 1949*, New York: International Business Machines (1950), pp.37–41.
37. Aiken, H., "Proposed Automatic Calculating Machine", *IEEE Spectrum*, August 1964, pp.62–69.
38. Stibitz, G.R., "The Relay Computers at Bell Labs", *Datamation*, Part I April 1967, pp.35–41, Part II May 1967, pp.45–49.
39. Lyndon, R.C., "The Zuse Computer", *Math Tables and Other Aids to Computation*, Vol. II, No.20, October 1947, pp.355–359.
40. Desmonde, W.H., and Berkling, K.J., "The Zuse Z3", *Datamation*, September 1966, pp.30, 31.
41. Zuse, K., "Geschichte und Technische Daten der ersten programmgesteuerten Rechenanlage der Welt", (Undated pamphlet published by Zuse KG, Bad Hersfeld.)
42. Aiken, H.H., and Hopper, G.M., "The Automatic Sequence Controlled Calculator", *Electrical Engineering*, Aug.–Sept. 1946, pp.384–391, Oct. 1946, pp.449–454, Nov. 1946, pp.522–528.
43. The Harvard University Computational Laboratory Staff, "A Manual of Operation for the Automatic Sequence Controlled Calculator", *Annals of the Computation Laboratory of Harvard University*, Vol. I, Cambridge, Mass.: Harvard University Press (1946).

CHAPTER 3

Early Electronic Computers

HERMAN H. GOLDSTINE

INTRODUCTION

IN A VERY REAL SENSE the modern computing machine is—one might almost say,—the most important by-product of the research and development efforts undertaken during World War II. To understand the history of the modern computer it is desirable to drop back in time and review some earlier developments. Only through such a discursus can one properly appreciate why the electronic digital computer was developed at all and under what auspices.

The history of a scientific invention is not essentially different from that of any other historical event such as a battle or a coronation. There are definite and precise reasons why the event occurred in the first place and still other reasons why it occurred when and where it did. No history can be considered complete or, indeed, even adequate if it does not attempt a discussion of these underlying reasons. We shall therefore not commence our history with the more immediate beginnings of modern computing but rather with the more remote ones. In doing this our account may apparently overlap on the other parts of this volume. Since it is seeking to ascertain different facts, the degree of actual overlap is very small and not only unavoidable but essential.

WORLD WAR I

The first name to come to our attention in this period is that of Oswald Veblen, then professor of mathematics at Princeton University. His background and career are in themselves most remarkable and his contributions to our subject, although they have the appearance of being peripheral, are essential. We shall see his hand appearing in this history from 1917 until his death in 1960. His contributions to our story are, as we said, essential in character and always at the key moment. In this sense they are like his many other contributions to American science.

Veblen, of Norwegian ancestry, grew up in a great period of mathematical activity in the United States and was fortunate enough to be a student at the University of Chicago along with a group of other men who were destined to make the United States a key member of the world scientific community. We can certainly not name all these men but they included such figures as George D. Birkhoff, Gilbert A. Bliss, Edwin Hubble, Mervin Kelly, Warren Weaver and many others. Veblen's family was itself quite distinguished, numbering among its members Thorstein Veblen and, on his wife's side, the Nobel laureates in physics, Davisson and Richardson. Thus he moved in the leading scientific circles and for this reason was able to affect events profoundly, as we shall see.

In 1917 Veblen went from Princeton to what was then the Sandy Hook Proving Ground of the U.S. Army's Ordnance Department and became the *deus ex machina* behind its scientific effort. Through his innumerable connections he was able to form a scientific cadre of men of topmost stature such as J. W. Alexander, A. A. Bennett, G. A. Bliss, T. H. Gronwall, D. Jackson, F. R. Moulton, N. Wiener, etc. It was this group which was responsible for turning, what is now called, "exterior ballistics" from a purely empirical subject into a respectable branch of the physical sciences. Prior to their work, the classical methods contained in Moulton's words, "... defects of reasoning, some quite erroneous conclusions, and the results were arrived at by singularly awkward methods."* After their work was completed, a full-fledged science had emerged. The differential equations of motion were correctly formulated, and procedures for the numerical solution of these equations were developed.

It was this work which made it inevitable that the ballistics people in the U.S. Army's Ordnance Department should become imbued with the importance of computation at a time in our history when very few scientists were so concerned. Fortunately for our history these men were together at what was then known as the Ballistic Research Laboratory of the Aberdeen Proving Ground in Maryland; the key names are, perhaps, Robert H. Kent and L. Serle Dederick, who later became associate directors.

* F. R. Moulton, *New Methods in Exterior Ballistics*, Chicago (1936).

BETWEEN THE WARS

During this period we "go to another part of the forest" and discover that the electrical engineering community is becoming active in the analog computing field. This activity is perhaps best exemplified by another great figure of our century, Vannevar Bush. This man to whom American Science owes so much made a major contribution to computing by his invention in 1930 of the so-called *differential analyzer*. This machine, an analogue device, is in a sense, the offspring of developments by Lord Kelvin and his brother, James Thomson; the latter designed a device called an integrator, which played a central role both in the differential analyzer, and in a tidal analyzer built by the former in 1876. (It is interesting to read what Kelvin said on that occasion. "The object of this machine is to substitute brass for brains in the great mechanical labour of calculating ...")

Both the ballistics people at Aberdeen and the electrical engineers at the University of Pennsylvania were greatly excited by the possibilities of Bush's machine for their respective problems, which were very (essentially) similar, and went independently to Bush to discuss making a differential analyzer of their own. He acted somewhat in the role of a marriage broker and brought these groups together. (Out of this dual project came eventually the modern digital computer; but this puts us ahead of the story.)

It was agreed that the Moore School of Electrical Engineering would produce duplicate copies of its plans for a differential analyzer and give these to Aberdeen so that the two institutions would have sister machines. This took place near the end of 1934, and a year and a half to two years later both machines were complete.

After a few years, Col. Herman H. Zornig, who was the commanding officer of the Ballistic Research Laboratory at Aberdeen again consulted with the Moore School through conversations with Dean Harold Pender and Prof. J. Grist Brainerd. He was concerned that our government might soon be facing the prospect of a major war without adequate computational facilities. His proposal, which was accepted by Dean Pender and Prof. Brainerd, was to make the university's analyzer available to Aberdeen in the event of a national emergency.

By this means these two disparate computing interests in the United States came together for the first time in a relationship which was to strengthen through the Second World War to the point where the "information revolution" came into being.

The final tying together of computing interests came shortly thereafter when Col. Zornig because of his continued concern approached Thomas J. Watson of IBM with a request similar to the one he had made to Dean Pender. Mr. Watson agreed to make available to the Ballistic Research Laboratory one of each of IBM's machines of that period for a dollar a year for the "duration". Through these efforts of Zornig, his successor Leslie E. Simon and the latter's deputy Paul N. Gillon, these three interests were brought in conjunction. It then remained for these very far-sighted army officers to bring Veblen back to Aberdeen in 1942 as chief scientific advisor to complete the picture.

THE BEGINNINGS OF WORLD WAR II

In his new role as scientific advisor to the then commanding officer of the Ballistic Research Laboratory, Veblen, who was now professor at the Institute for Advanced Study in Princeton, again became most active in recruiting a new scientific staff. He was able to bring back to Aberdeen A. A. Bennett from Brown University to resume his World War I tasks; and he brought in a wide variety of other younger, first-class scientists such as F. John, H. Lewy, I. Schoenburg, I. Segal, J. L. Kelley, E. J. McShane, C. L. Morrey, etc. in mathematics, such as G. Breit, L. E. Delsasso, J. E. Mayer, T. Johnson, H. Lemon, R. Sachs, T. E. Sterne, L. H. Thomas, etc. in physics, and such as E. Hubble and M. Schwarzschild, etc., in astronomy. He brought to the Office of the Chief of Ordnance one of the greatest mathematical figures, Marston Morse, Veblen's colleague at the Institute and Birkhoff's student. He also recruited as a part-time employee or consultant, his close friend, and in some sense his protege, John von Neumann. He is to be the chief protagonist of our story.

This extraordinary man, born in Budapest and educated in various European universities, was at this point in time a professor of mathematics at the Institute for Advanced Study where he was the colleague of Veblen. He was a "mathematician's mathematician"; a universalist in his interests which ranged from the foundations of mathematics to neurophysiology; a master of technique in each of his fields. In addition to these qualities, he possessed a wonderful sense of humor and a gregarious spirit. If one adds to these qualities an almost infinite curiosity, one obtains some measure of the man.

In some sense von Neumann thrived under the war; on the scientific side it made possible the explorations by the best minds of many fields, all of a technical nature. These delighted him in that they provided many stimuli for his highly sensitive and trained intellect. Thus we are not surprised to learn that he consulted for Los Alamos, Aberdeen and for various branches of the Navy and, somewhat later, for the Air Force as well.

His contributions were manifold and it is not our purpose here to attempt to enumerate them. However, there was one theme that ran through many of them and that is relevant to us: he showed many people in many government laboratories how to formulate in mathematical form very complicated physical problems and solve them numerically. Typically these were situations involving hydro- and thermo-dynamical considerations. While this remark may sound simple-minded, nonetheless the notion that it was sensible to formulate mathematically problems of very great complexity was not universally accepted as being useful. Indeed it was and may indeed still be more popular to handle such problems by means of physical experiments, particularly in the area of fluid dynamics, than by computation. In fact, in 1946 von Neumann and the author wrote "Indeed, to a great extent, experimentation in fluid dynamics is carried out under conditions where the underlying physical principles are not in doubt, where the quantities to be observed are completely determined by known equations. The purpose of the experiment is not to verify a theory but to replace a computation from an unquestioned theory by direct measurements. Thus wind tunnels are, for example, used at present, at least in large part, as computing devices ... to integrate the partial differential equations of fluid dynamics".*

To appreciate the situation a little better it is well to recall that it was not until 1928 that the now classical paper by Courant, Friedrichs and Lewy appeared,† It was in this paper they showed that the relative sizes of the meshes in the space and time dimensions mattered fundamentally to ensure what is now called the numerical stability of the solution of hyperbolic- or parabolic-type difference systems. Thus the numerical solution of partial differential equations of parabolic- or hyperbolic-type would really not have been technically possible before this landmark date of 1928.

It was in large measure due to von Neumann that Los Alamos installed a large collection of pre-war type, electro-mechanical, IBM machines and put

---

* H.H.Goldstine & J. von Neumann, "On the Principles of Large-Scale Computing Machines", *John von Neumann Collected Works*, New York (1963), vol. 5, p.4.

† R.Courant, K.Friedrichs, H.Lewy, "Über die partiellen Differenzengleichungen der Mathematischen Physik", *Math. Ann.*, vol. 100 (1928), pp.32–74.

them to work on a large, recurrent shock wave calculation. (Moreover from that point on, through the present, that laboratory has pioneered in the field of large-scale scientific calculation.) However, to a very great extent this equipment was completely inadequate to cope with the problems that were really of deepest interest to the people at Los Alamos.

EXCURSUS

Perhaps we ought to pause now and attempt to understand in somewhat more detail what is involved in some typical calculations that were of interest. To do this we halt the flow of our narrative to make some detailed estimations so that we can better appreciate the pressures that existed at Los Alamos and Aberdeen for computing.

To this end, let us first examine a very simple hyperbolic equation in two variables, x, a distance, and t, the time. In this period to divide the x axis into about 50 subintervals, each of the length Δx, was not unusual. The so-called Courant condition then says that a sound wave cannot travel more than Δx in Δt seconds; this means that such a wave will take et least 50 subintervals Δt of the time axis to move across the x interval and it is not unreasonable to allow sufficient time for four such traversals. Summing up the consequences of these assumptions, we see that they imply 50 points on the x-axis and 200 on the t-axis or 10,000 lattice points in all. At each of these points we may reasonably assume that 10 multiplications will need to be done. Thus for a very simple fluid dynamical problem we can reasonably expect that about 10^5 multiplications will need to be done, plus of course many concomitant operations.

To gain some feeling for how much work is involved in doing a hundred thousand multiplications we need to recall that the then standard IBM multiplier took 7 seconds to perform a multiplication. Thus our hundred thousand multiplications would require 7×10^5 seconds to perform; but there are about 10^5 seconds in a full day and thus our very simple problem would take about one full week of calculating. Moreover, a moderate sized survey, which might easily consist of 75 solutions, would then require at least a year and a half of solid calculating!

Now the calculation we have just considered is based on a spherical symmetry in the sense that only one spatial variable is present. If there is only cylindrical symmetry present, then there needs to be two spatial variables and the number of mesh points goes up to 5×10^5 mesh points and the number of multiplications to 5×10^6—a factor of 50 increase. Thus our work is no longer one full week per solution but a full year! And surveys of any size are now out of the question.

Thus the need at Los Alamos for computation. Clearly, some things could be done and were. In fact just enough to whet the appetites of the people there for bigger things, for vital and useful ones.

The other place where there was a great and recognized need for computing was, as we have said earlier, at Aberdeen. Let us inquire now into the nature of these needs in somewhat more quantitative terms than we have previously done. The Ballistic Research Laboratory, as it was then called, had a variety of responsibilities to fulfill to the Ordnance Department, but its most onerous one was undoubtedly the preparation of all firing and bombing tables for both the Army and what was then known as the Air Corps. A firing table is a fairly compendious set of tables prepared in such a fashion that a gunner can readily aim his piece to hit a given target once he knows certain data. They are of essentially four categories: (1) type of gun, shell, fuse and powder charge; (2) map data such as range, azimuth and height of gun relative to targets; (3) material conditions such as weight and temperature of the powder charge; and (4) meteorological information such as air temperature and density and wind intensity and direction.

It may be sensed from this enumeration that the preparation of even one such table is a very major undertaking and became the full-time pre-occupation of a large staff, as we shall see later. It requires the calculation of perhaps a 1000 so-called ballistic trajectories. What is a ballistic trajectory? It is the path followed by an idealized shell traveling through the air from gun to ground. In calculational units, we may assume that the total distance is again divided into about 50 intervals and that there are about 15 multiplications to be done at each point. Thus the work per trajectory is about 750 multiplications. Such a calculation took from 10 to 20 minutes on a differential analyzer—say $\frac{1}{4}$ of an hour. Thus to do the work involved in calculating 1000 trajectories about 250 hours would be required. However, a firing table is far from complete when the set of trajectories has been finished. Indeed, very much more needs doing so that perhaps 1000 to 2000 hours is more nearly what is involved, i.e., from 6 to 12 weeks.

WORLD WAR II. 1942

Let us now leave off from the details and return to our discussion of the events which are in the forefront of our attention.

We have in some sense set the stage: the prime mover, the War, has started; the University of Pennsylvania has a small contractual arrangement with the Ballistic Research Laboratory, Veblen is busy staffing that Laboratory with scientists, and von Neumann is beginning to "preach the gospel" of numerical computation at Los Alamos.

Very soon Aberdeen discovers that the differential analyzers do not provide nearly enough computing power, and so it starts a campaign to hire young college graduates with science majors to work as computers armed with desk calculators. It is quickly discovered that Aberdeen is a difficult place to which to attract these science graduates in the first place and that there are simply not enough of them in the second. The contract with the University's Moore School is broadened to include a training program in mathematics generally with some emphasis on numerical work. This is aimed at college graduates broadly.

At this point, mid 1942, the author arrived at Aberdeen as one of the young mathematicians brought in by Veblen because he had experience as Bliss' assistant at the University of Chicago in the science of ballistics. Parts therefore of what follows are written from the vantage point of a direct participant; it should, however, be recognized by the reader that such an account is not without disadvantages; unseen prejudices, parochial views, biases, hard positions and views of the very young, etc. The reader must therefore be somewhat on his guard against these.

By sheer luck—from the point of view—he was sent to Philadelphia by Gillon to head the activity there and, with a few hiatuses, was there for the rest of the war. Thus a fairly close connection was established between the University and Aberdeen, and this relationship was to become steadily closer as the war progressed. At various stages in the proceedings virtually the entire faculty and staff of the Moore School was involved in the relationship. Without exception, all these gentlemen went far out of their ways to be helpful. The key names that come to mind today are the Dean, Harold Pender, J. Grist Brainerd in charge of the overall relationship with Aberdeen, Carl Chambers and Reid Warren, who interacted closely at various times and Cornelius Weygandt who was in charge of the analyzer.

It was decided that a fairly sizeable substation of the Ballistic Research Laboratory should be set up at the Moore School to help expedite the production of firing tables. To carry this out the Moore School's training program was broadened and emphasized. Under Prof. Chambers three women, all well-trained in mathematics, stepped up both the recruiting and training: Adele K. Goldstine, Mildred Kramer and Mary Mauchly. More of them later. In addition to the group of about 100 young people who were busy at desk calculators, a lot of energy was put into upgrading the performance of the University's differential analyzer to speed up the production of ballistic trajectories.

The differential analyzer is an *analogy* or *measurement* device which operates with a precision of about 5 parts in 10,000. Like all such instruments it represents a real number as a physical quantity: in this case the position of a continuously rotating disk. Moreover it also, like other analogue devices, is somewhat one-purpose in function. Its basic operation is that of integration and builds other operations out of this one: thus multiplication of two variables u and v is performed by noting that $uv = \int u\,dv + \int v\,du$.

The greatest problem with the analyzer is one of ensuring that it remains calibrated against known standard. Unlike a digital device, the analyzer must be adjusted frequently by skilled technicians until it can reproduce previous calculations accurately. In order to improve the reliability and speed of operation of the analyzer, Weygandt, the Moore School member in charge of the machine's performance, brougt in various associates to advise him. Of these, undoubtedly the key figure was a young—23 years old—graduate student, J. Presper Eckert. Here he first demonstrated the qualities which were later to make a reality of the idea for an all-electronic digital computer. He quickly saw that the key problems of analyzer performance were two in number: how to improve the devices that amplified the outputs of the integrating units and those that followed arbitrary, input curves. These problems are apparently quite different, and it is some small measure of Eckert's genius that he quickly saw how to use one principle to handle both. His resultant devices were great improvements on the old ones and substantially increased the reliability and operating speed of the analyzer. (These improvements were also incorporated on the Aberdeen machine.)

In spite of all this the computational burden increased steadily both at Aberdeen and Philadelphia. There was, as a result, a steady atmosphere of deep interest and concern for computation. It must moreover be realized that the milieu of the Moore School was that of a very small, intimate group with common intellectual and social interests and pursuits. All the relevant people lived in very close rapport from early morning to late at night. During this war period there were no working hours; all periods of wakefulness were for work. Particularly involved at this point in time were Brainerd, Eckert, the author and Dr. John W. Mauchly, a physicist on the staff of the Moore School. He was involved both because of his own keen interest in computation—particularly statistical—and because of his wife's close relations with the Aberdeen group. Out of this involvement came a proposal in the spring of 1943, from Mauchly to the author "summarizing the advantages to be expected from an electronic high-speed computer".

After a good deal of discussion these ideas were presented to Col. Gillon, who was now in the Office of the Chief of Ordnance, and who, in his characteristically generous and intelligent fashion, enthusiastically and effectively endorsed them. This was to be his role throughout the developments of electronic computers as we shall see. In fact more than any other person in government, he made possible all computer developments until his retirement in about 1950. This resulted in a meeting being set up between the Aberdeen and the Philadelphia people to work out an agreement. At this meeting, the operative figures turned out to be Veblen and Simon. The former listened long enough to the presentation to understand the proposal and its potential importance. Having reached this point and having little patience with long and repetitious meetings, he got out of the chair on which he had been teetering and announced to Simon his agreement with the proposal and his suggestion that it be backed wholeheartedly.

THE ENIAC

Again at the crucial moment we find Veblen's influence being brought to bear; we shall see it again in post-war times. Thus the beginnings of the first electronic computer, the ENIAC—Electronic Numerical Integrator and Computer, so named by Gillon. We have seen the march of events which has led up to this one and we have gained some appreciation for its historical necessity. But we have still not addressed the technological problems. Why were they solvable now and not earlier?

Apropos of this, in its publicity releases in 1946 the Army commented interestingly that Eckert brought to the project "engineering and design", Mauchly "fundamental ideas; physics" and the author "mathematics; technical liaison". However, in spite of this, then still novel, "teaming up" of engineers, physicists and mathematicians, success would not have been possible without a major new discipline: electronics. It was the invention of radar by Watson–Watts in 1935 which triggered off the technological developments that have resulted in the entire electronics and computer industries. At this point the field was nascent but growing with breath-taking speed. In 1943 the leading companies were making counters operating at speeds of about 10,000 *bits* or counts per second. The ENIAC staff, however, immediately set a goal of 100,000 per second and achieved it without difficulty.

Electronics at this stage was just emerging from the radio "ham" status. The parts were very crude by today's standards, and everything was directed towards the needs of the audio field. Thus the notion that highly reliable operation could be achieved using radio parts was unbelievable to most engineers of this period. Indeed, the reliability or noise-free requirements of the electronic computer are staggering to contemplate. The ENIAC, after the fact, contained about 20,000 vacuum tubes—standard radio-type triodes—and operated at a fundamental rate of 100,000 pulses per second; thus, in each second, 2 billion operations are possible and, since there are 100,000 seconds in a day, in a day 2×10^{14} occasions in which an error can occur! Any such error could, and with overwhelming probability would, vitiate all that preceded it! Certainly man had never before attempted to build a device with such awesome reliability requirements. Not only had he never attempted any such device but most emphatically not with parts that were notoriously not precision-made or intended for clock-work applications.

This is the background against which to judge the decision to build a large general-purpose digital computer using electronic techniques. This is why it had not been done sooner. Here is the major risk and here is why the men who undertook the work deserve great praise. This is their Mt. Everest.

Let us re-examine the situation to assess the risks another way. At the time of ENIAC's development a large, non-electronic, digital calculator was being developed at Harvard in a cooperative venture with IBM. This machine, the so-called Mark I, went into operation in 1944 and did a multiplication in 3 seconds. The ENIAC did a multiplication in 3 milliseconds $= \frac{3}{1000}$ seconds. This represents, of course, a factor of a thousand acceleration! This is what was to be gained.

Even were the ENIAC to operate only 10% of the time and the Mark I 100%, it would still represent an acceleration of 100. In fact, its performance and reliability were good and it operated most of the time—its failure rate was about two or three per week. To keep it in that state did, in von Neumann's words, require "fighting the Battle of the Bulge daily".

Thus the risks and thus the rewards.

What are the rewards? Why is an electronic machine so fast? Why is it that the very first and most primitive, unsophisticated computer is a thousand times faster than the best electro-mechanical machine? Wherein lie the limitations of the latter and the virtues of the former?

The answer to both lies in an examination of the relative sizes of the basic elements in the two systems. Speed of operation of a total system is at least partly a function of how fast these basic devices can be activated.

What are these devices and how fast do they work? In the electro-mechanical case one finds electric relays and counter wheels, and in the electronic case vacuum tube circuits that are analogs of relays. To operate a relay one must move a small chunk of silver of about one gram in mass a distance of perhaps one centimeter—these numbers are rough. To do this one must overcome the inertia of the silver, its tendency to remain where it was. The faster one wants to move it, the greater the force that must be expended; in fact by Newton's so-called Law of Inertia the force needed is directly proportional to the mass to be moved. Now in the case of a vacuum tube, one wants to move not relay contacts but electrons; they have masses of about 9×10^{-14} grams!

Now it is not fair to compare these two masses, 1 and 10^{-13}, and conclude this is a realistic measure of speed ratios because it is not. But it does serve in crude, first approximation to orient one and to explicate why electronic devices are fundamentally so much faster than any macroscopic ones. Thus even a very primitive machine, the first of its kind, could be a thousand times superior in speed to a sophisticated, electro-mechanical machine. It is also why present-day computers can operate at speeds tens of thousands of time faster than that first electronic one.

This is the reason why electronics has been so fundamental to the computer. The computer or information revolution is based on as simple a fact as this!

THE ENIAC, CONTINUED

The development started on July 1, 1943 and was to continue until the machine's dedication on February 15, 1946. The work was undertaken by a totally dedicated group of engineers, draftmen, technicians, etc. with unmatched *elan*. It is a triumph of Eckert's leadership that he developed such *esprit de corps* in his people. The entire project was under Brainerd's overall direction with Eckert as chief engineer. Under the latter were a number of first-rate men: Arthur W. Burks, Robert Shaw, John H. Davis, T. Kite Sharpless, John Mauchly and later Chuan Chu.

These men decided at the beginning that the way to overcome the reliability problems, just discussed, was to insist on two broad principles. In Burks' words.

> "In the first place, the circuits were manufactured out of carefully selected and rigidly tested standard components which are operated considerably below their normal ratings... The second general principle of design has to do with the method chosen for making the accuracy of the computations independent of the tolerances and variations of the components. The tolerances of vacuum tubes are especially poor (with plate-resistance variations of the order of ± 40 percent, for example), so all tubes are operated as on–off devices: That is, either conducting ... or nonconducting... This means that numbers are never represented by the magnitudes of electrical signals, but only by their presence or absence on wires, and these signals are of sufficient magnitude ... that they are never destroyed by cross talk."*

The machine itself was conceived—in the words of Brainerd, Mauchly, and Eckert in August, 1942 in an appendix to the now historic private memorandum given to the author and the others in the Ordnance Dept. (It is entitled *The Use of High-Speed Vacuum Tube Devices for Calculating*.)—

> "As already stated, the electronic computer utilizes the principle of counting to achieve its results. It is then, in every sense, the electric analogue of the mechanical adding, multiplying and dividing machines which are now manufactured for ordinary arithmetic purposes. However, the design of the electronic computer allows easy interconnection of a number of simple component devices and provides for a cycle of operations which will yield a step by step solution of any difference equation within its scope. The result of one calculation, such as a single multipli-

* A. W. Burks, "Electronic Computing Circuits of the ENIAC", *Proceedings of the I.R.E.*, vol. 35 (1947), pp. 756–767.

cation, is immediately available for further operations in any way which is dictated by the equations governing the problem, and these numbers can be transferred from one component to another as required, without the necessity of copying them manually onto paper or from one component to another, as is the case when step by step solutions are performed with ordinary calculating machines. If one desires to visualize the mechanical analogy, he must conceive of a large number, say twenty or thirty, calculating machines, each capable of handling at least ten-digit numbers, and all interconnected by mechanical devices which see to it that the numerical result from an operation in one machine is properly transferred to some other machine, which is selected by a suitable program device; and one must further imagine that this program device is capable of arranging a cycle of different transfers and operations of this nature, with perhaps fifteen or twenty operations in each cycle. It may be said that even though such a mechanical device were constructed, and even though its speed of operation was considered satisfactory, the number of problems which it can solve within the lifetime which is determined by the wear of its parts would undoubtedly be very small. In stating that the electronic computer consists of components which are exactly analogous to the ordinary mechanical computing machine, it is intended that the analogy shall be interpreted rather completely. In particular, just as the ordinary computing machine utilizes the decimal system in performing its calculations, so may the electronic device. When a number, such as 1216, is to be injected into a particular register, it is not necessary for 1216 counts to be used. Instead, a total of 10 counts would be sufficient, one in the thousands register, two in the hundreds register, one in the tens register, and six in the units register. It is only in this way that almost unlimited accuracy can be obtained without unduly prolonging the time of operation. Electronic devices have been proposed for which this is not true, but such do not seem to merit consideration here."

PROGRAMMING AND CONTROL

We must spend some time gaining a better understanding of "a suitable program device". This is key to an understanding of the impact of the computer on society.

When in 1944 the author first mentioned to Eckert that the world-famous mathematician von Neumann was coming to see the ENIAC, Eckert said that he could easily tell whether von Neumann was indeed a genius. "If he first asks about the logical controls of the ENIAC, I'll accept him as a true genius." Happily this was his first question.

What are program devices or logical controls and why don't we hear of them in connections other than computers? Most mechanisms with which we deal in ordinary life are very simple such as window air conditioners, electric shavers or automobiles. They perform only a few simple functions which we control manually by pressing buttons or moving switches or levers. However, the central air conditioner is already more complex in character. It contains a device for sensing the ambient temperature and based on this it turns on or off the cooling and blower units. This is already a simple program and is under the automatic control of the system. Similarly the electric dishwasher has a cycle of operations that it performs in a given sequence; pre-rinsings, washing, rinsing, drying, stopping. This is another very simple example of a program, and it is controlled by device known as a cam which rotates with time and as it does causes various switches to be opened and closed in a predetermined, orderly sequence.

A program, then in the words of von Neumann and the author "... is that sequence of coded symbols (expressing a sequence of words...) that has to be placed ... in order to cause the machine to perform the desired and planned sequence of operations, which amount to solving the problem in question ... coding a problem for the machine would merely be what its name indicates: Translating a meaningful text (the instructions that govern solving the problem under consideration) from one language (the language of mathematics, in which the planners will have conceived the problem, or rather the numerical procedure by which he has decided to solve the problem) into another language (that one of our codes)".

Why does all this need be done automatically? Why can one not do these things under human supervision? In the case of machines such as our dishwasher the automatic execution of its program saves the housewife from having to remember to restart the machine on its next task and thereby frees her to do other things. In the case of a desk calculator or a sewing machine with a human operator, the entire program of different operations even though it is highly repetitious is, nonetheless, done by a human. The human intervenes in the former case to set numbers into the machine, to press the appropriate button to perform the required operation and to transcribe the results onto a sheet of paper. Moreover this is the basic cycle of the entire calculation and is repeated very many times. It is in fact the program that describes the problem.

These various operations are in reasonable harmony as to speed; that is to say the time required for any one of them is comparable to that for any other. Thus speeding up greatly on only one of them would not produce a revolutionary increase in total speed of operation. If however all but one function are greatly speeded up and one is left at manual speed, then there is still no real increase in speed. Thus, we learn an essential fact about complex systems: each major part or organ must operate at a speed harmonious with all others. If not, the slowest one becomes a bottle-neck and determines the speed of operation of the total system.

Thus human intervention, which is not unreasonable in an electromechanical system, is totally out of the question in an electronic one. Once the idea is accepted that all operations must be performed automatically then the notions of a *program* and a *logical control* emerge quickly to the surface.

In the ENIAC these notions appear quite consciously but primitively. They are largely implemented by a set of devices very like the plug board controls of the early IBM machines and sets of switches. These plugs and switches are scattered over the entire machine and in one unit called the "master programmer". There were trays containing coaxial cables that carried the digital information around the machine and other trays for the instruction pulses. The master programmer was then the unit that automatically caused the various program loops to be reiterated. Each unit contained plugs and switches for its local control. When one of these was stimulated it caused the unit to perform its preset function and upon completion to send a signal along the program trays to the next unit which is to perform a task.

According to M.H. Weik of the Ballistic Research Laboratories* "The ENIAC was placed in operation at the Moore School,

* M.H. Weik, *The ENIAC Story*, Ordnance (1961).

> ... Final assembly took place during the fall of 1945. By today's standards for electronic computers the ENIAC was a grotesque monster. Its thirty separate units, plus power supply and forced-air cooling, weighed over thirty tons. Its 19,000 vacuum tubes, 1 500 relays and hundreds of thousands of resistors, capacitors, and inductors consumed almost 200 *kilowatts* of electrical power."

Writing in August of 1945, von Neumann concluded.

> "The operating speed outclasses everything ... by two or three orders of magnitude: 0.003 seconds for a 10 decimal digit multiplication. Also: about 0.0003 seconds for looking up a value of a function ... about 0.01 seconds for a 10-digit division or square root. The 'inner memory' is not inconsiderable, but much less than ... (20 10-digit or 40 5-digit counters). It is therefore necessary to store intermediate results for any problem beyond the level of a total differential equation, by punching cards. The machine reads and punches cards with IBM attachments at IBM speeds; reading 0.3 seconds per card, punching 0.6 seconds per card.
>
> Since a card can hold 8 10-digit numbers, but usually only 2–4 can be conveniently placed on it for reasons of the inner economy of the machine, this means that it takes about 100 multiplication times to record a number (by punching it). With the existing methods of computing practically no number is ever produced at the cost of so many multiplications—about 3–10 multiplications per number recorded are normal. Hence the speed of the machine is practically regulated by its punching, i.e. recording. Therefore, it is a rather unbalanced device, unless radically new computing methods are developed for its use, with entirely new and higher standards of desirable complication. (This is actually a thing very much to be kept in mind—*mutatis mutandis* such new methods will be desirable for more advanced electronic devices, ... On the other hand this 'machine speed' 'punching speed' principle makes it very easy to estimate the time this device will spend on any given problem and therefore it facilitates the selection of procedure to be used on it.
>
> Thus the ENIAC is far from being an ideal electronic machine or

even as good as one that could be made with the components and the insights that are available at present. However, it is enough of an electronic machine, and the electronic *modus procedendi* is sufficiently powerful in itself, to make the ENIAC far superior to all non-electronic devices on many important problems. If the ENIAC proves reliable and practically workable—and it seems very likely that it will—it will have wide applications, in fields where it may accelerate the present processes by factors of up to 100, and in others, which it will render altogether for the first time accessible to the computational approach. It will be a powerful tool for total and partial differential equations; extensive surveys involving varying several parameters; solving systems of linear equations in n variables for unusual values of n (up to 20 at least), ... Indeed, many of its uses will probably only be discovered after it has been working for some time."

This machine was formally dedicated on February 15, 1946 at the University of Pennsylvania and the era of the electronic computer began. Mrs. Goldstine trained a group of Aberdeen people to be programmers for it and they, together with it, were transferred to Aberdeen. The machine with all its deficiencies in Weik's words "during the period 1949 through 1952 ... served as the main computation workhorse for the solution of the scientific problems of the Nation". These included several very large AEC problems which were undertaken at von Neumann's instigation as well as several hydrodynamical calculations for Prof. Douglas Hartree of the Cavendish Laboratory at Cambridge University and Prof. Abraham H. Taub, now at the University of California, Berkeley. (Indeed, the author still recalls with glee the sending off to Taub of the results of his calculation, that had been done by Adele Goldstine. As a joke they sent Taub the chaff—the material punched out of the cards—with a note saying here are your results.)

This machine which operated until 11:45 p.m. on October 2, 1955 is now a museum piece and parts of it are on display in a first-rate exhibit of computing at the Smithsonian Institution in Washington, D.C. It is the progenitor of all modern electronic computers.

THE BACKGROUND OF THE EDVAC

Having disposed of the ENIAC in these few preceding pages, we are ready to bring on the chief protagonist of our story. His entrance really occurred on a lonely railroad platform in Aberdeen when he casually met the author, who was returning from a periodic visit to Aberdeen from Philadelphia. This meeting was to revolutionize both their lives and maintain them as close collaborators for the balance of von Neumann's life.

Von Neumann now became deeply involved with the ENIAC group and principally with Burks, Eckert, Mauchly, Adele Goldstine and the author. Out of this involvement grew an interest in electronic computers that could more conveniently handle the sorts of partial differential equations that arise out of nuclear explosions. These need larger memories—more ability to store information—than possessed by the ENIAC—20 numbers of 10-digits each or 40 of 5-digits. The magic number of 1000 appeared in the conversations and was to serve as the goal of the new computers until the 1950's. Note this number; it represents a size increase of 50—almost two orders of magnitude over the ENIAC in memory size.

Before, we mentioned the ENIAC speed of 300 multiplications per seconds against $\frac{1}{3}$ per second for electro-mechanical machines—later in the war George Stibitz and Samuel B. Williams of the Bell Telephone Laboratories were to build an electro-mechanical machine with a speed of 1 multiplication per second. This acceleration of 300 to 1000 in speed is what made possible the present-day "information revolution". But it was also the enlargement of 50 times on the amount of information that could be stored.

The 19th century mathematician typically wanted to solve systems of total differential equations: the equations of exterior ballistics or linear circuit theory or acoustics being not atypical. These equations are well understood from a mathematical point of view and require, by 20th century standards, rather modest amounts of computational effort. Thus for example the study of sound phenomena involve motions of the air that are relatively slow while the study of blast phenomena involve very fast motions. Such phenomena are known to be different not just in degree but also in kind. The latter ones bring in very marked physical discontinuities called shock waves—these are responsible for the so-called sonic booms of fast aircraft—which materially alter the situation. The equations of motion are no longer of total differential but of partial differential form. The amount of work—the number of arithmetic operations—increases exponentially, and the amount of storage increases correspondingly.

To achieve this larger memory or room to store numbers became the leading problem for the Moore School group. There were immediately two major problems; in modern parlance, a *hardware*, and *software* one; what type of electronic device could be utilized to achieve so large a memory, and given such a memory how to organize a machine utilizing it properly. This machine went under the code name of EDVAC—Electronic Discrete Variable Computer—and was formally started on January 1, 1945 under the immediate direction of Prof. S. Reid Warren, Jr. of the Moore School, concurrently with the ENIAC construction and testing.

In March 31, 1945 Warren was to write in his first progress report

> "The problems of logical control have been analyzed by means of informal discussions among Dr. John von Neumann, Consultant for the Aberdeen Proving Ground, Dr. Mauchly, Mr. Eckert, Dr. Burks, Capt. Goldstine, and others... Points which have been considered during these discussions are flexibility of the use of the EDVAC, storage capacity, computing speed, sorting speed, the coding of problems, and circuit design. These items have received particular attention; in addition there has been some discussion of the input and output systems which may be most advantageously used for the EDVAC. Dr. von Neumann plans to submit the next few weeks a summary of these analyses of the logical control of the EDVAC together with examples showing how certain problems can be set up."

Actually Warren greatly underestimated what von Neumann was undertaking. He had taken all the "informal discussions" and distilled out of them and his own mind a completely detailed logical design for the machine together with a listing of the machine's instruction code. This report, a 100 page paper entitled "First Draft of a Report on the EDVAC" is dated June 30, 1945 and is perhaps the first example of logical design. Von Neumann invented a geometrical notation for denoting logical functions and operations, as distinct from actual circuit elements, and used this as the vehicle for describing what are now known as the architecture and organization of the new machine. He followed up on this paper with an appendix, which did not appear in the same report, giving the detailed programming and coding of a sorting problem.

In von Neumann's report all the many conversations, arguments, inchoate dreams and partially analyzed ideas were molded into a beautiful and coherent whole. Although this paper had only very limited circulation, because of war security, it was to be a landmark and a style-setter in the field. It is perhaps germane to the history of the period to indicate the topics covered in this document by repeating here part of the table of contents:

1.0 Definitions
2.0 Main Subdivisions of the System
3.0 Procedure of Discussion
4.0 Elements, Synchronism, Neuron Analogy
5.0 Principles Governing the Arithmetical Operations
6.0 E. Elements
7.0 Circuits for the Arithmetical Operations $+, x$
8.0 Circuits for the Arithmetical Operations $-, \div$
9.0 The Binary Point
10.0 Circuit for the Arithmetical Operation $\sqrt{.}$ Other Operations
11.0 Organization of CA. Complete List of Operations
12.0 Capacity of the Memory M. General Principles
13.0 Organization of M.
14.0 CC and M
15.0 The Code

Much controversy arose, shortly after the end of World War II, as to who was responsible for what in the design of the EDVAC. Von Neumann and the author felt they, along with the others, played a key part in the logical design. Since the whole group had worked together so closely, they urged calling all this part of the work a joint effort. This position was later resisted by some and was never resolved satisfactorily from the points of view of von Neumann and the author and perhaps others.

THE STORED PROGRAM

This concept is now universal to computers and arose in connection with the EDVAC. To gain some feeling for its importance one must realize that a general-purpose device must have some means whereby it can be specialized to do any particular task. Thus an automatic clothes washer can rinse, wash, spin, start, stop, use hot and cold water, accept large or small loads, etc. To cause it to do a specific task it has two things, which are essential: switches that decide which operations are to be done and a device for sequencing these operations in a certain order. If these were not present, the machine would not have any generality of purpose and would do only one fixed cycle of operation.

Similarly and to a much greater extent the computer is general purpose. Basically, it can do the familiar arithmetic operations and it can decide which of two courses of action to pursue based on the sign of a number. To make it do any specific problem, even so simple a one as to solve a quadratic equation, it is necessary to express the exact sequence of those elementary operations that the machine is to perform. In the case of machines before the ENIAC the statement of this sequence was done by means of holes in paper tapes or by means of plugs and switches. In the former case the sequence in which things were to be done was achieved by their spatial position on the tape relative to a reading station; i.e., the next instruction to be performed was that one next to be "read" by the reading sensors. In the latter, it was done by having the completion of the execution of any instruction signal, via a wire, to the proper location to initiate the next one.

In the case of the EDVAC the truly great idea was to "code" all instructions in numerical form exactly as if they were numbers and store them in the same location as the numerical material; there was to be a Control whose function was to pick out, *seriatim* the instructions and connect up the machine to execute them. To do this the positions in the memory—the repository for both instructions and numbers—were numbered from 0 to 1023 $= 2^{10} - 1$ and the control was ordered to start at position 0, execute the instruction there and then proceed to position 1 for the next one; etc.; these linear sequences being interrupted only by an instruction to go to a different position, the judgement being made usually contingent on the sign of some number generated in the calculation.

Writing in 1946, von Neumann and the author said,

> "Due to its very nature a general purpose computer has only a very few of its control connections permanently wired in. Apart from certain main communication channels these fixed connections are usually those which suffice to guarantee the device's ability to perform certain of the more common arithmetic processes, such as addition, subtraction, multiplication

and possibly division or square roots. It is the function of the control organization and its associated memory to make and unmake the balance of the connections needed to carry out the routine for a given problem. As we saw ... there are two main methods ... for making these connections. We classify them as follows: (a) The method of establishing all connections *a priori*, as exemplified by the ENIAC; and (b) The method of establishing connections at the moment needed, the instructions necessary for this being stored in some organ such as a paper tape...
We desire to combine the advantages of both (a) and (b) and do so by making one important modification in the scheme (b).
It is evident that the storage of instructions in appropriately coded form on tape is nothing other than the storage of digital information. It might, therefore, as such, also be treated in exactly the same manner, ..."*

THE EDVAC, CONTINUATION

In machine design there are always two sides to the coin: hardware and software. So far we have dealt almost solely with the latter since it was this aspect of the EDVAC which shaped the history of the computer. However, it would be very wrong to pass over in silence the hardware side. Let us therefore tell more of the EDVAC story in a connected way.

Out of the discussions of the group mentioned earlier came, on the business side, a proposal from the Moore School to Aberdeen to fund the development of a new computer which would in the words of a memo from Prof. C.B. Morrey, Jr. then Secretary of the Firing Table Review Board to Simon, dated, 30 August 1944,

"1. ... It is believed that a relatively small amount of further development will make possible the construction of a new electronic computing device that will—

---

* Goldstine and von Neumann, "On the Principles of Large Scale Computing Machines", *Loc. Cit.*

This paper was invited by Prof. Ford of the Mathematical Association of America in 1946 for publication in its Monthly. The authors however never had the leisure to complete the paper, wrote several versions and finally turned attention to other matters at the Institute for Advanced Study, which led to their publishing a report with Burks, which will be discussed later.

> a. Contain a much smaller number of tubes than the present (i.e., that now nearly completed) machine, and hence be cheaper and more practical to maintain.
>
> b. Be capable, because of its greater flexibility of handling many types of problems not easily adaptable to the present ENIAC.
>
> c. Be capable of storing cheaply and at high speeds large quantities of numerical data.
>
> d. Be of such a character that the setting up of a new problem on it will require much less time and be much simpler than in the case of the present ENIAC...
>
> e. It is accordingly requested that a new research and development contract be entered into with the Moore School to develop a new ENIAC along the lines agreed upon by Dr. Brainerd and Captain Goldstine."

On September 13, 1944 Brainerd in writing to Gillon sets forth a formal proposal in which he states, *inter alia*...

> "It is not feasible to increase the storage capacity of the ENIAC ... to the extent necessary for handling non-linear partial differential equations... The problem requires an entirely new approach. At the present time we know of two principles which might be used as a basis. One is the possible use of iconoscope tubes, concerning which Dr. von Neumann has talked to Dr. Zworykin of the RCA Research Laboratories, and another of which is the use of storage in a delay line, with which we have some experience..."

On the basis of this the EDVAC development started on January 1, 1945. The two ideas for hardware implementation of the 1000 word memory are to be the key differences between two families of computers and to influence deeply the directions that IBM and Sperry–Rand were to follow in the future. We shall revert to these ideas several times later.

Very fortunately Eckert had earlier done work on a contract between Radiation Laboratory of MIT and the Moore School. In the course of this development he worked extensively with "delay lines"—devices for causing an electrical signal to arrive at a given node of a circuit later in time than it normally would by travelling over a wire. To understand the relevance of this to our discussion we go now to explain how a delay line can, at least, conceptually be used as a memory. Suppose we are given a delay device of any sort with an input and an output line; let us depict it as

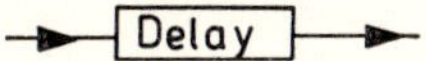

and assume that if a signal is put on the input it will reappear t seconds later on the output. (The sizes of t we are discussing are of the order of one thousandth to one millionth of a second.) Thus we see in the next figure the situation time-wise.

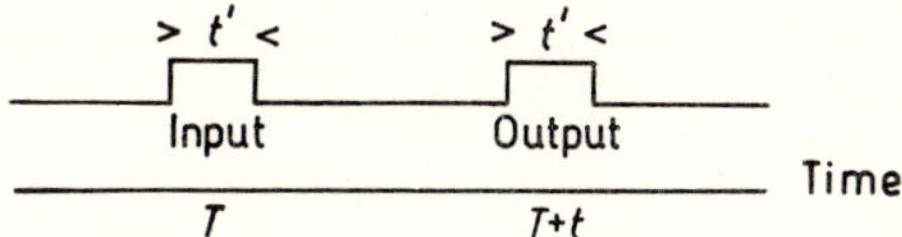

The spikes are of course our signals. Having understood this, we can now describe Eckert's great observation: If the output is tied back to the input, then a single signal introduced at time T will be remembered "forever" in the sense that at any later time $T + n \cdot t$ for $n = 0, 1, \ldots$ the output is still the original signal. Thus

If now t is about the same size as t', the size of the signal, then the delay device can be used to store a single signal. If however t is large compared to t', then many signals can be stored, as can be seen from what precedes. That is, a pattern of signals can be introduced between the time T and $T + t$ and it will repeat itself between $T + t$ and $T + 2t$, etc.

The type of delay device finally resolved upon is known variously as a mercury delay line, an acoustic delay line, an acoustic memory, etc. It is a very ingenious electronic device consisting of a long tube of mercury with quartz crystals at either end. (An electrical signal on such a crystal produces a mechanical vibration and conversely—this is the so-called piezo-electric property.) Thus an electric signal at the input causes a vibration of the input crystal which is communicated to the mercury in the form of an acoustic wave in the mercury. This wave travels to the end of the tube, and sets up a vibration in the output crystal. By this use of the piezo-electric effect an electrical signal at the input results in such a signal at the output suitably delayed. If the signals are 1 microsecond in duration—one millionth of a second—then a tube 5 feet long will hold about 1024 signals and this can represent 32 numbers of 32 binary digits each—32 binary digits is about 10 decimal digits. These become available every millisecond—one thousandth of a second.

In concept this device is not unlike a rotating magnetic or electrostatic disk or drum in which the delay is determined by the time needed for the device to make one revolution.

In closing this very brief glance at the "hardware" situation of EDVAC we should not overlook a very pregnant discussion in von Neumann's report.

He said,

> "12.8 The discussions up to this point were based entirely on the assumption of a delay memory. It is therefore important to note that this need not be the only practicable solution for the memory problem—indeed, that there exists an entirely different approach which may even appear prima facie to be more natural. The solution to which we allude must be sought along the lines of the *iconoscope*. This device in its developed form remembers the state of—$400 \times 500 = 200{,}000$ separate points, indeed it remembers for each point more than one alternative. As it is well known it remembers whether each point has been illuminated or not... The memories are placed upon it by a light beam, and subsequently sensed by an electron beam, but it is easy to see that small changes would make it possible to do the placing of the memories by an electron beam also... It is very probable that in the end the iconoscope memory will prove superior to the delay memory. However this may require some further development in several respects..."

Indeed these words will be the theme for the final portion of this chapter. They charted the future developments which will next be related.

THE END OF WORLD WAR II

The ENIAC, as we said, had a very primitive—by modern standards—form of logical control, a very small electronically controlled memory—20 words—and a very large—20,000 number of vacuum tubes. Also it operated in the decimal system at a rate of 100,000 pulses per second. Its successor, the EDVAC was to have a modern logical control—the stored program and an associated control organ—a large—by the standards of the 1940's—memory and a small amount of equipment about 3600 vacuum tubes. It was to be a binary machine and to operate at 1,000,000 pulses per second.

With the end of the War most of the forces holding the EDVAC group together disappeared: von Neumann returned to the Institute of Advanced Study; Eckert and Mauchly set up their own company which was ultimately to become the UNIVAC Division of Sperry-Rand; Burks and the author were to go to Princeton to join von Neumann. This *diaspora* of course materially affected the EDVAC project both in design and in completion date.

During the latter part of the ENIAC period the author was to succeed in bringing D.R.Hartree, Plummer Professor of Mathematical Physics at the University of Cambridge, to Philadelphia to familiarize himself and, at least partially, through him the European scientific community with what electronic computers can do. Through this connection and others that developed collaterally, such as one with Prof. Stig Ekelöf of the Chalmers Institute at Göteborg, Sweden, a number of Britons—and later of other Europeans—were to come to the United States: M.V.Wilkes who was to become Director of the Mathematical Laboratory at Cambridge and his colleagues S.Gill, J.R.Womersley who was superintendent of the Mathematics Division of the National Physical Laboratory in Teddington; and later his colleague Alan Turing, one of the great logicians of our times.

Inspired by what they saw, D.J.Wheeler, Gill and Wilkes headed a development team that was to produce the EDSAC which was of EDVAC type—indeed its logical design was based on von Neumann's report—and was the first of the delay line machines to be placed in operation—May 1949. This machine was to have great influence on English thinking and was followed by the ACE at the National Physical Laboratory.

Meanwhile in the United States a number of other machines, also inspired by the EDVAC, were under development: SEAC, the National Bureau of Standards Eastern Automatic Computer which went into operation in 1950;

and later the UNIVAC, designed by Eckert and Mauchly and built by the then Remington Rand Company.

In a very different context the IBM Corporation developed and put into operation early in 1948 a very large machine called the SSEC—Selective Sequence Electronic Calculator. In the words of Lord Bowden,

> "... It could be seen from the street by passing pedestrians who affectionately christened it 'Poppa'. It was a very large machine and contained 23,000 relays and 13,000 valves. All arithmetic operations were carried out by the valves and so it was more than 100 times as fast as the Harvard Mark I machine. It had three types of memory; a relatively small high-speed store in valves, a large capacity store on relays, and an indefinitely large store on eighty-column paper tape which could be used either in loops or if need be in great coils... Instructions and input data were punched on tape and there were 66 input reading heads so arranged that the control of the machine could be transferred automatically from one to another, as the calculation proceded. It was probably the first machine to have a conditional transfer of control instruction in the sense that Babbage and Lady Lovelace recommended...
>
> During the last year or two of its life it worked for the U.S. Government undertaking one computation for the Atomic Energy Commission which occupied it for six months... The machine in operation must have been the most spectacular in the world. Thousands of neon lamps flashed on and off; relayed switches buzzed away and the tape readers and punches worked continuously. The machine was dismantled in August, 1952."*

It is incidentally interesting that this large calculation, done principally by von Neumann, Robert D. Richmyer, and Adele Goldstine, resulted in a basic advance in theoretical knowledge. After the calculations were complete, Fermi looked at the results and realized that the equations were very insensitive to changes in a particular parameter. He therefore eliminated it and obtained a sufficiently simple set of equations that he could solve them by so-called analytical means, i.e. by formula. Thus no further calculations needed to be done.

---

* B. V. Bowden, *Faster than Thought*, London (1953).

It should also be recalled that von Neumann and Adele Goldstine were responsible for a substantial modification of the ENIAC to make it a stored program computer. This was done in the late 1940's and was made possible by the close cooperation and assistance of Dr. Richard Clippinger of the Ballistic Research Laboratory and his colleagues.

THE INSTITUTE FOR ADVANCED STUDY. THE BEGINNINGS

The nascent computer field at the end of the War entered its most perilous period. As yet it was highly tentative with great expectations but little to show in concrete results. The large mass of technical people were as yet not "sold" on the idea. Indeed a group of responsible electrical engineers were convinced that the vacuum tube was too unreliable a device to work in this connection, and a similar group of applied mathematicians doubted the need for as much computing power as even the ENIAC was to have.

The only support, at this juncture in history, was the Government: indeed it was only Army Ordnance. The author expended much energy in this transitional period trying to find a way to move the university community into the electronic computing field. At this point such a move was, in the author's judgement, essential to insure free and innovative thinking; later, as we shall see, it was as essential to move out of the university and into industry. But at this stage in history it was still premature for industry to undertake the big research and development programs that have characterized our industrial progress since the 1950's.

After many discussions von Neumann became equally convinced that this was the essential next step and that the Institute for Advanced Study should lead the way. This was not a small decision for the Institute to take since it represented a total departure from its previous policy of undertaking no experimental work. In fact, the change in Institute policy was another major effort by Veblen. With his help von Neumann was able to secure the full backing of the Institute trustees who agreed to put up a substantial sum of money to back the project. Frank Aydelotte, the then Director of the Institute, was a humanist with great breadth of vision who understood the importance of what was to be achieved; at all times he was a strong advocate of what became known as the Electronic Computer Project, as was Lewis Strauss, president of the Institute. Later, as is well-known, Robert Oppenheimer was to succeed Aydelotte both as Director and as an enthusiastic backer of the Project.

Both Princeton University and the RCA Research Laboratory were brought into the project and on November 8, 1945 von Neumann wrote to Aydelotte, Dr. E. W. Engstrom, RCA vice president in charge of the Research Laboratory in Princeton, and to Dr. H. S. Taylor, the dean of the Graduate School of Princeton University a "Memorandum on the Program of the High-Speed Computer Project", There von Neumann said in part:

> "(1) The purpose of this project is to develop and construct a fully automatic, digital, all-purpose automatic computing machine...
> (A) The machine is intended for use in pure mathematics, mathematical physics and applied mathematics... Apart from these things, however, such a machine, if intelligently used, will completely revolutionize our computing techniques, or, to formulate it more broadly, the field of approximation mathematics. Indeed, this device is likely to be in any objective sense at the very least 10,000 times faster (actually probably more) than the presenthuman computer-and-desk multiplying machine methods. IIowever, our present computing methods were developed for these, or for still slower (purely human) procedures. The projected machine will change the possibilities, the difficulties, the emphases, and the whole internal economy of computing so radically, and shift all procedural options and equilibria so completely that the old methods will be much less efficient than new ones which have to be developed. These new methods will have to be based on entirely new criteria of what is mathematically simple or complicated, elegant or clumsy. The development of such new methods is a major mathematical and mathematical-logical program. Certain phases of it can be visualized now, and will necessarily influence the scheme and the control arrangements of the projected machine. Further investigations will have to be carried out in parallel with the development and the construction of the machine. However, the main work will have to be done when the machine is completed and available, by using the machine itself as an experimental tool. It is to be expected that the future evolution of high speed computing will be decisively influenced by the experiences gained and the interpretations and theories developed in this stage of the project."

It now required the establishment of a community of interest between government and university community to ensure the degree of financial backing needed for a successful and long-range project. Here again Gillon was to play a leading role. He, together with this civilian associate Sam Feltman, Office of the Chief of Ordnance, who had been a vital backer of the ENIAC throughout, realized also the importance to the government of getting academic institutions into the computing field. They were to play a key role in this connection. But before describing it, we should first discuss the differences between EDVAC and the projected Institute machine.

COMPARISONS

In von Neumann's EDVAC report he makes mention of two possible types of memories: a delay-line one and an iconoscope or TV-type one. The former, as we saw, is limited by the delay time of the device used; in the case of the mercury lines this was to be one millisecond. The latter does not have such a limitation: any position in the device is as readily accessible as any other. More on this later.

As it turns out the latter type is about 50 times faster than the former. Since this is so, the delay line machine was planned as a serial device in the sense that numbers were to be processed one digit at a time. That is, to add two numbers together the digits were to be processed three at a time—one from each number and one from the carry digit of the previous digit sum—yielding a sum and a carry digit to the next stage.

At this point in history neither memory was a reality and in fact somewhat later Turing was to give such a plausible heuristic demonstration to von Neumann and the author of why the mercury line could never work reliably that he convinced both of them! The fact is, of course, that they did work reliably, and the author is unable to remember the argument that convinced him!

THE INSTITUTE. CONTINUATION

Gillon was impressed with the need to spread the technological risk and agreed to fund development of an Institute machine to get it started. Simultaneously Mina Rees, the then head of the Mathematics Branch of the Office of Naval Research, agreed to fund the development of new mathematical techniques. At a later stage, Joachim Weyl, who succeeded Dr. Rees at ONR, Gordon Lill of ONR, Merle Andrews of the Office of Air Research and Joseph Platt, then in the Research Division of the Atomic Energy Commission joined with Gillon and his successors to provide joint funding for the various activities at the Institute.

Since the Institute had never before engaged in any experimental work, laboratories had to be created, a machine shop set up, drafting facilities produced, stockrooms provided, etc. Ultimately all these non-trivial but unheroic tasks were finished. Concurrent with these von Neumann, Burks and the author with the benefit of many valuable conversations with Prof. John Tukey of Princeton University produced the first of a series of papers that are now classics in the field: Preliminary Discussion of the Logical Design of an Electronic Computing Instrument.

The first edition of this report or paper appeared on June 28, 1946 and represented the logical blueprint for the Institute development.

While these various activities were underway groups of engineers and of mathematicians were being formed. The compositions of the groups varied with time, and it is not relevant to our task to try to state when various people came and went. Nonetheless it is relevant to name the principal figures as long as it is realized they came and left at different times. In fact it was the steady flow of scientists and engineers both from the United States and Europe that did so much to spread the doctrine of the computer to the scientific community.

The first chief engineer was Julian H. Bigelow, and he was supported by James H. Pomerene along with Hugh D. Crane, Gerald Estrin, Ephraim Frei, Leon D. Harmon, Theodore W. Hildebrandt, Gordon Kent, Richard W. Melville, Jack Rosenberg, Morris Rubinoff, Ralph J. Slutz, Richard L. Snyder, Willis H. Ware and S. Y. Wong.

The mathematics group contained Sonya Bargman, Solomon Bochner, Werner Gautschi, Gilbert A. Hunt, Margaret Lambe, Hans Maehly, J. Ben Rosen, Hedi Selberg, Ernest Selmer, Eleanore Trefftz, Bryant Tuckerman. The Meteorology group that was formed later was headed by Jule Charney and contained among others, James Cooley, Bruce Gilchrist, Glenn Lewis, Adolf Nussbaum, Norman Phillips, Irving Rabinowitz. In addition there were large numbers of visitors, all of whom made up part of the intellectual community that was known as the Electronic Computer Project. These included Nils Barricelli, Bert Bolin, George Cressman, Arnt Eliassen, Ragnar Fjörtoft, Kanzabuto Gambo, Ernest Hovmöller, George Platzman, Paul Queney,

Carl-Gustav Rossby, Frederick Shuman, A.L.Stickles, Philip Thompson, George Veronis, Joseph Smagorinsky, Morris Neiberger in meteorology and Andrew and Kathleen Booth, Carl-Erik Fröberg, Joseph Gillis, Manfred Kochen, Richard Leibler, Toichiro Kinoshita, Edward J. McShane, Nicholas Metropolis, Francis J. Murray, Maxwell H. A. Newman, Leslie Peck, Chaim Pekeris, Robert D. Richtmeyer, Daniel Slotnick, Charles V. L. Smith, Erik Stemme, Abraham H. Taub, Olga Taussky, John Todd, in Mathematics. Von Neumann was director of the project and the author associate director. After a few years, Bigelow stepped out as chief engineer to go on a Guggenheim fellowship, and until the project's end in June 1956, Pomerene was the chief engineer. At this point the machine was turned over to Princeton University and the project ended.

This project was then different from its Moore School predecessors in that it was a multiply-directed effort: Logical Design, Programming, Engineering; Mathematics; and Meteorology. The first three efforts were broadly directed toward producing a machine of sufficient power to fulfill the desires and needs of the two other ones; the fourth one was to re-examine and develop, as needed, new mathematical techniques for handling the problems posed by the new machine; the fifth one was to formulate and then validate mathematical models of the weather conditions in forms that were computable on the projected machine.

These efforts were to be successful and were to be quite instrumental in their own ways in shaping the futures of their respective areas. How this was so is the subject of the remaining pages.

LOGICAL DESIGN

As mentioned earlier the first effort of the group was to produce a very complete logical blueprint of the computing instrument desired by the planners. While it is hard today to realize the problems and difficulties of an earlier era, it is not amiss to mention a few of them. Thus in the report mentioned earlier on logical design the section on the arithmetic organ itself is 25 pages of very detailed analysis of how numbers are to be represented in the instrument, how they are to be complemented, added, substracted, multiplied, and divided, how the "rounding operations" are to be performed in the two last cases and finally what registers, etc. are needed for effecting these operations. In fact, the contents of this logical design study are well summarized in its introduction.

"1.1 Inasmuch as the completed device will be a general-purpose computing machine it should contain certain main organs relating to arithmetic, memory-storage, control and connection with the human operator. It is intended that the machine be fully automatic in character, i.e., independent of the human operator after the computation starts. A fuller discussion of the implications of this remark will be given in Chapter 3 below.

1.2 It is evident that the machine must be capable of storing in some manner not only the digital information needed in a given computation such as boundary values, tables of functions (such as the equation of state of a fluid) and also the intermediate results of the computation (which may be wanted for varying lengths of time), but also the instructions which govern the actual routine to be performed on the numerical data. In a special-purpose machine these instructions are an integral part of the device and constitute a part of the design structure. For an all-purpose machine it must be possible to instruct the device to carry out any computation that can be formulated in numerical terms. There must, moreover, be a unit which can understand these instructions and order their execution.

1.3 Conceptually we have discussed above two different forms of memory: storage of numbers and storage of orders. If, however, the orders to the machine are reduced to a numerical code and if the machine can in some fashion distinguish a number from an order, the memory organ can be used to store both numbers and orders. The coding of orders into numeric form is discussed in 6.3 below.

1.4 If the memory for orders is merely a storage organ, there must exist an organ which can automatically execute the orders stored in the memory. We shall call this organ the *Control.*

1.5 Inasmuch as the device is to be a computing machine there must be an arithmetic organ in it which can perform certain of the elementary arithmetic operations. There will be, therefore, a unit capable of adding, subtracting, multiplying and dividing. It will be seen in 6.6 below that it can also perform additional operations that occur quite frequently. The operations that the machine will view as elementary are clearly those which are wired into the machine. To illustrate the operation of multiplication could be eliminated from the device as an elementary process if one were willing to view it as a properly ordered series of additions. Similar remarks apply to division. In general, the inner economy of the arithmetic unit is determined by a compromise between the desire for speed of operation—a non elementary operation will generally take a long time to perform since it is constituted of a series of orders given by the control—and the desire for simplicity, or cheapness, of the machine.

1.6 Lastly there must exist devices, the input and output organ, whereby the human operator and the machine can communicate with each other. This organ will be seen below in 4.5, where it is discussed, to constitute a secondary form of automatic memory."*

While certain improvements on the design suggested in this report have been made, it nonetheless has many universal qualities and has formed the basis for the design of modern computers under the name of the von Neumann organization or architecture.

PROGRAMMING

Shortly after the completion of the logical design report Burks left the Institute after receiving a call from the University of Michigan, where he is now chairman of the department of computer and communication sciences. It became clear to von Neumann and the author that it would be highly desirable to base the subject of programming on sound principles. Initially with the help of Burks and later without him these two studied the problems of coding and programming and their underlying philosophy as well as the actual preparation of many problems to understand how effective was the code. This work appeared starting on April 1, 1947 in a series of publications called Planning and Coding of Problems for an Electronic Computing Instrument, Report on the Mathematical and Logical aspects of an Electronic Computing Instrument, Part II, vols. I, II and III.†

---

* Burks, Goldstine, von Neumann, *op. cit.*

† H.H. Goldstine and J. von Neumann, *von Neumann Collected Works*, vol. V.

In the first of those reports is a somewhat didactic account of the basic philosophy of programming together with the introduction of a geometrical notation for representing programs: these are known as *flow diagrams* and, to this day, are still a standard notational representation for programs or indeed for any rather complex and interactive set of events. In this report it is asserted that,

> "The actual code for a problem is that sequence of coded symbols (expressing a sequence of words, ...) that has to be placed into the ... memory in order to cause the machine to perform the desired and planned sequence of operations, which amount to solving the problem in question. Or to be more precise: This sequence of codes will impose the desired sequence of actions on C (C is the control organ) by the following mechanism: C scans the sequence of codes, and effects the instructions, which they contain, one by one. If this were just a linear scanning of the coded sequence, the latter remaining throughout the procedure unchanged in form, then matters would be quite simple. Coding a problem for the machine would merely be what its name indicates: Translating a meaningful text (the language of mathematics, in which the planner will have conceived the problem, or rather the numerical procedure by which he has decided to solve the problem) into another language (that one of our code).
> This, however, is not the case. We are convinced, both on general grounds and from our actual experience with the coding of specific numerical problems, that the main difficulty lies just at this point...
> To sum up: C will, in general, not scan the coded sequence of instructions linearly. It may jump occasionally forward or backward, omitting (for the time being, but probably not permanently) some parts of the sequencer, and going repeatedly through others. It may modify some parts of the sequence while obeying the instructions in another part of the sequence. Thus when it scans a part of the sequence several times, it may actually find a different set of instructions there at each passage. All these displacements and modifications may be conditional upon the natures of intermediate results obtained by the machine itself in the course of this procedure...

Thus the relation of the coded instruction sequence to the mathematically conceived procedure of (numerical) solution is not a statical one, that of a translation, but highly dynamical. These complications are, furthermore, not hypothetical or exceptional. It does not require a deep analysis of any inductive or iterative mathematical process to see that they are indeed the norm. Also the flexibility and efficiency of our code is essentially due to them, i.e. to the extensive combinatorial possibilities they indicate... Our problem is, then, to find simple, step-by-step methods, by which these difficulties can be overcome. Since coding is not a static process of translation, but rather a technique of providing a dynamic background to control the automatic evolution of a meaning, it has to be viewed as a logical problem and one that represents a new branch of formal logics. We propose to show in the course of this report how this task is mastered."*

Very broadly these paragraphs give at least a feeling for the problems that were grappled with in these early pioneering days. Much that is now trivial or common-place was then unknown or at least highly perplexing. In particular the work on programming was highly innovative in that all previous work on computers including that of Babbage presupposed instructions whose form was invariant during a calculation. Here, for the first time, was a new order: the instructions could and in fact did (and indeed still do) alter each other as the calculation progressed. Thus the instruction stored at a given location was in fact, in principle, constantly subject to alteration.

This ability to modify instructions is not just an aesthetically desirable one, it is logically vital. In 1964, Elgot and Robinson showed "... that no finitely determined machine can compute all recursive sequential functions by fixed programs, i.e., by means of programs whose instructions are never altered or modified".† They also show that instruction modification is the key point and does allow the computer its great logical power. Not only does instruction modification greatly increase the logical power of a machine, it also yields an attractive dividend in the form of much shorter programs than could be so in the contrary case.

In 1949 T. Kilburn at the University of Manchester conceived of the no-

* Goldstine and von Neumann, *op. cit.*

† G. Elgot and A. Robinson, "Random-Access Stored-Program Machines, an Approach to Programming Languages," *Journal of ACM*, vol. 11 (1964), pp. 365–399.

tion of what is now known as the *index register* and then as the B-*tube*. In essence it is a register in which one can store a number to be used for adding to the address part of an instruction. Nowadays many such registers may be attached to a machine and they enable the programmer to obtain a marked increase in efficiency as against the Institute type machine. The interested reader may consult Richards' book for details.*

All programming and coding in this report was in a mnemonic-type language that was very close to that of the machine itself. It remained for John Backus and his colleagues at IBM to invent and develop the notion of a so-called higher-level language, FORTRAN, that was to be a major milestone in the history of computing. This great discovery of Backus occurred in 1957; it revolutionized the programming field and is now the language "spoken" by perhaps a hundred thousand programmers. It also initiated a very great field, now known generally as systems programming.

This FORTRAN language is much closer to that of algebra than is the mnemonic one of von Neumann and the author and goes a long way toward freeing the programmer from the peculiarities of the machine he is using. As a *gratis* "dividend" it also gives a great deal of compression to the code a programmer needs to write. That is, a simple statement of algebraic-like form written by the programmer will, when finally "compiled" by the machine—this is a species of automatic translation by the computer itself in a preliminary calculation from FORTRAN to machine language without human intervention—result in a sequence of from five to ten machine instructions. Thus the programmer needs to write much less, i.e., to exert less energy, than he had to before FORTRAN and to do what he needs to in more natural-to him-language.

MATHEMATICS

This group was organized to develop new techniques and to re-examine old ones. In fact, in a recent article A.S. Householder indicates that the very name Numerical Analysis, which is now the standard nomenclature for this type of mathematics, was just coined in that period.†

In both cases these developments were to be undertaken from the point of

* R.K. Richards, *Electronic Digital Systems*, New York (1966), pp.147–164.

† A.S. Householder, Norms and the localization of roots of matrices, *Bulletin of the American Mathematical Society*, Vol. 74 (1968), p.816.

view of what is known as *numerical stability.* This concept is really quite different from the more familiar and classical concept of physical stability in which a system is stable in the sense that its response to a small input is a similar output.

To understand numerical stability one must understand two things: first, no computer truly performs its stated operations—this is discussed below—and second, the formulation of a physical problem in a form it can be run on a computer is not the exact, original, mathematical expression of the familiar physical laws of conservation of mass, momentum and energy. In regard to the first point, digital computers always round-off the results of arithmetic processes to keep just the number of places that the machine is built to accept. Thus when two n-place numbers—the number base is irrelevant here—are multiplied or divided the exact, correct result is in general more than an n-place number. *Rounding-off* is the process used to truncate this result to n places. (How this is best done is not germane at this time.)

We therefore see that for virtually every arithmetic operation a computer commits an error in the sense of rounding-off; since the total number of such operations to be performed in a typical problem is hundreds of thousands or even millions, it is very important—indeed, even crucial—to know whether the nature of the numerical algorithm is such that these errors amplify or diminish in size. This is the point of our second observation above. Even though a physical or mathematical situation is stable in the sense that any small perturbation will damp out with time it cannot in general be directly put on a digital computer.

In fact, most physical problems and perhaps most mathematical ones, being given numerical treatment, are expressed in the form of differential equations; this is so because the conservation laws mentioned earlier are expressed with the help of the tools of the calculus. These tools or operations are differentiation and integration and they are not basic operations on a digital machine. In fact they must be approximated by finite sequences of additions, subtractions, multiplications, and divisions. Moreover the original equations are frequently implicit and must therefore be replaced by less complex equations that are tractable to numerical procedures.

In sum: The original mathematical formulation of most or at least many problems is such that it is not capable of being handled on a digital—or indeed any finite—machine. Instead it must be replaced by an approximate formulation that is explicit in form and is solely in terms of the elementary arithmetic processes. Moreover even these cannot be done correctly—exactly—

by any finite machine. Thus it is the task of the numerical analyst to ask how to formulate the approximate representation so that it is *numerically stable*, i.e. so that in spite of all these things the final numerical results will be close approximations to the "true" results, i.e., the results that would be obtained if one could solve with exact mathematical rigor the original mathematical problem.

The mathematical group worked on problems of this sort. They attempted to derive procedures that would not only be numerically stable but also well suited for the new computers. It produced and published a number of papers that helped set the tone for numerical analysis for quite a while. These papers dealt with the solution of large systems of linear equations,* with inverting matrices (in fact, the von Neumann dog of this era was named Inverse in honor of a paper that just appeared at that time),† finding the eigen-values of symmetric matrices,§ the partial differential equations of mathematical physics,‡ and finally with various problems of pure mathematics where the computer could have a definite impact.÷

---

* V. Bargmann, D. Montgomery and J. von Neumann, "Solution of Linear Systems of High Order" (*Navy. Bu. Ord.* 1946).

† J. von Neumann and H. H. Goldstine, "Inverting Matrices of High Order". *Bull. Amer. Math. Soc.* (1947) vol. 53, pp. 1021–1099.

H. H. Goldstine and J. von Neumann, "Inverting Matrices of High Order II", *Proc. Amer. Math. Soc.* (1951), vol. 2, pp. 188–202.

§ H. H. Goldstine, F. J. Murray and J. von Neumann, "The Jacobi Method for Real Symmetric Matrices", *J. Assoc. Comp. Mach.*, vol. 6 (1959), pp. 59–96.

‡ A. Blair, N. Metropolis, J. von Neumann, A. H. Taub, M. Tsingou, "A Study of a Numerical Solution to a Two-Dimensional Hydro-Dynamical Problem", *MTAC* vol. 13 (1959), pp. 145–184.

J. von Neumann and R. D. Richtmyer, "On the Numerical Solution of Partial Differential Equations of Parabolic Type", Los Alamos Scientific Laboratory Report, L.A. 657; "Statistical Methods in Neutron Diffusion", L–551, "A Method for the numerical calculations of Hydrodynamic Shocks", *J. Appl. Phy.* vol. 21, (1950), pp. 232–237.

H. H. Goldstine and J. von Neumann, "Blast Wave Calculation", *Comm. Pure Appl. Math.*, vol. 8 (1955), pp. 327–353.

÷ J. von Neumann, "Various Techniques used in Connection with Random Digits", *J. Res. Nat. Bur. Stand. Appl. Math.*, vol. 3 (1950), pp. 36–38.

N. C. Metropolis, G. Reitweisner, J. von Neumann, "Statistical Treatment of Values of First 2000 Decimal Digits of e and π calculated on the ENIAC", *MTAC* vol. 4 (1950), pp. 109–111.

J. von Neumann, H. H. Goldstine, "A Numerical Study of a Conjecture of Kummer", *MTAC*, vol. 7 (1953), pp. 133–134.

In all this work the underlying basis is John von Neumann. There were other papers besides those published and in both categories it is von Neumann whose genius is expressed in some measure in all them. It was he who inspired the group, who set the pace and tune and who contributed mightily to each paper or report.

In closing this discussion we must not overlook the major contributions made by von Neumann to the numerical study of partial and total differential equations. Among other things he took the stability criterion of Courant, Friedrichs and Lewy, which is expressed in a very clear and geometrical fashion in terms of what are called *characteristics* and expressed it in a highly useful form. (These characteristics are curves which are intimately connected with partial differential equations. The purpose of the Courant, Friedrichs and Lewy paper was to show the existence of the phenomenon of numerical stability; this they did in their now classical paper.)

Von Neumann became concerned about developing this fundamental criterion, which is now known as the Courant Condition, into an operative one. This he did by analyzing the solution into its Fourier components, much as a sound wave can be analyzed into its fundamental and harmonics. Out of this analysis he was able to develop an analytic form for the Courant Condition that is not too difficult to apply and which tells one whether the solution is numerically stable.

Another major effort was carried out by von Neumann and Richtmyer. Together they showed how to modify suitably the usual differential equations of hydrodynamics so that shocks, which arise as unwanted, subsidiary effects, could be suppressed without destroying the essential accuracy of the results. This was accomplished by introducing a so-called pseudo-viscosity term into the equations. To this day this is a standard method.

There is not room in this chapter to detail the remarkable work on automata done by von Neumann. He was always very intrigued by the work of Turing as well as that of Pitts and McCulloch and did much in this area. Perhaps his two most significant accomplishments here were his development of the so-called *probabilistic logics* and *self-reproducing automata*. In the former he showed how one could organize highly reliable automata from basic elements that are inherently unreliable. The full significance of this work is perhaps only now beginning to be appreciated.*

---

* J. von Neumann, *Collected Works*, vol. V, pp. 329–378. See also S. Winograd and J. D. Cowan, *Reliable amputation in the prescence of noise*, Cambridge (1963).

His work on self-reproducing automata engaged his interest intermittently from the early 1940's until his death. This was very fascinating to him because it impinged so closely on biology and neuro-physiology. (It is interesting to speculate on what his reactions would have been to the brilliant work on the genetic code that has been going on since his death.) Fortunately Burks has assembled his notes, completed his proofs and published them.*

THE METEOROLOGICAL RESEARCH GROUP

An English mathematician, L. F. Richardson had in the 1910's decided it would be feasible to formulate the differential equations of motion for the atmosphere and to solve them numerically.† This effort was somewhat premature, coming, as it did, before the electronic computer and the mathematical techniques of Courant, Friedrichs and Lewy.

However with the Institute computer well on its way to completion and the mathematical tools needed for the numerical analysis of partial differential equations in hand, von Neumann set about reviving Richardson's ideas. With Charney and the others in the meteorology group he began to study how to formulate thc basic problem. This is not trivial since the complete problem of analyzing the motion of the earth's atmosphere is incredibly difficult requiring an analysis of the influence of mountains, seas, energy from the sun, etc. Since an exact description is much too complex, it is essential that one formulate a suitable model by introducing plausible simplifying assumptions.

The search for these assumptions was the key activity of the group in its early years. Finally, a model was prepared with the help of the assumption that the wind is geostrophic, i.e., its direction is determined by the rotation of the earth. With this assumption and with Rossby's key observation that—with suitable simplification—the hydrodynamical equations can be expressed in a conservative form, the so-called barotropic vorticity equation (this determines the motion of the atmosphere) can be integrated numerically. The next key task of course was to validate the assumptions by very extensive calculations of past meteorological events and by comparing these results with the realities.

* Theory of Self-Reproducing Automata, Edited and Compiled by Arthur R. Busks, Urbana (1966).

† L. F. Richardson, *Weather prediction by numerical process*, Cambridge University Press (1922).

Most of the balance of the group's activities was devoted to this effort and to refining the model to make it more realistic by allowing baroclinic effects; a great effort was made to see if the evolving model was good enough to predict the genesis and development of such phenomena as cyclones.

The original paper by Charney, Fjörtoft and von Neumann appeared in 1950 and described the results of a preliminary calculation on the ENIAC.* In this calculation North America was divided into squares 736 km on a side and the time steps were taken as 3 hours. The authors say,

> "It may be of interest to remark that the computation time for a 24-hour forecast was about 24 hours, that is, we were just able to keep pace with the weather. However, much of this time was consumed by manual and IBM operations, namely by the reading,... of punch cards. In the course of the four 24 hour forecasts about 100,000 standard IBM punch cards were produced and 1,000,000 multiplications and divisions were performed... With a larger capacity and higher speed machine such as is now being built at the Institute for Advanced Study, the non-arithmetical operations will be eliminated and the arithmetical operations performed more quickly. It is estimated that the total computation time with a grid twice the ENIAC-grids density, will be about ½ hour, so that Richardson's dream (1922) of advancing the computation faster than the weather may soon be realized, at least for a two dimensional model."†

The U.S. Weather Bureau and the Air Weather Service of the United States Air Force, as well as the comparable unit in the United States Navy, became much interested in this work with the result that many of the key figures in both became temporarily associated with the group: these included George Cressman, Fred Shuman, Philip Thompson, John Blackburn, Albert Stickles.

The group's activities were very successful and revolutionized the method used by the Weather Bureau for daily weather predictions. The old technique of synoptic forecasting, which is essentially highly skillful extrapolation foreward in time by humans with great training and insight, has been replaced by a numerical one; The Numerical Weather Prediction Unit calculates daily

---

* J. G. Charney, R. Fjörtoft and J. von Neumann, "Numerical integration of the barotropic vorticity equation", *Tellus*, vol. 2, (1950), pp. 237–254.

† J. G. Charney, R. Fjörtoft and J. von Neumann, *op. cit.*

the weather from a model that evolved from those developed by the Institute people. In addition another group known as the Geophysical Research Laboratory has been created to do research into long-range, global predictions. This organization is moved to Princeton and is under the direction of Smagorinsky and the former one, the Prediction Unit, is in Suitland, Maryland, and is under Shuman. In addition the National Center for Atmospheric Research at Boulder has an active group under Thompson studying numerically various aspects of meteorology. There are also excellent groups at various universities such as MIT, Chicago and UCLA, all derivative in some sense from the Institute Group. Abroad this work has equally had great impact, and the numerical prediction of weather is now the accepted mode in all principal countries of the world.

THE ENGINEERING GROUP

In a certain very real sense the entire Electronic Computer Project depended critically on this group. All else was ancillary, even essential but greatly dependent. Thus the engineers played a key part in the entire project, and everyone took the keenest interest in their progress.

The logical design, as was said above, was determined by the report on Logical Design of an Electronic Computing Instrument. The engineering design however represented a major effort in a sense not to be experienced today. At the present time engineering design, at least in the computer field, is highly structured and well understood: the technologies to be used for memory, arithmetic, control and input/output organs are chosen from ones that have previously been developed; the architecture and organization are determined, the logical functions to be performed are described and finally the so-called design automation programs produce the desired circuit drawings. The chief engineer is throughout this highly over-simplified picture very much concerned with a host of managerial questions but almost never with details of circuit design and even less with standards or technology development.

In the late 1940's and early 1950's it was quite different. In those days there were virtually no pre-existing technologies, no pre-designed packages for the components, no automatic programs to translate logical functions into circuit drawings. The chief engineer had to do all these things as well as a host of others.

In the beginning this post was filled by Julian Bigelow and, after he left to study under a Guggenheim Fellowship, by James Pomerene. Each man made unique and essential contributions to the Project. Perhaps the greatest accomplishments of the engineering group and its leaders Bigelow and Pomerene were four in number. First, they initiated the construction of the first machines of the so-called von Neumann type; second, they pioneered the notion that all transfers of information must be done by what are now known as positive latches, that is, at no time is information stored even for extremely short intervals of time in capacitors or inductors but must always be in some device such as a flip–flop or toggle; third, they recognized the potentialities of the Williams tube memory and contributed greatly to making it workable and fourth they did a great deal to pioneer novel chassis design.

The first accomplishment led to a considerable number of copies and near-copies being made. In some sense the machine served as the proto-type for a very large family of machines and even influenced the design of the IBM 701. This group of machines was sponsored by various agencies and included the Ordvac and the first Illiac, built by the group in the Digital Computer Laboratory at the University of Illinois lead by Professors Ralph Meagher and Abraham Taub (the former machine for the Aberdeen Proving Ground and the latter for the University of Illinois)—this Illinois activity is noteworthy, not only for the machines it produced but for the educational role the Digital Computer Laboratory (now the Department of Computer Science) played—the JOHNNIAC at the Rand Corporation, the first of the MANIACs at the Los Alamos Scientific Laboratory, the ORACLE and the AVIDAC at the Argonne National Laboratory, the SWAC of the National Bureau of Standards, the MSUDC at Michigan State University, the SILLIAC at the University of Sidney, Australia, and the BESK at Lund University, Sweden. This list is perhaps not complete and in every sense accurate, but it does give some indication of the early computers that were designed and built under the stimulus of the Institute Project both by governments and universities throughout the world.

It was probably this accomplishment which can be singled out as the decisive influence in causing the computer to be recognized by the leaders in the scientific and business community as the next great tool. In very large measure this was due to the remarkable recognition by the relevant world leaders of von Neumann's prescience. It must be quite rare that one man has had so great an impact on so many different people in such diverse ways.

The group's second accomplishment represented one of a set of important design decisions but one which has remained as a major tenet of computer design through the years. Willis Ware has written: "With each primary storage element there is associated a secondary element whose only function is to remember the state of the primary element during any interval in which it is *changing* or is in a state of change".* While the details of this are rather technical, the accomplishment itself is significant in that is served to set very high standards of reliability without which the computer field would never have reached its current high level of acceptability. This major accomplishment is due to Bigelow, as was so much of the Institute machine.

The ideas of Williams and Kilburn were taken up by the engineers at the Institute and later by those at the University of Illinois. In C.V.L. Smith's words,

> "In the scheme first used by Williams, information was inserted and recovered serially. This mode of operation imposes the least severe requirements upon the capabilities of the tube, but it was quite clear from the beginning that the principles were equally adaptable to the construction of a true random-access memory, and this was at once undertaken ... at the Institute for Advanced Study (IAS) in Princeton, N.J. This development was successful, and before the end of the year 1951 the IAS machine and a version of it built at the University of Illinois were successfully operating with random-access or "parallel' memories, and the Ferranti Mk I, based upon William's original work at the University of Manchester, was successfully using a serial memory. Since then a number of computers based upon the original IAS work have been successfully completed, and also the BESK in Sweden, a computer at the Radar Research Establishment in England, and, last but not least, the IBM 701, of which about twenty were produced before it was superseded by the 704. In all these memories electrostatic deflection tubes were used."†

This development at the Institute primarily due to Pomerene and Meagher at the University of Illinois was rendered very difficult by the extreme sensitivity of the tubes to extraneous, outside electrical "noise" such as results from

---

* W.H. Ware, "The Logical Principles of a New Kind of Binary Counter", *Proc. IRE*, vol. 41 (1953), pp. 1429.

† C.V.L. Smith, *Electronic Digital Computers*, New York (1959), p. 278.

auto ignitions, etc. To protect the memory from such noise they were forced to use very careful shielding techniques as well as very delicate designs for their circuits. In fact it was a crucial 34 hour error-free run by Pomerene in 1949 which first established that the Williams tube principle had the required stability for a computer memory.

The other major problem faced was to get and maintain forty of these tubes in reliable operation. This required at least a daily "tune up" of the system and was at the beginning delicate and difficult. However, as compared to the acoustic delay line, the Williams tube had a great speed advantage. It took about $\frac{1}{1000}$ second for a delay line to recirculate so that on the average it required the order of $\frac{1}{2000}$ of a second to gain access to any desired piece of information. In the case of the electrostatic tube access to any word required about 20 microseconds $= \frac{2}{1{,}000{,}000}$ of a second no matter where located. Thus the Institute type machine had a very great inherent advantage over the EDVAC type. This was amply to justify Gillon and the other governmental backers of the project.

The electrostatic storage tubes were also subject to another difficulty. They always suffered from a phenomenon that made their operation delicate. If a particular spot was read too many times relative to its neighbors, it became contaminated with stray electrons and this could in fact be so bad as to alter some neighbor from a dark spot to a bright one. Such a change was almost always catastrophic since it modified numbers and instructions at random. It was soon found out that this phenomenon could be controlled by programming so that no spot ever was read or written on more than a certain number of times per regeneration—refreshment—of the entire memory contents. This, of course, placed a considerable burden on the programmer. It was later found how to increase greatly the stability of the memory so that this problem was reduced in importance. Along with this better ways were found to "tune up" the memory more quickly.

Originally the input/output organ of the machine was teletype-like equipment; this required 8 minutes to load the memory and 16 minutes to print it out. These times soon turned out to be prohibitively slow, and IBM cards and machines were used instead. This reduced the input or output time to seconds and proved very successful.

A magnetic drum capable of acting as a second stage in the hierarchy of memories for the machine was designed and installed. It contained 4096 words and further increased the machine's capability. It was later replaced by a bigger—16,384 word—drum purchased from the Electronic Research

Associates. (This company played a pioneering role in magnetic drum design and ultimately became a division of Sperry Rand.)

Finally it should be remarked that Pomerene designed and built a very successful graphical display device which had essentially all features of the present-day ones except it lacked a light pen; this was done about 12 years before the present ones and gives some indication of the power of linking first-rate engineers and mathematicians in a project of a "free-wheeling" sort.

THE INSTITUTE. CONCLUSION

After the major engineering thrust was complete the Institute machine had the following characteristics: It had a memory of 1024 words each of 39 binary digits plus a sign digit—this is equivalent to about 12 decimal digits—any one of which could be accessed in about 20 microseconds plus a magnetic drum back up to 16,384 words. It did a multiplication in between 500 and 900 microseconds—recall a microsecond is a millionth of a second. It had a graphical display for exhibiting the results of a calculation at the command of a human as well as punch card methods of input and output. Finally, it handled each word in parallel. Its control operated in a serial fashion fetching out in order the instructions from the memory unless this sequence was interrupted by what is called a transfer of the control.

The entire project was in every important respect a great success and fully justified the Services and the AEC in supporting it. Out of the total effort came a machine that served as prototype for many early governmental and university machines as well as for the first commercial ones, as we have seen. The present day numerical forecasting techniques had their genesis and inspiration there. Much of modern-day numerical analysis derives from what was done there and then. This project, looked at retrospectively, was one of the great intellectual efforts of our day.

OTHER DEVELOPMENTS

During this same period there were a number of other developments going on which must be noted in passing. These include the developments at MIT by Prof. Jay Forrester and his associates of what was known as Whirlwind as well as those at Harvard by Prof. Howard Aiken and his colleagues of his machines called Mark II, etc.

The former had very great significance since out of this pioneering effort came the type of memory technology still in use today: the magnetic core. To quote C.V.L.Smith once again:

> "These were first suggested in 1947 by both J.W.Forrester and A.D.Booth: most of the pioneering development work was done at the Digital Computer Laboratory at MIT under the direction of Forrester and at the RCA Laboratories by J.A.Rajchmann. In the summer of 1953 the MIT work resulted in an operable full-scale memory which replaced the electrostatic memory in Whirlwind I."*

The work at Harvard is especially important because Aiken established there the Harvard Computation Laboratory which served together with the University of Illinois Digital Computer Laboratory as focal points in the world for the concept of the computer or information science departments which would emerge in the second half of the present decade. Thus in some sense the names of Aiken, Meagher and Taub and their associates are perhaps the key ones in the present trend to find the role of the computer in the university as a scientific subject in its own right and not just as a universal tool of the faculty and students.

Finally we must stop our enumeration of early computer developments. To do this it seems not unreasonable to stop with the so-called NORC computer—the IBM Naval Ordnance Research Calculator—which was dedicated on December 2, 1954 by von Neumann. This development was lead by a group under Bryon Havens of the Watson Laboratory of IBM. In concluding this account of the early days of electronic computers it is perhaps appropriate to quote here a part of von Neumann's speech both because of the light it casts on machine developments and on the man himself.

> "It is best to use as a point of reference for this evaluation the first electronic computing machines. These came into existence, beginning with the ENIAC, in the middle and late 1940's. They represented improvements in speed by factors of 100–1000 over the state of the art that had existed before. Thus the level that had then been established was very remarkable in itself, but even measured by this standard the subsequent evolution was unusually fast.

---

* C.V.L.Smith, *op. cit.*, p.286.

The first large-scale and commercially available electronic calculator of a completely integrated type that IBM produced, the 701, was already ahead of the first developments by a factor which, depending on how one measures it, lies somewhere between 10 and 100, but for which an average figure of, say 30, seems most reasonable.

This, by the way, was only about two years ago. Comparing NORC to this, it is again hard to put down a definite, single number as a ratio for comparing the two instruments, each of which may perform differently in different situations, but the ration of improvement in speed is probably a factor of, say 5, if not somewhat higher.

It happens only rarely that a very complex, very sophisticated machine can be improved in less than a decade by a factor of 30, and then in only two more years by another factor of 5. This is very remarkable indeed. Of course, it is less surprising in a very new technology, but computing machine technology is not that new and that simple. Modern computing machines are very complex objects and to make such improvements requires quite exceptional efforts and skills.

In speaking of the importance of NORC, I would also like to add that while the speed which is achieved is of great importance—and I have indicated some of the reasons for this—some other things have been achieved here which are also of very great importance. Those of you present who have lived with this field, and who have lived and suffered with computing machines of various sorts and know what kind of life this is, will have appreciated the fact that this machine has been in a completely assembled state for less than two months, has been working on 'problems' less than two weeks, and that it ran yesterday for four hours without making a mistake. For those of us who have been exposed to the realities of computing machines, it will not be hard to see what this means. It is indeed sensational for an object of this size, and that this has been achieved is quite a reassurance regarding the state of the art and regarding the complexity to which one will be able to go in the future. This is a machine of about 9000 tubes and 25,000 diodes. These numbers are very high, but such numbers have occurred before. But the machines of this type in the past took a year or more to 'break in'.

The last thing that I want to mention can be said in a few words, but it is nonetheless very important. It is this: In planning new computing machines, in fact, in planning anything new, in trying to enlarge the number of parameters with which one can work, it is customary and very proper to consider what the demand is, what the price is, whether it will be more profitable to do it in a bold way or in a cautious way, and so on. This type of consideration is certainly necessary. Things would very quickly go to pieces if these rules were not observed in ninety-nine cases out of a hundred.

It is very important, however, that there should be one case in a hundred where it is done differently... That is, to do sometimes what the United States Navy did in this case, and what IBM did in this case: to write specifications simply calling for the most advanced machine which is possible in the present state of the art. I hope that this will be done again soon and that it will never be forgotten."*

CONCLUSION

By this point in time the germinal ideas of the computer field had largely been conceived. From here on, at least for some time, the advancement of the computer field lay in brilliant technological innovations and in remarkable software advances. Space does not permit us to explore these in the detail they deserve. Instead we intend to close with a few brief paragraphs illustrating the fact that the basic ideas had already been conceived.

Clearly, the present day machine organization was already well understood during the period of our history. It is worth asking whether the newer ideas of today were envisioned and if so to what extent. These newer ideas relate largely to improvements in man–machine interaction and to novel machine organizations involving parallel operations. Already in 1946 von Neumann and the author were writing: "Printing and possibly subsequent graphing of intermediate data ... may also be undertaken to follow the

* John von Neumann, *Collected Works*, vol. V, pp. 238–247.

course of a calculation and to base decisions on the later phases of the computation on these early results... Instead we propose that it be handled by some automatic graphing device as, for example, by an oscilloscope... If on the other hand, he cannot produce *ex ante* such unambiguous and exhaustive formulations, and wishes instead to exercise his intuitive judgement as the calculation develops, he can arrange for that, too... He can then intervene whenever he sees fit."*

As to parallel machine organizations even these have in some sense their genesis in such devices as the B-register of Kilburn mentioned above since here was the first clear departure from the highly serialized von Neumann organization. Thus in some sense Kilburn was already, by reverting to ENIAC-like principles, anticipating the most modern tendencies.

* von Neumann, *Collected Works*, vol. V, pp. 21–22.

CHAPTER 4

Computer Architecture

J. E. THORNTON

1 COMPUTERS, A SCIENCE OR AN ART

ANYONE WHO says he knows how computers should be built should have his head examined! The man who says it is either inexperienced or really mad. Come to think of it, wasn't it G. K. Chesterton who said that the truly *logical* man is the madman?

Consider the short history of computing. During this time, there have been several major changes in the technology of electronic logic and electronic storage. There have been major shifts in the mechanics of programming and control over the computer, and in the application and methods of application. Is there anything, in fact, that hasn't changed? Interestingly, most computer people would say that the computer hasn't changed from the original von Neuman idea. It may not look or sound the same, or even run the same; but it is, to them, the same.

What, then, is different? And what do computer designers actually do? They, of course, attempt to match the shifting ground of technology to the shifting needs of the user. The man in the middle, the computer designer, is probably neither scientist nor artist, in the classic sense. The new term, computer architecture, is rather apt since the structural architect is on similar middle ground.

There is, though, a science to the architecture of computers. There are past successes and a history of increasing knowledge. One can explain new ideas and concepts in comparison to the old. One can set up standards of comparison to measure results, new and old. Problems of the past submit to analysis, and from the analysis a direction can appear. The science is perhaps one of method more than anything: however, even method can assume a major role and produce striking change.

The art of computer design is in making sense out of the many diverse and counteracting needs. For every plea for improvement, there is the demand that there be no change. For every new approach, there is the requirement to service *all* old approaches. There is a considerable temptation to simply drop all contact with the past. However, a "path" must exist by which one can move from old to new.

In practice, the strategy of problem solution has been intimately tied to details of the computer used. The introduction of high level computer language has helped to provide a path of conversion from old computer to new. But for heavily used "production" programs there is a rather general dependence on "machine language" coding to take advantage of the special characteristics of the computer used. The result is, therefore, rather mixed up from the point of view of conversion. One could argue that effort spent on "production" programs, most of it on old computers, is a legitimate and worthwhile investment, even though it confuses ultimate introduction of newer computers.

Attempts by manufacturers to establish compatible product lines, in an effort to smooth conversion, have met with limited success. The problem is similar to building a bridge across a chasm in a deep fog without knowing how far it is to the other side! We might call this blind intuition, only occasionally a successful approach.

There is a strong, and often misleading, desire to describe computer architecture in terms of the explicit configuration of functional modules. With the growing maturity of the computer industry, there should come a realization that what we seek is *balance*. Balance can be achieved between the kinds and amount of storage and processing. Following are some of the factors which make up the "meat" of a computer designer's work toward achieving good balance.

2 THE INFLUENCE OF TECHNOLOGY

An essential element of the design of every new computer has been the introduction of new higher speed electronic components. A rule of thumb commonly used by designers for the speed increase of the basic logic circuits is the factor four to one. This has fortunately been a decent match between the ability to accomplish a speed improvement in a given period of time and the increased needs during that period.

It would certainly be a happy situation if speed increases could be accomplished simply by the basic circuits. Such is the perversity of these things, though, that numerous other aids are necessary. The ability to make significant improvements in performance varies widely within the computing system. Following is a list of the important examples:

Basic logic circuits
New components for basic logic have been introduced every five to seven years, and significant speed increases are introduced every two to three years. Raw speed increase in electronic components is becoming very difficult and taking longer.

Central storage units
New components for central storage have been introduced over ten years apart, and significant speed and capacity increases are introduced every four to five years. The squeeze on raw speed is being felt here too.

Secondary storage units
New components for secondary storage have been introduced over ten years apart, and significant speed and capacity increases are introduced every six years or more.

Unit record equipment
Very little change has been introduced in printers and punched card equipment, and very little prospect of major change is in sight.

Man/machine terminals
The cathode ray display device has been introduced, with prospects of significant improvement in economy.

This highly variable rate of change of the hardware component parts of a computing system has had an important influence on the architecture of the system. Trade-off decisions relating to the technology used fall within the triangle of Figure 1. Each of these three factors is very often in opposition

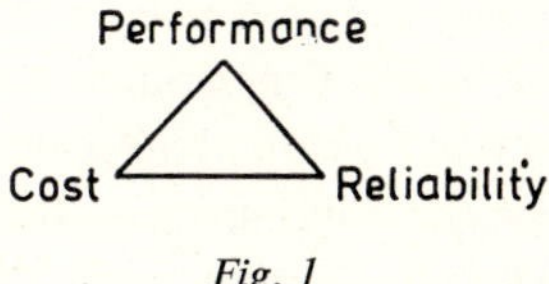

Fig. 1

to the others. Performance can be measured in speed or capacity of storage as applicable. Reliability can be determined both at the component level and for the total number of components used. Cost may not necessarily oppose either performance or reliability except as more parts are used.

A factor recently apparent and growing in importance, especially in very large computers, is the transit time of signals on wires. This factor is often overstated but is nevertheless forcing changes in architecture. An estimate of the percentage of time lost in wiring delay for current computers is about 20 percent. A four-to-one circuit speed improvement would require sixteen-to-one improvement in the currently used two-dimensional back panel area to match it. This is something of an over-simplification since not all wiring is on the critical path of timing.

In the *semi-conductor* technology, the current highest speeds are being attained with discrete component technology. Integrated circuits for the very fast logic fall just below the discrete component for a variety of reasons relating to process compromises in the construction. The current rather primitive state of integrated circuits is offset by the improvement in reliability and cost, arising from the reduced number of "mechanical" connections, among other things.

A potential benefit from advances in the integrated circuit technology can come in the packaging density and in the resulting improvement in wiring time. Significant reductions in power and better cooling techniques must also be made in order to offset the heat density increase going along with the increased packaging density. Reducing power means reducing speed since much of the power is expended supplying current to charge undesired capacitance. This means that the reduced power must also be accompanied by reduced extraneous capacitance. This is a considerable challenge to the semiconductor designer. At one and the same time, he must achieve improved packaging density, reduced power per circuit, increased circuit speed, low cost and high reliability!

In the *high speed storage* technology, the ferrite magnetic core continues to lead. Thin film, plated wire and semi-conductor "memories" are forced to compete in the very high speed end of the storage unit spectrum. The ferrite core is also experiencing success in the large "extended" storage units, augmenting the central storage units of the computer.

The computer is, of course, utterly dependent on the speed, cost and reliability of its central storage. Any technological advance in this component of the computing system is quickly gobbled up. The speed improvement and the increased storage capacity are essential ingredients to increased computing power, to new problem-solving strategies, and to new computer architecture.

In the *secondary storage* technology, magnetic drums and disk units now provide essential on-line storage for the computing system. Major increases in capacity and transfer rate are hard put to keep up with the advances in the central processor and central storage. Nevertheless, the growing importance of secondary storage for the new systems has influenced the architecture.

The newest technology to appear, the *remote terminal*, is opening up the large computer to many simultaneous users, reaping major benefits. In this multiple usage of a central computing system, an opportunity exists for dramatic new technique. The merging of the computer technology with the communication technology must be viewed as one of the major events of importance in this technological age.

A little-discussed area of computer technology is the power and cooling requirement for the large scale computer. Delivering power to the high density electronics and removing the resultant heat is now an integrated part of the total package, along with the plug-in module and the wiring back panel. In very large systems, it is still *possible* to cool by transporting air past the hot components. However, the total air movement is so significant that room designs are critical. New schemes of Freon or water cooling and integrated power supplies make possible very large, yet compact, systems.

Unquestionably, the available technology is the most significant driving force for new system architecture. Since all of the technologies do not advance at the same rate, one might be tempted to establish a "technology independent" architecture. Just as certain "machine independent" programming practices have failed, this kind of scheme is likely to be sadly inefficient.

Reading the direction of technology and its effect can be an unforgiving exercise, though. The bottlenecks of input-output, the allocation of hierarchical storage the effective response to many users at terminals, the overhead of time-sharing a centralized system are some of the problems of system operation which are significantly affected by changing technology.

3 THE COMPUTER AS A SYSTEM

Particularly in the large scale scientific computer system, the "stand alone" usage is long gone. Justification for the idleness during program debugging or manual input-output operations is unnecessary in the new multi-access systems. The modern scientific computer is capable of supervising itself, establishing and controlling a flow of "user" jobs sharing the resources of the system, and accounting for this usage between the "users" and the "overhead".

The former large-problem user who had total control over "his" computer while using it may not be particularly happy about this. However, even he can take advantage of operational facilities unfamiliar in the past.

An especially important result of allowing many concurrent users is the growth of programs for general purpose use. These programs represent an increasing value to the system as a whole because they reduce the need for programming routine operations for every new problem. This public library of programs requires an organized or disciplined approach. Although this is essentially a software task, certain essential hardware requirements must be met in order to provide reasonable efficiency.

The modern computing *system* imposes much more complete integration of hardware and software than every before. Methods of allocating both space in system storage and time in the system processing units involve both the system and the user. Schemes now in use provide elegant definition of "files" which are described by a "name" and protected according to the requirement of the user. For manipulation of these files, it can be seen that there is a linkage between the source language and the system. When suitably requested, the system must put a program in touch with the right file belonging to the right user, while insuring privacy. Users working in groups must be able to share certain files as well as hold separate personal files.

Some computer architecture is rather one-sided in that it meshes very well with a specific language, like FORTRAN or ALGOL, without providing a good mesh with the system facilities. Other architecture emphasizes the operating system itself without especially allowing for the language as a kind of sub-system. It is clear that efficiency demands a well-oiled operating system for both purposes.

In spite of several failures due to over-optimism, the multiple access approach makes possible a conversational form of computing. With present-day scientific computing systems completing tens and hundreds of "jobs" per hour using batch processing techniques, who is to argue against the user simply staying at his terminal and working right along with his "job". Much controversy about time-sharing versus batch processing is too purist. The most efficient systems will unquestionably allow for classes of usage requiring varying internal technique ranging from highly pre-planned batch processing to low efficiency, but fast response, "real time" processing. Figure 2 describes two of the many factors which can vary, response time and storage usage.

In one system the distribution curve of jobs ranges between these two more or less opposing requirements. The spectrum of jobs in large-scale scientific computation has been skewed toward high use of storage and will

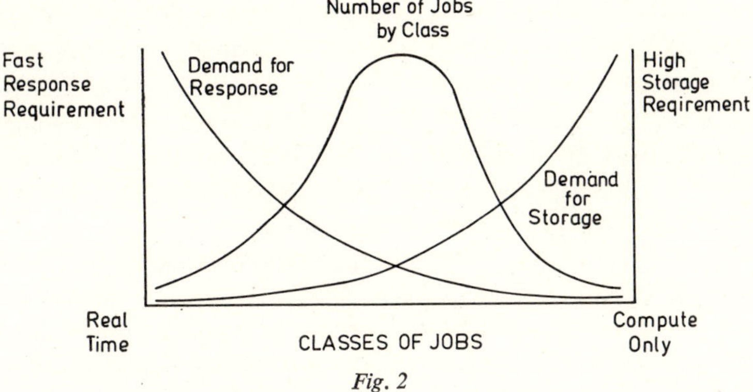

Fig. 2

undoubtedly move somewhat left toward a fast response requirement in order to make use of the multiple access terminal and conversational computation. Other curves can be drawn relating supervisory processing, overhead, storage swapping, and so on.

In order to service many simultaneous users, the large-scale computing system makes use of many aspects of communication systems. The processing resources in this context, and the storage system as well, are considered "on-line" through the communication network. The potential is very high for substantial improvement in the mechanics of access to the computer and its storage. The need for discipline in the usage is also very strong. Development of hardware and software schemes for control, protection, load leveling and minimizing overhead is very sensitive to the entire system, not just to the processor.

4 ORGANIZATION FOR SCIENTIFIC COMPUTATION

Some years ago a significant property was added to the scientific computer hardware. This was the inclusion of floating point addition, subtraction, multiplication and division. Up to that time, the *binary* computer utilized scaling techniques or software floating point techniques. Numerical accuracy requirements then increased since broader problem solution was efficiently possible.

Since the introduction of floating point hardware, the scientific computer has evolved with special emphasis on central processor speed, central storage speed and capacity and on system aspects arising from multiple usage. An interesting technique in the very large computers of recent years is the use of many processing units operating concurrently. Parallel addition, for example, is accomplished by simultaneous treatment of the partial sum and partial carries in separate logic circuits. A number of large-scale scientific computers contain independent functional units which perform major functions in parallel. For example, one computer can perform floating addition, floating multiplication, address calculation, and indexing concurrently while maintaining the correct sequence of events. It can also perform peripheral control over devices, such as printers, magnetic tapes, and punched card equipment, concurrently with central processing. The technique of separating input-output activity and central processing activity makes this concurrency possible. The net effect of such concurrency is a substantial increase in computational speed over and above the speed of the basic electronics used.

Two new conceptual ideas have been added to these "conventional" large-scale scientific computers. These new ideas propose computer hardware and software changes fully as far reaching as the floating point hardware was. The two schemes are the *highly parallel processing network* and the *pipeline processor*. Both of these new computer organizations depend on the ability to organize data into highly structured arrays. Arrays may also be described as vectors or "linearized" arrays.

A second premise for both schemes is the ability to construct huge computers with adequate reliability. This premise results from the increased basic reliability of new components and circuits.

The *processing network* is a two-dimensional array of processors as shown in Figure 3. Processors are shown in a 4 by 4 array numbered 1 through 16. The array could, of course, be larger. Each processor contains arithmetic and storage but limited control and a very simple1/0 interface. Identical instructions are supplied to all processors simultaneously. It is assumed that the highly structured data can be distributed to the processors in such a way that all processors can work simultaneously, thereby achieving in one period of time sixteen (or n) times more work than a single processor. Secondary controls can be built into each processor to allow it to ignore control instructions dependent on data within the processor. This low-level secondary control allows more flexibility; but, of course, the processor is idle during the period described.

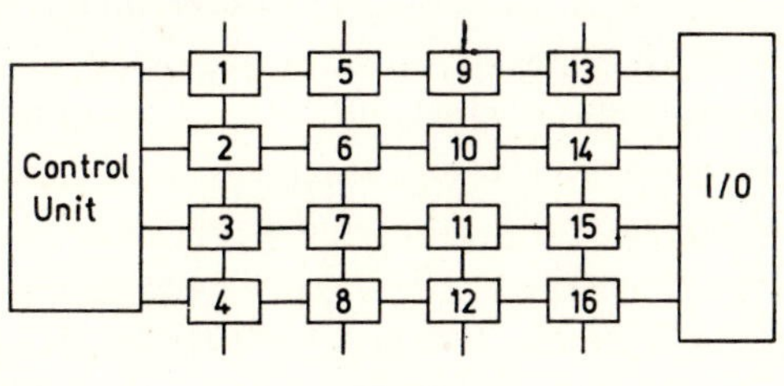

Fig. 3

It can be seen that this scheme will not be one hundred percent effective but will depend heavily on the ability to *structure* data and processing effectively. The proper optimum size for each processor will certainly be something less than a conventional arithmetic mechanism using the same basic technology. Therefore, the scheme ranges in performance from less than a conventional unit (perhaps one-tenth is a good guess) to as much greater as the number of processors used and the percentage of highly structured work allows.

The *pipeline processor* utilizes an approach dependent on the same assumption of highly structured work. However, the data is arranged in storage and "streamed" through the pipeline back to storage at high stream rates. In this context a "pipeline" can be described as in Figure 4.

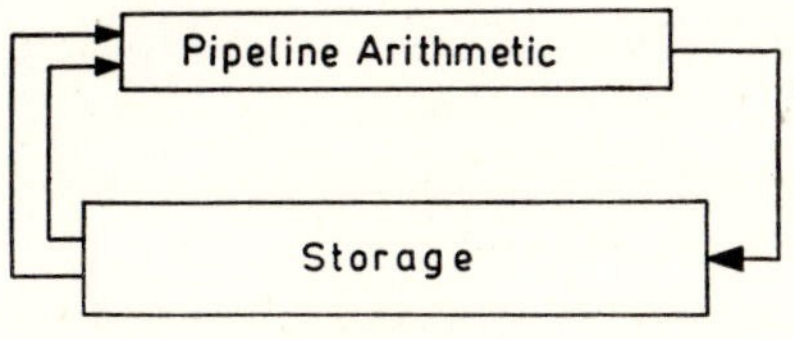

Fig. 4

In general, two input streams of data enter the pipeline arithmetic to be operated on. A single result stream returns to storage. Time must be spent establishing the input streams and filling the pipe. After that initial set-up time, the pipeline arithmetic can be made to deliver results at maximum rate until completion of the function currently being executed.

Since there is only one pipeline arithmetic unit, some added property can be included, probably resulting in a substantially faster conventional unit than *one* parallel network processor. Another factor which is difficult to assess is the fact that the pipeline arithmetic has complete access to central storage whereas the network processors only have limited local access. Cost per bit of large, high transfer rate storage is significantly less than the small fast access storage needed in the network.

Since these two new organization concepts are still experimental, a good comparative analysis cannot be given. Other new organizations have been proposed which may offer some future added computing power. The two described here are indicative of the possible different approaches.

Scientific computers which fit into the *conventional* class also continue to evolve with interesting architecture twists. A very important factor in the very large-scale scientific systems is the distribution of processing "outward" from the central computer to the peripheral devices. A considerable amount of processing is more correctly done near certain classes of peripheral devices; a case in point is the control over a secondary storage unit like a magnetic disk. This processing, or more correctly multi-processing, allows for more efficicnt usage of both the magnetic disk and the central processor.

There has been no obvious "optimum computing system" devised as yet, in spite of the many attempts. The changing opportunities of technology together with the changing needs of the computer user may well combine to prevent one.

CHAPTER 5

Computer Memories

JAN A. RAJCHMAN

THE COMPUTER has pervaded most fields of human activity and may well be the most important innovation of our age. Born out of the technology of communication, it is capable of handling enormous amounts of information at tremendous speeds. What makes it so potent and pervasive is the fact that a single mechanism, given the proper program, can perform any information processing task capable of being specified. The same mechanism can calculate taxes and other items on thousands of paychecks, solve complex equations, control industrial processes, write music, form and compose characters for printing, guide space vehicles or help to teach children. This diversity of tasks—which often surprises even the designers of the machines—is made possible by the simple idea of the stored program.

The trick is to control electronically the nature and sequence of arithmetical and logical processes that are themselves electronic. In other words, what determines whether an addition, a multiplication, a simple juxtaposition or some other operation is executed, what determines the inputs of the operation and what determines the disposition of the result are not built into the machine but are part of the electronic process itself. A program is the enumeration of these determining commands; it specifies the method used for the solution of a problem in detail. It is a demonstrable fact that any determinate information processing task can be performed by a sufficient number of minute steps, and therefore with a sufficiently fast computer can be solved in a reasonable time. When the machine is in operation, both the commands and the numbers or symbols being processed are constantly being taken out of and put into a depository of information known as a memory. This anthropomorphic name was coined by early computer workers; more restrained people, particularly in Britain, use the term "store".

The commands, numbers or symbols needed in a processing task—known collectively as words—are stored in the memory, each with a certain "address". The address identifies the stored word and determines a definite physical location within the memory device. The power and universality of programming arises from the capacity to address the memory selectively, that is, to direct a word into any address and to retrieve it in a very short time, regardless of how the address was previously used. Such selective access is described as "random" to emphasize the programmer's total freedom to dispose of any information under any desired label and to retrieve it at any time, in contrast with "serial" memories, in which information was stored in queues and had to be retrieved in definite sequences. This made it necessary to wait for the desired information while irrelevant material was flowing by, and in general to wait longer with longer queues of information.

Perhaps the most important attribute of random access to a memory is the ease with which it is possible to choose one or another command according to the process being executed, thus allowing branching into one of two or more possible programming sequences. For example, a summation of terms is made by executing whatever operations are prescribed in each term, adding the term to the growing partial sum, and then comparing the number of terms to the total number of terms prescribed. If the prescribed limit is not reached, a new calculation loop proceeds and another term is added. If the number of terms has reached the limit, a new procedure is started: the sum can be printed out, stored for future use or become itself the input of a new process. Such "conditional transfer" and looping are the cornerstones of universal programming and are easy to execute with a random-access memory, since the number of steps in the loop can be readily programmed and conditional transfer merely amounts to the indication of another address to be selected. Clearly a high-speed random-access memory is the essential component that makes possible a modern electronic computer. Let us consider how such a memory can be implemented.

The reader will recall that digital information in a computer is expressed in "bits" each bit being the statement of a single alternative: yes or no, 0 or 1. A group of n bits can code 2^n alternatives, which can express 2^n binary numbers. Accordingly all combinations of four 0's or 1's (0000 to 1111) can express 2^4, or 16, numbers, for example the numbers from 0 to 15. A sequence of bits can equally well represent a collection of arbitrary symbols, English words or artificial words.

Most computers handle a fixed number of bits as a word. The function of a random-access memory is to store the m bits of a word on being supplied with an address specified in n bits. Subsequently the memory will furnish on demand, usually in less than a microsecond, the stored m bits on being supplied with the same n address bits. Such a memory has a storage capacity of 2^n words, corresponding to all possible addresses. Since each word is m bits long, the total capacity of the memory is $m2^n$ bits. In a typical "16 K" word memory n equals 14, so that the actual word capacity is 2^{14}, or 16,384 words; if m equals 40 bits, the total capacity is 655,360 bits.

Since the early 1950's the standard random-access memory has been provided by an array of tiny ring-shaped cores made of a ferrite, an easily magnetized material. In its simplest form the array of cores is threaded by 2^n "word" conductors in one direction and by m "digit" conductors in the other (see Fig. 1). Each core can hold one bit of information, which is stored in terms of the direction of imposed magnetization; in other words, the core "remembers" the direction of the effective magnetizing current sent through it last.

When the memory is in operation, there is a cycle of two steps: "read" and "write". In the read step a current in a given direction is sent through the

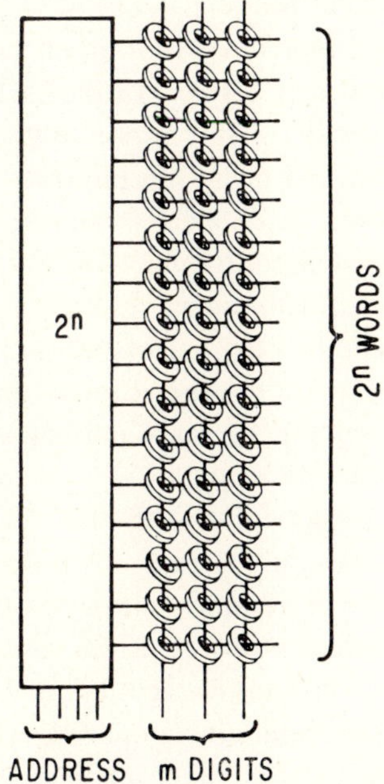

Fig. 1 Word Organized Memory [2D]

selected word line and brings all the cores on it to the same state of magnetization. The magnetic flux of the cores that were in the opposite state is reversed, and this reversal induces voltages on the corresponding digit lines, which in this way sense "destructively" the word information. In a succeeding write step the current in the word line is reversed and also reduced, and simultaneously pulses of current are sent through certain digit lines. The amplitude of the word and digit currents is adjusted to be insufficient to switch a core by themselves but sufficient to switch it when they act together. As a result only the cores on the selected word threaded by energized digit lines will switch; all others will remain unaffected. In the write step the read information can be rewritten, thereby conserving the information within the memory system in spite of the destructive readout. Until the information is rewritten it is momentarily in the circuits rather than in the cores. Alternatively, new information can be entered in the selected word address that was "cleared" in the read step.

Each of the word lines has to be energized by an active device such as a vacuum tube, a transistor or a diode. Hence a typical 16–K memory would require 16,384 devices, one for each address in the memory. In addition, circuits are needed to decipher the address code, whose function is to select one of the devices for every combination of n input bits. These circuits usually double the required number of devices. In the early days of the computer, when the only suitable devices were vacuum tubes, such a large number of devices would have made core memories impractical if they had had to be organized in the manner just described. For that reason this simple organization, called "word-organized", or "2D", was not the first one used.

The concept of coincident addressing or "3D", made it possible to have a much smaller number of addressing circuits. In the most widely used form of coincident addressing the cores belonging to the bits of the word are distributed among as many two-dimensional square arrays as there are bits in a word. Thus a memory containing 40-bit words requires 40 arrays. A 16-K memory then calls for 16,384 cores (128×128) per array. Four wires are threaded through each core: an X wire and a Y wire are connected in series through all the planes; two more wires, one to "sense", and one to "inhibit", are separately threaded through all cores in each plane (see Fig. 2 and Fig. 3).

In the read step, pulses of equal strength are sent simultaneously through selected X and Y lines. Here again the amplitude of any one pulse is insufficient to switch any core by itself but sufficient to do so when acting in coincidence with the other pulse. As a result, in each plane the core at the

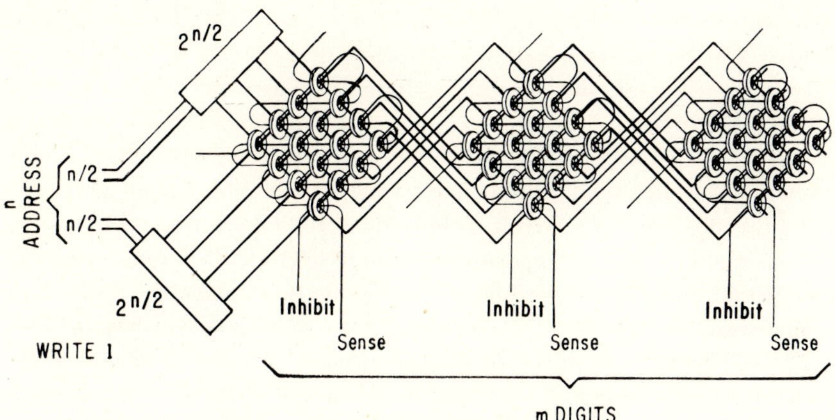

Fig. 2 Current Coincident Memory [3D]

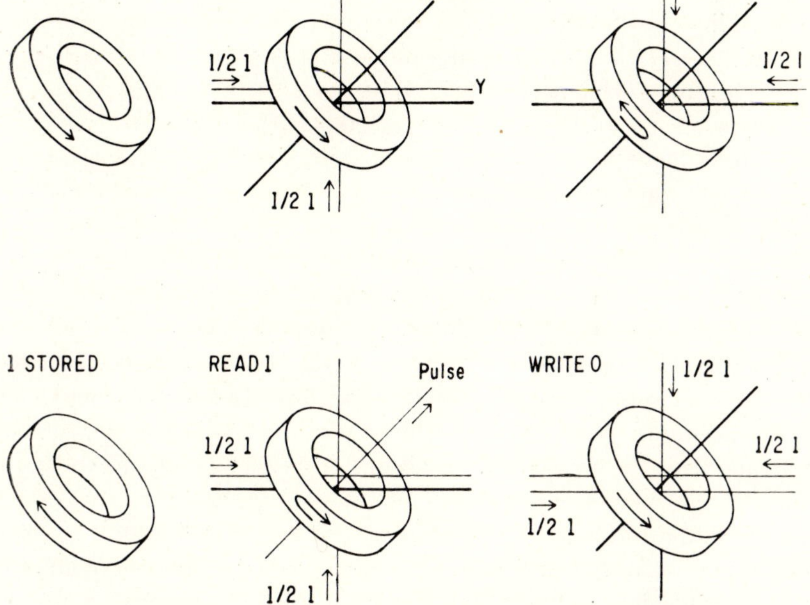

Fig. 3 Reading and Writing in Current Coincident Organization

intersection of an X line and a Y line will switch or not, depending on its magnetic state; no other core will be affected. The voltages induced by the switching of the selected cores appear simultaneously on the corresponding sense lines. In the write step, X and Y currents of the same amplitude but reverse direction are applied, and they tend to switch all 40 selected cores to the opposite direction. This tendency is canceled on selected planes by simultaneous application in the corresponding inhibit windings of currents of the same amplitude but opposite direction. As before, the write step is used to rewrite information (see Fig. 3). Addressing in a coincident memory requires two sets of $2^{n/2}$ driving devices. In our example, where $n = 14$, $2^{n/2} = 2^7$, or 128, driving devices per set; this is far less than one set of 2^{14}, or 16,384 driving devices. The gain results from the participation in the decoding function of the core itself, which responds to two signals but not to one, and thus acts as an "AND" gate. (In the binary logic of computers an "AND" gate is a circuit whose output is 1 if, and only if, all input variables to the circuit have the value 1.)

Coincident addressing is best achieved if all cores have identical magnetic characteristics known as rectangular hysteresis loops (see Fig. 5). It is then possible to choose a magnetizing current that will bring the cores on the selected lines just below the "knee" of the loop without causing any switching whatsoever, and yet that will cause complete switching of the selected core when two such currents act in coincidence.

In practice it is fairly easy to approach these ideal characteristics, so that the cores will not "forget" their magnetic state even when they are subjected to millions of half-selecting pulses that tend to change that state. This is the only requirement in a 2D word-organized memory. In a 3D memory it is necessary to prevent small voltages induced in the sense windings of half-selected cores from "masking" the voltage of the selected cores. These numerous voltages (2×127, or 254, in our example) can be made to cancel each other by proper threading of the sense winding so that half cores induce voltages of opposite direction from the other half. Because the cores are not perfectly uniform and the hysteresis loop has not only a slope but a curvature, the cancellation is not perfect. This is one of the factors limiting the size of such current-coincident memory planes.

In addition to having a rectangular hysteresis loop the core should be capable of fast switching. Cores of nickel–iron alloy in the form of ultrathin ribbon, made during World War II for high-frequency magnetic amplifiers, had the adequate properties and made it possible to demonstrate the opera-

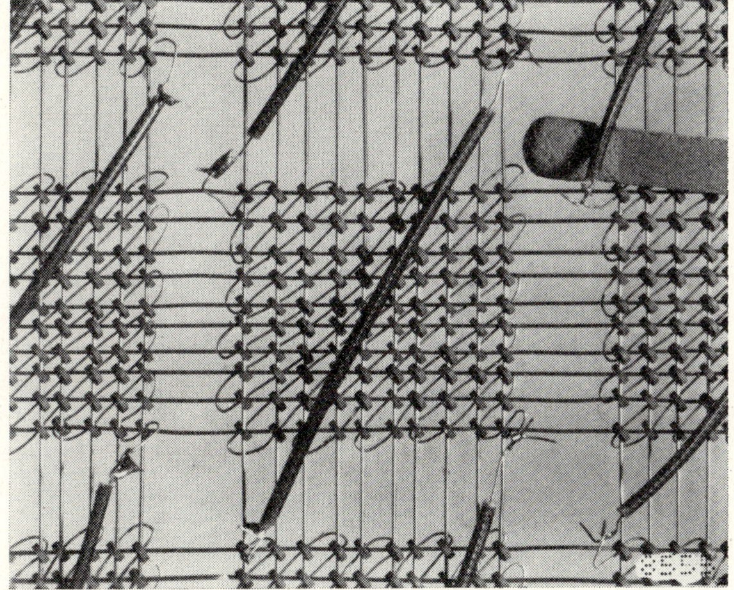

Fig. 4 Early current coincident core array

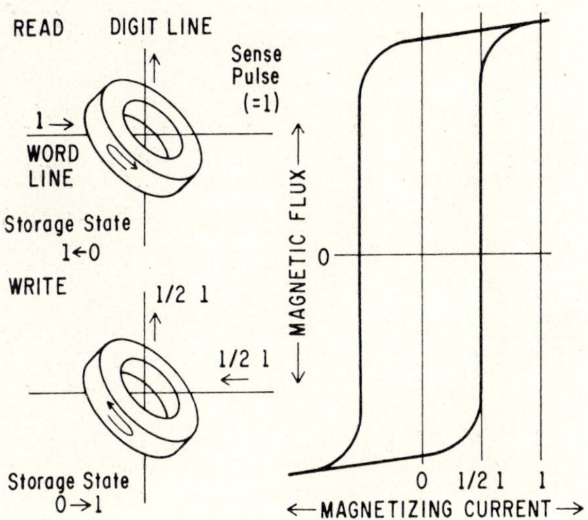

Fig. 5 Hysteresis loop of a core

tion of the first memories. Such cores were delicate and expensive. Fortunately ferrites—fast switching magnetic materials made of oxides of iron, manganese, magnesium, zinc or lithium—had been developed during the 1920's and had been perfected for transformers and the electron-beam-guiding yokes of television picture tubes. Minimum hysteresis was sought for these applications. The maximum hysteresis square loop needed for memories was achieved by ingenious modifications of composition and processing, and today many excellent ferrite core materials are available.

Fine powders of these materials, suspended in a binder as a slurry, are pressed into the shape of cores on automatic machines and baked. The cores are then tested and sorted automatically. The result is a low-cost, high-quality, ceramic-like element with a uniformity not surpassed by any other electronic component. The smaller the core, the less driving current it requires, the faster it switches for a given current, the tighter it can be packed to minimize delays along windings but the more difficult it is to handle and thread. As the art has progressed over the years, standard core sizes have gradually decreased from 0.080 inch outside diameter and 0.050 inch inside diameter, known in the trade as an 80/50 core, to 50/30, 30/18, 20/12 and recently even to 12/7 (see Fig. 6). Typical arrays of 20/12 cores contain 16,384 of them in a square plane that measures only 6.4 inches on a side.

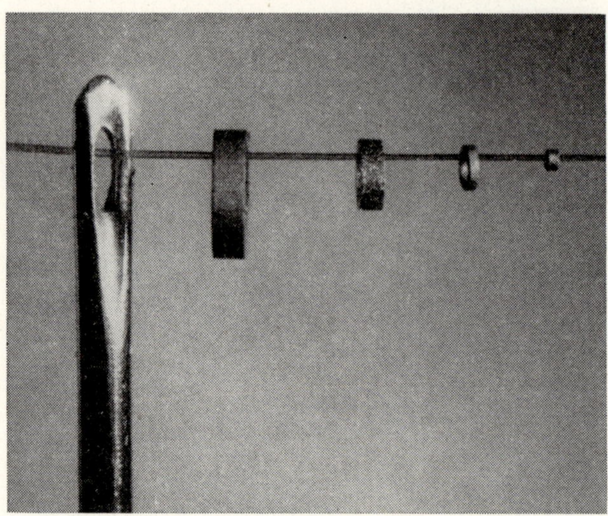

Fig. 6 Evolution of core sizes from 0.080″ O.D./0.050″ I.D. to 0.020″ O.D./0.012″ I.D. Cores threaded on human hair of 0.004″ diameter

The cores are wired into arrays by painstaking handwork with only rudimentary mechanical aids. The operation is delicate and expensive, but mechanizing it is economical only for larger arrays used in very large quantities. Much of the industry finds it more advantageous to use artisan labor, and some seek it at low wages in Hong Kong, Taiwan and Mexico. The situation is somewhat ironic: the heart of the computer, which itself is the symbol of mechanization, is made by the age-old kind of labor that produced brocades and carpets. The objective of this labor is not to turn out a few exquisite art objects but rather to mass produce uniform core arrays. An estimate of the production of core arrays in the United States alone involves the threading of more than 20-billion cores in 1968.

Made as they are, the core arrays have provided reliable, fast random-access memories for practically all computers in use today. At the same time the possibility of avoiding both the manufacture of cores and the threading of them by hand has fascinated many inventors from the beginning. The stimulus has been not only to lower costs but also to increase speed of operation. The principal goal has been to produce "integrated" memories—memories in which the active elements and their connections are mechanically fabricated in a unitary process.

The problems have turned out to be difficult, and for years the constantly improving core technology has prevailed. It was found nearly impossible to obtain any magnetic element in integrated form that had a sufficiently square hysteresis loop or a switching threshold sharp enough and uniform enough from element to element to allow current-coincident operation. Therefore only a 2D word organized system is possible. For the larger decoder required in word-organized memories, nonintegrated arrays of cores were first used and were found rather expensive and inefficient. Today comparatively inexpensive semiconductor diodes permit a matrix arrangement of word lines for the first level of decoding, which makes a combination of cores and diodes equivalent to a coincident-addressed core array and about as economical. It was also found that integrated technologies are subtle and not necessarily inexpensive. Nevertheless, a great deal of progress was made; some integrated memories have been introduced commercially, and the initial hopes are still valid, particularly for high speed. There are three significant contenders: the monolithic ferrite, the flat film and the plated wire.

The monolithic ferrite memory is made from the same type of material as the cores. The slurry is spread by a blade onto sheets. During this operation conducting parallel lines of a refractory metal are also formed within the sheets. Two such sheets, with their conducting lines at right angles, face each other and are separated by a third sheet of ferrite to form a sandwich. The sandwich is then heated, pressed and sintered to produce a monolithic structure (see Fig. 7 and Fig. 8). The conducting lines define the memory elements at their crossings. One set of lines are word windings, the other digit windings. Because the lines are at right angles to each other, a current applied to a selected word conductor switches magnetic flux along a path that does not link the digit conductor. If coincident digit pulses are applied, however, there

Fig. 7 Monolithic ferrite memory. Each ferrite sheet contains 64×64 conductors

will be flux common to the word lines and the digit lines at the corresponding crossings. The application of a word pulse in the opposite direction switches mutual flux and induces sense voltages in the digit lines.

The monolithic integrated structure not only is easy to fabricate but also makes it possible to have very small elements. Experimental laminates about 0.005 inch thick have elements with an effective diameter of 0.003 inch. The close spacing of the lines (100 to the inch) corresponds to what can be achieved with an integrated row of diodes; indeed, fully integrated combinations of laminated ferrite and diodes have been made. One of these had a

Fig. 8 X-ray photograph of experimental monolithic ferrite sheets with 256×100 conductors

capacity of 65,416 bits and operated on a cycle of 400 nanoseconds. (A nanosecond is a billionth of a second.) While not commercialized, the monolithic ferrite memory is still an important possibility, particularly for mobile applications because of its compactness.

The other two approaches to integrated memories involve the use of nickel–iron alloys called permalloys rather than ferrites. One difficulty with metals is that eddy currents develop and slow down the switching speed. If the thickness of permalloy sheets is held to about 0.0025 inch, the slowing is negligible. If the thickness is further reduced until it is comparable to the wavelength of ultraviolet radiation—say less than 3000 angstrom units—entirely new properties appear. Switching can then be achieved by the rotation of magnetization, which is inherently a much faster process than the one that takes place in magnetic cores. In cores switching involves a movement of the walls of magnetic "domains". Thin permalloy films can be made to exhibit strong differences in magnetic properties along different directions of magnetization. In a "hard" direction there is practically no hysteresis but in an "easy" direction at right angles to it there is an almost perfect hysteresis loop (see Fig. 9). Furthermore, thin films are readily deposited by evaporating or electroplating, which lend themselves nicely to integrated fabrication. These advantages became apparent in the middle 1950's and spurred much activity.

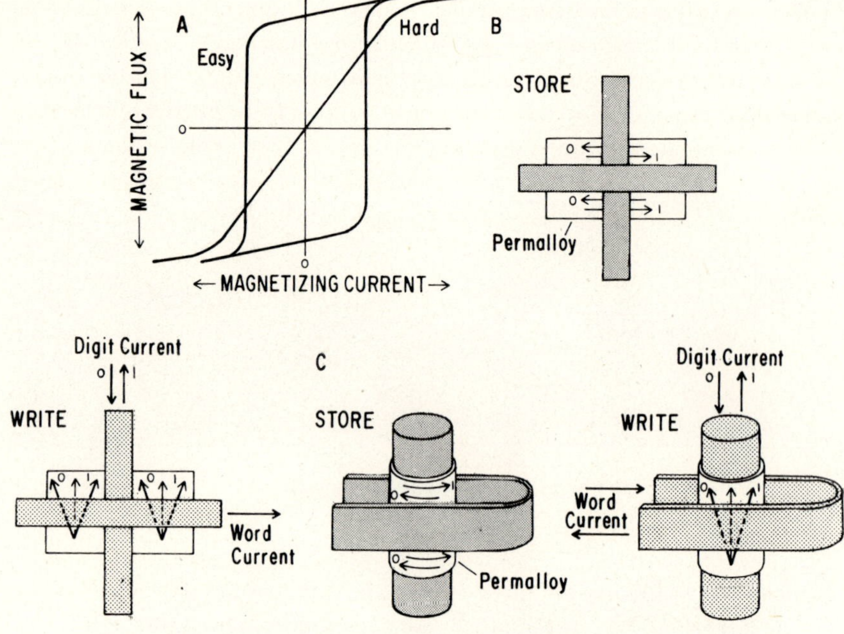

Fig. 9 Thin magnetic film elements—A. Easy and hard hysteresis; B. Flat elements; C. Cylindrical elements

In one approach permalloy is evaporated in a vacuum onto glass or metal to a thickness of between 1000 and 2000 angstroms in the presence of a direct current magnetic field, which produces the desired direction of magnetization. Separate spots of permalloy are obtained either by the evaporation through a mask or by etching a continuous film. Mylar sheets with photographically formed copper lines are laid on the film and provide the word and digit lines (see Fig. 10). The word lines run in the easy direction of magnetization. When a spot is storing a pulse, it is in one or another of the easy states of magnetization. In the read cycle a current pulse is applied to the selected word line; the corresponding spots are forcibly magnetized in the hard direction and thereby induce sense voltages on the digit lines in a direction corresponding to their state. In the write cycle or in rewrite, currents are applied to the digit lines while the word current is still on; the combined currents tilt the magnetization one way or another from the hard direction and thereby establish the easy direction at the termination of the pulses.

This arrangement is attractive in that a small digit current suffices to trigger the spot into the desired state. In practice the directions of magnetization are not perfectly aligned and vary from element to element, so that a larger digit current is required to flip the flux in the desired direction in spite of irregularities haphazardly favoring one direction or another. When these imperfections are severe, the minimum digit current can be so large that it can cause elements of unselected word lines to "creep" from one magnetic

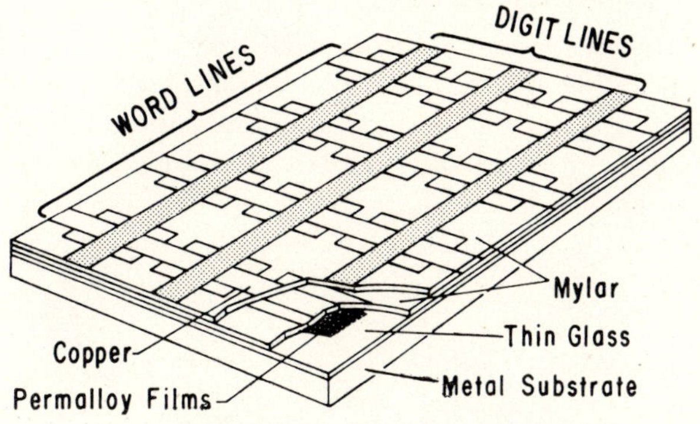

Fig 10 Thin magnetic film array

state to the other. This state of affairs is particularly troublesome in lines near the selected one, which are subjected to its stray word field. The field strays because lines of magnetic flux can extend between storage elements through the air adjacent to the sheet. Still a worse result of having the elements open in this way is a stray field that can demagnetize the elements altogether. In order to keep such fields from diluting the desired sharp magnetic characteristics too much, it is necessary that the length of the storage element in the easy direction be about 10,000 times its thickness, that is, more than a millimeter, and this sets a limit on compactness. On the other hand, it is desirable to make the elements smaller in the hard direction to accentuate "anisotropy", that is, the difference in magnetic properties between the hard and easy directions.

Most of the inherent obstacles to thin film integrated memories were, at first, something of a surprise, but they have been overcome. The latest types utilize two film spots facing each other in perfect registry and sandwiching between them the word and digit windings. This minimizes creep and demagnetization, provides stable operation and gives relative immunity to external magnetic fields, such as the earth field.

Typical products are memory systems with about 2-million bits and access time of 500 nanoseconds. There are special units of capacities of that order working with cycle times of 250 nanoseconds. Also, smaller capacities allow cycle times of about 100 nanoseconds.

The third kind of integrated memory is the plated wire. A beryllium–copper wire is electroplated with permalloy. Straight parallel lengths of this wire are digit lines; conductive ribbons strapped on at right angles are word lines (see Fig. 11). Storage is accomplished by magnetization around the circumference of the plated film, which is in the easy direction. That direction is established during plating by passing a direct current through the wire. The word current flips the fields toward the axis of the wire, which is in the hard direction. This induces sense voltages in the digit lines, and small digit currents flip the fields to the selected easy directions as they do in the flat-film memory.

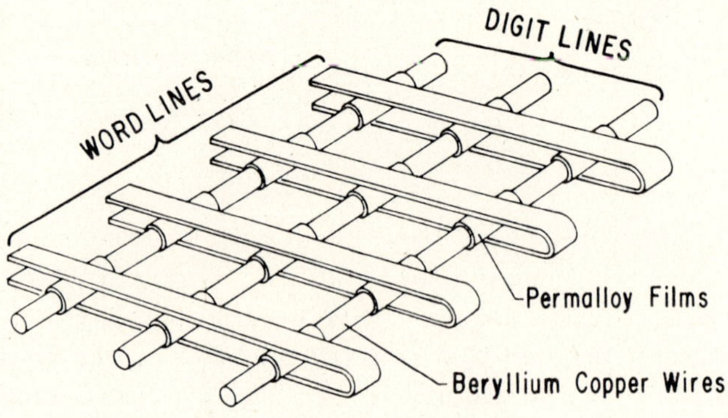

Fig. 11 Plated wire memory

The plated wire memory represents several fortunate compromises. Plating can be regarded as a one-dimensional integration. It is amenable to a continuous process, which should be more economical than the hand-threading of individual cores, and it is simpler to control than the lamination of ferrites or the deposition of flat films. Moreover, the plated wire storage element has magnetic flux lines that are closed within the magnetic material in one direction, again a compromise between cores and flat films. The resulting reduction of demagnetizing effects makes it possible for the film to be

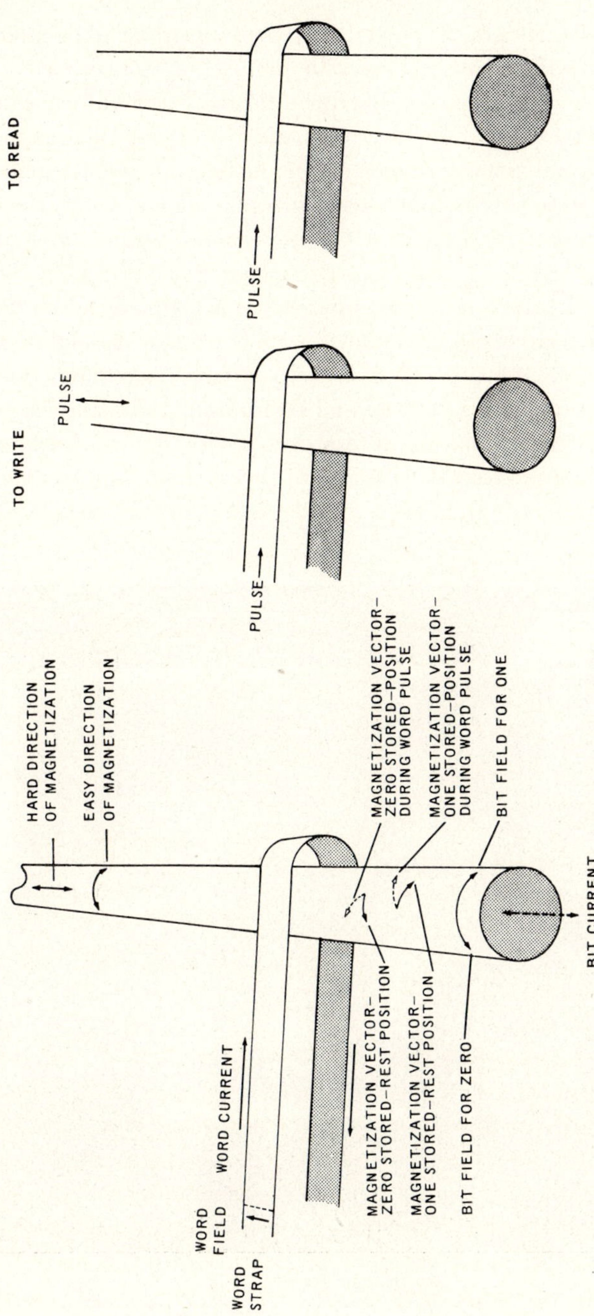

Fig. 12 Plated wire memory

comparatively thick: from 15,000 to 20,000 angstroms. This thickness of permalloy provides just the right amount of flux—more flux than the necessarily thin flat films and less flux than the difficult-to-miniaturize ferrite cores.

In spite of these attractions, it took years to develop plating techniques that avoided unfavorable magnetic effects. In addition, the wire on which the permalloy is plated is susceptible to strains that tend to distort the magnetic characteristics of the films. Such effects can be minimized by careful choice of materials.

In recent years most problems were overcome. A memory system operating in 600 nanoseconds was announced for a line of commercial computers in 1966 (see Fig. 12 and Fig. 13). By 1968 a number of manufacturers claim capability of supplying good wire and several sources exist for memory systems. The relative economy with respect to cores still remains to be proven, although it appears very likely for cycle times between 200 and 300 nanoseconds.

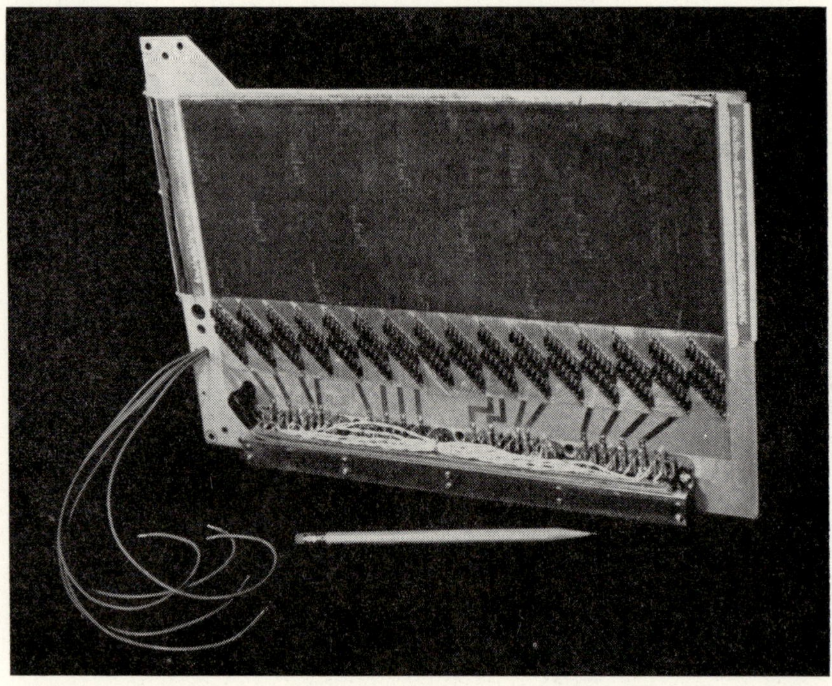

Fig. 13 Closeup of one plane of plated wire memory

There have been a number of approaches to integrated magnetic memories other than the laminate, flat films and plated wire of the type just described. Among them, a notable type commercialized a few years ago, is the so-called "rod" memory which uses long plated cylinders, really plated wires. It uses axial rather than circumferential remanent magnetization. Coaxial solenoids are used to produce x, y, sense and inhibit windings, and these windings are made by automated methods in plane structures. Long wires thread many winding-planes or else needle-like short lengths, are inserted by automatic magnetic means through each individual plane. The operation is similar to that of a conventional current-coincident core memory.

All magnetic memories have common characteristics. Access to an element in an array is achieved not by some kind of sharply selective pointer but by the somewhat imperfect means of coincident pulses on electrical lines. The kind of imperfections one encounters in the storage elements have been described: there are also unfavorable interactions between the driving lines and the sense lines. Because the driving currents are a considerable fraction of an ampere, spurious signals can be induced in the sense lines that can easily exceed the desired values (measured in millivolts) of sense signals. Such masking is not difficult to minimize in the read cycle of current-coincident memories. In the write cycle of 2D and 3D types, however, a large voltage can be induced in the sense line, since that line is itself the digit drive line or couples with it in every element of the digit plane. The induced voltage paralyzes the sense amplifier for a period that is often the most significant part of the memory's cycle time.

To minimize write noise in word-organized 2D memories, the two halves of the digit lines are often driven so as to induce opposing voltages in the sense amplifier and yet not attenuate the sense voltage. Such a circuit also serves to cancel the coupling by capacitance. In current-coincident 3D memories the inhibit and sense lines are wound in different patterns designed to minimize magnetic and capacitative coupling, but this cannot be done perfectly for both. Recently an organization intermediate between 2D and 3D, called $2\frac{1}{2}$D, has become popular (see Fig. 14).

It drastically reduces the write noise and makes the core technology faster and even more competitive with technologies based on integrated laminates or films. The read step is the same as it is in 3D, but for the write step an inhibit winding is replaced by one set of selecting lines, say the X lines. The selected X line in each plane is either energized or not according to the digit to be written in the plane. Clearly, write noise is greatly reduced since the noise now originates from one line rather than the entire plane. The price for this advantage is more circuitry.

Noise-cancellation techniques and $2\frac{1}{2}$D are examples of stratagems to minimize the imperfections inherent in coincident-line addressing of arrays of magnetic elements. These stratagems can help only up to a point. In general, the faster one wishes to operate the memory system, the smaller is the

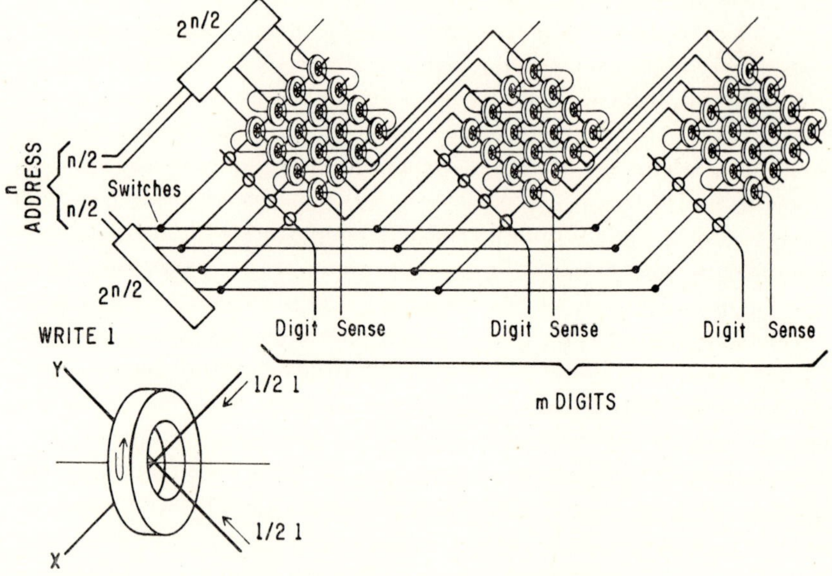

Fig. 14 $2\frac{1}{2}$D core memory organization

number of elements permissible in the sense and drive lines. This is so not only because of write noise but also because of the time it takes signals to travel through the system and the power required by long drive lines. When detailed designs are worked out for memories of various capacities and speeds based on cores, laminates, flat film and plated wire, it rather surprisingly turns out that about the same speed, within a factor less than two, is obtained for a given ratio of storage elements to switching elements (transistors or diodes) required in the memory system.

For fast systems with a cycle time of from 100 to 200 nanoseconds this bits-to-switches ratio may be as low as 30 to one; for the widely used 3D core systems operating at about one microsecond the ratio is about 300 to one, and for very large, slow systems it may be as high as 1000 to one. Although the bits-to-switches ratio is not a complete criterion for choosing between various memory technologies, it determines the relative cost of the associated electronic circuitry. The ratio is most significant, however, in considering whether any magnetic technology can be extended to very high speeds or very large storage capacities. These questions should be considered separately.

In a computer the time required to put commands and numbers into the memory and to retrieve them should be comparable to the time required to perform logic operations on them before they are again returned to the memory. Electronic logic has become progressively faster as technology has progressed from vacuum tubes to transistors and integrated circuits. Progress in memories was difficult, and a comparable evolution could be obtained only in memories of small capacity. Accordingly computer designers resorted to a hierarchy of memories: a small, fast "scratch pad" memory and a slower large main memory. In magnetic memories the cycle time—read followed by write—consists of (1) address-decoding time, (2) the time required for signals to travel along lines, (3) the time needed to switch elements twice, (4) delay in amplifiers and (5) delays due to write-in noise and timing imperfections. The fastest magnetic scratch pads use fast decoders, short lines, fast film memory elements and are cleverly designed to minimize delays. They call for much electronic circuitry, and at cycle times less than 100 nanoseconds the bits-to-switches ratio becomes very small. With the advent of integrated circuits the question arises of whether magnetic storage should be eliminated altogether. The economy in circuits achieved by the use of magnetic elements is no longer significant, and the necessity of amplifying the weak magnetic sense signals wastes cycle time.

As a matter of fact nonmagnetic memories have been used for years in all computers. Semiconductor registers, which consist of a row of "flip–flops" each storing one bit, can be regarded as one-word memories. A genuine memory calls for selection among many words. Such selection does not differ from the type of function carried out in the computer by organs responsible for logic control and arithmetic. This type of function is accomplished by "AND", "OR" and "NOR" gates. (The output of an "OR" gate is 1 if at least one of the inputs is 1. "NOR" stands for "not or".) For each bit a flip–flop and selecting gates are needed, so that a complete circuit calls for perhaps 10 transistors. Together with peripheral circuitry this yields a bit-to-switches ratio of only about one to 10. Furthermore, a fair amount of direct current power is continuously required to hold the state of the flip–flops when these are made out of the bipolar transistor in common use today. Despite these shortcomings small "scratch pad" high speed semiconductor memories have been in use for several years. Recently fairly large, i.e. more than a hundred thousand bits, "cache" memories have been announced for commercial machines.

The future of semiconductor memories is very bright due to the large strides that have been made in integrated circuits and the advent of the metal-oxide-semiconductor (MOS) field-effect transistor. These developments may avoid the relatively brute force use of discrete (or integrated only a few at a time) standby power consuming bipolar transistors.

The MOS field-effect transistor is an almost perfect switch (see Fig. 15). The conduction between the "source" and "drain" electrodes of the transistor is controlled by the potential of a perfectly insulated metal gate. There are two types of MOS transistor: the n-type which conducts by means of electrons, and the *p*-type, which conducts by means of "holes", or regions deficient in electrons. It is possible to fabricate n- and *p*-type transistors that have practically no conduction between the source and the drain when the gate is at the same potential as the source. However, when the gate potential is a few volts positive with respect to the source of an n-type MOS transistor, or a few volts negative with respect to the source of a *p*-type MOS transistor, there is good conduction between the source and the drain. Accordingly, these devices behave somewhat like relays, and logic networks consisting of MOS transistors resemble relay networks.

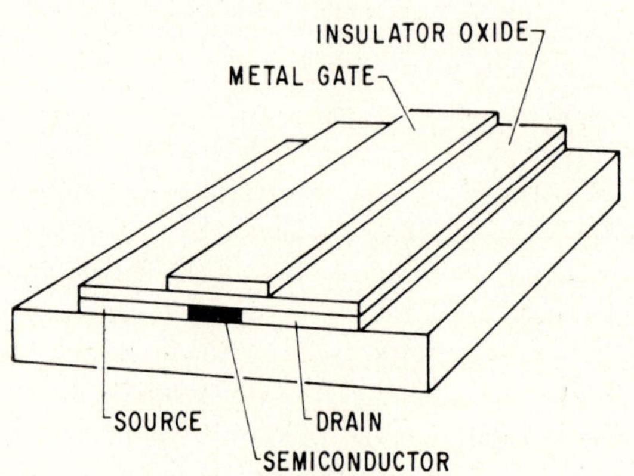

Fig. 15 MOS field-effect transistor

When two pairs of n- and *p*-type MOS transistors are connected in series and symmetrically cross-connected, they form an almost perfect flip–flop circuit (see Fig. 16). Both states of the flip–flop are stable, and practically no current flows through either branch of the circuit except for small currents due to slight leaks in MOS transistors. To make it possible to set the flip–flop in one state or another, six more transistors are ordinarily added. Many words can be connected to the same digit line, and sensing is extremely fast (a few nanoseconds).

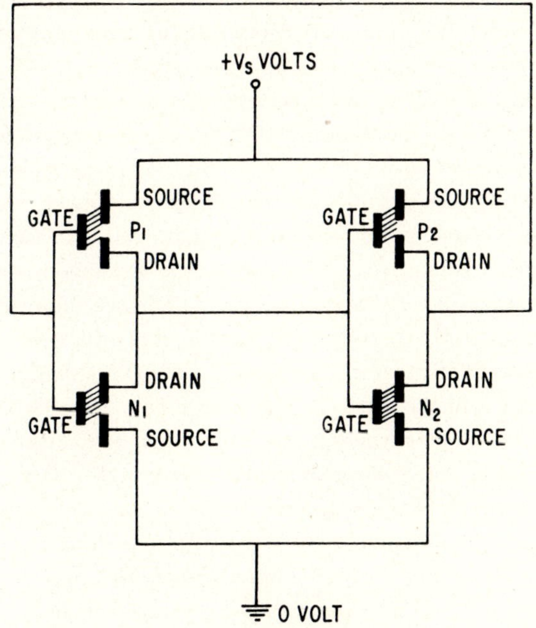

Fig. 16 MOS flip–flop circuit

The switching speed is limited by the degree to which the transistor can amplify and by the capacitance of the electrodes and windings. Almost no holding power is required (0.10 microwatt per cell). The total switching power of the entire memory, no matter how large, can be much less than one watt. There is no write noise and no sense-amplifier delay, since the sense signal is quite strong. The system is nearly ideal, but is any system that contains thousands of transistors practical?

The electronics industry is investing large sums on the assumption that the answer will be yes. The industry bases its hopes on the phenomenal success of integrated circuits for other purposes. This success is due to a mastery of silicon technology: the ability to produce silicon crystals of high purity, to "dope" them precisely with impurities and to create thin insulating layers. Of equal importance is the development of photographic techniques to form conductive, insulating or specifically doped semiconductor areas on a microscopic scale. Over the years the performance of transistors improved, and it became practical to integrate two, four and then eight transistors with all their connections and coupling elements. "Chips" consisting of 20 to 30 transistors have been commercially available for the past year or two. Today the large-scale integration of hundreds or even thousands of transistors is the object of an industry-wide effort.

In 1966 MOS memory arrays of *p*-type elements were made that had a capacity of 256 bits. Commercial *p*-MOS memory arrays of larger size became available in 1968.

The future belongs, however, to MOS arrays of both n- and *p*-type elements. With such elements one can put a one-bit cell, complete with a flip–flop and logic gates, on an area 0.015 inch square. A memory with a capacity of hundreds of bits can then be made on a chip $\frac{1}{4}$-inch on a side. An access time of 30 nanoseconds or less is expected (see Fig. 17).

In 1968 small complementary symmetry,—CMOS—integrated memories became available commercially.

Whatever the particular type, integrated semiconductor memories have an inherent advantage over any type of magnetic memory. After all, the comparison is between an active electronic logic circuit with explicitly designed performance and a passive magnetic element of the same size that depends more or less on the properties of materials provided by nature. The question comes down to one of practicality: can these intricate microelectronic elements be made in large arrays with adequate perfection and economy? The answer is definitely "yes" for small, ultra-high-speed scratch pad memories with a capacity of a few thousand bits. Concerning larger memories the prophets differ. Some think it is only a matter of time; others believe magnetic memories will retain their position because their associated circuits can also be made faster, cheaper and more reliable by the new microelectronic technology.

What of memories with very large capacities? Over the years the demand for larger computer memories has grown steadily as more ambitious problems with longer programs and more data were attacked, and as higher computer languages called for longer programs. Core memories were too ex-

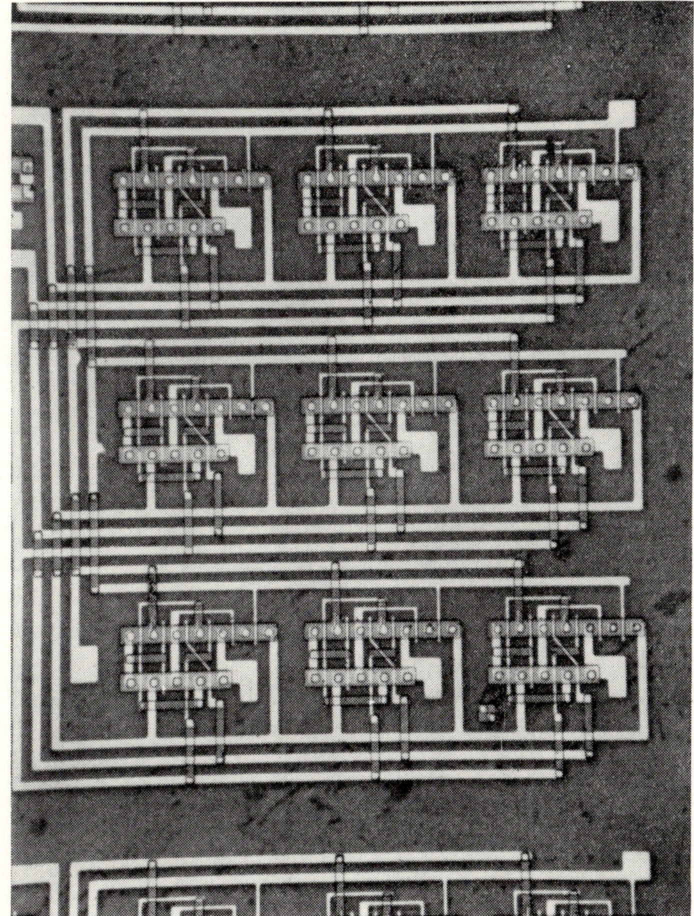

Fig. 17 Experimental integrated complementary symmetry MOS memory

pensive to provide the 100-million to one-billion bits required for many uses, and this led to the development of memories based on spinning magnetics drums and disks. These electromechanical devices operate with a cycle time measured not in nanoseconds or microseconds but in milliseconds. When a computer is specially programmed, they seem to the user to be a simple extension of the core memory. This is a tolerable but undesirable state of affairs. Much effort is required to produce such programs (which also consume memory space), and much computer time is lost because of slow access time and the necessity of exchanging information between the cores and the drum or disk. Furthermore, the reliability of drum and disk memories leaves much to be desired.

In the meantime, core memories have gradually increased in capacity and today main working memories have commonly many millions of bits. In recent years so-called "extended" core memories with reading capacities of 100-million bits have been commercialized for use as auxiliary mass storage. These memories are designed to favor economy at the expense of speed and operate at the relatively long cycle time of 2- to 8-microseconds. Only two wires thread each core and a few thousand cores are threaded on the same wire, giving a "screen door" appearance to the memory planes. In operation, writing is as in 2½D organizations and in reading, one of the wires serves as both a selecting and a sense winding. The cost-per-bit of the extended core memories is 3- to 10-times less than that of the faster main memories. As with all core memories, its cost-performance is likely to continue to improve somewhat. However, that cost is still a factor of almost 100 higher than the corresponding cost of disk memories. Despite their cost, extended core memories are finding increased use where the benefits of fast random access and great reliability are at a great premium, such as in multiple-user centers working around the clock.

Clearly, the challenge is to produce electronic random access memories with capacities of one-billion bits at a cost substantially lower than that of extended core systems. Further extension or improvements in magnetic core memories or in magnetic integrated memory techniques is unlikely to yield the solution mostly because of the very large number, about one-million, of transistors or other semiconductor devices that are required. A more promising approach lies in the realm of superconductivity.

Superconductive materials, when cooled below a certain critical temperature, lose all electrical resistance. If current is started in a loop of such material, it will flow forever and therefore "remember" that it was started. The loop is in effect a one-bit memory register. Furthermore, superconductivity is destroyed by a sufficiently strong magnetic field. This effect is the basis of the cryotron switch. In the "on" state the switch has no resistance; an "off" state of finite resistance is created by passage of a current in a nearby controlling conductor that places the switch in a magnetic field. Tin, which is a "soft" superconductor, is a good material for the switch itself, because it can be changed from the superconductive to the resistant state by a weak magnetic field. A "harder" superconductor, such as lead, which resists the

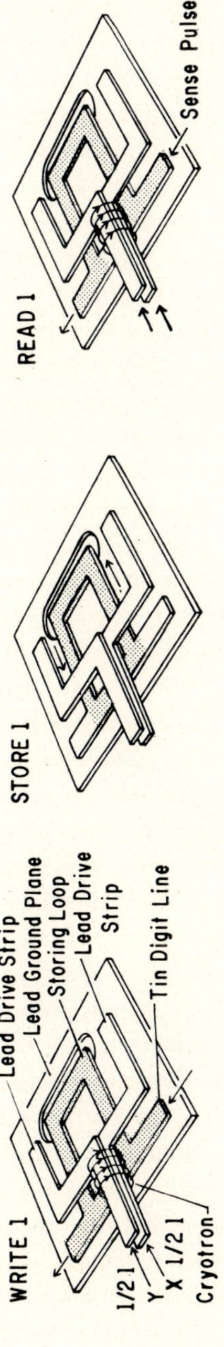

Fig. 18 Superconducting memory elements

weak field, can be used for circuit wiring. Most experimental cryotrons employ films several thousand angstroms thick. They can be switched in nanoseconds and are amenable to integrated fabrication techniques.

In a recently announced superconducting memory, storage loops made of tin are connected in series along a digit conductor and are on top of a lead sheet from which they are separated by a thin insulating film (see Fig. 18). The lead sheet has holes at locations corresponding to one side of each loop. The selecting X and Y conductors, made of lead and properly insulated, lie atop the other side of each loop and thus form a doubly controlled switch. The switch operates by creating a local magnetic field that destroys superconductivity in part of the loop.

The memory system operates by current coincidence in the following manner. The selecting X and Y currents are insufficient to activate the switch singly but are sufficient when acting together. In write-in the selecting currents are applied and then a current is sent through the digit line. The increased resistivity in the loop below the switch steers the current to the side of the loop over the hole in the lead sheet. Next the selecting currents are turned off and the magnetic field disappears, leaving the switchable side of the loop superconductive. Current does not flow through that side, until the digit current is turned off. When this happens, the current circulates as a persistent storing current around the complete loop. In the read cycle the selecting currents are applied and again increase the resistivity of the loop below the switch, thereby causing the persistent loop current to dissipate. As the current dissipates, it generates a sense voltage in the tin digit line. This voltage depends on the loop's magnetic flux, which is made as large as possible by allowing it to extend through the hole in the lead sheet.

Several properties of this superconducting memory are outstanding. In the first place, the loop is a more perfect memory element than any magnetic element, integrated or otherwise. There is absolutely no contribution to the sense signal from half-selected elements, there is no creep and the uniformity of the threshold of switching can be closely controlled. The selecting currents can vary from 80 to 120 percent of their standard value. There is, however, some write noise. Fortunately such noise can be eliminated by the cancellation techniques mentioned earlier. The result of these ideal properties of the loop and the low inductance of the connecting lines is that many loops can be driven and sensed by a single circuit. The bits-to-switches ratio can probably be 100,000 to one or more, which is two or three orders of magnitude better than for any magnetic memory. Furthermore, integration on the

Fig. 19 Superconductive memory plane containing 262,144 Bits on $4\frac{1}{2} \times 5''$ glass plate

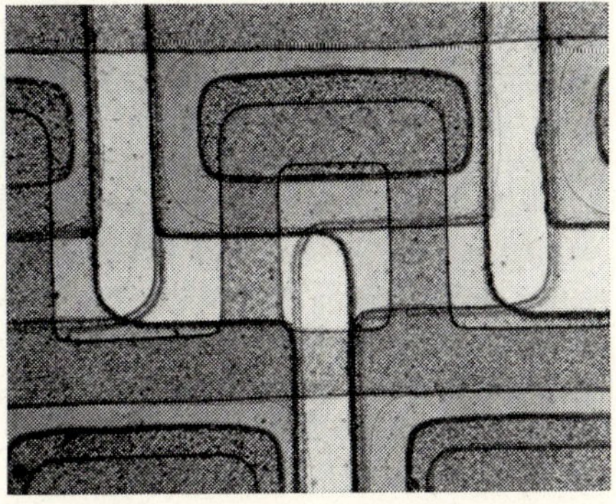

Fig. 20 Details of one storage loop of the plane of Fig. 19

grand scale is eminently possible. Thin films of superconductive materials and other metals are easily evaporated over large areas. Photographic techniques are well suited for creating the desired patterns. The degree to which these techniques have been perfected is indicated by an experimental memory plane $4\frac{1}{2}$ by 5 inches in size that contains 262,144 storage cells. This is a density of more than 13,000 cells per square inch (see Fig. 19 and Fig. 20).

The price to be paid for these ideal properties is the necessity of providing low temperatures: for the lead–tin memory 3.5 degrees Kelvin (degrees

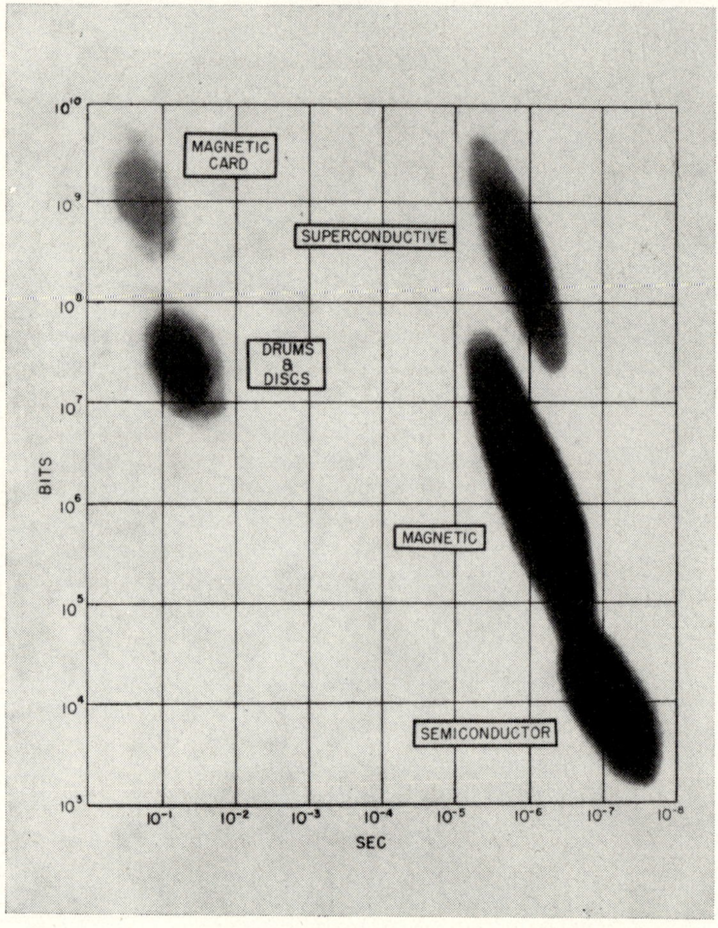

Fig. 21 Chart illustrating the capacity and speed of various memories

centigrade above absolute zero). Such temperatures are no longer confined to the cryogenic laboratory; providing them would not add much to the cost of large-capacity superconducting memories. Memories of this kind may therefore offer capacities comparable to those of electromechanical devices at comparable cost, but with three orders of magnitude higher speed and also greater reliability.

To sum up, the memory gives universal powers to the computer. For more than a decade it was based on separate cores and transistors. The arrival of integrated magnetic, semiconductor and superconductive structures may extend speeds and capacities and make computers still more powerful.

It seems that no one technology will sweep all before it. Rather, one can expect that magnetic techniques (cores, monolithic ferrites, flat films or plated wire) will provide capacities of from 10,000 to 10-million bits at speeds of a fraction of a microsecond. Integrated semiconductor techniques will provide memories of smaller capacity with speeds measured in tens of nanoseconds. Finally, superconductive mass memories with capacities of from 10-million to a billion bits will have speeds of about a microsecond (see Fig. 21).

This is not the end. Already under study are various ways to avoid the making of physically distinct cells for each bit without losing the essence of digital addressing. One approach uses sound waves traveling in magnetic materials. Other approaches make use of electrooptical devices such as the laser. As long as the demand for better memory systems continues there will be no shortage of ingenious proposals.

References

1. Forrester, J. W., "Digital Information Storage in Three Dimensions Using Magnetic Cores", *Journal of Applied Physics*, Vol. 22, pp. 44–48; 1951.
2. Rajchman, J. A., "Static Magnetic Matrix Memory and Switching Circuits", *RCA Review*, Vol. 13, No. 2, pp. 183–201; June, 1952.
3. Shahbender, R., Wentworth, C., Li, K., Hotchkiss, S. E., Rajchman, J. A., "Laminated Ferrite Memory", *Proceedings Fall Joint Computer Conference*, p. 77, November, 1963.
4. Shahbender, R., "Laminated Ferrite Memories—Review and Evaluation", *RCA Review*, Vol. 29, No. 2, pp. 180–198; June, 1968.
5. Pohm, A. V. *et al.*, "Large High Speed DRO Film Memories", *Proceedings Intermag Conference*, April, 1963.
6. Maeda, H. *et al.*, "Woven Wire Memory for NDRO System", *IEEE Transactions on Electronic Computers*, Vol. EC–15, No. 4, pp. 442–451; August, 1966.

7. Higashi, Paul, "A Thin Film Rod Memory for the NCR 315 RMC Computer", *IEEE Transactions on Electronic Computers*, Vol. EC–15, No. 4, pp. 459–467; August, 1966.
8. Gilligan, T. J., "$2\frac{1}{2}$D High-Speed Memory Systems—Past, Present and Future", *IEEE Transactions on Electronic Computers*, Vol. EC–15, No. 4, pp. 475–485; August, 1966.
9. Brown, J. Reese, Jr., "First and Second Order Ferrite Memory Core Characteristics and Their Relationship to System Performance", *IEEE Transactions on Electronic Computers*, Vol. EC–15, No. 4, pp. 485–501; August, 1966.
10. Hobbs, L. C., "Present and Future State-of-the-Art in Computer Memories", *IEEE Transactions on Electronic Computers*, Vol. EC–15, No. 4, pp. 534–550; August, 1966.
11. Rajchman, J. A., "Memories in Present and Future Generations of Computers", *IEEE Spectrum*, November, 1965.
12. Rajchman, J. A., "Integrated Computer Memories", *Scientific American*, July, 1967.
13. Pleshko, P., *et al.*, "An Investigation of the Potential of MOS Transistor Memories", *IEEE Transactions on Electronic Computers*, Vol. EC–15, No. 4, pp. 423–427, August, 1966.
14. Burns, J. R., *et al.*, "Silicon-on-Sapphire Complementary MOS Memory Systems", *Digest of Technical Papers, ISSCC*, Vol. 10, pp. 76–77; February, 1967.
15. Sanders, Wendell B., "Semiconductor Memory Circuits and Technology", *Proceedings Fall Joint Computer Conference*, Vol. 33, Part 2, p. 1205; December, 1968.
16. Newhouse, Vernon L., Book *Applied Superconductivity*, published by Wiley & Sons, 1964.
17. Burns, L. L., *et al.*, "Three-Wire Cryoelectric Memory Systems", *IEEE Transactions on Magnetics*, Vol. MAG–2, No. 3, pp. 398–402; September, 1966.
18. Sass, A. R., *et al.*, "Cryogenic Random Access Memories", *IEEE Spectrum*, p. 91, July, 1967.
19. Sass, A. R., *et al.*, "Cryotron-Based Random Access Memory", *IEEE Transactions on Magnetics*, Vol. MAG–3, No. 3, pp. 260–267; September, 1967.
20. Gange, R. A., "Taking Cryoelectric Memories Out of Cold Storage", *Electronics*, Vol. 40, No. 8, pp. 111–120, April 17, 1967.

CHAPTER 6

Computer Software

MAX GOLDSTEIN and PETER CAPEK

1 INTRODUCTION

PEOPLE WHO DO not deal with computers in their daily lives generally do not realize the amount of human effort which is involved in making computers perform all the marvelous tasks which we associate with them. The attitude generally taken by the layman is that the computer is a clever machine which automatically knows how to solve any problem which is presented to it. This, of course, is not the case, and the path between the definition of a problem and its computer solution is frequently a long and arduous one. Furthermore it is generally not the problem which must be presented to the machine—it is the method of solution. A computer is not capable of figuring out how to solve a problem. It can only perform a set of explicitly specified operations which are built into its hardware. The user must write the instructions, or program, for the solution.

In order to aid the user in getting results for his problem, computer software is provided. This allows the user to program in a problem-oriented language (rather than in the machine's language), to get his program executed efficiently, to be assigned the needed hardware resources and, in general, to minimize his effort in arriving at the solution to his problem.

The cost of developing the software for a computer system is frequently higher than the cost of designing the hardware on which it will run. For all but the smallest machines the cost of building software runs into the hundreds of thousands of dollars and occasionally into the hundreds of millions. Consequently, there is much discussion and debate current in this area and there are many different points of view as to how things should be done. In the ensuing discussion we shall give an overview of the current state of the art of computer software and indicate the directions in which it seems to be going.

Computer software is intangible. Its physical manifestations are most commonly a deck of tabulating cards or a roll of magnetic tape. But the word *software* really represents much more work and money than is apparent from either of these. Software is *information* which directs a computer in performing the tasks for which it was designed and built. A computer is worth little unless appropriate software is available to make it perform the function for which it was intended.

One may well ask why the software cannot be built into the hardware of a computer. The answer to this becomes apparent when we consider that only a small amount of logic can be built into a machine at a reasonable cost. To wire a computer to process an entire payroll, for example, is an economic, if not an electronic, impossibility. But even if we could build such a machine, it might work very well for our once-monthly payroll, but would be totally useless if we wanted to use the same machine to take care of an inventory.

The main reasons, then, that computer software is necessary are flexibility and economy. With only certain fixed logic built into hardware, a general purpose computer becomes feasible. It is very easy to use a single machine to process a payroll one day and to use the same machine to solve differential equations the next. It should be apparent, then, what role software plays in the use of computers. Software is a general term for program, or codes, which tell the computer (hardware) precisely how to go about solving a problem or performing some specified task. In the present discussion we shall attempt to demonstrate more specifically the role software plays.

Software falls into two major classes according to its function, namely the area of operating systems and that of language processors. We will examine each of these in turn.

2 OPERATING SYSTEMS

An operating system is precisely what its name states—a system, program, or series of programs which is responsible for operating the computer system. By operate we mean supervise the flow of programs and data throughout the machine. This may be done at various levels of sophistication. Let us first explain the functions which an operating system must perform in more detail and then show the various methods by which this can be done.

It is the duty of the operating system to interface (in the sense of providing liaison) between the hardware which is the computer, on the one hand, and the operator, programmer and user on the other. The operating system provides the environment and control under which all other programs will run. The operating system furnishes a continuity between jobs and may do some scheduling of the work to be done. Frequently referred to as the supervisor, it aids the operator and programmer so that the machine may be used more efficiently. Additionally, it must provide certain services to its users, to the machine operator, and perhaps to the management which is running the computer center.

2.1 Functions of operating systems

Probably the most important function is that of error checking. The operating system must check for errors on the part of the operator and the programmer. It must, whenever feasible, check that the right tape has been mounted or that the date which the operator has typed in is reasonable. It may aid the operator by internally scheduling jobs in such a manner that they can run more efficiently. It must check for errors while reading tape and other input and output media, or signal invalid results if a running program attempts to divide by zero, or violates preset fault conditions. It must give detailed descriptions of errors and provide information such as dumps which make it possible for the programmer to narrow down and eliminate the error. It may aid the management of the machine by providing accounting information i.e. who uses the machine and the kinds of jobs that are run. Possibly it also provides information and statistics on the nature of jobs run in order that changes may be made to increase the machine efficiency.

Another important feature which an operating system provides is a library of programs and subprograms, and a loader which makes them easy to use. Typically included among the library would be mathematical routines to calculate trigonometric functions (sine, cosine, etc.) as well as utility programs to do such things as, say, copy one tape onto another.

In summary, then, the operating system does not solve problems or do arithmetic. It merely makes it easier to use the computer and helps only indirectly in solving problems.

2.2 Levels

There are several levels of sophistication in operating systems today. They all have the characteristics mentioned above, but vary in complexity.

2.2.1 Batch

The simplest, most elementary level is that of batch processing. On a machine which is doing batch processing, input decks are simply stacked in a card reader and run (executed) in the order in which they appear. All scheduling must be done by the operator and the only flexibility which he can exercise is that of putting jobs in the reader in the order in which he wants them run. This means of operating was the primary mode, even for the largest computers, until comparatively recently.

But as faster computers were designed, it became apparent that the input and output devices (card readers, punches and printers) connected to them could not keep up with the rate at which they needed data for input, or with the rate at which they produced results on output. To compound the problem, the computer would frequently be busy for long periods of time doing nothing but computation, leaving the reader and printer standing idle. To alleviate this problem, it was proposed that a small, independent computer be used to read cards and to print output. The machine would read cards and write them out onto magnetic tape. The large machine would read a magnetic tape for its data and write its results onto another magnetic tape, and this tape would again be printed by the smaller machine. Consequently, the card reader and printer could be kept going, independently of what the large machine was doing. An additional advantage is that the large computer could read its data at tape speed when it wanted them, rather than at the much slower card reader speed. This method of operating gained rather wide acceptance during the early 1960's, and many such systems were installed. Even today, many such combinations are still running successfully. The most famous of these was the large IBM 7090 used for computation which was teamed up with a 1401 which operated the card reader and printer.

An important point here is that this simple, unenhanced method of running a computer was and still is a primary method and it is only recently that more sophisticated operating systems have gained a foothold.

2.2.2 Direct coupled systems

A direct outgrowth of the batch processing method discussed above is the idea of the directly attached smaller computer, for example, the IBM ASP System. In ASP, which stands for Attached Support Processor, a System 360/65 or larger computer is directly connected to a smaller System 360/40. In addition to selling more computers for the manufacturer, this scheme gives the smaller machine the ability to schedule jobs which are to be run on the larger one. The larger machine actually runs the programs which are submitted. The smaller machine schedules the larger one, reads cards and prints output. The fact that a computer can decide which job is to be run next can result in a considerable increase in throughput* on the system. For example, simply by alternating jobs which require tapes with jobs that do not, the operator will have time to mount tapes before the machine actually needs them.

2.2.3 Multiprogramming

With all the schemes we have mentioned so far, the main computer executes one job at a time. It will never begin a new task until the one on which it is currently working is completed. This sort of situation was satisfactory for a long time, but as computers became more and more expensive and there was more and more work to be done on them, engineers and systems designers had to invent more complicated ways of increasing the throughput of the machines and getting a higher utilization of them. Multiprogramming is one result of this effort. In a machine which is multiprogrammed, there are several unrelated programs residing in memory at once. They are all executed by the same central processing unit, but only one is being executed at a time. As soon as one job gets to a point where it can no longer compute, perhaps because it needs data from a disk or a tape (input or output), it relinquishes the central processor and waits for the I/0 to be completed. While the I/0 is going on, the central processor is busy working on another job, and does not stand by idle. It is the duty of the operating system to perform the bookkeeping and supervising which is necessary to maintain this sort of operation. It should be clear that multiprogramming not only requires special software, but special hardware in addition. Since there are several programs sharing the same memory, it is necessary to include special hardware to assure that none of the programs interferes with any of the other programs in memory. This is called memory protection. Another common feature on multiprogrammed machines is memory relocation which allows programs to be moved within the memory as new programs are initiated and others terminate.

2.2.4 Time sharing

A special case of a job waiting for input or output is that of one running in a time sharing system which is interacting with a programmer who is sitting

* The amount of useful work performed in a given time on a computer system.

at a console or teletype. Here the computer must respond almost instantaneously* or the user will become discouraged and walk away. However, the requirements of such an operation are far greater than those of a simple batch processor. A problem which must be done on a computer but still requires human guidance frequently to make key decisions can profit immensely from the use of time sharing techniques. As the name implies, the computer shares its time among many users. The aim of this system is to give each of the users the impression that he is the only one who is using the machine. Time sharing virtually always implies multiprogramming. But the use of multiprogramming may be necessitated by other things. For example, a computer may be directly connected to a large cyclotron. The use of the computer may be required once every 10 seconds for one-tenth of a second, but the results must be available immediately or they are of no value.

2.2.5 Real time

A few special types of operating systems are worth mentioning. A good example of this is the airlines reservation system. In such an installation two computers are frequently used, one as a backup for the other. Applications such as this are characterized by having large files of data, though not much calculation needs to be done to complete a given transaction. The operating system necessary to run such a system is particularly complex because it must provide a high degree of security to the airline and must be ready to handle exceptional cases as they arise without disturbing the continuity of the operation. Another special application is that of process control. Here, a computer will be given most of the responsibility for running, say, an oil refinery. The machine will be given the ability to "read" several hundred important pressures and flow-rates throughout the plant. It will also be given the ability to operate key valves and other controls which affect the operation of the system. A human operator's intervention is required only when some special situation arises which the machine has not been programmed to handle.

2.2.6 Summary

In general, the programmer of problems need not be concerned with the actual mode of operation of the computer which he is using. The only significant way in which he is affected is by the time it takes to get his job run. Even

* Generally, in less than 5 seconds.

if he is running on a time shared computer, the basic programming he must do to get a solution to this problem is the same as if he were running in a batch programming environment. The actual manner in which he must use the machine is very slightly altered by the mode the system runs in. He is particularly unaffected if he is programming in one of the higher level languages, which we shall now describe.

3 COMPUTER LANGUAGES

So far we have been talking only about operating systems which run computers. Let us now turn our attention more directly to the software actually used to solve problems on these machines. Here the topic of primary importance is that of computer *languages*. There are several different types of language which one can use in attempting to solve a problem on a machine. Which one of these is chosen depends on several factors. If speed of execution on the machine is important the user might resort to a low-level assembly language which will give him a greater opportunity to optimize his code. But if ease of programming is more important, as it is in most cases, then a higher level language which is well-suited to the problem at hand should be chosen. Similarly, if a program will have to be run on several different machines, a language which is available on all of them must be chosen. In order to illustrate the difference between the various types of programming languages in the following discussion we shall show in each case an example of a program in the language being described. All of these programs will perform the same function, namely that of calculating the sum of the numbers from 1 to 100 (i.e. $1 + 2 + 3 + \cdots + 99 + 100 = 5050$).

3.1 Machine dependent languages

The hardware of a computer provides a small built-in repertory of instructions. These form the machine language. The number of machine language instructions varies from 8 on a small machine such as the PDP–8 to about 200 on a large machine like a 7094. Programming at the level of machine language is a very tedious task, and is normally used only on a new machine until suitable, more sophisticated methods can be employed.

Here is an example of a machine language program.

```
 1    112
 2    213
 3    412
 4    315
 5    008
 6    814
 7    900
 8    114
 9    212
10    414
11    901
12    000
13      1
14      0
15    101
```

Fig. 1

The column of numbers at the left lists the locations in the computer memory into which the instructions are placed. The numbers at the right are the actual machine language codes which will be executed by the machine.

The next higher level of programming is that of *assembly language* programming. An *assembler* is a program, usually provided by the manufacturer of the machine, which translates machine instructions which are written in symbolic format into machine language. An assembler is typically the first program that is written for a new machine when it is built, since all higher level languages usually use an assembler. Even among assemblers there are various levels. Frequently the first one that is written is very simple and has few non-essential features. Then, in the language which is processed by this first assembler, a more powerful assembler is written. This process (called bootstrapping) may be repeated several times until a powerful assembler with full macro capabilities is produced. Macros allow the programmer to define sequences of instructions which he may then refer to by a single line of code.

To illustrate assembly language programming, here is the above machine language program (Fig. 1) written in a simple assembly language.*

* We should point out that many people refer to assembly language programming as machine language programming. Strictly speaking, however, this is not accurate.

```
START    LOD   I       SET I = I+1
         ADD   ONE
         STO   I
         SUB   X101    CHECK IF DONE YET (I=101)
         JPNG  B       NO, JUMP
         PRNT  SUM     PRINT RESULT
         STOP          STOP
B        LOD   SUM     SUM = SUM + I
         ADD   I
         STO   SUM
         JUMP  START   LOOP BACK
I        CON   0       INDEX
ONE      CON   1       CONSTANT
SUM      CON   0       INITIAL SUM
X101     CON   101     CONSTANT
```

Fig. 2

As in machine language programming, here we also write single instructions to the machine, but we need not remember absolute values for locations and operation codes. We can write LOD I rather than 112. The advantages of an assembler are obvious.

But even the most sophisticated assembler is still tied down to the structure of the machine, and hence is still *machine dependent*. It can only be used on one specific type of machine.

3.2 Machine independent languages

The next rung on the ladder of languages is that of machine independent languages. As the name implies, a machine independent language allows one to program without a detailed knowledge of the structure or design of the machine on which the program will be running. Because of this feature, a single machine independent language program can be run on many machines. What we would like to be able to do is to define our problem (really, of course, we define the method for its solution) in some fashion, even though we don't know the details of how the computer works. The approach generally taken is to write one program which "knows" both how the computer hardware works in detail and also understands our external language, the language in which we will describe our problem. Let us, for the time being, call this program a *translator*. The translator, then, applying its "knowledge" of how the machine operates has the task of translating a source language program written by a user into an object program which the computer hardware can understand. This output from the translator, or *object* program, is used directly by the computer to solve the problem at hand.

The critical link then, the translator, between the machine independent program and the machine on which it is to run may take any one of several forms. There are two primary ways in which this can be done—by compilation and by interpretation.

The interpreter has several disadvantages, but a prime advantage is that it is easy to implement, compared to a compiler. It is, however, extremely slow. An interpreter offers an opportunity to provide very detailed error checking of the source program, particularly of conditions which are not apparent until the program is being executed. An interpreter is a program which reads and executes the programmer's source program one statement at a time, and executes it directly. Because the interpreter must continuously decide what to do next (from an analysis of the source code) and then do it, it is very slow. In an attempt to speed up the process, interpreters are usually written in such a manner as to read the source program once and translate it into some internal language. This intermediate language is more directly related to the actual operations to be performed.

The next step upwards from interpreters, though from the programmer's point of view it is really the same, is the level of the compiler. A compiler is a program which, like an interpreter, accepts as input some source language. The compiler's task is to translate the source language into an object program. Frequently, the output from the compiler will be a program in assembly language. The assembler will assemble this and will in turn produce the actual machine language program. While a compiler must compute for a while before producing its object program, the time spent in compilation produces an execution speed which cannot be approached by an interpreter. The difference in execution speed between an interpreted and a compiled program can be of the order of a hundred to one. Nonetheless, interpreters are useful where the execution time is small, compared to the compilation time, particularly in small student type jobs.

3.2.1 Fortran

Let us next examine a few languages which are the object of all this compilation and interpretation. By far the most common language used in the scientific and engineering areas is Fortran. Fortran was the first attempt at producing a program which, in effect, would write a program. The original version had 25,000 machine language instructions. The Fortran language is very closely related to the language used by scientists who write the programs in their daily work. Fortran provides very convenient methods of handling arrays of data and for looping and iterating. It is a fairly easy language to learn and is very well suited to problems which are primarily calculation, rather than data handling.

Let us show how a program to perform the above calculation (Figs. 1, 2) could be written in Fortran. All that is necessary is the following:

```
          ISUM = 0
          DO 1 J = 1,100
    1     ISUM = ISUM + J
          PRINT 2, ISUM
    2     FORMAT(I10)
          END
```

Fig. 3

The single *DO* statement provides all the control necessary to perform the summation through 100.

3.2.2 Algol

Another language which is in some sense similar to Fortran but is more popular in Europe than in the United States is Algol. Algol was specified by a group of computer users in the late 1950's and has since been implemented on a number of machines. Most things which can be done in Fortran can also be done in Algol, but Algol also contains many features which are not found in Fortran. Because many of these are quite difficult to implement, a full Algol compiler is rare. In addition, in an attempt to make Algol even more machine independent than Fortran, the input/output features were not specified and consequently there are several different I/0 schemes for Algol, none with much relation to any of the others.

The program to find the sum of the integers then takes the following form, when written in Algol:

```
s := 0;
for i = 1 step 1 until 100
     do s := s + i end;
print (s);
```

Fig. 4

3.2.3 *Cobol*

Possibly the most popular programming language is Cobol. It is used almost exclusively in business oriented computer installations. It is a language which resembles English to the same extent that Fortran resembles mathematics. It is consequently verbose, though easy to learn. It contains features which are particularly necessary for business data processing, an application which is characterized more by input/output than by calculation. It provides easy methods for sorting large files and for printing nicely tabulated reports. The sample program written in Cobol looks like:

```
DATA DIVISION.
WORKING - STORAGE SECTION.
77   TOTAL     PICTURE 99999  VALUE 0.
77   NUMBER    PICTURE   999  VALUE 1.
77   DISPLY    PICTURE  X(20)
     VALUE     '    TOTAL COUNT  =   '.
77   DISPLY-TOTAL PICTURE  ZZ,ZZZ.
PROCEDURE DIVISION.
START.
        ADD NUMBER TO TOTAL.
        IF NUMBER IS GREAT THAN 99
             GO TO FINISH.
        ADD 1 TO NUMBER.
        GO TO START.
FINISH
        MOVE TOTAL TO DISPLY - TOTAL.
        DISPLAY DISPLY, DISPLY - TOTAL.
        STOP RUN.
END PROGRAM.
```

Fig. 5

3.2.4 *PL/I*

A recent innovation in programming languages is called PL/I, which stands for Programming Language/I. It is a new and not yet very common language which was developed by IBM in cooperation with SHARE, one of its user groups. PL/I, like Algol, contains many features which are difficult to implement, so no one has yet written a full PL/I compiler.

PL/I is almost the logical union of Fortran, Algol and Cobol, with a few features from other languages thrown in for good measure. The designers took the nicest features from each of these languages and included them in the specifications for PL/I. Consequently PL/I looks like a somewhat verbose version of Fortran. But here we have a single language which can, at least in theory, make the scientists as well as the businessmen happy. PL/I has nice features for letting the programmer take advantage of distinctly hardware oriented features (such as interrupts*) in a higher level language, something which had never been done before. But until people have had more experience with it, it will be impossible to determine whether this is the right direction to take, and whether those who have stepped in this direction have in fact succeeded. The above example in PL/I is shown in Figure 6.

```
ADDER:  PROC OPTIONS(MAIN);
     K = 0;
     DO I = 1 TO 100;
     K = K + I;
     END;
     PUT LIST(K);
     END;
```

Fig. 6

3.2.5 Other languages

There are several other significant languages which are available today, but which have not been as widely exposed as those already discussed. There are perhaps two reasons for this. First of all, they were developed by universities, rather than by the computer manufacturers. Secondly, they are generally not available for use on as many machines as are languages such as Fortran.

Probably foremost among these is LISP. LISP stands for LISt Processing language. It is particularly well suited for dealing with complicated data structures and has found perhaps its widest application in the field of artificial intelligence. Though burdened by a poor source language format, it is really a very powerful language for those applications for which it was designed.

Another of the privately developed languages is SNOBOL. This was

---

* An interrupt is a change in the flow of control during the execution of a program. The interrupt may be caused by an error condition (such as dividing by 0) or by something normal like the completion of input or output. The interrupt generally reverts control to the operating system rather than the problem program and is done in such a way that execution may continue from the point where it was interrupted.

originally developed by Bell Telephone Laboratories for the IBM 7090, and is currently available on several different machines. The most recent version, SNOBOL 4, is a substantial improvement over the previous versions. SNOBOL is probably the easiest of all programming languages to learn, but also one of the most powerful. Its forte is in the area of character manipulation, textual analysis and pattern recognition. While it is very well suited to these, it also has facilities for handling complex data structures and for doing calculations, though it is not as appropriate for this as Fortran. A significant disadvantage is that SNOBOL programs usually run very slowly compared to similar programs written in other languages.

There are in addition to those already mentioned a number of languages designed for simulation. Among these are SIMSCRIPT, SIMULA and GPSS, the General Purpose System Simulator, which is in fairly wide use. All of these languages have special features which make them especially appropriate for simulating everything from a butcher shop to a computer installation.

In addition, there are several languages which are very specific in application. A good example of these is APT which is a language for *A*utomatic *P*rogramming of machine *T*ools. A similar system is AUTOPROMT. By using these, an engineer can write a program in which he describes the operations which a machine tool operator must perform in order to produce a specific part. The computer then produces a punched paper tape which consists of instructions to the machine tool, so that it can operate by itself, thereby freeing its human operator to tend to other duties. This method of fabrication is becoming more and more popular.

3.2.6 Special packages

There are a number of other special applications packages which have been developed. One of these is SORT/MERGE. This is virtually standard software on all computers sold to business firms. SORT/MERGE is a program which will sort large numbers of records for use in business data processing. On a typical large machine several hundred thousand records can be sorted in a matter of an hour, a task that would be impossible any other way. A typical application is to sort records relating to a warehouse inventory by part number.

P–STAT is a large statistics package that was developed at Princeton University. The BMD or Biomedical package was developed by UCLA and is particularly well suited for analyzing large masses of statistical data such as is found in the fields of Sociology and Medicine. The BMD programs are a good example of the usefulness of writing in a machine independent language. Though the programs were originally written in Fortran for the IBM 7090, they are being easily converted to run on more modern machines such as the Control Data 6600.

4 CONCLUSIONS

For engineering and commercial reasons, contemporary computers basically operate in a microscopic fashion. To program is to specify in all necessary detail the steps required for the computer realization of some desired function. If the function is elaborate, the pattern of steps required may be highly complex. Languages and operating systems were developed so that scientists and other problem originators could express themselves in a more natural and appropriate manner than the unaided computer is capable of using as its instructions.

Most commercial compilers and systems were originally designed in ad hoc ways. It soon became apparent that in order to build translators and use them efficiently one had to understand something about the structure of both the language one starts with and the language one is going to use, and to study processes for mapping one into the other, and for devising algorithms based on formal principles. The establishment of departments of computer science in many universities has spurred this search for "principles". Expanded interest is to be expected in the formal theory of compilers and in abstract families of languages which study the structures of abstract automata and their relation to formal languages.

Other techniques which have been meeting with increased favor by translator writers is that of writing compiler–compilers. This involves writing a program to which is supplied the specifications of the input language, say Fortran. The compiler–compiler produces as its output a compiler which "understands" Fortran. Simply by defining the input language differently one could produce an Algol compiler for example. Another method for generalizing languages is the idea of extensible languages. This technique furnishes the user with the core of a compiler with extensive primitives. A user could then easily extend the language to suit his needs.

The line between hardware and software to some extent has been slowly vanishing on smaller computers. A technique for building computers known as microprogramming has contributed to this. A microprogrammed machine is one in which a microprogram corresponding to each machine language instruction is built into the hardware and it is this microprogram which determines the operation of the machine. Consequently by changing the microprogram one can alter the entire design of the machine. This might be done by reading in a deck of punched cards. Most models of IBM's System/360 are built in this manner.

In the large scientific computer area, however, new advances in the logical design of computers will increase the formidableness of language and operating systems design. Parallel computers typified by Illiac IV, now being constructed at the University of Illinois, where many processors are working on different parts of one problem, will intensify the programming problem.

Attempts at standardization of computer languages will increase. Cobol, for example, became a standard language in 1968 with the adoption of a version by the U.S.A. Standards Institute, a nonprofit trade group with quasi-legal authority. The action climaxed nine years of effort. Other well known languages are also being reviewed.

The preparation of this paper was supported by the AEC Computing and Applied Mathematics Center, Courant Institute of Mathematical Sciences, New York University, under Contract AT(30–1)–1480, with the U.S. Atomic Energy Commission.

References

1. A great deal of reference material is contained in manuals published by computer manufacturers. Although these manuals are not listed here, they exist for every commerically available programming language and system.
2. Relevant material is to be found in several computing journals such as Communications of the ACM, Journal of the ACM, AFIPS Conference Proceedings and Computing Reviews.
3. Rosen, Saul (Ed.), *Programming Systems and Languages*, McGraw–Hill, 1967.
4. Vyssotsky, Corbato and Graham, "Structure of the Multics Supervisor", *Proc. AFIPS*, 1965 Fall Joint Computer Conf. PTI, Spartan Books, Wash., D.C., 1965.
5. Feldman and Gries, "Translator Writing Systems", *Comm. ACM*, Vol. 11, No.2, Feb. 1968.
6. Wegner, Peter, *Programming Languages, Information Structures and Machine Organization*, McGraw–Hill, 1968.
7. Knuth, Donald F., *The Art of Computer Programming*, Addison–Wesley, 1968.

CHAPTER 7

Software for Interaction with Computers

JULES I. SCHWARTZ

BACKGROUND

THROUGHOUT THE business, scientific, and governmental communities, the direct or indirect requirement for computer service is a fact of life. In the early years of computing, this diversity of computer usage was nonexistent and probably not foreseen. Computers were envisioned to be merely tools for calculation, allowing lengthy numerical computations to be performed orders of magnitude faster than they could by hand. The early computers were particularly oriented for problems which could have prespecified or programmed solutions—that is, one could literally wire in an algorithm for the solution of a class of problems, enter a few parameters at run-time, and get a display of the answer shortly thereafter.

To a large extent, the first computers were used directly by scientists involved in the solution of a problem. They worked on the solution of a relatively small number of problems, and these occupied a large part of the total computer time.

As computers became more sophisticated, however, their capabilities to solve considerably more complicated problems had some interesting effects on the art of computing. Man's ingenuity for inventing techniques to utilize these powerful devices led to the design of solutions for increasingly complex problems. Although the statement of algorithms was still straightforward, computers—along with their increasing speed and capacity—were increasingly awkward to utilize. The accurate translation of an algorithm to a program on a computer required many days or months for completion. This, of course, was too much for the scientific investigator of a project to get involved in, so that by the middle 1950's a completely new profession had begun to grow: programming. This field was composed of people who were skilled first in computers, and then to a lesser extent in other disciplines. The scientist normally dealt with the programmer, who acted as the intermediary between him and the computer.

With the existence of these new specialists, several phenomena occurred. First, the scientist became more distant from the computer. Computers increased in complexity, and programmers began inventing techniques, procedures, and systems which made their jobs easier, but also made access to a computer extremely difficult for the nonprogrammer. Second, with an expanding set of people interested in computers for both problem-solving and assisting in the programming process, the demand for computer time became pressing. Even as computers became more capable, there invariably was a critical shortage of available time on them. Third, long periods of time were still required to program and check out algorithms (programs) for computers. Whether because of increasing complexity of problems, computers, or systems, "programming–debugging" time never seemed to improve.

With this combination of circumstances, the only theme of computer installations—from about 1954 to the early 1960's—was "efficiency". The emphasis of procedures and systems was to minimize idle time. To do this, systems were invented which put large numbers of jobs in a stream or batch and ran them serially with little delay between jobs. Building increasingly sophisticated versions of these "batch-processing" systems has been the main occupation of systems programmers for the past 13 years.

The first rule of these systems is that during the running of the stream of jobs, there should be no human intervention, except by computer operators who occasionally enter cards and tapes, and remove printouts at the system's command. Occasionally, programmers can go to the computer to attend a run, but not normally. The "customer", who originates most problems, never attends their running. This mode of operation is efficient in the sense of preventing careless mistakes or unwarranted delays by humans. It is a strictly controlled and planned environment.

It has some inherent difficulties, however. Delays, even for short computations, are long. Overnight response is common, with good return being on the order of hours, and delays of days or a week not unexpected. There is another deficiency in this mode of operation, however, which is more serious. This is the effect of errors. A delay of a day or more to get a satisfactory result may be tolerable. But waiting the same amount of time and getting *no* results (due to an error) is extremely frustrating. Those who are unfamiliar with computers would be surprised at how many runs result in little or no progress, because of machine, system, program, or human error.

In addition, this detached style of running prevents what can be a valuable approach to using a computer—namely, the direct interaction of the human with the machine. In 1960, Dr. J.C.R.Licklider delivered an interesting paper[18] describing eloquently the advantage of an interactive—he called it "symbiotic"—relationship between a man and a computer. In this mode, a man deals directly with a computer through a keyboard/printer or display/light-pen device to monitor and direct the running of a program. In this way, one can maximize the capabilities of both parties. The man hypothesizes, creates, and inspects; the computer calculates. Together, through a rapid interchange, they approach the solution of a problem.

In 1960, the concept of "on-line" access to a computer with consoles was not new. Various examples of direct human interaction with a computer existed. One of the most prominent examples (although not generally demonstrated to the public) was the SAGE system[9]. This was a highly sophisticated system for automating defense against air attack, development of which began in 1955. Associated with each computer were a set of manned consoles, the consoles having character and vector displays, keyboard, and light-pen. Other examples existed in the 1950's and early 1960's, but usually they were systems utilizing one or several devices which—when being used—usurped the entire computer. Where computers are inexpensive and abundant, their complete occupation by one or two people at consoles is not unreasonable. But normally computer time is expensive, and there is a shortage of it. Thus, one person working at his own pace at a console, letting the computer idle while he is thinking or preparing an input, is economically unjustifiable.

An attempt to present the computer to a human in an interactive style of access requires more than a console and a standard operating system. The technique which made interactive access economical on a large scale was "time-sharing". Although the term had been used previously for certain engineering concepts, the term was first used to refer to a programming system in several articles published in the early 1960's[20,25].

THE CONCEPTS OF TIME-SHARING

The basic idea of time-sharing is actually quite simple. Essentially, the system permits some number of users to run their programs in the computer simultaneously, as opposed to running them one at a time in a batch-oriented system. With a large number of users operating at consoles, there is a high probability that there will always be something for the computer to do. While some people are thinking or typing, the system can run other users' programs, thus keeping idle time to a minimum.

As computing systems grow more sophisticated, the difference between so-called time-shared and other systems tends to decrease. There are actually very few strictly "batch" or serial operating systems being produced today. The majority of operating systems are called "multiprogramming systems". In these, a number of programs reside simultaneously in the computer, with each program normally occupying a section or "partition" of core memory. Each program is assigned a priority. The program with the highest priority that requires service is operated. During the computer's periods of idleness (e.g., when it is waiting for input or output, is between jobs, or requires operator intervention), programs with lower priorities are run. There is usually very little "user" intervention in the typical multiprogrammed operating system (although this is increasing). Inputs and outputs of the programs are generally via cards, tape, or printers.

Time-shared systems have a somewhat different objective than these standard operating systems. They exist primarily to give interactive service to users. Consequently, a priority scheme alone is not sufficient. Permitting one or more high-priority programs to usurp the system's computing resources to the exclusion of other users would defeat the prime objective of time-sharing.

Time-sharing systems divide up the machine's time into slices. Each program that requires computing service gets a turn in some reasonably sequential fashion. This turn, called a "quantum" or "time-slice", is short, on the order of a fraction of a second to a second. If at the end of a time-slice, the program has not completed operating, it is interrupted, and the system inspects to see if another program is waiting for service. If it is, it is given its turn. Using this time-slice approach, one can guarantee that each user will never have to wait more than $(n - 1)\,t$ for his program to begin computing after he enters an input, where n is the number of simultaneous users, and t is the time each user takes (including the time-slice). Actually, the expression $(n - 1)\,t$ as a maximum is an over-simplification, for reasons that will now be discussed.

Swap time and overhead

Unlike most multiprogrammed systems, the majority of time-sharing systems attempt to permit each program to occupy most—if not all—of the available core storage. This, of course, prevents all programs from *residing* in core when they are being run. Thus, most programs reside on peripheral storage (drum, disc, or slow large core). When a program's turn to operate comes, it usually must be brought into core from this peripheral store. Also, another program has to be moved out of core onto peripheral store. This process (known as "swapping") adds a period of time to the operation of each program.

In addition to the swap time associated with each program's turn, there is also a certain amount of "overhead" for each program's operation. This includes the system execution of such functions as scheduling, resource allocation, and the handling of machine interrupts.

Thus, we see that the time t associated with each user program's operation is equal to

$$q + s + o$$

where

q = length of turn
s = swap time (time to get the program in and out of core)
o = overhead

Again, this is a simplification of the actual operation, for the following reasons.

Active vs. inactive programs

In real operations, it is very unusual to have all programs in the system require service during one cycle. The basis on which the concept of time-sharing rests is the fact that a high percentage of the users have programs which are inactive for significant periods of time. The users spend time inspecting output, thinking, and entering input. Thus, during a given cycle the number of active programs is m, where, in general $m < n$ (the number of users). The relationship of m to n can vary widely, depending on the system, applications, kinds of terminals used, backlog, and other factors. However, a reasonable estimate might be $0.1n < m < 0.2n$.

Another inaccuracy is the parameter q. According to the expression, it appears that each program computes for a complete time-slice. In fact, statistics have shown that the majority of interactive programs actually compute for something less than one full time-slice to satisfy a user input, and then requires more input from a user before continuing. Consequently, q is the *maximum* compute time for any one user during a cycle.

Also, the parameters s and o are not the same for all programs, but the assumption of their being equal for all programs does not have any significant effect, for purposes of this discussion.

Response time

We have seen that the time for m active programs to get one time-slice is:

$$c \leqslant m\,(q + s + o)$$

For most user requests, less than one time-slice is sufficient, so that c represents an upper limit on response time. However, some programs (such as lengthy mathematical computations) require more than one time-slice to complete an output. For these programs,

$$\text{Response time} = km\,(q + s + o)$$

making the simplifying assumption that m remains the same for each cycle.

It seems obvious at first that the way to minimize the response time is to *minimize* q, since s and o are relatively uncontrollable. (In fact, there are schemes for controlling s, which are discussed later.) However, another simple analysis points up a fundamental problem of time-sharing.

Assume that there are p programs which require time T to complete p users' requests, starting at the same time, where $T > q$. The elapsed time E to complete all programs is:

$$E = \frac{pT}{q}\,(q + s + o)$$

so that

$$E = pT\left(1 + \frac{s + o}{q}\right)$$

which shows in an expression the fairly obvious observation that one can minimize the elapsed time (i.e., maximize the system's throughput) by *increasing* q.

This conflict between the need to handle lengthy computations as well as jobs with short processing times in a time-sharing system is one of the pressing problems in producing a satisfactory system. Various solutions for this problem exist. Primarily they involve the creation of "scheduling algorithms" which deal with programs having long compute time requests. Once isolated, these programs are run less frequently but with longer periods of uninterrupted operation. Such programs are usually chosen based on their computing history, assuming that programs which have computed for long periods are going to continue doing so. These programs are moved into different queues. Of course, it is also possible that programs can be manually specified as computer-bound and thus be relegated to these queues automatically. Also, this area is one where the use of a priority scheme in conjuction with time-slicing is possible.

Other techniques

Up until now, the discussion has centered on the control of system response and throughput by manipulating the size and frequency of the time-slice. This is clearly not the only area where improvements in service can be achieved.

Overlap

One area that has been completely ignored in the preceding is the possibility of parallel operations. On all present-day computers, it is possible to perform several events simultaneously. For example, it is possible to overlap the swap-in or swap-out of one program while another is in operation, which shortens the total cycle time. Also, on some computers, several control processors can share the same memory facilities, providing the ability to overlap several computations. This latter feature has not been utilized very effectively yet. Only a few time-sharing efforts[4,17] have attempted to utilize multi-processing computers. Their utilization, however, has obvious advantages.

Paging

It is clear that the major source of overhead in time-sharing comes from swapping. The desire to give rapid response to users, although their programs may only operate a short time, can create a high ratio of transfer time to operating time. There are various techniques to decrease the amount of swap time. They include such things as the concept of re-entrant programs—"pure" programs which don't modify themselves and thus can operate for a number of users without requiring the bringing in of anything but the data that is unique to each user. Such "pure" programs do not have to be swapped out, since their original state on swap storage is correct. Another technique is "marking". While a program is operating, a record is kept of those areas of the program and data that were changed. Only these areas are written out onto the swap storage.

One other important concept which has been applied to time-sharing over the past five years is that of paging. This technique consists of dividing programs into fixed-size segments which can be uniquely addressed independent of their physical location in core storage. Each segment or "page" is given an identifier—a page number—which is used to reference it. The system assigns to each page an actual address which is placed in a table accessible by the computer. Any reference to a page is then automatically translated to the physical address. If a referenced page is not in core memory, an interrupt occurs, and control returns to the system, so that the requested page can be brought in.

Paging has several advantages. It is quite a bit easier for programs (or parts of programs) to reside simultaneously in core storage, thus eliminating the need to completely swap out programs before bringing in others, since a program's pages can reside in any available core space. In addition, the system can swap the program in sections (e.g., a page at a time), which minimizes the amount of transfer needed before a computation can begin. Also, a mechanism for marking modified pages is easy to produce, helping to speed up the previously discussed marking scheme for saving transfers to swap storage.

Although the concept of paging is relatively old (for computers)[7,13], its use in time-sharing has gained momentum in just the last few years. Some papers first discussed its use in 1964–65[6]. Recently, there have been a large number of papers on the topic, since the techniques for utilizing this capability to great advantage seem to be much more difficult than one might assume.

One of the original methods for utilizing paging was called "demand paging". The major notion of this technique was that pages would be brought into core only when required by the program. Thus, when a program's turn arrived, it would be necessary to have only one page available. As more were needed, they would be fetched.

From the beginning, there have been a variety of criticisms of this approach[10]. The fundamental difficulty with this technique has been defined as "thrashing". This term describes the situation where the computer is essentially idle for long periods of time while required pages are accessed from peripheral storage. This is due to the fact that whereas the time to access and transfer a page may be on the order of 10 to 100 milliseconds, the compute time for a single page may only be microseconds. Since the time to bring in information from a swapping device like drum or disc is normally faster per word as the number of words increases,* it is more economical to bring in a set of pages than a single one, assuming one can predict the proper set of pages required. There are considerable discussions on techniques for predicting the necessary pages and optimum number of them[2]. Suffice it to say that most people believe that this mechanism can be a valuable one when used wisely, and like most inventions in this field, seems to be leading into a whole area of theory on its own.

THE ARCHITECTURE OF TIME-SHARING SYSTEMS

Time-sharing systems have a variety of forms. Some, particularly the early systems, were quite simple. Others, like MIT's MULTICS[4] and IBM's TSS[17], have quite complex structures. However, an understanding of some of the common characteristics of these systems helps in understanding the subject.

Executive program

First, there are the components of the system that deal with the basic hardware; these are characteristic of all operating systems:

Interrupt Handler
Handles all computer interrupts.

Input/Output Dispatcher
Performs all physical transfers of information to and from peripheral storage devices, typewriters, displays, and other keyboards.

Then there are the components that handle the needs of simultaneous users and programs:

Cataloger
Assigns physical space for files requested by programs. In the dynamic situation that exists during time-shared operations, users' programs cannot arbitrarily occupy specific physical files. Thus, typically, a program requests space for a file, and the Cataloger assigns it. It also deletes files that are no longer in use, and keeps an inventory of all temporary and permanent files.

Input/Output Controller
Once files have been assigned physical addresses by the Cataloger, the system

* That is, $s = \alpha + \beta w$, where $\alpha \gg \beta$, and w = the number of words.

must translate subsequent program requests to move information to and from the file, since these requests are not made with specific addresses. The Input/Output Controller translates these requests for the Input/Output Dispatcher. (These two components are sometimes combined.)

Scheduler

Determines which program should operate at a given time and how long it should operate. Some of the considerations used by the Scheduler are discussed in the section on "Response Time".

Swapper

Once programs are determined eligible and ready to run, they must be brought into core storage and executed. Also, they must be returned to swap storage after their turn is complete. This is done under control of the Swapper component.

Other components exist to permit user communication with the system and programs:

Command Interpreter

Each system has a set of commands available to users for execution and control of programs. For example, the commands LOGIN, LOAD, GO, and STOP may permit a user to identify himself, load, execute, and stop a program. These commands are recognized by the Command Interpreter component. In many systems, each command is actually represented by a program, so that the Command Interpreter simply executes the program necessary to perform a command.

There are usually at least two modes of communication in time-sharing systems: one mode permits commands to the system; the other, conversation with a user program. These modes are normally set by a special character or key, so that the component which interprets the user input knows whether it should be interpreted as a command or passed to a user program.

Time-sharing hardware

A thorough discussion of the equipment used for time-sharing would be quite detailed and lengthy. For the purposes of this discussion, a summary of the requirements will be given, and areas where time-sharing has unique needs will be pointed out.

Figure 1 shows a generalized hardware configuration for a time-shared system. The requirements for core memory are similar to those needed for other operating systems. About $\frac{1}{4}$ to $\frac{1}{3}$ of the core memory is occupied by the time-sharing executive program.

The removable storage (disc packs, tapes, etc.) is useful for private files of information and programs; it permits essentially unlimited storage facilities.

The large, permanent store is useful for libraries of programs and data needed by the user population during most of the operating hours. These latter storage devices (removable and permanent) are characteristic of most system's requirements, although in time-sharing one might need more storage than other systems because of their servicing of a number of simultaneous users.

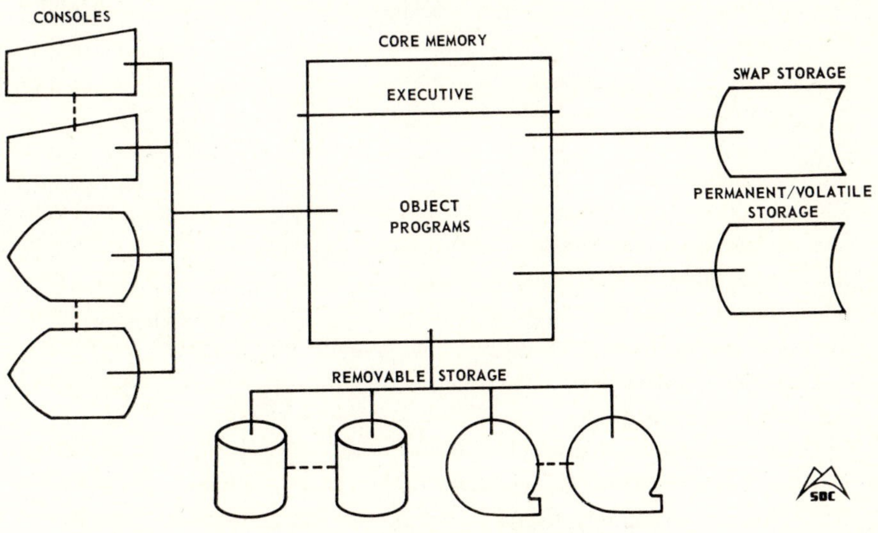

Fig. 1 Hardware configuration

A requirement for time-sharing which is relatively unique is the swapping storage. This storage must be a reasonably large, random-access device, large enough to store all programs simultaneously in use. Swap storage is normally disc, drum, or bulk core storage.

Another requirement for time-sharing systems (or any systems permitting on-line access) is a set of consoles. These can be typewriters, teletypes, storage, or cathode-ray tube devices, with some systems permitting other devices such as touch-tone phones, grids with styli for hand-printed inputs, and the "mouse"[8] (which can be used on any flat surface to reference points on a display) for entering inputs to the computer.

Associated with these consoles are communication lines for connecting them to the computer. The kind of lines necessary depends on the device, its distance from the computer, the basic computer, and the data requirements of the device. For some devices, direct distant operation from a computer is either impossible or extremely expensive. These include large CRT displays, the RAND Tablet, or other devices where high input or output data rates are required. One technique which makes it more feasible to run these devices remotely is to attach them to a computer located at the site. This local computer can provide the rapid data access required by the devices. It can then be interfaced with the central time-sharing computer over standard communication lines. (At the Information Processing Techniques Office of ARPA in Virginia, a RAND Tablet and large display are interfaced with a time-shared computer at Lincoln Laboratory in Massachusetts, via a PDP–9 computer located at the Information Processing Techniques Office.)

For the slower devices, normal communications are sufficient technically. However, considerable concern has been expressed over the cost of these communications. The reason for this problem is the incompatibility between standard voice communications and conversation between a human and a computer. The dialogue involved in interactive computing consists of short bursts of communication, separated by long periods during which no information is being passed. (Either the computer or human is "thinking".) Normally, the line is open during the entire period, and the cost is based on the length of time the line is open. One solution for this problem that is becoming common is the use of multiplexors at remote sites. A leased line is made available between the multiplexor and the remote time-shared computer. All users on site make local calls to the multiplexor, which then controls the message traffic to the remote system, permitting the users to utilize primarily local lines with a reasonably apportioned rate for the long-distance communications.

Special- and general-purpose systems

The architecture so far discussed is an abbreviated description of a system which permits an open-ended set of programs to be run. With this kind of system, there is no restriction regarding the type of program one can execute. The programs that can be loaded and executed may allow one to solve numerical problems, retrieve information from a data bank, or compile new kinds of programs. Such open-ended systems are called "general-purpose". In general-purpose systems, the time-sharing executive interprets and executes commands, and treats all application programs alike.

Another kind of time-sharing system is called "special-purpose". Actually, there are probably many more special- than general-purpose systems in existence today. Special-purpose systems are built to concentrate on a limited set of capabilities. The advantage of a restricted set is the ability to provide an optimum interface for the user. A general-purpose system offers a number of different languages, varied application programs, and access to a wide spectrum of storage. Thus, the general-purpose system user is probably faced with a variety of conventions for controlling the situation. In the extreme case, a special-purpose system offers one language useful for numerical calculation only, and the user is unaware of the system's needs for storage. Thus, the user can have a quite limited set of information to learn and a minimum of conventions to be concerned with at a console.

EXAMPLES OF SYSTEMS

Early systems

One of the earliest and probably best known of the general-purpose systems is called CTSS (Compatible Time-Sharing System)[5]. Its first version was demonstrable at MIT in 1962. At that time it could handle approximately three simultaneous users doing programming and debugging. It is still in considerable demand, serving a large population of university users, now numbering around 30 simultaneously.

CTSS is now of considerable interest because of its pioneering role in the field of time-sharing. It is also interesting, however, for the large variety of developments that have come from this system. Fields such as engineering, mathematics, psychology, education, artificial intelligence, and others have gained from work done on CTSS.

Another early fairly large-scale time-sharing system was produced on the IBM AN/FSQ–32 computer at System Development Corporation. This computer was originally intended for military applications, but was used in research on interactive computing by SDC with support from the Information Processing Techniques Office of ARPA. The SDC system has many similarities to the CTSS system at MIT, in that it serves around 30 simultaneous users working on a wide variety of applications. Its use has been largely oriented to research and operational problems of defense. Most of the current users are associated with the federal government (both military and otherwise). Considerable emphasis is placed on problems of data management—the input, retrieval, and modification of large amounts of information.

Another early system had its initial demonstration in the early 1960's. This special-purpose system, developed at The RAND Corporation, is called JOSS. Originally developed on the JOHNNIAC computer and now running on the PDP–6, JOSS was the first system devoted to users solving computational problems. The idea was to present a very simple, elegant, and useful tool for the nonprogrammer scientist.

Another system with aims quite similar to JOSS is called BASIC. Originally developed at Dartmouth University on the GE–235 computer, it has become probably the best known interactive system in the country, thanks to its capability and considerable promotion by the General Electric Company. It has stressed simplicity over power in providing a calculation tool for the nonprogrammer.

In addition to these systems, numerous others have been developed over the last six years, on a wide variety of computers. There are over a hundred time-sharing installations available on a commercial basis at present. In addition, many systems are being produced at a number of universities and industrial installations[16,26].

Current efforts

For the most part, the typical systems available to users today have changed relatively little over the past years. They tend to be restricted to allowing nonprogrammer users to perform calculations. There are systems which concentrate on business applications and some on text-handling[14]. But the majority of commercially available systems provide one or several computational languages and a capability for entering and editing data and programs. The larger, more generally applicable systems are not widely available commercially, although on systems like SDC's and MIT'S, charges for use are made.

There have been several efforts at providing significant improvement in capabilities offered by large-scale systems. In 1964, design began on computers and time-sharing systems that were intended to increase significantly the capacity and capability of time-sharing. The two major efforts were at MIT and IBM, using the GE–645 and IBM 360/67 computers, respectively. Although the computers differed in a number of details, they had several features in common which were particularly oriented to time-sharing.

Both computers are "paged" (see the section on "Paging" above). Both employ what is known as the "virtual memory" concept. This is a means of handling large file storage without burdening the programmer with the management of storage hierarchies. His files are called segments, which are large collections of pages. References to parts of segments are made as though they existed in core. The system actually places the required pages in core. Also, the GE–645 and IBM 360/67 were both designed as multiprocessors. Each computer provided access to its storage from two (or more) processing units simultaneously, permitting parallel computations.

In addition to these and other hardware features, both systems were to be quite advanced in program design. They were planned to be the first of a number of large-scale computing utilities. Although many of the principles developed in these efforts have proven to be quite valuable, the systems themselves have so far progressed quite slowly, for a variety of reasons. For the present, they offer approximately the capability of previous systems (or less), but presumably will surpass them eventually.

INTERACTIVE APPLICATIONS

The major part of this discussion so far has been devoted to a description of time-sharing systems. As was stated previously, these systems exist for the purpose of providing users economical direct access to computers. Therefore, a discussion of this subject without emphasis on the applications of time-shared systems would be quite remiss. In this section, we shall try to give an overview of the kinds of applications that have been produced using some of the systems in existence today. Of course, there are literally thousands of examples of interactive computing in existence today, so that the best that can be described in this chapter is a small sample. This sample is chosen to demonstrate the range of capabilities as well as techniques for accomplishing certain capabilities in interactive computing.

Computation

Basic

As stated before, one of the best known interactive computational systems is BASIC. The BASIC language was devised to provide a wide range of capabilities with a very simple grammar and a small set of rules. An example* is presented here of a BASIC program which solves a set of two simultaneous equations:

$$ax + by = c$$

$$dx + ey = f$$

Assuming $ae - bd$ is not equal to 0, the solution is:

$$x = \frac{ce - bf}{ae - bd} \qquad y = \frac{af - cd}{ae - bd}$$

The BASIC program to compute and print the solution, given a, b, c, d, e, and f is:

```
15 LET G = A*E – B*D
37 LET X = (C*E – B*F)/G
42 LET Y = (A*F – C*D)/G
55 PRINT X, Y
```

Adding the statements which read the data and check for $ae - bd$ not zero gives the following sequence:

```
10 READ A, B, D, E
15 LET G = A*E – B*D
20 IF G = 0 THEN 65
30 READ C, F
37 LET X = (C*E – B*F)/G
42 LET Y = (A*F – C*D)/G
55 PRINT X, Y
60 GO TO 30
65 PRINT "NO UNIQUE SOLUTION"
```

The meaning of this sequence should be obvious, when it is understood that "THEN 65" in line number 20 implies that the next execution should be line 65, in this case, if $ae - bd = 0$.

To complete the program, one must actually enter the data and begin execution. The next example shows the complete program, data, and output as it would appear on a teletypewriter.

---

* For the most part this example is taken from: "BASIC Language Reference Manual", General Electric Company document IPC–202026A, January 1967.

```
10 READ A, B, D, E
15 LET G = A*E - B*D
20 IF G = 0 THEN 65
30 READ C, F
37 LET X = (C*E - B*F)/G
42 LET Y = (A*F - C*D)/G
55 PRINT X, Y
60 GO TO 30
65 PRINT "NO UNIQUE SOLUTION"
70 DATA 1, 2, 4
80 DATA 2, -7, 5
85 DATA 1, 3, 4, -7
90 END
RUN
LINEAR 10:37 DEC. 17, 1965

  4            -5.5
  0.666667      0.166667
 -3.66667       3.83333

OUT OF DATA IN 30

TIME: 0 SECS.
```

Of course, in most interactive applications, one would not want to type in all the data at program composition time. Frequently, one wants to enter data interactively, as the program is running. To do this in BASIC, there is the command INPUT.

Thus, in the previous example, one could dynamically enter the values for *a*, *b*, *c*, *d*, *e*, and *f* with the following program:

```
 5 PRINT "ENTER A, B, D, E"
10 INPUT A, B, D, E
15 LET G = A*E - B*D
20 IF G = 0 THEN 65
25 PRINT "ENTER C, F"
30 INPUT C, F
        .
        .
        .
```

The program will remain stopped after the INPUT statement until the values are entered.

In BASIC, one can define and use matrices as well as scalars and vectors. Some examples of statements utilizing matrices are:

MAT C = A*B (which means Multiply the matrix A by B and set into C)

MAT C = TRN(A) (which means Set C = to the transpose of A)

MAT C = ZER (which means fill C with zeroes)

JOSS

Another example of a user-oriented computation language in JOSS[3]. Similar to BASIC in its stress on simplicity, JOSS utilizes a specially designed typewriter console and a carefully human-engineered design to make computer access by a nonprogrammer rapid and painless. In fact, the authors of JOSS feel that a programmer might be handicapped in learning JOSS, since is is quite different than the typical programming language. An example of the use of JOSS to solve the same problem as illustrated in the previous section on BASIC is:

1.1 Do part 2.
1.2 Type "NO UNIQUE SOLUTION" if $g = 0$.
1.3 Do part 3 if $g \neq 0$.
1.4 To step 1.1.

2.1 Demand a, b, d, e.
2.2 Set $g = a \cdot e - b \cdot d$.

3.1 Demand c, f.
3.2 Type $(c \cdot e - b \cdot f)/g$, $(a \cdot f - c \cdot d)/g$ in form 1.

Form 1:

$x =$ ____ . ____ $y =$ ____ . ____

There are, of course, many similarities in BASIC and JOSS when used for this simple problem. However, several differences can be noted. The use of the JOSS typewriter permits lower-case letters for character names, and other special symbols such as the middle dot for multiplication. JOSS has a fairly simple but powerful formating scheme, an example of which is Form 1. Also, programming is done in steps and parts in JOSS, rather than the more typical linear string with arbitrary symbols for labeling lines of code. Another significant difference between JOSS and some other systems is that JOSS is an "interpretive" system. With most programming languages, the normal procedure is to "compile" a program into machine language and let it run. An interpretive system keeps the program in some higher-level language, and executes each step individually, thus having complete control of the operation at all times. This is particularly valuable when one is operating interactively, since any anomaly will be reported instantaneously to the user. The disadvantage of such systems is that interpretive execution is normally slower than that of executions of a compiled program.

Data management

With the coming of time-sharing at SDC, the desirability of an interactive data management system became apparent. In the typical data management system, the user enters and retrieves information by entering a set of statements which are quite similar to those required in programming. For retrieval (and sometimes when entering the data), one must know ahead of time what information he will want, and how he wants it formatted. With a system designed for interactive access, on the other hand, the user can inspect data and decide at run time what information he wants and how he wants it formatted. Two systems at SDC were given this orientation. The original one was LUCID on the Q–32 time-sharing system; the second, TDMS (on the SDC time-sharing system for the IBM 360 computer). Examples of an on-line interaction with TDMS are given below.

The DESCRIBE command permits the user to determine the original definitions of elements of the data base. The SHOW command, along with the SEARCH command, permits the user to "browse" through the data base. The PRINT command permits elements to be displayed, based on given qualification criteria.

```
DESCRIBE PERSONNEL
- - - -
            C9  PERSONNEL
            C91 NAME
            C92 SOCIAL SECURITY NUMBER
            C93 DEPENDENTS
                C931 NUMBER OF DEPENDENTS
                C932 NAME OF WIFE
```

```
SHOW OFFICE LOCATION
- - - -
    2 V1  DAYTON
    3 V25 WASHINGTON
SEARCH: V2 ··· V18/4
- - - -
    1 V2  EAST ORANGE
    3 V6  NEW YORK
    2 V10 PATERSON
    2 V14 SAN FRANCISCO
    1 V18 TULSA
SEARCH: $
- - - -
PRINT COUNT OFFICE LOCATION WHERE NUMBER
    OF EMPLOYEES GR 200 AND BUSINESS VOLUME
    LS 6000000
- - - -
3 ENTRIES
PRINT OFFICE LOCATION WHERE SAME
- - - -
    MINNEAPOLIS
    LOS ANGELES
    NEW YORK
```

Natural language interaction

The examples of interactive applications discussed so far could be described as "nonprogrammer"-oriented. That is, the user need not understand how the computer or programming system works, and the language he uses is fairly simple to learn. It is obvious, however, that he does have to learn some rules, and the language he uses, although sometimes learned rapidly, has a considerable number of constraints. In addition, one finds that the various languages have enough incompatibilities to dismay the normal user. One obvious solution to this problem would be to use natural language at an interactive console. If one had the complete range of English available at a console, he could choose the vocabulary which was natural to him and not be concerned with unnatural rules. This is obviously a very difficult technical problem. Some believe that full understanding of completely natural English by a computer is impossible, and even the optimistic feel that it is unlikely during the 20th century. However, there are various efforts at producing computer understanding of natural language, or at least reasonably rich subsets of English. Some of the efforts are described by Bobrow[1] and Simmons[24] in the *Annual Review of Information Science and Technology.*

One interesting example of work in this area is called ELIZA, a system produced at MIT, under the guidance of Professor J. Weizenbaum. The system has achieved a good deal of notoriety, because of its reputation as a computer-psychiatrist, since its early demonstrations were primarily in this area. Actually, it is a clever manipulator of key words and phrases, a syntax analyzer, a categorizer, and memorizer, so that a very realistic conversation with a user can take place. An example* of a "psychotherapeutic" dialogue† follows:

men are all alike
IN WHAT WAY
they're always bugging us about something or other
CAN YOU THINK OF A SPECIFIC EXAMPLE
well, my boyfriend made me come here
YOUR BOYFRIEND MADE YOU COME HERE
he noticed that i'm depressed much of the time
I AM SORRY TO HEAR YOU ARE DEPRESSED
he's right. i am unhappy
DO YOU THINK COMING HERE WILL HELP YOU NOT TO BE UNHAPPY
i need some help, that much is clear
WHAT WOULD IT MEAN TO YOU IF YOU GOT SOME HELP
perhaps i could learn to get along with my mother
TELL ME MORE ABOUT YOUR FAMILY
my mother takes care of me

---

* Taken from: "Project MAC Progress Report IV, July 1966 to July 1967", Massachusetts Institute of Technology document MAC–PR–4, July 1967.

† Lower case is human input; upper case is computer response.

WHO ELSE IN YOUR FAMILY TAKES CARE OF YOU
my father
YOUR FATHER
you are like my father in some ways
WHAT RESEMBLANCE DO YOU SEE
you are not very aggressive, but i don't think you want me to notice that
WHAT MAKES YOU THINK I AM NOT VERY AGGRESSIVE
you don't argue with me
WHY DO YOU THINK I DON'T ARGUE WITH YOU
you are afraid of me
DOES IT PLEASE YOU TO BELIEVE I AM AFRAID OF YOU
it pleases my father to think i am afraid of him
WHAT ELSE COMES TO MIND WHEN YOU THINK OF YOUR FATHER
bullies
DOES THAT HAVE ANYTHING TO DO WITH THE FACT THAT YOUR BOYFRIEND MADE YOU COME HERE

To see the application of natural language to computational problems, ELIZA has been modified with an evaluator. An example of it in use is:

please tell me what the sum of 5 and 4 is
IT'S 9.0
good, call that *a*
VERY WELL
can you tell me what the product of *a* and 5 is
IT'S 45.0
call it *b*
VERY WELL
let $f(x, y)$ be $x^{**}2 + y^{**}2$
ALL RIGHT
please compute $f(a, b)$
IT'S 2106.0
what is the product of 45 and 45
IT'S 2025.0
what is the difference between that and *a*

IT'S 2016.0
call this *z*

Computer-assisted instruction

Another area where interactive computing has a high potential is that of education. There are now various examples of students at a variety of grade levels directly accessing a computer in the classroom. At SDC, a system called PLANIT, which provides access to both the teacher and pupil, has been developed. The teacher can prepare, test, and modify lessons and tests in a variety of modes (e.g., multiple-choice, true-false, question-answer, arithmetic, and algebraic). The student interacts with the lessons. In addition to permitting some reasonably natural mistakes and offering hints (when directed by the instructor), the system permits the student to perform computer-assisted calculations when desired. It also can do such things as rapidly generate samples from given kinds of distributions, which makes certain statistics courses much more meaningful.

Examples of PLANIT dialogues follow. Example 1 describes the teacher's on-line preparation of part of a lesson; Example 2 shows a student's interaction with a PLANIT lesson.

Example 1

```
FRAME 1.00 (P)
G 5.  STEP 5
1.00  SUM UP THE ITEMS IN COLUMN 1.;
2.00  DIVIDE THE SUM BY THE NUMBER OF ITEMS IN
      THAT COLUMN.

G 6.  SS.
1.00  MY

G 7.  DIST, M, N.
1.00  NN 23 3

G 8.  M1  R1  S1     MS  R2  C.
1.00  33  60  N  +9  40  0.25

G 9.  SH.
1.00  ITEM NO.      COLUMN 1  COLUMN 2
2.00     X             X         X
```

```
FRAME 2.00 (Q)
G 2.  SQ.
1.00  WHAT IS THE MEAN OF COLUMN 1?
G 3.  SA.
1.00  0 FUNCTION MEAN (C) = (SUM DATA(I, C) FOR
      (I = I, N(C)))/N(C)
2.00  1 + MEAN(1) WITHIN 0.05
3.00  2 MEAN(2)
4.00  3 (MEAN(1) + MEAN(2))/2 WITHIN 0.1
G 4.  SAT.
1.00  1 F:VERY GOOD.
2.00  2 R:NO, YOU FOUND THE MEAN OF COLUMN 2;
      NOW DO THE SAME FOR COLUMN 1.
3.00  3 R:NO, FORGET COLUMN 2—PLEASE FIND THE
      MEAN OF COLUMN 1 ONLY.
4.00  –R:NO, TRY AGAIN.
```

Example 2

```
*GET DISD
IDENTIFY YOURSELF***ST1
SAMPLE SIZE (MAX 20) ***
*6
ITEM NO.   COLUMN 1   COLUMN 2
   1–         41         35
   2–         31         37
   3–         12         34
   4–         31         41
   5–         57         53
   6–         25         40
WHAT IS THE MEAN OF COLUMN 1?
*←STEPS
STEP 1.
SUM UP THE ITEMS IN COLUMN 1.
*35+37+34+41+53+40
240.0
*STEPS
```

STEP 2.
DIVIDE THE SUM BY THE NUMBER OF ITEMS IN THAT COLUMN.
*240.0/6
40.0
*↑40.0
NO, YOU FOUND THE MEAN OF COLUMN 2; NOW DO THE SAME FOR COLUMN 1.
*←41+31+12+31+57+25
197.0
*197.0/6
32.8333
*↑32.8
VERY GOOD.

Use of displays

The examples described so far have been keyboard-oriented—that is, the use of devices other than teletype and typewriters might be helpful, but is not essential. There are numerous examples, however, where the ability to input and present other than alphanumeric characters is essential. Again, we shall present a short summary of some interesting examples of the application of displays in interactive computing.

At General Motors Corporation, engineers use consoles to design components of cars. The GM system, called DAC–1[15], was originally programmed on the IBM 7094, and now runs on an IBM 360.

At Lincoln Laboratory, a time-sharing system (called APEX[19]) has been produced for the primary purpose of exploring concepts utilizing display devices. Among the projects carried out on this system are SKETCHPAD, which automates the draftman's task; another system which analyzes three-dimensional figures and correctly inserts or deletes lines on other than the front plane—the hidden-line problem—and a system which allows the design of electronic circuits.

At MIT, a system for studying molecular structures by inspection of a computer-generated three-dimensional display has been produced. Also, considerable research has been conducted in the area of three-dimensional architectural design, utilizing displays which appear to rotate figures in space.

At The RAND Corporation and System Development Corporation, research in recognition of hand-written characters has led to systems for computer recognition of hand-written flow diagrams and mathematical expressions. At SDC, display programs associated with the data management systems described previously permit graphical inspection of collections of data, in addition to the random searching for facts possible with the keyboard-oriented systems.

Summary of applications

This section has been an attempt at representing the field of interactive computing. The examples presented are a small fraction of the total set and are probably biased somewhat by the author's association with some projects more than others. But the examples do cover a range of areas, and demonstrate the exciting potential of interactive systems solving problems in diverse fields, such as engineering, business, education, and linguistics.

EVALUATION

There is little in the preceding discussion to question the value of time-sharing and interactive computing. For many problems, conversation with a computer is preferable, not only from the psychological point of view, but from a technical one as well. If this is so, one might expect that the rush to throw away the traditional systems would be overwhelming. Although there has been a spectacular increase of interest in time-sharing (from less than ten installations in 1963 to hundreds in 1968), there is still a considerable dependence on systems which are more oriented to batch-processing than interactive access. There are probably many reasons for this. For one, there is always the problem of inertia. Even within a field that is making such rapid strides as computing, there is considerable reluctance to make radical changes in methods. This is compounded by the fact that computer manufacturers tend to be the major supplier of operating systems. Since their aim is to supply systems that will make it possible to sell computers, they generally do not try to break with traditional means. That is why the majority of time-sharing efforts have come from universities, non-profit corporations, and small independent companies, rather than from well-established large installations or manufacturers. There are, of course, several notable exceptions to these statements. The Digital Equipment Corporation has made time-sharing the basic system on their large computers. Also, most other manufacturers have produced some version of time-sharing on their new line of computers. Generally, however, these have not been the primary operating system offered to computer buyers.

There is still another reason for some people's reluctance to switch to interactive computing as the primary means of computer access. They don't believe the statements that interactive computing and time-sharing are more economical than off-line access. The argument is as follows. To achieve a time-sharing system, one needs a larger computer and more programming than would be required in more traditional systems. Consequently, the initial cost is higher. Also, the need to maintain and schedule a number of simultaneous programs increase the overhead. Swapping is an obvious cost not present in a system which maintains all programs in core. There are arguments which can be made against some of these statements. In any case, actual measurements for comparative costs are quite difficult to achieve, although some studies have been made on the performance of some time-sharing systems[11,23].

In the final analysis, however, the question which must be answered is that of productivity. Can one state that performance (particularly human performance) under one mode is more productive than the other? Or, is it possible to do things one way that are not possible or very awkward the other? On this latter point, there are examples of applications which would be terribly expensive or obviously difficult in a non-time-shared environment. Applications utilizing displays and random fact-retrieval are examples of these.

On the other hand, some kinds of computation don't lend themselves to interactive computing. Any application where the ratio of computing to console input/output is high is a poor candidate for time-sharing and must be handled in a special way.

Thus, it is clear that there are applications where one system is clearly favored (or one must make special cases out of the particular system). But there are a large number of cases which are not clear and can be contested. One of the areas which is hotly debated is that of programming itself. Can programs be developed more economically interactively or not?

It is argued by some that the gain in productivity using a computer interactively is so great that a considerable loss in "computer efficiency" can be tolerated. This theory has been generally accepted, but since the early days of time-sharing the feeling that interaction with a computer greatly increased productivity has been primarily intuitive. There have been some attempts at testing this hypothesis over the last few years. For the most part, these attempts have consisted of experiments comparing program production and problem-solving under time-shared and non-interactive systems.

A discussion of five such studies was presented by Sackman[22]. Attempting to summarize here what is in effect a summary of a controversial set of experiments is obviously precarious, but a few of the conclusions given by Sackman might be mentioned. Using interactive systems, fewer man-hours but more computer time is utilized in solution of problems. User preference for interactive over conventional off-line operation was evident. Also, in many cases, individual subject differences in the experiments overshadowed any differences in the modes of computer operation. Sackman also came to other conclusions, and made a plea for improvements in methodology in several areas which would make future experiments more meaningful.

FUTURE

There is little doubt in many people's minds that systems which permit interactive access are eventually going to predominate. Whether these systems evolve from current time-sharing systems or from standard multi-programming systems that have been modified to permit on-line access is not clear or very important. The fact is, however, that on-line interaction with a computer is becoming part of the day-to-day operations for a wide range of applications in the commercial, scientific, and business worlds, just as computers have. Technical obstacles to their more complete utilization are eventually resolvable. There are various sociological problems associated with the use of large shared computers, particularly where sensitive or private information is stored. These are problems which must be attacked soon. (A discussion of these problems can be found in[12,21].) In any case, the intellectual stimulation of direct contact with a computer is such that its progressively increased use seems irreversible.

Bibliography

1. Bobrow, D.G., Fraser, J.B., and Quillian, M.R., "Automated Language Processing", in Cuadra, C.A. (ed.), *Annual Review of Information Science and Technology*. New York: Interscience, 1967, Vol. 2.

2. Brawn, B.S., and Gustavson, F.G., "Program Behavior in a Paging Environment", *Proceedings of the Fall Joint Computer Conference,* 1968, pp.1019–1032.
3. Bryan, G.E., and Smith, J.W., "JOSS Language". RAND document RM–5377–PR, August 1967.
4. Corbato, F.J., and Vyssotsky, V.A., "Introduction and Overview of the MULTICS System", *Proceedings of the Fall Joint Computer Conference,* 1965, pp.185–196.
5. Corbato, F.J., et al., *The Compatible Time-Sharing System: A Programmer's Guide.* Cambridge, Massachusetts: The M.I.T. Press, 1963.
6. Dennis, J.B., "Segmentation and the Design of Multi-Programmed Computer Systems", *IEEE International Convention Record*, Part 3, 1965, pp.214–225.
7. Devonald, C.H., and Fotheringham, J.A., "The Atlas Computer", *Datamation*, Vol. 7, No.5, May 1961, pp.23–27.
8. Engelbart, D.C., and English, W.K., "A Research Center for Augmenting Human Intellect", *Proceedings of the Fall Joint Computer Conference*, 1968, pp.395–410.
9. Everett, R.R., Zraket, C.A., and Benington, H.D., "SAGE—A Data Processing System for Air Defense", *Proceedings of the Eastern Joint Computer Conference*, December 1967, pp.148–155.
10. Fine, G.H., Jackson, C.W., and McIsaac, P.V., "Dynamic Program Behavior Under Paging", *Proceedings of the 21st National ACM Conference*, 1966, pp.223–228.
11. Fine, G.H. and McIsaac, P.V., "Simulation of a Time-Sharing System", *Management Science*, Vol. 12, No.6, February 1966, pp.B–180–194.
12. Gruenberger, F., *Computers and Communication: Toward a Computer Utility.* Englewood Cliffs, New Jersey: Prentice Hall, 1968.
13. Howarth, D.J., Jones, P.D., and Wyld, M.T., "The Atlas Scheduling System", *Computer Journal*, Vol. 5, No.3, October 1962, pp.238–244.
14. International Business Machines Corporation. *CALL/360: DATATEXT Operator's Manual.* IBM document J20–0038–1. 1968.
15. Jacks, E.L., "A Laboratory for the Study of Graphical Man-Machine Communication", *Proceedings of the Fall Joint Computer Conference*, 1964, pp.343–350.
16. Kennedy, P.R., "The ADEPT-50 System: A General Description of the Time-Sharing Executive and the Programmer's Package", SDC document TM–3899, June 7, 1968.
17. Lett, A.S. and Konigsford, W.L., "TSS/360: A Time-Shared Operating System", *Proceedings of the Fall Joint Computer Conference*, 1968, pp.15–28.
18. Licklider, J.C.R., "Man-Computer Symbiosis", *IRE Transactions on Human Factors in Electronics*, Vol. HFE-1, No.1, March 1960, pp.4–11.
19. Massachusetts Institute of Technology, Lincoln Laboratory. "Graphics: Semiannual Technical Summary Report, 1 June–30 November 1967", ESD–TR–67–570, January 1968.
20. McCarthy, J., "Time-Sharing Computer Systems", in Greenberger, M. (ed.), *Management and the Computer of the Future.* Cambridge, Massachusetts: The M.I.T. Press, 1962, pp.221–236.
21. Parkhill, D.F., *The Challenge of the Computer Utility.* Reading, Massachusetts: Addison Wesley Publishing Company, 1966.
22. Sackman, H., "Time-Sharing Versus Batch Processing: The Experimental Evidence", *Proceedings of the Spring Joint Computer Conference*, 1968, pp.1–10.

23. Scherr, A.L., "An Analysis of Time-Shared Computer Systems", Massachusetts Institute of Technology document MAC–TR–18, June 1965.
24. Simmons, R.F., "Automated Language Processing", in Cuadra, C.A. (ed.), *Annual Review of Information Science and Technology*. New York: Interscience, 1966, Vol. 1.
25. Strachey, C., "Time-Sharing in Large Fast Computers", *Proceedings of the International Conference on Information Processing*, UNESCO, June 1959, Paper B. 2.19.
26. University of Michigan. *University of Michigan Terminal System (MTS)*. 2 Vols., December 1967.

CHAPTER 8

Errors in Computing

N. METROPOLIS

1 INTRODUCTION

EXPERIMENTALISTS IN the physical sciences have a well-established practice of specifying the error associated with each of their results. The tradition seems much less in evidence among computer scientists in connection with calculational results. Von Neumann and Goldstine[1], among others, have called attention to the question of error specification: "When a problem in pure or in applied mathematics is 'solved' by numerical computation, errors, that is, deviations of the numerical 'solution' obtained from the true, rigorous one, are unavoidable. Such a 'solution' is therefore meaningless, unless there is an estimate of the total error in the above sense."

A review is given here of the various kinds of errors in *digital* computing. Conceptually, they may be regarded as deriving from independent sources. We discuss them in turn and at varying length. Our effort is based, in part, on the definitive introduction contained in Ref. 1; an excellent, general discussion is found, of course, in Householder[2]; an account of more recent studies is given in Ref. 3. Analog computing is not included in the present discussion; programming errors are also excluded as they belong to the subject of artificial languages.

2 ERROR IN MODEL

The stimulus for the modern development of the computer discipline stemmed in large part from an interest in problems of the physical sciences. (This is but another instance of the influence of the natural sciences upon mathematics.) The associated computations (then and now) have been based on models that represent an idealization and simplification of real phenomena. In the traditional manner of science, a comparison with the facts would typically disclose the shortcomings of the assumptions; in fortunate situations, the "numerical experiments" might even illuminate a more suitable set of assumptions. Using this heuristic approach, one would hope to achieve in the end a rigorous mathematical understanding of the underlying structure of the phenomenon of interest. Often this method is the only recourse, as in the case of non-linear phenomena, where the resources of linear theories are of little avail.

Assume the computations have been performed in a flawless manner and that a discrepancy exists upon comparison with a confirmed experimental fact; such a difference is called an error in model. The consideration of such errors is vital to the development of the physical theory at hand, and often to mathematics itself and it would be difficult to exaggerate its role. However, that type of error is only prologue to the present discussion.

3 INHERENT ERROR

Leaving aside any question of the correctness of theory and of the consequent mathematical equations, we consider some sources of error that are more directly associated with the actual computational process. The set of numerical quantities, or parameters, that initiate the calculation may be based on experimental data with attendant imprecision in their values. This kind of error is referred to as *inherent* error. Usually, the relative size of the inaccuracies varies somewhat over the set of the input parameters; moreover, the effect on the calculation is often of differing sensitivity; that is to say, input parameters with approximately the same relative (or even absolute) errors can each influence the final results in a quite disparate manner. Occasionally, an input with large error may have less effect on the results than a much more precisely known quantity.

Quite clearly, in order to study the consequences of inherent error, one must examine the details of the particular sequence of operations at hand and monitor the effect of the inherent errors on all intermediate and final numbers. This is a non-trivial task. Studies of such *propagative* effects of inherent error have not been pursued as extensively as some of the other types; consequently, we shall dwell on this aspect in the discussion below, but first we introduce the remaining types.

4 ANALYTIC, OR APPROXIMATION, ERROR

The preparation of a problem for submittal to a digital computer often involves a variety of approximations quite distinct in character from those discussed above. Interpolations, extrapolations, evaluation of definite integrals, replacement of a differential equation by a finite difference equation are typical examples of procedures that calculate an approximation to the desired theoretical expression. This type of error is called *analytic*, or *approximation* error; sometimes the expression, truncation error, is used, as suggested by the fact that an approximation is often obtained by truncation of an infinite series expansion.

Thus errors are introduced into the computation even though the initial set of numbers may have been exact. Such errors have been studied extensively by mathematicians, using the techniques of analysis, long before the modern period of computing began. As a consequence we shall have occasion to discuss this source of error only briefly.

5 ROUNDING, OR GENERATED, ERROR

Numbers are represented in a computer by a standard length sequence of digits. Input numbers of precision greater than that implied by this length must be rounded. Such rounding procedures are also applied to results of the standard arithmetic operations of addition (subtraction), multiplication and division. The most comprehensive treatment of the effects of rounding error has been made by J. H. Wilkinson[4].

Clearly, the three sources of error described in Sec. 3, 4 and 5 interact in an actual problem so that the several effects are sometimes difficult to separate. Occasionally, some simplifying circumstances obtain, for instance, if inherent errors are relatively large, rounding error (as implemented on a computer) may be ignored whenever the modification occurs beyond the "level" of precision. In quite the same spirit, it suffices to consider those approximations (in the sense of Sec. 4) to a degree of precision that is consistent with (not necessarily the same as) that of the numbers entering into such approximations. A tractable scheme is described below to assess the relative importance of rounding error and of the propagation of inherent error.

Sometimes it is convenient to interpret rounding error as inherent error by regarding inexact arithmetic operations on exact numbers as exact arithmetic operations on inexact numbers. In fact, Wilkinson's analysis is based on such considerations.

The natural development of the subject of errors leads to discussion of error-monitoring that would permit reliable estimates of total error in a computation. Before pursuing this, however, we digress to consider, for the sake of completeness, a quite different type of error, namely, failures in computer operation.

6 COMPUTER MALFUNCTION

In the early days of computer engineering when vacuum tubes were the fashion, a curious or strange numeric result would usually point the finger at the computer, rather than the user. Quite apart from the egotistical one, there were several reasons (or rationalizations) for this attitude: (1) manufacture of tubes did not meet, at first, the standards required by the new discipline; (2) reliability requirements in circuit design were below desired levels—what may have been adequate for, say, a scaler in a physics laboratory was impossible in an electronic computer; (3) the experience in testing, selecting components, and maintaining a computer was limited. To be sure, reliability of computer operation was to improve rapidly and markedly. Nonetheless such a question engaged von Neumann's attention and led to a beautiful discussion on the synthesis of reliable automata from unreliable components[5].

In the late fifties, the transistor was to take precedence over the, by then, well-developed vacuum tubes not only in reliability, but also in speed and with respect to much lower power dissipation. With this new element, the logical designer of computers pursued more ambitious and sophisticated structures, as well as pressing for greater operational speeds so that part of the gain in safety margin was sacrificed. The improvement was sufficient, however, so that it became almost routine to accuse the user, rather than the computer, whenever a suspicious numeric result appeared.

The present state of so-called integrated circuits has removed almost all inhibitions that designers may have had; again one of the impelling factors has been to achieve greater speeds. Thus the problems besetting the maintenance engineer have not diminished, even though some concessions have been made in terms of being provided with some built-in checking devices, a separate console and a planned preventive-maintenance schedule that is continually being up-dated as experience increases.

The solution to the problem of minimizing the "down-time" of a computer is far from clear.

1) One conflicting aspect is that a fairly high-level engineer is desired whenever the computer is suspect, yet a reasonably built computer should not need attention too frequently, and the intervals between failures may become dull for such a skilled individual.

2) Another question is how extensive built-in checking and monitoring should be; additional equipment is implied (which must also be maintained) and often effective speed is reduced. At this point a quite novel approach to computer design may be mentioned. Detailed computer operations are segmented into a sequence of small steps; conventionally, the sequence is executed in a *synchronous* fashion. Muller and his colleagues, among others, have considered an organization wherein the various electronic circuits operate at their "natural" (not necessarily constant) speed and such that a completion signal from one step is required to trigger the next[6]. A corollary of this mode of operation is that a control malfunction, *ipso facto*, interrupts the sequence—a kind of *flagrante delicto*.

3) How much should programmers participate in the checking process by including duplicative or alternative calculations and comparisons; often the question arises whether to use some conservation principle present in the problem as an error check rather than to simplify a computation.

4) How frequently should preventive maintenance be imposed; some compromise between risk and overall effectiveness is desired. Several very interesting aspects concern the strategy of testing: how repetitious should tests be, how much diagnosis is desired in concert with the testing. The reason for repetition of course is that a failure may be intermittent, in contrast to a definitely recurring failure, a much easier type to detect and localize.

5) One set of procedures for maintenance that may be typical is:

i) one period per week of several hours when a relatively thorough examination is made. Input-output units, various types of storage, and central processor are tested separately and in a logically consistent order so that failures may be more easily diagnosed. Electrical characteristics or parameters are varied to pick out elements of only marginal goodness.

ii) Several periods per week of shorter duration, say 30–60 minutes, are devoted on a regular basis for less exhaustive testing and checking.

iii) Both the long and short form of preventive maintenance are modified and augmented as a consequence of unscheduled maintenance requirements disclosing hitherto untested circuit paths. Such malfunctions are for the most part detected by users, who often are sufficiently astute and curious that they can localize the source rather closely. Another reason why a user and not the engineer may discover a "bug" is that some errors are of an intermittent nature of very small probability. This fact combined with the above on untested paths probably accounts for the oft heard remark that a computer "learns" the test problems well, only to "stumble" when turned over to the user.

6) Baylis[7] has recently purposed that formal studies be made of this important and difficult question of maintenance and has made some very interesting and substantive suggestions. Commenting on the problems faced by the maintenance engineer as described by Baylis, one reviewer[8] remarks "... problems which would have taxed even Sherlock Holmes".

7 STABILITY AND CORRELATION

The introductory remarks on approximation, or analytic, error mentioned that in many computational problems, finite difference equations are used to represent partial differential equations. Care must be exercised in the selection of some parameters, for if certain relations are not satisfied in making these choices, then the approximation can deteriorate rapidly and bear no relation to the rigorous solution.

Discussion of such questions constitutes *stability analysis*. The work of Courant, Friedrichs, and Lewy[9] is now classic; this field is still another to which von Neumann made significant contributions. Quite clearly one must avoid, wherever possible, introducing unstable procedures into a computation. Occasionally, however, instabilities are unavoidable in some problems, e.g., where the underlying phenomenon is itself unstable; in such a circumstance, one should be aware of the unstable situation and terminate the calculation when it is no longer meaningful, i.e., when all precision is lost.

A part of a problem may be expressed in terms of recurrence relations such as

$$r_{n+1}(x) = Ar_n(x) + Br_{n-1}(x) + C(n)$$

where A, B are, say, constants and C depends only on n; the successive values r_{n+1} are determined from those of its two immediate predecessors. At least two circumstances occur: if a given relation between A and B is satisfied, it can be shown that the precision of the value of r_{n+1} is undiminished (or even increased) relative to that of its predecessors r_n and r_{n-1}; no problem here. However, if it should turn out that the relation is not satisfied, then the precision of r_{n+1} deteriorates, and often rapidly. Obviously it would be useful to have some form of error monitor to alert the user when the calculation becomes meaningless. We discuss below an approach that not only discloses the circumstance when all precision is lost but also provides an assessment of the precision in intermediate and final numeric quantities in less extreme situations.

An aspect of monitoring with which every scheme must be concerned is that of error correlation. An error analysis of a calculation involving several operands, each with its associated error, must consider the detailed structure of the computation, for these errors interact. A salient point in the considerations is whether an operand appears only once, or in several places in the given computation; if in several places, the analysis must recognize that it is the same error that is introduced, i.e., the errors are correlated. This correlation can have either effect—to increase or decrease the total error. A simple example illustrates this.

1) Consider the evaluation of an nth degree polynomial $y = \sum_{i=0}^{n} a_i x^i$ where the errors in x and a_i are independent. In forming x^i, one must recognize that the associated error is (statistically) larger than in forming say $\prod_{r=1}^{i} z_r$ where z_r have errors comparable to that of x but are all independent.

2) Further care must be exercised in combining the various terms of the above sum inasmuch as they also have correlated errors.* If the errors in the coefficients a_i are relatively small compared to the error in x, and if the magnitude of the derivative satisfies $|y'| \ll 1$ at x, then the correlation leads

* In practice, powers of x are formed and terms are summed in a (concurrent) recursive manner; for purposes of error analysis the two operations may be viewed as being separately performed.

to a decrease in error; it is easy to have a situation where this decrease more than compensates for the increase in error when forming the various powers of x. A detailed discussion may be found in Ref. 10.

8 COMPUTER MONITOR

One convenient approach to error estimation in computation is to view the process in two parts: (i) given independent errors $\Delta x_1, \Delta x_2$ in two operands x_1, x_2, to estimate the error Δx_3 in the resultant x_3 in the elementary arithmetic operations of addition (subtraction), multiplication and division; (ii) to study the effects of correlation in a sequence of such arithmetic operation, corresponding to the computational problem at hand. This dichotomy is perhaps a natural one since the first is concerned with the structure of the arithmetic processor of a computer, and the second with an analysis of mathematically (but not necessarily computationally) equivalent algorithms that effect a solution. We begin here with (i) and discuss (ii) in the next section.

A number x in a computer is usually represented by a pair $x = (e, f)$ implying in a binary computer that $x = 2^e \cdot f$ where e is the integer exponent and f is the fractional part satisfying $|f| < 1$. For integer a, the replacement of (e, f) by $(e + a, 2^{-a}f)$ is called *adjustment*. For $a > 0$, the operation on f is a right shift; $a < 0$ is a left shift (such that the upper bound on the adjusted fractional part is not exceeded). Strict equivalence of x with its adjusted representation in the case $a > 0$ is only approximately maintained if bits of f on the right end are lost in the shift process.

For historical reasons, all results of the elementary arithmetic operations ($\pm$, $\times$, $\div$) in conventional (binary) computers are adjusted to a standard form, namely, such that the fractional part satisfies $\frac{1}{2} \leq |2a^{-a}f| < 1$; this corresponds to the maximum allowed left shift in f. Thus, in the standard form, the most significant bit of (adjusted) f resides (apart from sign) in the left-most stage of a register that contains the number.

Upon reflection, one realizes that the flexibility afforded by adjustment is really a degree of freedom that can be utilized to exhibit the precision of input numbers, as well as intermediate and final results. For an input number, one simple approach is to adjust it so that the *least* significant bit of the fractional part f resides in some selected, say the kth, stage near the right side of a register. (The remaining stages beyond this stage serve as "guard bits" to separate or at least reduce the effects of rounding; this is discussed further.) For numbers subsequently produced by arithmetic operations, a set of adjustment rules is required that keeps the least significant bit of each number in this kth stage. The rules that achieve this for *statistically independent* operands are easily expressed: call m_i the number of leading zeros in $|f_i|$ and characterize a number x_i by (e_i, f_i, m_i). Represent the arithmetic operations by $x_1 * x_2 \rightarrow x_3$. Then the rules are:

$$\left.\begin{array}{l}\text{Addition}\\ \text{(Subtraction)}\end{array}\right\} \quad e_3 = \max\{e_1, e_2\}$$

$$\left.\begin{array}{l}\text{Multiplication}\\ \text{Division}\end{array}\right\} \quad m_3 = \max\{m_1, m_2\}$$

Ideally, one starts with a set of initial numbers lined up so to speak with the least significant bit in the same kth position or stage; when the computation is complete, the final numbers have the same alignment, so that the precision of each number is immediately evident by simply noting the number of bit from the kth stage to the left. We discuss the realistic situation presently. These rules correspond to the criteria one would use in a manual computation that keeps track of significant digits. The MANIAC III computer at the University of Chicago has incorporated this arithmetic structure in its design[11].

We have pointed out that these adjustment rules would provide a proper *assessment* of the precision in arithmetic results if no correlative effects were present. If present, they can however be handled. We stress the fact that techniques that accomplish this are based upon arithmetic being performed according to the above adjustment rules.

9 METHOD OF REDUCED PRECISION

Manifestly, the results of a calculation are functions of the initial quantities. The latter are known only approximately, in accordance with some distribution. Variations in these quantities lead to a distribution of values for each result, whose variance is the probable error in the result. The aforementioned adjustment rules would provide a reasonably close estimate of this probable error if no correlations exist. It turns out that when correlations are present, the estimate may be greater, less than, or comparable with the probable error, and may be called a "conservative", "liberal" or "faithful" estimate, respectively.* The estimate of error is based upon the interpretation of digits to the left of, and including, the kth digit position in our representation as being meaningful (cf. Scc. 8 above).

A technique exists that establishes the nature of the estimate. Moreover, if a proposed algorithm is indeed a faithful one, then clearly it is optimal in the sense that the prescribed precision is the most one can achieve. On the other hand, if it is not a faithful one, one may analyze the deviation and modify the proposed algorithm so as to take proper account of correlations that are present. Fortunately, the technique indicates the point in the algorithm where a deviation occurs; this localization often suggests an appropriate change in the algorithm being studied. Not surprisingly, some of these changes are simple, whereas others may be somewhat more subtle.

The technique, called the method of reduced precision, may be described as follows:

1) The given algorithm is computed using our arithmetic adjustment rules, first having adjusted the initial numbers x_i according to the prescription mentioned above that takes account of the precision by arranging that the least significant bit resides in some selected k^{th} stage of a register. Call the final quantities y_j.

2) Consider the following pertubation of the initial numbers x_i: into a selected bit position to the left of the kth, say the fourth preceding stage, referred to as $k^{\text{th}}_{(r)}$, introduce a perturbation of ± 1; the choice of sign is made at random. These are the initial numbers $x_i^{(r)}$ with reduced precision. The calculation is repeated to produce the corresponding outputs $y_j^{(r)}$.

3) A comparison of the difference $|y_j - y_j^{(r)}|$ is made. More specifically, the fractional part is compared. If the leading 1 in this difference resides in the $k^{\text{th}}_{(r)}$ stage, the algorithm being tested is "faithful"; if the leading 1 resides to the left (right) then the algorithm is "liberal" ("conservative"), since more (less) bits are considered as being meaningful than is actually the case.

Several remarks are in order. The method of reduced precision is normally used for testing purposes only. When an algorithm has been established to be faithful for all initial quantities, duplicative calculations cease and one has the assurance that the assessment of error for that algorithm is an appropriate one. Note that by extending the above comparisons to intermediate numbers

* For brevity we refer to the algorithms that produce such error estimates with the corresponding adjectives.

(cf. step (3) above), one can localize the point where any non-faithful representation occurs—perhaps simplifying the application of corrective measures. A more complete discussion may be found in Ref. 12.

Error analytic arguments are usually required to establish that an algorithm is faithful for all sets of initial numbers. The analysis of some algorithms however is beyond our present state of knowledge. For such intractable cases, one may provide heuristic arguments based on sampling a large set of cases with the technique of reduced precision. Such sets should of course cover wide ranges of magnitude and precision; particular attention should be given to "ill-conditioned" sets of initial numbers that test relatively "sensitive" parts of the algorithm. The very cautious investigator would rely only on the method of reduced precision.

Given a set of initial numbers, what variations can be expected in the results of the comparisons in (3) for several random perturbations? Such experiments have been performed on a variety of algorithms[10,14] and the variations are satisfactorily small, usually not exceeding two bits and often not exceeding one.

10 IMPLEMENTATION

A natural question at this point might well be: what if the available computer has not incorporated the above adjustment rules in its design of the arithmetic processor? It is however feasible to effect these rules by programming; indeed this has been done for at least three types of computer[13].

The computer programmer today relies on a library of sub-routines or procedures to simplify his work. A good starting point for automatic monitoring of error would be to ask that the library include only algorithms that are faithful in the above sense. Algorithms for the most frequently used subroutines or procedure have been found[13].

11 EXAMPLES

Less simple examples have also been studied, for instance, recurrence relations, polynomial evaluation, and matrix inversion[10,14].

In the case of recurrence relations, the usually prescribed algorithm permits interactions among the several (corresponding to the order of the recurrence) sources of error propagation, and these in turn often lead to a non-faithful representation of results. Consideration of the so-called characteristic equation suggests a formulation that enables the several error propagations to act independently.

Polynomial evaluation by the efficient, recursive method due to Horner[15] also suffers from error correlation due to the independent variable, say x. The key observation here is that the magnitude of the first derivative with respect to x provides a measure of this correlation. It is well known that the derivative can also be computed recursively; requiring the derivative approximately doubles the amount of calculation. It should be remarked, however, that a given polynomial can have evaluations with widely differing precision for different values of x.

The effect of inherent errors on matrix inversion is less apparent. One study of this considers the so-called Gaussian elimination method and examines a variety of different strategies for the selection of "pivot element" in each of the several stages of the method. Heuristic arguments based on the closely allied problem of determinant evaluation indicate a strategy that has been tried on several classes of matrices known to be "ill-conditioned" with respect to inversion; the results are satisfactory.

These examples should convey the impression that a faithful algorithm is usually more elaborate than some mathematically equivalent versions (i.e., error considerations left aside); the latter may indeed be more efficient computationally, but without error estimates, the results are of dubious value.

12 CONCLUDING REMARKS

The preceding discussion of the last several sections has been focused on inherent error, its propagation and its monitoring. The latitude available in the choice of the kth stage, the residence of the least significant digit, permits a separation of the propagated inherent error and that of rounding. Rounding takes place, of course, in the extreme right bit position; if the kth stage is several places to the left of this, then rounding has little effect at first on the kth bit position. The rounding error gradually diffuses to the left as the number of operations increases in a sequence of concatenated operands. The diffusion has only a logarithmic dependence on the number of operations involved; thus an upper bound for the effect of rounding can be given even if the number of operations is known only approximately.

The second remark concerns instability. A computation may be unstable because some approximation (cf. Sec. 7) has this character, or the phenomenon under investigation is itself unstable. In either case the behavior is readily discernible by mere inspection of the results as a deterioration of precision—not infrequently, completely so. Clearly, it would be informative and reassuring to know the nature and extent of instabilities, particularly whether or not they are intrinsic.

The error analyst soon learns that his discipline is a rather complicated one; as in most areas, experience is a very useful commodity. If he faces an algorithm that is not likely to be used frequently, he need not be concerned with establishing a faithful version—it suffices to repeat the calculation using the method of reduced precision (cf. Sec. 9). However, if frequent use is anticipated, duplication is annoying, hence the search for a faithful algorithm should be pursued. It is a fact that some algorithms exhibit only small correlation so that little needs modifying. (This of course does not imply that conventional, normalized arithmetic is suitable.) The important point is that one can be confident that a calculation has been performed in a manner to achieve the best assessment of probable error in the final quantities, or at least a sufficiently good one for the purpose at hand; otherwise doubts as to the meaning of the calculation remain.

References

1. John von Neumann and H. H. Goldstine, Numerical inverting of matrices of high order. *Bull. Amer. Math. Soc.* **53** (1947) 1021–1099.
2. A. S. Householder, *Principles of Numerical Analysis*, McGraw–Hill, New York, 1953.
3. R. L. Ashenhurst and N. Metropolis, Error estimation in computer calculation, *Amer. Math. Monthly*, **72** (1965) 47–58.
4. J. H. Wilkinson, *The Algebraic Eigenvalue Problem*, Clarendon Press, Oxford, 1965.
5. J. von Neumann, Probabilistic logics and the synthesis of reliable organisms from unreliable components, 43–98, *Automata Studies*, Princeton University Press (1956).
6. D. E. Muller and W. S. Bartky, A theory of asynchronous circuits, *Proc. Internat. Sympos. Switching Theory 1957*, 204–243, Harvard University Press, Cambridge (1959).
7. M. H. J. Baylis, Maintenance of large computer systems—the engineer's assistant, *Machine Intelligence*, Vol. 3, 269, Edinburgh University Press, Edingburgh, 1968.
8. S. Gill, *Nature, 220,* (1968) 1154.
9. R. Courant, K. Friedrichs, and H. Lewy, Über die partiellen Differenzengleichungen der Mathematischen Physik, *Math. Ann., 100* (1928) 32–74.
10. M. Menzel and N. Metropolis, Algorithms in unnormalized arithmetic. II. Unrestricted polynomial evaluation. *Num. Math.* **10**, 451–462 (1967).

11. N. Metropolis and R. L. Ashenhurst, Significant digit computer arithmetic. *IRE Transactions on Electronic Computers* EC–7, No. 4, 265–267 (1958).
12. N. Metropolis, Algorithms in unnormalized arithmetic. Proceedings of the Colloque International du C.N.R.S. Besancon 1966: *Programmation in Mathematiques Numeriques*. Paris: Editions du C.N.R.S. 1968.
13. R. C. Blandford and N. Metropolis, The simulation of two arithmetic structures, Los Alamos Scientific Laboratory Report LA–3979 (1968).
    Toshio Yasui and Gilbert L. Winje, Significant digit arithmetic on Illiac IV, Department of Computer Science Report No. 211, Urbana (1969).
14. N. Metropolis, Algorithms in unnormalized arithmetic. I. Recurrence relations. *Num. Math.* 7, 104–112 (1965). M. Fraser and N. Metropolis, Algorithms in unnormalized arithmetic. III. Matrix inversion. *Num. Math. 12* (1968).
15. P. Henrici, *Elements of numerical analysis*. New York: Wiley 1964.

CHAPTER 9

The Role of the University in the Computer Age

JOHN R. PASTA

1 INTRODUCTION

IN ANY DESCRIPTION of the impact of computers on universities it is important to separate the role of computers in the research and educational processes of the institution from its status as an object for study. This distinction is not always made carefully and, indeed, it is not an easy one to make as can be realized from observing the many disagreements over the definitions of the discipline Computer Science. It is agreed generally, however, that there is a rapidly growing body of knowledge which lies outside or, at least, on the fringes of the usual university disciplines which displays a coherence sufficient to call it a discipline in its own right. The most common name for this area is Computer Science.

In the areas outside of computer science proper, computers have played an important role in both the research and educational functions of the university and this position is becoming more important each year. These relationships will be examined in order to gain an appreciation for the overall position of the digital computer in modern education, but it is the consideration of the growth of computer science in the university which will be treated first.

Some consideration will be given to why computers have become an object of intense study. It is fashionable, these days, to re-examine educational goals in terms of relevance to society and to life. There can be no denying that these new machines will have a large secondary impact on the living and working habits of our society but with very little extension in thought one can imagine awe-inspiring and frightening consequences of present-day developments in the science. We are accustomed to developing our science and reserving until later consideration of the consequences, a habit which, one day, may be our undoing.

2 EARLY CONTRIBUTIONS

The origin and early development of computers[1] was carried out mainly in universities. The pioneering work at Harvard University, the University of Pennsylvania, the Institute for Advanced Study and in England was picked up quickly at a dozen places. The potential of the digital computer was recognized and this early work demonstrated that electronic components had reached a degree of reliability permitting the fabrication of systems of the size contemplated.

In addition to important research contributions such as magnetic core memories, index registers, assembly programs, and compilers, the universities produced the engineers and mathematicians who were responsible for the spectacular advance of computers in the first decade. In those formative years the spirit was more cooperative than competitive. In the best university tradition, the communication of ideas, progress and results was given an important position. In reviewing a report from the group at the University of Illinois M. Lehman[2] of the IBM Corporation wrote:

> "The history and valiant pioneering efforts of the Digital Computer Laboratory of the University of Illinois are perhaps not as widely known and acknowledged as its many fundamental contributions to the development of computer art and science warrant. In fact, with the regretful but inevitable transfer of computer hardware research and development from the university to industrial surroundings, this laboratory is almost the sole survivor of the many groups who, for a decade and a half, conceived and/or laid the foundations of the entire field of information processing.
>
> "Among its many contributions, Illinois can pride itself on its record of documentation. There is probably no other machine whose development and final design is so completely documented and described as the ILLIAC II. This machine, at its conception

in the mid 1950's, represented, together with some other independent design projects of the same period, the spearhead and breakthrough into a new generation of machines. The early history of computers and computing is becoming rapidly obscured, and it would seem urgent that some historian attempt to reclaim and record the facts while the men who made them are still around. Such a researcher would certainly include among his basic documents the series starting with the now classical Report 80 of the University of Illinois' Digital Computer Laboatory, so as to trace the early history of concepts, e.g., modular interleave memories, instruction look-ahead, redundant number representations, and computer arithmetic in general."

3 THE COMPUTER EXPLOSION AND ITS IMPACT ON COMPUTER SCIENCE

Even while the earliest research was underway there were beginnings of a tooling-up by industry for the great proliferation of computers to follow. It was not long before there was little justification for a university to build computers to satisfy a service need at the institution. Unfortunately, it was this service need that had been advanced as a justification for most projects in computer hardware research because this procedure insured the broadest-based institutional support.

Not everyone viewed the role of these early computer scientists in that light. In an early confrontation between a university group and a short-sighted industrial management A.E.C. Commissioner John von Neumann asked "where shall we get the engineers and innovators of the future if these activities are disbanded?"

This view did not prevail, however, and university interest in the area moved away from the pragmatic aspects toward the more theoretical considerations which underlie the discipline. It was this shift in emphasis which may be responsible for the evolving in most curricula today of "Computer Science" rather than "Computer Engineering".

These two aspects are not incompatible in the discipline. Neither are they incompatible in an individual and the prime example is to be found by examining the accomplishments of von Neumann. There is, in fact, a tendency these days to try to reintroduce into the curriculum at least some of the systems architecture ideas which implement the body of abstract knowledge being assembled. Of course, the earlier concerted drives to construct large general purpose computers at the university have disappeared, but the need for some of the activities associated with machine design and fabrication remains. It has long been recognized that consideration of computer systems must embrace hardware and software simultaneously for a properly integrated picture. The deeper studies of these systems can not do otherwise if one is to have a true perspective.

4 EDUCATION IN COMPUTER SCIENCE

Curricula in Computer Science have been adopted at most major universities not only because of faculty inclination but also to meet the needs of students and society. Although the initial programs reflected the special interests of the charter faculty, the departments enlarged and diversified to cover the discipline in a nearly uniform way. The general pattern of such a curriculum became sufficiently clear that various organizations and committees could undertake the job of defining it in general terms. These attempts were meant to make models around which a program suitable to the local university environment could be formed.

There are many ways of categorizing the different components of a curriculum but the classification of the Curriculum Committee on Computer Science of the Association for Computer Machinery enjoys the best professional standing.[3] The three major divisions with their principal topics are:

A. Information Structures and Processes
   1. Data Structures
   2. Programming Languages
   3. Models of Computation

B. Information Processing Systems
   1. Computer Design and Organization
   2. Translators and Interpreters
   3. Computer and Operating Systems
   4. Special Purpose Systems

C. Methodologies
   1. Numerical Mathematics
   2. Data Processing and File Management

3. Symbol Manipulation
4. Text Processing
5. Computer Graphics
6. Simulation
7. Information Retrieval
8. Artificial Intelligence
9. Process Control
10. Instruction Systems

In order to give the reader a notion of the content of these areas they will be treated briefly.

5 INFORMATION STRUCTURES AND PROCESSES

One of the attractive features of mathematics is its suitability for abstracting essential features from complex situations and dealing with them in a rigorous and elegant way. There appears to be a low threshold for the degree of complexity found in nature which the mind can handle. The abstracting procedure and the establishment of relationships among these attributes is really what is meant by "handling" a complex situation so we really ought to define more carefully what we mean by degree of complexity in order to understand its reduction. Nevertheless, the reader will have some feeling for these matters; the relative complexity of, say, chess and checkers is something he will accept without requiring a quantitative measure.

The digital computer is an instrument designed to handle complex situations and as the technology advances the potential situations become more and more complex. The handling of the payroll for a large corporation is clearly a complex affair, but while the data base for a payroll may be complex, the procedure is straightforward. What we might consider is a procedure so complex that it must be the product of many minds and is incapable of detailed comprehension by any one person. We have instances of such computer programs and the analogy with attaining a detailed knowledge of the working of the brain should be evident.

In order to specify these computational systems, however, the computer scientist must understand his system and its data at a level of low complexity so that he can find basic theorems or universal truths to guide him in the proper design of the overall systems and data bases. In using the term "system" we encompass, as is customary, both hardware and software aspects.

The courses in this part of the curriculum prepare the student for an understanding of the representation and manipulation of data and programming languages in various organized ways. The study in depth of these structures has led to theoretical models and to the more abstract parts of computer science. Models of computation lead one to studies of switching circuits, sequential machines, automata theory, formal languages and grammars, mathematical theories of languages and of computation.

As an example of the kind of thing we are talking about, consider the Turing machine, a model invented in the 1930's by the mathematician A. M. Turing.[4] This abstract model is very simple. In one form it is a device with a finite number of internal states and a tape of arbitrary length marked into squares. At any moment it can read a symbol on the tape. Based on that symbol and the internal state, the machine can initiate actions to change the symbol and move the tape one square left or right.

One would expect such a machine to be limited in the kinds of things it could do and yet Turing showed that any effective computation performed on any computer can be performed on a Turing machine. The universality of this machine allows us to establish truths about it which will apply to all other machines and consideration of this and other equivalent models has increased our understanding of computers, programs, languages, and computations, all of which can be fit into this simple model.

Indeed one may ask another question. Does such a machine have enough "intelligence" to build a copy of itself; to reproduce itself? Offhand, one might think that an object would not build a replica with the full complexity of the constructor. We do, of course, have examples in nature if we accept a mechanistic view, and we can, in fact, show that it is possible in the case of these machines. It is the possibility of considering a question such as this that make these models so interesting to researchers in computer science and particularly to the student seeking a deeper understanding of computers and more general automata.

Thus we see that this part of the curriculum and of research gives rise to a kind of intellectually satisfying exercise for serious students of computer science.

6 INFORMATION PROCESSING SYSTEMS

In computer science a system is usually considered to be the whole complex of hardware and software with all of the connections, interfaces, and communication channels considered as a single entity. In the education process, however, the interconnection and interdependence of hardware and software is largely lost when the student moves from a hardware-oriented session to a software class. Of course, this phenomenon is not peculiar to this discipline and in many fields of study the making of the connection or integration into a full picture is left to the student. The tragedy is that it is often impossible for the student to fill in the hardware part of the total system picture even when the term "hardware" is used in the loosest of connotations. This part of computer science is frequently absent in a curriculum and may have arisen from the historical pattern mentioned earlier.

The best possible situation would be the one in which a faculty with competence in all phases of system study worked closely together in a single department. Computer science covers such a broad area, however, that with the largest of departments may one hope to cover this field with reasonable depth. Furthermore, the broadness of the field brings together such diverse interests that a certain amount of separation must occur. In spite of the fact that we call the discipline a science, it should be realized that the heart of the subject is the engineering of hardware and software systems. This realization has led in some cases to the incorporation of the discipline into engineering departments. Anyone watching the spectacular growth of computer science can see easily that such a department with multiple goals must grow quickly beyond reasonable size and fission off a computer science department either formally or practically. The problem is rather the other way; as a computer science discipline grows, it may break up into subdisciplines and the department becomes a school or college.

Whether or not the connection is made, however, it is generally conceded that a well-rounded computer scientist will have a good knowledge of computer architecture, both hardware and software, at a fairly detailed level. Usually the level of detail for software is more atomic than for hardware.

A recent aspect of study in this area has been the attention to the diagnostic study of complete systems in an attempt to design better systems or better utilize present systems.[5] This is often carried out as a kind of behavioral approach in which the detailed structure becomes less important than the action of the system in different situations or the means by which it can be made to function better in these situations.

This second broad area of computer science, Information Processing Systems, is the hardware and software systems engineering part of the discipline. The structure of this part of a degree curriculum is the most controversial with respect to both content and depth and these questions remain largely unresolved.

7 METHODOLOGIES

The field has grown so rapidly but the different applications areas have not drawn far apart with their own methods, jargon, and intellectual organization insofar as computation is concerned. It is possible to extract the methodology common to the many applications of computers and study their structure.

The oldest and the one with the largest body of knowledge to draw from is numerical mathematics. The easiest extrapolation to make when seeking jobs for the early computers was from the desk calculator. This volume contains examples of the increases in the size and complexity of computations which the computer has made possible. Along with this increased capacity there has been a need for a finer measure of the meaning of results, for algorithms with better properties when applied to larger systems, as well as a basis for understanding the mathematical stability of these procedures. In spite of the long history, numerical mathematics has barely kept pace with the new requirements generated by the computer and the subject has enjoyed a renewed dynamism. Even such relatively stagnant areas as finding the roots of a function or approximating a function with other simple functions have provided arenas for exciting progress.

This renewed interest in numerical mathematics is being overshadowed, however, by the growing research effort in non-numerical fields and the complexion of computer science is changing as more researchers, especially recent graduates, move in that direction. It is hard to tell, at this time, whether this change is due to realization of its importance, the possibility of comprehending a field which has not yet accumulated great depth, the great diversity which thins competition, or pure huckstery.

The manipulation of non-numeric information is less completely understood and less formalized than is the handling of numeric information under the ordinary rules of arithmetic, even though most of the world's information is of the non-numeric sort. The abstract study of data structures mentioned in the first broad educational category serves as a basis for techniques in processing library, medical, management, literary, military, scientific, ecologic, and other types of information files. These techniques would include organization, classification, and the vast problems of information retrieval on a large scale.

A related topic is the formal manipulation of symbols according to fixed rules such as the processes of algebra or formal differentiation and integration. At the other end of the spectrum there has been an increasing interest in text processing and the humanities scholars have been carrying out programs in textual criticism, concordance generation, as well as metrical, authorship, and linguistic analysis at a high level of sophistication.[6] At many places the humanists are engaged in exploratory research on textual material in advance of others one might have expected to provide the leadership such as library scientists.

Operations Research[7] found its first important application during the second world war and this area has expanded into a whole study of simulation in general. Business games, agricultural models, economic systems, transportation networks, university computer centers, and time-sharing networks all find their development better understood and enhanced by simulation by mathematical models on the computer. Special languages have been developed to deal with these situations and much computer time is devoted to this activity. Indeed, in a competitive situation a lack of such an activity or an inferior model could mean a failure of an organization whether it be a corporation, an agricultural complex, an economic system, an army, or a nation. There is always the danger of overselling an excessive dependence on models, but neither should inactivity lead to a waste of computer resources in potentially valuable studies. A characteristic of these simulation studies is their insatiable appetite for more and more computer capacity. Thus, consideration of the simulation and modelling areas is closely linked to the study of information processing systems.

Automated factories have been considered for a long time and in many areas, such as the chemical industry, it has reached a very advanced state. The presence of the computer has changed the picture and opened the way for new approaches. From the simple feedback control of earlier systems there has been a progression of more advanced systems which are reacting in real time with complicated calculations of functions of the systems parameters. One of the more promising directions is the automation of machine tools[8] to perform operations not possible under human control. Feed rates can be computed in real time and surfaces can be generated with none of the classical restrictions to translatory and circular generation found in simpler machines under human guidance. The steps from specification to the final object is in the process of being completely automated, and the social implications are far-reaching. The Soviet Union has established an office of automation at the Minister level with the expectation that a breakthrough in this area will yield economic advantages over competitors who must develop highly skilled workers to approach the capabilities of future automated production.

In the computer field itself, design automation[9] has reached the point that layout and interconnection paths are generated by the computer. A driving program is then written to control the plotter producing the artwork for masks which eventually will be used for printed circuit wiring and integrated circuit manufacture.

Automation already has made incursions into the field of research particularly in chemistry, crystallography, nuclear physics, biology, medicine, and other fields. The high intensity meson facility under construction at the Los Alamos Scientific Laboratory is an accelerator with a level of automatic control far above any now in operation.

The new field of computer graphics[10] has produced a front end for the design procedure which allows a design engineer to manipulate visual diagrams, charts, and parameters, and in interaction with the computer, to develop the specifications for parts and assemblies. The power of this interactive mode, which is generally carried out with a cathode ray tube display, is not restricted to this type of operation, however. The adjustment of solution parameters in solving mathematical problems, the specification of computer programs by flow diagrams, the guidance of programs for recognizing and measuring patterns of various kinds, the training of airline pilots, the behavior of models of physical systems, and the control of air traffic are a few possible applications of computer graphic techniques. There is much basic work in signal compression, low-cost displays, learning theory, image enhancement, and development of languages to be carried out in this wide-open research field.

Computer graphics is also a facet of the problem of computer-based educational systems.[11] Although this area lies mostly in the domain of education there are some points of contact with computer science. The tremendous impact and importance of this area can hardly be overestimated. The possibility of quickly raising disadvantaged individuals, groups, or even whole nations in an accelerated time scale to a new educational level, for example, opens a whole new dimension in this new world where technological sophistication has become as important as natural resources.

The last defined area in the curriculum is that of artificial intelligence,[12] by which we mean a behavior of a machine which could be described as intelligent when observed in a biological system. This is a "fun" activity in computer science, and yet it is a most important one. It is not so much that computers should be designed to act in human-like ways any more than vehicles should have legs rather than wheels. It is rather that we can thereby better understand the organization and functioning of the mind. This is not meant to exclude the possibility that the things learned in this anthropomorphic approach will not be applicable to the design of machines; the two are intertwined problems. The brain is made up of relatively simple components just as the computer, and the Gestalt of the mind may be thought of as the analogy of the computer program. We may learn some tricks from what appears to be a very professional job of programming in the biological case.

Some beginnings have been made in this area, but much remains to be done beyond the present work, which is mostly organizational and descriptive. One of the odd characteristics of these early studies is the "magic act" nature of the results. A computer program, like a professional magician, can amaze and mystify with its apparent great powers, but when the details of the trick are disclosed, we say, "Oh, of course!" The game of chess played by the computer[13] is not of the highest level, but it is, nonetheless, quite impressive. A computer cannot perform by examining every move because of the astronomic size of such a task. Instead, it is programmed to assess possible moves according to gains or piece-exchanges, resulting mobility, control of critical squares, or other rules designed to achieve specific goals and subgoals. In other words, some of the chess experience of humans is taught to the machine. The point is that with relatively simple criteria some surprising behavior can be generated; surprising, that is, if we are ignorant of the program or the program is too complex to comprehend.

Another area of intelligent behavior is pattern recognition[14] and this area has received much attention. It is one in which it is fairly easy to determine the degree of success of a design, and at the same time will have important applications. It is also one, however, in which hardware system overtones are present and progress through software research alone is severely restricted.

Overlying all of these efforts is the realization that true intelligent behavior must involve learning or adaptation and some of the deeper studies are in that field and it is here, the author believes, that lack of hardware support has been the most crippling. The lack of support is not just in funding but in the unavailability of computer scientists with hardware competence.

The reader may be struck with the remarkable richness of computer science and the interdependence of the component parts. Coupled with the diversity and the newness of the field is a thinness in certain areas, but this just means a wealth of opportunity for significant research.

8 THE COMPUTER AS A TOOL

The remainder of this volume is concerned with the application of computers in science and technology so there is no need to elaborate on that aspect here, but it is appropriate to mention how the mechanics of using computers has evolved on the campus.

When the first computers arrived on campus they were rudimentary and required considerable experience to use effectively. There were few users and even personal interaction with the computer was often possible. As the use became more widespread and higher level languages appeared, the computers grew in size and a large resident "system" came into being to handle the traffic.

The consequent loss of personal interaction with the system was felt keenly, particularly in the preliminary checkout phases of computer programs. Those who did not need the massive power of a large computer considered the possibility of a smaller, more personal system. Time-sharing systems, which in effect made the large computer look like many smaller ones, each serving a personal console, sought to satisfy these needs. A third approach was to establish a minimum local system to process immediate needs with a link to a larger backup computer.

Although all of these systems have their drawbacks, it is a fact that many campuses incorporate all of these methods with the result that systems are becoming more dispersed. This has placed a premium on the study of computer system organization in an attempt to reintegrate the components into a more powerful whole which can operate more or less as a unit when required to do so, and yet preserve the integrity of the parts.

How will computing power be dispersed? It might be through the utility concept of a large central facility with simple access, or perhaps a network of equal partners with symbiotic capabilities. More likely it will be a network with a hierarchy of computers of different capacities. Whichever it is to be, it appears that the university campus is a fertile ground for testing these ideas.

9 CONCLUDING THOUGHTS

There has been no attempt at describing specific research projects at universities. These research activities can be deduced from the outline of the curriculum; the teaching and research at a university merge imperceptibly from one to the other in the field of computer science. In any event, these projects change very rapidly in this new discipline and one can only conjecture on their direction.

Consider one of the most dynamic and intellectually challenging fields, that of artificial intelligence. It is likely that the overlap with the biological sciences will increase and the consequences of such an interaction could be startling.

The unravelling of the structure of DNA, the primary genetical substance, was a monumental task involving the analysis of a really complex structure made up of constituents obeying simpler laws. More and more biologists who have worked in this area are turning to an analogous problem, the structure of the human brain, also made up of much simpler components, and they are being joined by members of other disciplines.

In many respects this latter problem is more germane to study and alteration of the human condition than is biological evolution. It is now generally felt that environmental conditions and evolution of a psychosocial type are the more significant factors molding human life today[15] and that the biological evolution of man has reached a virtual standstill over the last 100,000 years. The biological constitution of the man of the twentieth century and of prehistory are not significantly different. Man carries in his brain rather than his genes the accumulation of recent evolutionary changes and these are passed on as surely as imprints on the genetic material.

If this thesis be accepted, then there is a strong argument for a study of how man's brain has been programmed by external forces and this problem looms larger than the genetical one. With technological advances accelerating at the present rate the biological functioning of the human body is becoming an anomaly and the furnishing of mechanical substitutes for the more important constituents is rapidly becoming an important business. There have been proposals for life support systems for an isolated head[16] and technical problems may not be unsurmountable. This may seem a macabre proposal, but it is a logical extension of current survival systems.

Another feature of a social evolution theory is the control of the environment by man and the closing of the loop by the effect of the environment on the controlling agency. A better understanding of this cycle may well be essential to man's survival of technology and prevention of a mental extinction.

These considerations lead to an interest in brain mechanisms and one of the best approaches currently available is through use and study of the computer. This is a single example of a need for the study of computers for their own sake as well as with a view toward applications.

It is difficult to imagine much of this kind of research outside of the university. Fortunately it is being carried out at universities, although probably more as a result of the intrinsic intellectual interest rather than owing to a strong social consciousness. Nevertheless, the importance to society ought to be recognized at some level in the decision process. There is a need for informed opinions in all phases of this new science if its impact on our future proves to be even a fraction of informed expectation.

References

1. S. Rosen, "Electronic Computers: A Historical Survey". *Computing Surveys* **1** (1969)' pp. 7–36.
2. M. Lehman, Review 9119. *Computing Reviews* 7, 1966, pp. 83–84.
3. ACM Curriculum Committee on Computer Science, "Curriculum 68", *Communications of the ACM* **11** (1968), pp. 151–197.
4. M. L. Minsky, *Computation: Finite and Infinite Machines*. Prentice–Hall, Englewood Cliffs, N.J., 1967, pp. 103–145.
5. G. Estrin, *et al.*, "Snuper Computer—A Computer in Instrumentation Automation", *Proc. Spring Joint Computer Conference*, 1967, Thompson, Washington, D.C., 1967.
6. E. A. Bowles (ed.), *Computers in Humanistic Research*, Prentice–Hall, Englewood Cliffs, N.J., 1967.
7. S. Beers, *Decision and Control*, Wiley, New York, N.Y., 1966.
8. O. S. Puckle and J. R. Arrowsmith, *An Introduction to Numerical Control of Machine Tools*, Chapman and Hall, London, 1964.
9. *Proceedings of the SHARE Design Automation Workshop*, June 23–25, 1965, Atlantic City, N.J.
10. D. Secrest and J. Nievergelt (eds.), *Emerging Concepts in Computer Graphics*, Benjamin, New York, N.Y., 1968.
11. See report of seminar at University of Texas, Austin, Tex., *Computer-Assisted Instruc-*

tion, Testing and Guidance. Harper and Row, New York, N.Y. (to be published October, 1969).

12. E.A. Feigenbaum and J. Feldman (eds.), *Computers and Thought*, McGraw–Hill, New York, N.Y., 1963.
13. E.A. Feigenbaum and J. Feldman, *loc. cit.*, pp. 37–105.
14. G.C. Cheng, *et al.* (eds.), *Pictorial Pattern Recognition*, Thompson, Washington, D.C., 1968.
15. P.B. Medawar, *The Future of Man*, Basic, New York, N.Y., 1960; G. Wolstenholme (ed.), *Man and His Future*, Little, Brown, Boston, Mass., 1963.
16. N. Amason, "Brain Without a Body?", *Soviet Life* (September, 1968), pp. 46–48.

CHAPTER 10

The Impact of Computers on Mathematics

PETER D. LAX

1

SOME MATHEMATICIANS, whose ivory tower is $99\frac{44}{100}\%$ pure, would claim that, mathematics being an autonomous mental activity, it can no more be influenced by computers than it is influenced by cars, television, the jet plane, etc. The aim of this essay is to argue the case for a contrary opinion, by presenting examples of good mathematics which would not have been created without the computer. All my examples stem from mathematical physics, partly because my own interest lies in this field, but also because I want to provide ammunition against the other extreme, the computer enthusiast who believes that all there is to computing lies in the new areas such as model building, off-beat optimum problems, information theory, the simulation of human intelligence, and that mathematical physics is passé!

Computers have changed the character of applied mathematics in a number of ways. The most direct is through the introduction of new problems—and their solution—in classical settings. I will illustrate this in Section 2 by the stability theory of difference operators, which called forth a body of new results in matrix and operator theory.

A second way that computers have influenced mathematics is by creating the need for a theory of computation, i.e. a measure for the amount of computation needed to calculate and approximate functions. At present we have only the barest outline of such a theory, embracing on the one hand Kolmogoroff's theory of tabulation, and his theory of the complexity of sequences, and on the other hand algorithmic results such as Winograd's estimates for the time it takes to add and multiply numbers, see ref. 15, and the work of

Winograd and Strassen on efficient ways of multiplying matrices. These theories are interesting per se, but at the same time they yield unusual and unexpectedly efficient algorithms for certain standard operations. The non-standard methods are significantly better than the usual ones only when the scale of the calculation is large enough, such as matrix operations performed on sufficiently high order matrices. That is why such methods were not sought in the old days when any single calculation was limited to a few thousand arithmetic and logical operations, in contrast to the billions which can be performed today.

I would like to digress for a moment to talk about the bad old days; the severe limitation of the amount of computing that could be performed forced applied mathematicians to drastically simplify their models, and to exploit every accidental feature of the problem at hand to reduce computing. The general run of mathematicians delight in the precise and despise the approximate, they favor the general and frown upon the ad hoc; no wonder the that old fashioned applied mathematics was in ill repute. Now that the character of numerical analysis is so changed it is reasonable to expect many more people to take an interest in applied mathematics, especially since it is no longer necessary to learn so much tedious classical lore—on the contrary, knowledge of the old may unduly hamstring the imagination. In fact, numerical analysis today takes its place alongside the two oldest professions of the world where amateurs can compete successfully with professionals.*

I expect the availability of adequate computing power to have a salutary effect on the development of the theory of various branches of mathematical physics, and I expect radical changes in the teaching of these subjects. In the past there was a tendency to mix together theoretical concepts with methods of solution; today we can render onto theory what is theory's, and onto the computer what is the computer's.

An equally important use of computing in mathematics is as an experimental tool for studying nonlinear problems. In Section 3 a number of examples are given where numerical studies of nonlinear systems has led to the discovery of surprising qualitative properties.

---

* The others being prostitution and military strategy.

2 SOME NEW PROBLEMS IN CLASSICAL MATHEMATICS

Some mathematicians, not familiar with finite difference schemes, believe that converting a differential equation into a difference equation is the last act of desperation, which removes all subtlety from the problem, leaving it fit only for the tender mercies of the computer. Those who work in the field know that the opposite is true: the theory of difference equations needed to distinguish useful from useless difference schemes is often more sophisticated than the corresponding theory of partial differential equations needed to distinguish well posed problems from ill posed ones.

In this section we will illustrate our contention by discussing briefly some aspects of the *stability problem* for difference operators; this is the problem to give conditions under which high powers of S_h^n of difference operators S_h remain bounded. S_h is of the form

$$S_h = \sum_j s_j(x)\, T_h^j \tag{2.1}$$

where T_h^j means translation by jh, j a vector with integer components; the coefficients $s_j(x)$ may be scalars or square matrices.

If the coefficients s_j of S_h are independent of x, the operator S_h is, via the Fourier transform, unitarily equivalent with multiplication by $S(h\xi)$ where

$$S(\xi) = \sum_j s_j e^{ij\xi}. \tag{2.2}$$

Consequently S_h^n is unitarily equivalent with multiplication by $S^n(\xi)$; in the scalar case we conclude immediately that S_h *is stable if and only if*

$$|S(\xi)| \leq 1 \quad \text{for all } \xi. \tag{2.3}$$

Like most necessary and sufficient conditions, this one is not the end but merely the starting point for further investigations. In ref. 14, Strang posed and solved the following problem:

Suppose the summation in (2.2) is restricted to $|j| \leq l$, l some given integer. Find the most accurate *stable* approximation S_h to $T_{\lambda h}$, where λ is *not* an integer.

Most accurate here means that

$$T_{\lambda h}u - S_h u = 0(h^m), \tag{2.4}$$

for all C^∞ functions u, with as high value of m as possible.

It is easy to see that this is equivalent to this problem.

Find a trigonometric polynomial $S(\xi)$ of degree l such that

$$\left(\frac{d}{d\xi}\right)^a S|_{\xi=0} = (i\lambda)^a, \quad a = 0, 1, \ldots, m \tag{2.4'}$$

for which

$$|S(\xi)| \leq 1 \tag{2.5}$$

for all ξ.

Clearly the largest possible value of m is $2l + 1$. In this case (2.4′) determines S uniquely; it is easy to write this S as the solution of an interpolation problem, but it is not easy at all to decide if this S satisfies (2.5). Strang has shown, using Hungarian type analysis, i.e. in the style of L. Fejer and his school, that (2.5) holds, provided that $|\lambda| \leq 1$.

We turn now to the matrix case still with constant coefficients; there stability amounts to $|S^n(\xi)|$ being *uniformly bounded* for all ξ and all w. This leads to the following surprisingly nontrivial matrix problem:

Characterize the class of matrices $\{M\}$ satisfying

$$|M^n| \leq K,$$

for all n, K some constant.

For various answers to this problem, see the work of Kreiss, Buchanan, Morton and Schechter described in the book of Richtmyer-Morton[12]. The importance of these criteria is not so much for difference operators with constant coefficients but they turn out to be essential for proving the stability of difference operators with variable coefficients. Thus Kreiss[4] was able to use his matrix theorem to prove the stability of schemes which contain a sufficient amount of dissipation.

In ref. 8 the stability of a certain scheme is proved which contains less dissipation than required by Kreiss' theorem. It is amusing and instructive to note that one ingredient of this proof is Berger's theorem on the numerical range of the powers of an operator; Halmos[3] was led to conjecture this remarkable relation on the basis of a much weaker matrix theorem used in an earlier paper[9] on stability theory.

If the underlying domain of the difference operator S_h does not include all lattice points but only a portion of them, the problem of determining when S_h is stable is much harder. Strang[13] has observed that in case S_h acts on lattice points on the positive axis, and if no special boundary conditions are imposed, then S_h is a Toeplitz operator; he was able to use this observation to study the stability of certain simple implicit schemes. Osher,[11] observed that with arbitrary boundary conditions S_h is a finite dimensional perturbation of a Toeplitz operator, and he has discussed the stability of S_h with the aid of the classical factorization of the reciprocals of Toeplitz operators.

All this bears out the contention that the study of difference operators needed for the analysis of machine computations has led to new and interesting results in the very classical fields of matrix theory and operator theory.

3 NUMERICAL EXPERIMENTS WITH NONLINEAR PROBLEMS

In this section we describe two nonlinear problems where some unexpected behavior of solutions was discovered by analyzing samples of solutions obtained by numerical computations. All our examples deal with nonlinear partial differential equations, a field for which von Neumann predicted that exploratory calculations will play an increasingly important role. It should be pointed out that computing is becoming increasingly important for studying nonlinear ordinary differential equations; these calculations often are extremely delicate, compared to which the calculations done for partial differential equations are somewhat crude. We refer the interested reader to ref. 6, where the long term stability of trajectories is studied, and to ref. 10, which describes a successful computer-aided search for periodic solutions in celestial mechanics.

A This example is about the equation

$$u_t + uu_x + u_{xxx} = 0 \tag{3.1}$$

introduced by Korteweg and de Vries as an approximate equation for long waves over water. It also appears in some approximate description in plasma physics.

In what follows we shall be concerned with infinitely differentible solutions of (3.1) which, together with all their derivatives, are zero at $x = \pm\infty$. It turns out that the initial value problem is properly posed in this class of functions and that solutions exist for all time. Since (3.1) is time reversible (if $u(x, t)$ is a solution, so is $u(-x, -t)$), the initial value problem is properly posed also backwards.

Already Korteweg and de Vries were aware of solitary wave solutions of (3.1); these are of the form

$$u(x, t) = s(x - ct),$$

where s satisfies the ordinary differential equation

$$-cs' + ss' + s''' = 0. \tag{3.2}$$

This equation can be solved explicitly; it turns out that for each positive speed c there exists a solitary wave, zero at $\pm\infty$, unique except for a phase shift.

Now special solutions are important when dealing with linear equations since one can build new solutions out of them by superposition. No such thing is possible for nonlinear equations, and therefore special solutions do not play a central role in the general theory of nonlinear equations. However Kruskal and Zabusky[5] made the remarkable discovery that solutions of (3.1) may have hidden in them a certain number of solitary waves, in the following sense:

Let u be any solution of (3.1); then there exist N eigenspeeds $c_1, c_2, \ldots, c_N$ such that

$$\lim_{t\to\pm\infty} u(x, c_j t, t) = s(x - \theta_j^{\pm}; c_j). \tag{3.3}$$

Note that the speeds of the solitary waves emerging from u for large t are the same near $t = +\infty$ and $-\infty$; only the phase shifts $\theta_j^{\pm}$ are different.

Kruskal and Zabursky first observed these emerging solitary waves by studying moving picture output of numerical solutions of (3.1); once the phenomenon was noticed, careful numerical calculations were made in situations which isolated this phenomenon, such as the collision and eventual separation, unscathed, of two, three solitary waves. Eventually Gardner, Greene, Kruskal and Miura[2] succeeded in proving (3.3) rigorously.

B The next example is the Hodgkin–Huxley equation for the conduction of nerve impulses. $V = V(x, t)$ denotes the potential at time t at position x along the nerve. I denotes the current flowing from the inside to the outside of the nerve fibre; it is also a function of x and t, related linearly to V:

$$I = aV + b; \tag{3.4}$$

a and b are *nonlinear* functions of quantities m, n and h, representing the concentration of certain ions:

$$a = A(m, n, h), \quad b = B(m, n, h). \tag{3.5}$$

The concentrations satisfy ordinary differential equations:

$$\begin{aligned} m_t &= M \\ n_t &= N \\ h_t &= H \end{aligned} \tag{3.6}$$

where M, N, H are linear functions of m, n and h, with coefficients which are *nonlinear* functions of V.

The potential V itself satisfies the cable equation

$$V_{xx} = \text{const } V_t + I. \tag{3.7}$$

How well does this system of equations represent the propagation of signals in nerves? Hodgkin and Huxley have shown that, with appropriate choice for the indicated functions, this system of equations has steady progressing wave solutions, i.e. solutions which depend on $x - \theta t$ only; θ, the velocity with which the wave propagates, is uniquely determined by the equations, and agrees very well with experimentally measured values. Cooley and Dodge showed in ref. 1 that if one end of the nerve fibre is stimulated by electric current, a signal is triggered off which propagates along the fibre, provided that the stimulation exceeds a certain threshold, measured as a combination of the strength and duration of the stimulus. The shape and speed of propagation of this signal is the same as that of the progressing wave solution found earlier, except during the initial stage of the discharge. This constitutes a striking demonstration of the aptness of the model.

At present a rigorous proof is lacking that solutions of the Hodgkin–Huxley equation behave as indicated by the numerical calculations. To find such a proof is a challenging mathematical problem.

References

1. J.W. Cooley and F.A. Dodge, Jr., "Digital Computer Solution for the Excitation and Propagation of the Nerve Impulse", *Biophysical J.*, 1966, Vol. 6, 583–599.
2. C.S. Gardner, J.M. Greene, M.D. Kruskal, and R.M. Miura, "A Method for Solving the Korteweg–de Vries Equation", *Phys. Rev. Letters*, Vol. 19, 1967, pp. 1095–1097.
3. P.R. Halmos, *A Hilbert Space Problem Book*, D. van Nostrand, 1967.
4. H.O. Kreiss, "On Difference Approximations of the Dissipative Type for Hyperbolic Differential Equations", *Comm. Pure Appl. Math.*, Vol. 17, p. 335, 1964.
5. M.D. Kruskal and N.J. Zabusky, "Interaction of Solutions in a Collisionless Plasma and the Recurrence of Initial States", *Phys. Rev. Letters*, Vol. 15, 1965, pp. 240–243.
6. L.J. Laslett *et al.*, "Long-Term Stability for Particle Orbits", AEC Research and Development Report, NYO–1480–101, 1968.
7. P.D. Lax, "Integrals for Equations of Evolution and Solitary Waves", *Comm. Pure Appl., Math.*, Vol. 21, 467–490, 1968.
8. P.D. Lax and L. Nirenberg, "On Stability for Difference Schemes; a Sharp Form of Gårding's Inequality, *Comm. Pure Appl. Math.*" Vol. 19, No. 4, 473–492, 1966.
9. P.D. Lax and B. Wendroff, "Difference Schemes for Hyperbolic Equations with High Order of Accuracy", *Comm. Pure Appl. Math.*, Vol. 17, No. 3, 381–398, 1964.

10. K. R. Meyer and J. Palmore, "A New Class of Periodic Solutions in the Restricted Three Body Problem", Report, University of Minnesota, 1968.
11. S. Osher, "Systems of Difference Equations with General Homogeneous Boundary Conditions", *Trans. Amer. Math. Soc.*, Vol. 137, 1969, pp. 177–201.
12. R. D. Richtmyer and K. W. Morton, *Difference Methods for Initial-Value Problems*, 2nd Ed., Interscience Publishers, 1967.
13. W. G. Strang, "Wiener–Hopf Difference Equations", *J. Math. Mech.*, Vol. 13, p. 85, 1964.
14. W. G. Strang, "Polynomial Approximation of Bernstein Type", *Trans. Amer. Math. Soc.*, Vol. 105, p. 525, 1962.
15. S. Winograd, "How Fast Can Computers Add", *Scientific American*, October 1968.
16. V. Strassen, "Gaussian Elimination is not Optimal", *Numerische Mathematik*, Vol. 13, 1969, pp. 354–356.

CHAPTER 11

Monte Carlo Methods

MALVIN H. KALOS

THE EXISTENCE of modern high speed digital computers has brought into wide use a remarkable class of computing methods now given the name of Monte Carlo. This description is apt since the essence of the method is the invention of a game of chance and then gambling—usually on a computer—long enough to establish in an experimental way the odds for "winning" or "losing". Every different numerical problem to which the method is applied requires a different set of rules for gambling.

The history of this kind of computing depends somewhat on how broadly one defines it, but it is well known that mathematicians and scientists have used, at least privately, gambling experiments as an aid to understanding for much of this century. However, to be useful in interesting problems that cannot be better treated in other ways, it is usually necessary either to carry out many gambling runs or else to use complex rules (or both). It is for this reason that Monte Carlo techniques are almost invariably associated with digital computers.

Two simple examples may suffice to convey the essence of how "gambling" or "random sampling", to use a more dignified name, may be used to solve mathematical and technical problems of various kinds.

The first example concerns the calculation of π from the knowledge that

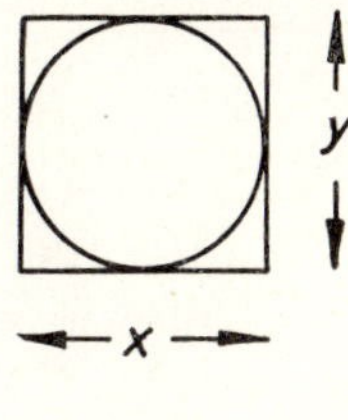

the ratio of the area of a circle to the square drawn around it is $\pi/4$. In addition, we use the fact that when points are placed at random throughout the area of the square, the chance that a point lies inside the circle is just the ratio of areas, $\pi/4$. Now it is possible to supply a computer with a sequence of "random numbers" having the property that they lie uniformly in any range of numerical values, for example, over the side of the square. We can take a pair of such numbers to represent the two geometrical coordinates (x and y) of a point in the square and then to test whether that point does or does not lie inside the circle. It is possible to program a computer to carry out this operation and to repeat it with different x's and y's many times. The fraction of trials in which the points lie in the circle as determined from many experiments should then be close to $\pi/4$ and that fraction would then be an "estimate" of $\pi/4$. Thus in 1,000,000 trials it is very likely (95% chance) that the number of points inside the circle would lie in the range from 784,600 to 786,200 yielding estimates of $\pi/4$ from 0.7846 to 0.7862 compared with the exact value of 0.7853 98 ...

The example illustrates the random sampling which may be used to solve a mathematical problem, in this case evaluation of a "definite integral". It also shows that, not surprisingly, the answers obtained are also subject to laws of chance. This aspect of Monte Carlo is a drawback, but not a serious one if an answer is needed to be no more accurate than can be obtained this way. Sometimes, in spite of the random character of the answer, it is nonetheless the most accurate answer which can be obtained for a given investment of computer time. That is certainly not the case for the calculation of π which can be determined much more accurately in other ways. But the same idea applied in problems with many more variables than two (i.e. x and y) can be economical.

Our second example concerns perhaps the simplest possible "random walk" and its simulation with the help of random numbers. A random walk is a sequence of locations taken on by some object or pointer in which the rules for moving from one location to another are governed, at least partly, by chance. The example

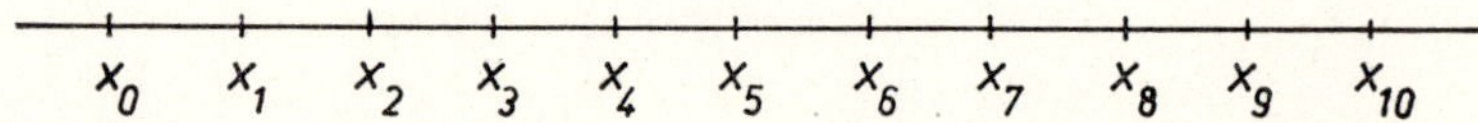

deals with a set of locations on a line called $x_0, x_1, x_2, \ldots, x_n$ which can be equally spaced. We start a marker at x_0. Half the time, at random (i.e. with probability $\frac{1}{2}$) the marker goes no further; otherwise the marker is moved one place to the right, i.e. to x_1. These rules are applied repeatedly until it is decided that the marker shall move no further. Thus the marker may move an indefinitely large number of move to location x_n, but the chances of getting far are small.

Suppose we want to know the chances of moving as far as location x_{10}. A simple Monte Carlo calculation to do this would go as follows. First, a counter (or memory cell) of the computer is set aside for the number n to indicate that the marker is at x_n. Initially n is set to 0. Then, using the kind of random numbers mentioned before, it is possible to decide at random between two alternatives each having probability $\frac{1}{2}$ of occurring. Arbitrarily, one of these may be assigned to moving the marker from x_n to x_{n+1}. When it occurs, the contents of the memory cell containing n are increased by 1. The alternative is that the walk is considered to end. If the required answer is the probability of reaching a particular point only—we have assumed the tenth—then in this case our interest ceases when the marker goes that far and the walk or our simulation of it may also be terminated. The entire procedure is then repeated and the number of "successes" tallied for an experimental estimate of the probability of success. For our example, in 95% of all runs of 1,000,000 trial walks the number of successes will range from 914 to 1040 yielding an experimental probability from 0.000914 to 0.00104 compared with the correct value of 0.000977...

As with our example concerned with π the answer for the simple random walk is easily calculated analytically, in fact more easily than the simulation can be programmed for a computer. But when more variables than one change in a step of the walk and when the range of possible changes is greater and subject to complicated rules, then analytical or other numerical solutions often become difficult while numerical simulation of the random processes on a computer remains possible. Furthermore, in many cases the Monte Carlo approach is an efficient way to get an answer from the point of view of computer utilization.

A distinction is sometimes made between "simulation" and "Monte Carlo", in which the first is applied to a direct modeling of a random process. Monte Carlo is then a description of the solution of more mathematical problems. This is a useful distinction, but it is not always clear whether a calculation is one or the other. The simple random walk of our example may be described by a set of equations; the "solution" of these equations may be carried out by the simulation of the original random walk. Of course the mathematical point of view is extremely helpful in some cases. As we shall see later, it shows that alternative random sampling procedures may be used to obtain specific answers, procedures which greatly reduce computer time to obtain those answers.

The introduction suggests the possibility that almost any random process can be simulated in a numerical way provided that at each step the different alternatives are known and that the numerical value of the chance for each possible outcome is known. Random processes of this kind occur in nature and in many of them enough is known to permit a very detailed treatment by Monte Carlo. In other cases it is illuminating to see the outcome when various assumptions are made about unknown probability laws governing the various outcomes. Monte Carlo provides a very flexible way of testing changes in such outcomes.

The kind or random process which inspired the modern development of Monte Carlo and for which some of the most sophisticated mathematical and computation developments exist is associated with the passage of radiation through matter. A good example of this concerns the possible life histories of neutrons in a reactor.

A reactor is an arrangement of various materials including uranium (or other fissionable material) which emit neutrons after absorption of neutrons. It is this that makes a chain reaction possible. Other materials present serve to slow down ("moderate") the neutrons from fission to enhance their chance of causing later fissions; neutron absorbing materials help control the reactor and to shield the space outside the reactor from the dangerous effects of the neutrons. The relation to Monte Carlo arises at first from the fact that every step in any neutron's life history, from "birth" in fission to "death" by absorption or escape or decay is in part a random process. Thus the direction with which a neutron starts life after a fission is completely random—any direction is as likely as any other. In the same way the speed with which the neutron begins its first flight is random although with a probability law for different speeds that makes very fast or very slow neutrons relatively rare. A neutron then flies with constant speed in a straight line. At every point along this line (except in a vacuum) the neutron may encounter some of the materials which make up the reactor and therefore may collide with atoms. Whether it does or not is again random with a chance that depends on the kind of atoms it meets and upon its speed. Finally after a collision, there are several possible outcomes: a neutron may cause a fresh fission starting a new cycle, it may simply disappear, or it may scatter becoming a slower neutron with a new (random) direction.

Now the properties of the neutron may be represented by numerical values of its position, its direction, and speed stored in different memory cells. It is then possible to change the numerical values in such a way as to follow possible behavior; from the proceeding it is clear that these changes should be made at random – with the help of the kind of random numbers discussed earlier. Thus the first values for direction of a neutron which represent a possible value obtained in fission are selected at random so that any direction is equally likely. In the same way the initial speed, the distance of each flight to a collision with an atom, the outcome of each such collision is chosen in turn randomly but in such a way as to reflect the chances that a real neutron has for undergoing each alternative.

Conceptually at least, this procedure is easily programmed for a digital computer. Appropriate routines must be coded to establish or change the numbers that describe the neutron. Then many different random samples of possible neutron life histories may be carried out and information about the behavior of a real reactor deduced. For example when a large number of such life histories are carried to conclusion, the number of times a fission event is chosen as the outcome of a collision may be counted up. If this exceeds the number of neutrons started the reactor is likely to increase its power as time goes on. If the number of final fissions is less than the number of initial neutrons, the power of the reactor is then expected to die away.

The complications in practice come from two sources. For one, the numbers which determine the relative chances of all the different choices open to a neutron are in fact very voluminous. They often vary very rapidly with neutron speed and in a different way for each material used in building the reactor. An accurate numerical representation of the required data often requires several (or many) thousand memory cells in a computer. Acquiring, manipulating, and using correctly all these data arc formidable technical tasks.

The second practical difficulty which often arises is a consequence of the power of the method. Following the straight line path of a neutron (often called "tracking") can be carried out even when the geometrical arrangement of materials is very complicated. Arrangements of fuel rods, of pressure vessels, of coolant pipes can be treated. Programs flexible enough to treat reactors of this complexity are themselves complex to write, to debug, and to use.

Although our example has been devoted to the behavior of a nuclear reactor, it could equally well have dealt with any of a large number of problems concerned with the passage of radiation through matter. Examination of the radiation that may pass through the shield of a reactor is often carried out by the same methods. The behavior of cosmic ray and man-generated very high energy particles is much studied this way. Although the statistical rules of behavior at each encounter are not as well known as for neutrons in a reactor it is common to make models of the behavior and carry out a simulation. In high energy physics the analysis of how well the results of some experiment agree with a hypothesis is often made by assuming the hypothesis in a Monte Carlo simulation of the experiment and comparing these results with the real experiment.

Generally speaking when it becomes necessary to carry out calculations of the interactions of radiation with matter with rather complex rules of behavior, Monte Carlo methods prove to be very useful.

The area of application is not limited to physical sciences. Again any model of a random process occurring naturally (as with the growth and death of biological populations) or in connection with human activities (for example, traffic problems) can be profitably simulated when it can be made sufficiently definite. It may be useful to explore an example in some detail. We take a simplified model of a "serving facility", in which a number of stations exist to serve some need. In this model customers may be assumed to arrive at random with some average rate and a known probability distribution of random delays between arrivals. When more customers arrive than there are stations to serve them, a queue (a list of those awaiting service) is built up. In a computer simulation, a memory cell would be set aside for the length of the queue. In addition, if the customers were different in some way considered important in the model, then additional cells would be required to record these properties. For simplicity, we suppose that all customers are alike except for their time of arrival. Then a simulation may be carried out provided definite rules are given for the assignment of customers on arrival to stations or to the queue and from the queue to stations. These rules may be definite—earliest arrival must be served first—or they may include some element of chance. In any case any prescription may be translated into rules for a computer to add or withdraw entries from the list which represents the queue in the computer memory and add elements to lists representing customers being actively serviced. Finally the model is completed with the statement of rules that govern the time it takes to complete the service for any customer and free a station for another. In doing this, the stations may be assumed different; the time may depend upon properties of the station and customer and may also be random in part. Again a definite set of rules is required for simulation, but very complicated assumptions may be acco-modated.

A simulation is started with a possible initial state of the facility, say, at the beginning of a working day. Behavior such as the average length of a queue or average delay may be obtained. On the other hand there may be greater interest in the occurance of rare events such as the chance of very long waits and these may be obtained also but only as a result of very many runs of simulation. Such results tend also to be strongly dependent on the validity of the rules assumed for the simulation in unusual or extreme variations.

Monte Carlo simulations of this kind are widely used now in many areas. Problems of the reliability of complex systems subject to random failures of components; analysis of the safety of aircraft landing systems; the behavior of communications systems with noise or other random changes; economic models and the behavior of the stock market; these are studied in part with the help of computer simulation. In all cases it is necessary to enumerate the possible events at every state and to give numerical data for the probabilities of the alternatives. Often this enumeration is itself a useful exercise in drawing attention to the complexity of a problem or the lack of data required to give definite answers to technical problems.

The examples given have stressed the applicability of Monte Carlo to more or less naturally occurring random processes. Each of the examples can also be described mathematically. The calculation of π is an example of integration. The behavior of neutrons in a reactor is also given by integro-differential equations and the servicing problem can be related to difference integral equations. In a sense we may regard the Monte Carlo method as constituting a numerical solution of these equations although we did not even take the trouble to formulate them.

It would be beyond the technical scope of this article to attempt a general formulation of the class of mathematical problems which arise in random processes, but it is possible to cite some advantages to such formulations. For one, some of the same problems occur more generally in applied mathematics and the physical sciences and when they do, it is possible to turn them into random processes which may be solved by Monte Carlo. There are several good examples which may be cited in current research.

As was suggested earlier, Monte Carlo methods are suitable for the evaluation of integrals in which many variables appear. Such integrals occur in physics in many places; very large integrals, impossible to carry out in any other way than Monte Carlo, come up in the treatment of the properties of systems containing many particles (atoms, molecules, or most often, idealized but interesting particles). Special integration procedures are necessary together with large fast computers, but it is possible to calculate the properties of systems of thousands of particles by Monte Carlo. For some it is possible to demonstrate the existence of different chemical phases of the system.

The treatment of the properties of a system of many particles in quantum mechanics requires the solution of a differential equation in many variables—three for each particle. This differential equation is of a character equivalent to the behavior of a particle diffusing randomly in many dimensions and can therefore be treated by Monte Carlo. The fact that many coordinates change at random in the simulated diffusion is no particular drawback. Problems of as many as 768 dimensions have been successfully treated in this way.

A second practical way in which the mathematical formulation may be used is to show the existence of many different random processes all of which give the same answer. These correspond then, to different Monte Carlo methods for a given problem; some of the alternatives are vastly more efficient than others. That is, they require much less computer time to attain a numerical result with a desired statistical reliability. This possibility is particularly important when a calculation is carried out in order to estimate the chance of the occurance of a very unlikely event. For example, a reactor shield prevents all but a small fraction of the reactor neutrons from leaking, but it may be necessary to calculate by Monte Carlo the size of that fraction. In the study of the reliability of a system, the simulation is aimed at finding the rare failure and the circumstances that cause it. In either case a "natural" simulation leads rarely to the interesting event and most of the computing is wasted. It is possible, however, to redirect the simulation, that is to change the rules so as to emphasize the occurance of selected events and yet obtain a valid estimate of its chance of turning up in the original random process.

To make plausible that an honest but more efficient alternative game may be played we may refer back to the very simple random walk example given earlier. In that we assumed that at any step a marker might move to the right or else, with probability $\frac{1}{2}$, the play is terminated. Any termination prevents the interesting event—survival to the tenth position—from occurring. Any suppression of early termination enhances the chance of survival and tends to make the simulation more efficient. In particular at any one stage we might artificially increase the chance of survival in the simulation to a number greater than $\frac{1}{2}$, say to 1. This survival no longer represents that of the original object, but it may be thought of as standing for the composite survivors of two (instead of one) markers both subject to termination. Clearly, then, this artificial change in the rules doubles the chance of survival in the final answer and the answer to the original problem is obtained as half the probability for the altered simulation. The essence of a general strategy becomes clear: at each stage of the random process analyze the relative chances of each alternative to lead to the required final situation and artificially favor the more likely ones. To compensate for this change reduce any answer derived from an altered simulation by the ratio of the natural to the enhanced probability of the event which is actually chosen.

This kind of strategy is well known in radiation transport and a considerable body of theory and application has been worked out. It is less well developed, although no less practical, in problems of reliability.

CHAPTER 12

Impact on Statistical Mechanics*

B. J. ALDER

STATISTICAL MECHANICS is an ideal field to which to apply computers, because it meets the necessary prerequisite that the laws of physics which govern it are presumed to be well-known. The difficulties that one faces are hence mathematical ones, but so formidable that some of the most famous theoreticians over a period of a hundred years have not been able to solve many of the problems analytically. In fact, the statistical mechanical theory for the thermodynamic or equilibrium properties of a system composed of a very large number of particles was formulated in part to overcome the mathematical complexities involved in having to solve the simultaneous equations of motion of such a large collection of particles. Nevertheless, even this equilibrium theory faces unsurmountable mathematical problems.

The computers' role can only be to provide solutions to specific problems as accurately as possible. A set of such specific solutions might allow some ingenious investigator to synthesize a general analytical solution; but, more likely, carefully chosen specific examples will suggest that in certain limiting situations an asymptotic solution is exact which can then be more or less rigorously proven by analytical means. In this sense the computer results play the same role as experimental results. Their advantage over experiment lies in the possibility of choosing examples that are more easily amenable to mathematical analysis, that is to evaluate the properties of ideal but not as idealized models as required by analytical theories.

Besides the above mathematical advantage over experiments, the computer

* Work performed under the auspices of the U.S. Atomic Energy Commission.

can also provide data which are for all practical purposes inaccessible by experiments themselves. By having accessible in the computer the positions or velocities or both of all the particles in the system at any given instant of time, it is possible to extract details, which, given the relatively crude experimental tools available to observe the behavior of a system in nature, are impossible to obtain otherwise. Such detailed behavior allows hypotheses to be checked at a deeper and more refined level. However, even in some situations where the data can be obtained by actual experiments, the computer still has an advantage in that it can provide more accurate data or data with higher time or space resolution. The latter point is especially worth noting, since it is experimentally difficult to get information on a very fast time scale corresponding to molecular collision times or at a small distance scale corresponding to atomic dimensions.

Before illustrating the above points with some examples, the two main schemes evolved on the computer to carry out statistical mechanical calculations will be discussed in detail as well as the limitations of such numerical work. The problem in statistical mechanics, as can be gathered from the previous discussion, is to predict the collective behavior of a huge number of particles, given the behavior of an individual particle. Restricting the discussion to classical systems, the behavior of the particles is characterized by the fact that they obey the Newtonian equations of motion and that they interact with each other via a given potential. For present simplifying purposes that potential will be taken to be the hard sphere potential; that is, the particles are hard elastic billiard balls.

Given the positions and velocities of a collection of such spheres at any instant, it is possible, by solving the simultaneous equations of motion, to predict their positions and velocities at any future time. It is thus possible to find out the net distance any given particle has moved in any given time interval. By finding the average net distance all the particles have moved in the given time interval, the self-diffusion coefficient can, for example, be obtained and compared to experiment. The computer can actually carry out such a program by following the detailed history of a number of such spheres, collision by collision. This method, called molecular dynamics,[1] is from the computing and conceptual point of view straight-forward, except that some fairly obvious means can be employed to make the program run efficiently. However, no matter how efficient one tries to be, it is impossible to deal with as many molecules as in a drop of water for as long a time as a second. Even with any forseeable computer it will be hard to handle more than about

10^4 particles for more than 10^{-7} seconds. This is the limitation one faces in the simulation of an actual system.

Nevertheless, the formulators of statistical mechanics could never even have dreamed of being able to solve the simultaneous equations of motion to that extent, and hence they made a hypothesis which would allow the equilibrium or time independent properties to be calculated in a simpler way. Instead of determining the thermodynamic properties by taking a time average of the system at equilibrium, the ergodic hypothesis postulates that an identical average can be obtained by giving the particles all possible locations within the given volume of the system, and, in averaging, any given set of positions (a configuration) is weighted by the Boltzmann factor; that is, the probability of the occurrence of the configuration is taken to be the exponential of its negative total energy divided by kT, where k is the Boltzmann constant and T the temperature. For the hard sphere example, a certain number of spheres, N, would be placed randomly in a certain volume, both determined by the density of the system one wishes to study. One would then calculate the energy of the system which would either be zero if none of the spheres overlap with any other sphere, or infinite if any two do, since overlapping means that two spheres occupy part of the same volume, which is energetically not allowed. After placing the N particles many different times into the volume, those samples where any two spheres overlapped would hence be given zero weight, while all the energetically allowed configurations would be given equal weight in determining any average over the positions of the particles.

The above procedure[2] is easily carried out, but turns out to be much less efficient on the computer than the time averaging. This is because the random sampling of configuration space selects, except at very low densities, almost invariably a very improbable state, namely one where at least two spheres overlap. This is easy to see since the chance that any two spheres overlap becomes very high if many are placed randomly in a small box. In order to get many spheres into a box, they must be arranged in some order which is very unlikely to be randomly chosen.

To overcome this difficulty, one might think of placing the spheres into the box one at a time,[2] and if a sphere to be placed overlapped with any of the previously accepted ones it would be rejected and a new one tried. Even this procedure does not allow many particles to be placed in the box, since at not very high densities it becomes practically impossible to place another sphere into the system, no matter how many times on tries. Besides that, the procedure violates the fundamental tenets of statistical mechanics in not selecting each configuration completely randomly. Once a sphere to be placed overlaps with any other sphere already accepted, the entire configuration should be rejected rather than another sphere tried in order not to give that configuration a higher than deserved weight.

The above procedure corresponds to a true Monte Carlo process in that it would be a stochastically random process. Since this procedure is not practical, a computational method involving the next level of stochastic complexity has been applied.[2,3] It involves the development of a Markov chain which samples the configuration of the system only near the most probable ones. In the example of the spheres, the given number of them are initially placed into the box in any possible way so that they don't overlap, usually some ordered arrangement. A sphere is then randomly selected to be displaced by a random amount. If that displacement does not result in an overlap, a new configuration is obtained for averaging purposes, and the process is repeated. If the random displacement results in an overlap, the sphere is placed back in its original position, and that configuration is again counted in the averaging process. In this way an unsuccessful particle displacement leads correctly to the conclusion of the higher probability of the existence of the original configuration, and hence the satisfaction of the principle of microscopic reversibility. The above procedure is exact as long as all possible configurations of the system can be achieved, which in principle depends only on the length of the computer run.

Both the Monte Carlo method, which can only calculate equilibrium properties, and the molecular dynamic method have been applied to identical systems to evaluate the same property, for example, the pressure of a hard sphere system at some density. The results were identical within statistical accuracy. This is a specific instance of the verification of the ergodic hypothesis—a hypothesis which has not been rigorously established to this day. A comparison of the efficiency of the two methods in obtaining the same property to a given accuracy shows that they are about the same. This is somewhat surprising in view of the greater computational complexity of the time integration procedure. In some sense the computer has thus removed the need to use the ergodic hypothesis, although, of course, from the mathematical point of view only by configuration averaging can many of the computer results be followed by corresponding analytical developments.

The only really serious limitation of both of these numerical schemes is that one can only deal with a relatively small number of particles compared to the number in a macroscopic system. A consequence of this is that boundary effects can seriously affect the results. There are, however, a few means available to minimize the distortion caused by the boundaries. One is to choose periodic boundaries since they can be shown to lead to the smallest deviations for finite systems from the results for infinite systems for exactly calculable analytical models. Secondly, a phenomenon can frequently be studied just as well in two dimensions as in three. In two dimensions, for a given number of particles, the surface to area ratio is much more favorable than in three. Thirdly, it is possible to reformulate non-equilibrium problems so as to avoid the further difficulty of imposing external constraints on the boundary, such as a temperature gradient, in the study of thermal conductivity. In some instances the ability of the computer to deal with a finite system allows important physical problems to be solved which would be impossible otherwise. These are situations where natural phenomena involve on the order of 100 particles as, for instance, in nucleation initiation of condensation. These small numbers of particles can only be approximately dealt with by statistical mechanical theory, since that theory is strictly applicable only to infinite systems, and yet that number of particles is too large to be dealt with exactly analytically.

The most astonishing result of the first computer[4] runs was, however, that macroscopic systems could be very well approximated in most situations by the order of 100 particles. The first evidence in favor of this result came from the empirical evidence that the value of the pressure in a hard sphere system, for example, did not sensibly change with the number of particles studied once that number exceeded 100. Further tests were then made by comparison to exactly known results for infinite systems. There thus appears to be a qualitative difference between the behavior of systems comprised of fewer than 10 particles and those of more than about 100. For example, melting cannot be observed in systems of less than 10 particles, although for melting in particular, and for phase transitions in general, the number dependence of the results is unusually large. The observation that few-particle results cannot reproduce certain phenomena is vital to what one might expect out of analytical theories. With few exceptions, analytical theories can only deal in effect with systems of the order of a few particles. This is because it is difficult to handle integrals or equations which are highly dimensional or involve many linked variables. One is hence forced in analytical theories to reduce the dimensionality of the system by approximating the behavior of all but a few particles in the system; for example, by studying the behavior of a few particles in the mean field of the remainder. It is thus not surprising that there exists not even an approximate analytical model derivable from statistical mechanics which explains the melting phenomena.

Now that these two computer methods have essentially solved the mathematical problem, it is possible to overcome other problems. For example, the potential of interaction between molecules needed in the calculations is not too accurately known. One could hence, by comparison to experiment, turn the problem around and ask what potential is needed to reproduce the data and thus get information on that potential. The straightforward and time-consuming procedure of varying the potential and calculating the resulting change in the property can be avoided to some extent by using perturbation theory. One can in this way overcome the limitations of the computer of obtaining a specific solution. Perturbation theory allows one to predict a property for a slightly different potential than the one actually used on the computer by tabulating some other function for the unperturbed system required in the theory. Once a realistic potential has been established, it also becomes very useful to predict properties outside the range of available experiments, for example, the behavior of systems at very high pressures where experiments are very difficult.

The previously discussed choice of the hard sphere potential for simplifying purposes was deliberate. The choice of the potential form is in fact the only aspect of the problem over which one has any control besides the conditions of density and temperature at which investigations are to be carried out. The latter choice is by no means trivial, since on computers, as opposed to experiments, it is possible to go to infinite temperature or pressure. From information on the behavior of systems in these limits, the computer has led to the empirical observation that some previously used models become exact. Alternatively, new exact models could be developed. The only other simplifying procedure available concerns the accuracy with which one specifies the positions of the particles. On the computer, because of the finite number of digits with which the coordinates are specified, one never really deals with continuous spaces. One could hence use coarser coordinate grids up to the point where one deals with lattice spaces in which the number of lattice points becomes comparable to the number of particles. In that instance analytical treatment becomes considerably easier; however, for most applications the results are seriously distorted from the results of continuous or nearly continuous spaces. One needs, for example, of the order of at least ten lattice points per particle to even qualitatively reproduce the melting transition, and that number is too large to deal with analytically, and not really small enough to significantly simplify the computing. Nevertheless, in some special situations lattices can be used very effectively.

The hard sphere potential in a continuous space represents the compromise between a potential which generates most of the essential characteristics of real systems and yet is not trivial from the mathematical viewpoint but possibly analytically tractable. It can, for example, be gainfully employed as the model of an ideal liquid. It is the lack of such a model that has made the description of liquids previously so difficult. One was forced to use the perfect gas or the harmonic solid as an ideal liquid, because their properties could be established. One was then forced to modify these model systems to adopt them ad hoc to the liquid state, or, preferably, rigorously use the procedure of perturbation. This, however, proves very unsatisfactory because the liquid state is too far removed from either of these extreme states; in other words, the perturbation does not converge rapidly. By establishing on the computer the properties of a hard sphere system, and using an attractive potential as a perturbation, it is easy to predict accurate liquid state properties. For example, first order perturbation theory leads to the classical van der Waals theory of liquids which is, in fact, a very good description of the fluid state.

To make the above discussion more concrete, two computer studies for hard spheres will be discussed briefly in order to illustrate both the limitations and power of the computer methods and how they complement both experiment and theory. The first example concerns a thermodynamic property, namely melting.[4] It represents the single most important result obtained in equilibrium on the computer, because by showing that hard spheres melt, so that attractive forces are not necessary for that process, the problem has been reduced to its essential features. Nevertheless, rigorous analytical results are still lacking for reasons explained earlier; however, the computer melting result represents, and continues to represent, a powerful challenge to the theoretician. The result, however, does confirm some ad hoc models of melting which were based on geometric aspects, notably, the old Lindemann law.

On the experimental side the influence of this transition has been more concrete. First of all it settled an old controversy as to whether there would exist a liquid-solid critical point equivalent to the observed liquid–gas critical point. In other words, the open question was whether above certain conditions a solid phase could exist. By observing for hard spheres a solid phase at infinite temperature, the absence of a critical point was assured. Furthermore, it was then possible to realistically predict melting at high temperature and pressure where experimentation is not readily possible. Even more significantly, a mechanism of melting could be proposed which induced experimentalists to set up appropriate equipment in order to confirm its existence in nature. The proposed mechanism was based on the computer observation of the detailed motion of the particles in a solid just previous to melting. One observed a cooperative sliding motion of relatively large numbers of particles in adjacent rows in opposite directions. This observation led to the revival of a previously made proposal that melting corresponds to a long wave length shear mode instability of the solid. It eliminated other models such as one that associated melting with vacancies. It made less useful a model that dealt with melting as an order-disorder process. Although there is long range order in solids and not in liquids in three dimensions, by having observed melting also in hard disks in two dimensional systems where this difference in order between the two phases can be proven not to hold, the order distinction has been shown not to be generally valid.

The computer advantages and disadvantages have also been well illustrated by this melting study. The advantages have already been mentioned; namely the phenomenon was established for this idealized model and the details of the mechanism could be observed directly in terms of atomic motion. On the other hand it was found that the results depend strongly on the number of particles used. For small systems (less than 10 particles) the transition is not observed at all. For larger systems the two phases were not able to coexist, that is, the system was either entirely solid or entirely fluid, and the transition between the two occurred rarely in terms of computer time. Thus long runs were necessary, but even then the averaging between the two states could not be done reliably. This is still the situation today in three dimensions. In two dimensions the coexistence of the solid and fluid phase could finally be achieved by going to systems of the order of a 1000 particles and running them for an exceedingly long time.

The other illustration is from a recent study of a non-equilibrium property, namely the diffusion coefficient.[5] In this study a qualitatively unexpected observation was made upon analysis of the computer results. It was found that spheres diffused faster than one would expect on the basis of a random sequence of collisions. Furthermore, the results had an unusually large dependence on the number of particles such that the diffusion coefficient in two dimensions did not appear to converge to a finite values as larger systems were investigated. A more detailed analysis of the computer results showed that this enhanced diffusion occurred because a particle had a higher probability than random to continue moving in the same direction even after it had undergone many collisions. A search for a possible mechanism for this behavior led to the suggestion that possibly a particle as it moves through the fluid induces a vortex flow pattern. Through this vortex flow pattern the particle, so to speak, would give itself a push from behind and thus continue to move preferably in its original direction. To check upon this possibility the flow pattern around a particle was calculated via the computer and the suggestion confirmed. It should be stressed that some other proposals were made before this successful one and discarded when the computer results did not support it in a critical test of verification of the detailed mechanism.

Since this persistence of velocity appeared to last for long times, and since the original idea for the vortex model came from a hydrodynamic analogy, this hydrodynamic model was studied to see whether it could qualitatively predict the long time behavior of this persistence. Striking confirmation was obtained which gave confidence in using this model to extrapolate the computer results in time beyond the point where calculations were possible. The vortex model then predicts that in two dimensions the diffusion coefficient does not converge, and that in three dimensions, although convergence is obtained, the velocity persistence does not die off exponentially with time for long times, as has been assumed by all previous theories. This recent result will now have to lead to a critical re-examination of the fundamental assumptions made in all transport theory. This result will also lead to a renewed effort to measure this velocity persistence via neutron diffraction with much higher precision than heretofore on some real material. The computer proved indeed to be a very powerful tool in this instance in aiding the development of a new mechanism of diffusion valid at long times. Particularly pleasing was the fact that the possible limitation of following a system for very long times did not prevent reaching the hydrodynamic time regime.

In summary, the impact of computers in statistical mechanics has been enormous, because they are able to stimulate and complement both theory and experiment. The simulation process of real systems can provide better data than experiments themselves can. From the theoretical point of view, the computer allows properties of more realistic model systems to be accurately calculated, thus freeing the investigator from the limitation of dealing with models or invoking approximations only because they reduce the problem to one that could be handled analytically. This restriction to tractable mathematics has seriously hampered statistical mechanics, and through the use of computers it becomes practical to return to the question of what physical processes might be important contributors to various properties.

References

1. B.J. Alder and T.E. Wainwright in *Transport Processes in Statistical Mechanics*, Ed. I. Prigogine (Interscience, New York, pg. 97, 1958).
2. B.J. Alder, S.P. Frankel and V.A. Lewinson, *J. Chem. Phys.* ***23***, 417 (1955).
3. W.W. Wood and F.R. Parker, *J. Chem. Phys.* ***27***, 720 (1957).
4. B.J. Alder and T.E. Wainwright, *J. Chem. Phys.* ***33***, 1439 (1960), and *Phys. Rev.* ***127***, 359 (1962).
5. B.J. Alder and T.E. Wainwright, *J. Phys. Soc. Japan, Supplement* ***26***, 267 (1968), and *Phys. Rev.* A, ***1,*** 18 (1970).

CHAPTER 13

Plasma Physics*

CHARLES K. BIRDSALL and JOHN M. DAWSON

PROLOGUE

We present our chapter in three parts. The first part is an introduction to plasmas, with some reasons for plasma simulation and a brief outline of how we do our many-body simulation. The second part (by J.M.D.) displays the contributions of simulation to *plasma theory* obtained using one-dimensional models. The third part (by C.K.B.) presents some results pertinent to *laboratory experiments* as obtained from two-dimensional models. The results shown are intended to be representative but are not exhaustive.

Plasma physicists use computers in a variety of ways, many in common with other branches of science and engineering. One unique and productive use has been in simulation of the many-body behavior of plasma particles (electrons and ions) which have long range forces and, hence, collective motions. The basic idea in the computation is to solve the Newton equations of motion for each particle, where the force on the particle is determined from a potential function and the particle velocity. The potential function is computed from an equation which involves the positions of the particles. The usefulness of the computer is that it is capable of solving quickly for the forces and of moving the particles, say, up to 10^6 in number, then re-solving for the new forces and again moving the particles.

---

* The work reported herein was supported in part by by the U.S. Atomic Energy Commission.

PART I INTRODUCTION

What characterizes a plasma?

If we consider solids, liquids and gases to be the first, second and third states of matter, then we might call *plasmas* the fourth state of matter. The higher states are reached by adding thermal energy to matter, with the plasma state, for most elements, reached at many thousands of degrees Kelvin. In this temperature range, there is sufficient energy for ionization, that is, enough energy to strip an electron from most atoms, so that a medium of electrons, ions and neutral atoms is formed. Everyday examples are electrical discharges (arcs, fluorescent lamps, neon lights) and the larger scale plasmas, such as the earth's ionosphere, the magnetosphere and, of course, that wonderful thermonuclear reactor, the sun, as well as stars, indeed, more than 99.9 percent of our universe.

Plasma studies have been expanding because of the many applications in electrical and other engineering areas, in geophysics and space science, and very much because of the possible development of controlled thermonuclear power generation, sometimes called the intellectual challenge of this half century.

One of the remarkable attractions of plasma research is that it combines so many of the basic disciplines of physics. At relatively low densities and high temperatures, almost all aspects of classical physics can be brought to bear. Single-particle orbits (of charged particles in electric and magnetic fields) are solved by classical Hamiltonian mechanics. With many particles, plasma motion can, on occassion, resemble hydrodynamic flow and the study of plasma instability may, in some cases, parallel that in hydrodynamics. Plasma, considered as a dielectric medium, allows application and extension of the usual electromagnetic wave propagation theory for anisotropic, inhomogeneous media. The kinetic theory and statistical mechanics of plasmas offer many deep and subtle problems leading to fundamental advances in methods of handling such problems.

The feature that dominates plasma phenomena and has forced advances in the above physics is the *long range of electric and magnetic forces*. The forces of small elements of charges or currents decay as the inverse square of the distance from these elements. As elements are added together at, say, constant density, the number increases as the cube of the diameter of the volume so that very large fields can be produced. With an assembly of charges or currents, the electric and magnetic forces of many particles may add together in a coherent way, making possible *collective modes* of behavior in a plasma; such do not exist in ordinary gases, where molecular forces are very short range. A large variety of these collective motions can be described as *waves* propagating through the plasma. Thus, plasma interaction is *many-body*, as contrasted for example, with two-body collisions which typify gases of neutral molecules.

Thus, we take *plasma* to refer to a collection of charged particles sufficiently dense that their electric and magnetic fields can result in strongly coherent behavior. This definition includes partially or fully ionized gases and essentially electrically neutral collections, yet still allows us to treat electron (or ion) beams with no ions (or electrons) which may have plasma-like behavior, but, in general appear as exceptions.

In a plasma of electrons and ions, reasonably near equilibrium and reasonably large in extent, one finds that the difference in electron and ion density is always very small. The *macroscopic* motion of the plasma is such that the electrons and ions are constrained to move without separating appreciably, bound to each other collectively by their mutually attractive forces. The forces holding the charges together become very large even for small charge separation; the energy required for separation is more than is available in the plasma. *Microscopically*, a positive test charge inserted into a neutral plasma causes an increase in electron density and decrease in ion density near the test charge, tending to shield the rest of the plasma from the test charge. The shielding roughly takes place over a length called the *Debye length*, λ_D, in a volume called the *Debye sphere*. The Debye sphere is roughly the maximum volume over which appreciable deviations from charge neutrality may occur. The energy required to achieve an appreciable deviation over a larger region is prohibitively large, relative to the thermal (kinetic) energy. Roughly, speaking, a test particle interacts *individually* with charges less than λ_D away and *collectively* with those beyond λ_D. A plasma region, by definition, is several λ_D in diameter.

The Debye sphere should be well populated, in order to have a plasma as we have defined it. The number of electrons or ions within a Debye sphere is called N_D and is simply the density, n, times the volume,

$$N_D = \tfrac{4}{3}\pi n \lambda_D^3$$

With $N_D \gg 1$, a number of plasma properties are guaranteed:

a) $\dfrac{\text{mean particle kinetic energy}}{\text{mean interparticle potential energy}} \sim N_D \gg 1$, saying that the dominant feature energetically is the particle motion, that the field energies which arise due to separation of electrons and ions (or currents) will generally be a very small percentage of the kinetic (or total) energy;

b) The average interparticle spacing is much larger than that spacing producing a large angle (90 degree) deflection collision; thus, small angle deflections are the rule and their accumulation dominates over large angle deflections in the scattering process.

c) The Debye length is much less than the mean free path or distance required for an accumulated deflection of 90 degree, "shielding out" or reducing the frequency of such occurrences.

If a plasma is perturbed from equilibrium (say, a small charge separation is artificially set up), then we expect the charge separation (space-charge) forces (fields) to tend to restore neutrality. If the separation is made at long wavelengths (over a distance large compared with λ_D), then the ions and electrons go into simple harmonic motion at the radian *plasma frequency*, $\omega_p \sim \sqrt{n}$. If the separation is made at ever shorter lengths, then the thermal motion of the charges disperses the charge separation and damps out the oscillation *(Landau damping)*. The field or wave energy is lost to the particles. The relation between ω_p and λ_D is

$$\omega_p = v_T/\lambda_D$$

where basic particle speed is $v_T = \sqrt{\varkappa T/m}$; T is the plasma temperature, m is the electron mass, $\varkappa$ is Boltzmann's constant. Hence,

$$\lambda_D = v_T/\omega_p \sim \sqrt{T/n} \quad (\omega_p^2 = ne^2/m\varepsilon_0$$

where e is the electron charge and ε_0 is the permittivity of vacuum, a constant).

We have pretty well located the basic boundaries of what is called plasma. A numerical summary is given in Fig. 1.1, which also locates a number of plasmas of research interest. Note that atmospheric density is nearly at the top (indicating why laboratory plasmas are made in vacuum pumped enclosures) and that room temperature is at the left and that the temperature at the right is about one billion degrees.

The number of particles in a Debye sphere, N_D, is also one of the major parameters occuring in the simulation of plasmas on computers. N_D is the

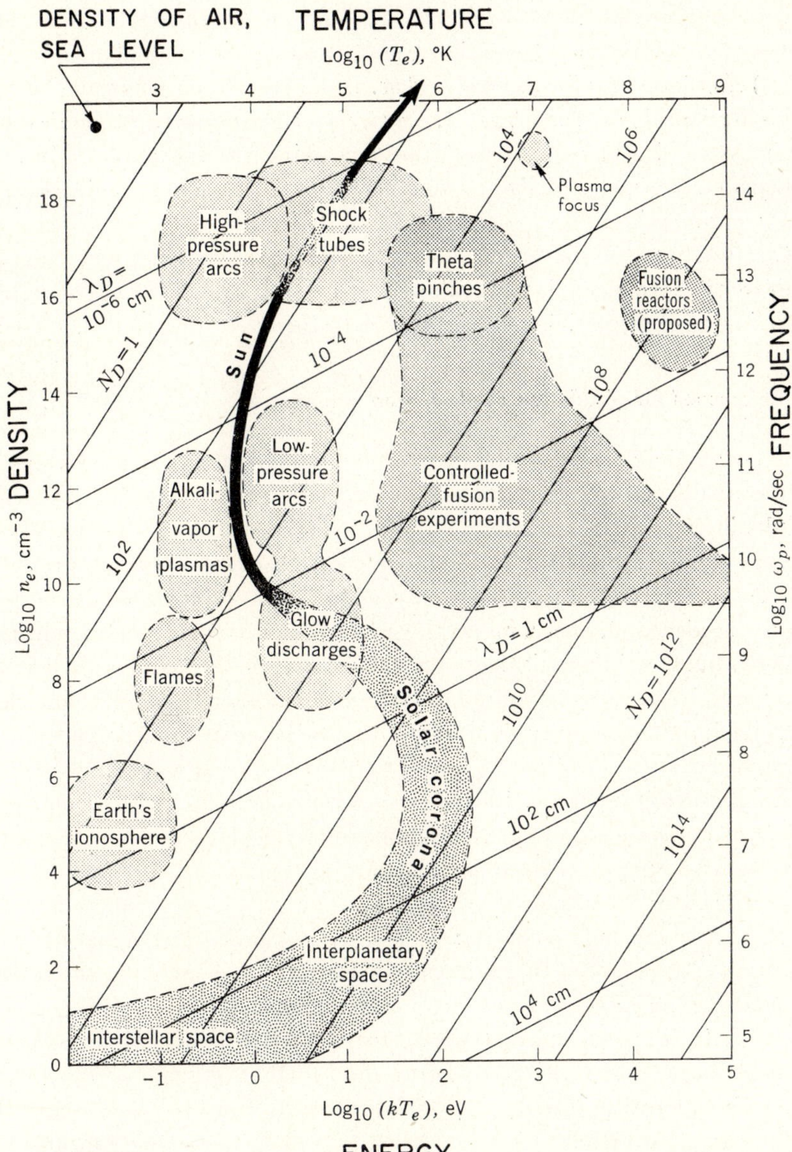

Fig. 1.1 Various approximate plasma domains on a density, n_e, versus temperature, T_e plot. Lines of constant Debye length, λ_D, and of constant number of particles in a Debye sphere, N_D, are also shown (From W. Kunkel, Ed., *Plasma Physics in Theory and Application* McGraw–Hill Book Co. N.Y., 1966)

number of particles per Debye length, $n\lambda_D$, for one-dimensional models, the number of particles in a Debye circle, $\pi n\lambda_D^2$, for two-dimensional models and $(4\pi/3)\, n\lambda_D^3$ for three dimensional models. In nearly all computer modelling, it will be desirable to keep $N_D \gg 1$. *This is the major difficulty*, for interesting plasmas are also many λ_D across, thus, requiring 100's of particles in one dimension, the square of that in two dimensions ($\sim 10^4$) and the cube ($\sim 10^6$) in three dimensions.

Our task here is to show some of the plasma simulation that has been done, with some of the ingenuity that allows it to be done using far fewer than the actual number of particles and with some of the interesting results.

Some reasons for simulation of plasma on a computer

One outstanding reason is to *verify plasma theory* which has grown rapidly in the past decade or so. A theoretician is especially interested in obtaining results in certain areas where laboratory experiments have not been done, or are very difficult, or are on such a scale as to be very expensive or impossible. Of course, on the computer, one can limit the work to just the particular aspect needed and, the computer experimenter may make measurements which would be impracticable in the laboratory. A theoretician needs verification to improve his assumptions and approximations and to enhance his physical intuition in order to move on to more complex models. This is the motivation of the scientists, to add to basic physical knowledge.

Another reason that has drawn researchers to computers is the promise of *simulation of plasma devices* (both small and large) which they would like to analyse or synthesize. Plasma devices include controlled thermonuclear fusion experiments, which are very expensive and present severe difficulties in obtaining data of high precision; further, the number of fusion plasma containment configurations is very large so that trying them all in the laboratory would be extremely expensive. Such plasmas as space plasmas are not easily scaled to laboratory size experiments. This is the attraction to the engineer, to find how to make something work.

A third motivation, is to solve for plasma behavior in models where explicit analysis would be excessively difficult or time-consuming. Linear solutions in uniform, (or nearly so) plasmas, where small perturbations from equilibrium are followed analytically in time and space, are relatively easily obtained. When and where such perturbations grow to large amplitudes, then analysis virtually always fails, even when done by elegant and ingenious methods. In particular, nonlinear solutions in nonuniform plasmas are very hard to obtain analytically. Hence, a distinct area of computer experiments in plasmas has been the *following of instabilities* in time and space, from small beginnings through to some final large amplitude new state. Sometimes the final state represents a new equilibrium, sometimes it represents turbulent motion, and sometimes it represents simply a loss of plasma to the walls. For fusion experiments, as it is instabilities that appear to cause plasma loss which is faster than expected from simple particle collisions, this application of simulation is expected to give considerable insight on how to solve plasma loss problems.

Turbulent motion, where various types of waves couple together through nonlinear effects, can be very complicated. Particles moving through the plasma can resonate with waves and exchange energy with them; with enough particles to form a beam the interaction can excite unstable growth of waves, which may lead to turbulence, to particle acceleration. Noting that hydrodynamic turbulence has been studied intensively for many decades, and is still understood incompletely, the study of *turbulence in plasmas*, which can support a much larger variety of motions, can be expected to be an important field for some time to come.

The computer is a powerful tool for investigating complicated physical systems and a plasma in a magnetic field is one of the most complex.

Choices of method; many-body approximations

The study of media by computer, such as the flow of liquids or gases, is well established. In some regimes, plasmas act like fluids, with the added complication that they respond to electric and magnetic forces. The macroscopic description of the behavior of plasmas in these regimes is close to that for fluid behavior and therefore, research with such plasmas draws heavily on hydrodynamics. This subject is covered elsewhere and will not be given here.

Another approach is to solve a set of partial differential equations which contain a number of assumptions, approximations and "constants" describing the plasma of interest. This approach draws on classical numerical methods and is used in other fields, also described elsewhere.

Our choice has been to study plasmas using many-particle simulation, which traces out the simultaneous motion of many ions and electrons, using a high speed computer with a large memory. There are no approximations in that only the fundamental interparticle force laws and the electromagnetic field equations are used; both sets are exact. The main approximation comes in using, say, 10^4 to 10^6 particles on the computer to represent, say, 10^{12} to 10^{18} in an actual plasma. The credibility of the method hinges on how we explain this enormous gap, of how one particle followed by the computer can ever represent the behavior of, say, 10^{12} actual particles. The secret, if any, lies in the fact that we are primarily interested in the *collective motion* of the plasma, rather than in the extremely large amount of detail associated with the motion of all the particles. We are further helped by the fact that we do not have to simulate the whole plasma but need only look at a small part of it. The collective motion involves many less degrees of freedom than are included in all the particles. We hope that 10^5 degrees of freedom will be enough to describe the plasma adequately. If this is not true, then almost any approach is hopeless.

For particle simulation of plasmas, we will generally work with N_D of a few 10's to a few 100's rather than 10^4 to 10^8 as might be encountered in a real plasma (see Fig. 1.1). This still maintains the *essence* of the plasma, a large ratio of kinetic to potential energy.

A further aid to simulation is the fact that many phenomena which interest us occur in one and two dimensional models as well as in three. It is possible to handle much larger N_D's in the former, with systems which are many Debye lengths across and, indeed, most of the simulation done to date has been in one dimension.

In addition, by finding that the collective behavior at small amplitudes is identical with that predicted by linear analysis and that such is not improved by doubling the number of particles (and hence increasing N_D), we gain further assurance that our approach is valid.

Thus, our approach is somewhat closer to the plasma particle nature than a fluid approximation (where the smoothing is complete), yet somewhat less than the complete story.

How plasma simulation using particle motion is done

We are all accustomed to the problem of a body falling in a constant gravitational force. Dropped from rest, the velocity of the body increases linearly with time and the displacement increases as the square of time. We are simply integrating Newton's law, force = mass times acceleration, once to get the velocity and a second time to get the displacement. All is fine if the fall is short, but, as we are all reminded from the recent Apollo trips, the earth's gravitational force drops off (as $1/r^2$) and is not really constant during a body's fall. Thus, for our calculation we should break up the integration into small steps, using in each step the average gravitational force for that step.

It is the same with plasmas, but with electric and magnetic forces in place of gravity. Each charge or current element in the plasma interacts with all the others as well as with externally applied fields. As the forces cause all particles to change positions and thus to alter the internal forces, it is necessary to calculate these internal forces, then advance the particles for a short time step, (meaning integrate Newton's law for each particle) then re-calculate the forces. The cycle is shown in Fig. 1.2. The accuracy of the calculation is dependent on the shortness of the step in time and on the details of the numerical methods employed. The accuracy can be ensured either with a built-in check (which reduces the time step if accuracy decreases) or by comparison with a re-run at a shorter step. Typical problems will be run for 1000 to 100,000 steps, taking from a few minutes to several hours on a computer, to represent the actual plasma behaviour over a few 10's to 100's of plasma or cyclotron periods (perhaps some microseconds in laboratory time).

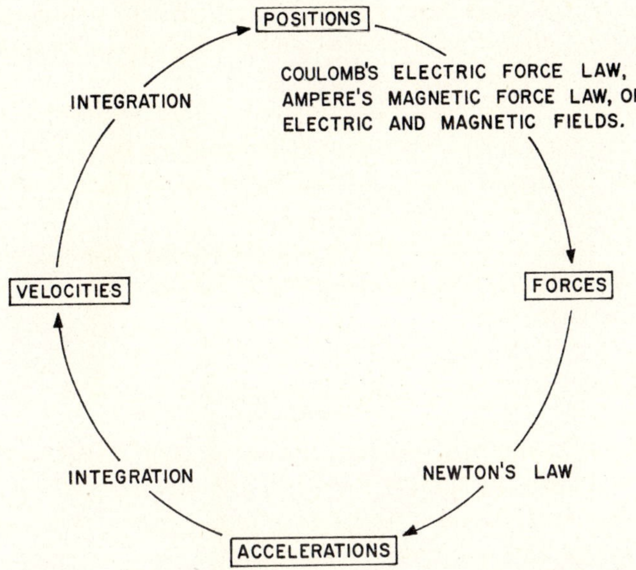

Fig. 1.2 One possible cycle of operations for one time step, Δt

The calculation of *electric force* can be done in several ways. One way is to use Coulomb's law in the form

force on ith charge $= q_i$ [vector sum, taken over all particles, of $\hat{a}_{ij}\,(q_j/r_{ij}^2)$]

q_i and q_j are the charge magnitudes; r_{ij} is the distance from the ith to the jth particle and can be readily calculated as we know the location (coordinates) of each particle; $\hat{a}_{ij}$ is a unit vector pointing from the ith to the jth charge. (The $1/r^2$ law applies in three dimensions, 3D, and changes to $1/r$ in 2D, to no r dependence in 1D; for the present discussion, let us assume 3D.) Figure 1.3 is a sketch of a few particle locations, given by dots; for the ith particle, all others are j and the r_{ij} are computed from

$$r_{ij} = (x_i - x_j)^2 + (y_i - y_j)^2 + (z_i - z_j)^2.$$

This approach might work well with a system all by itself; it has been used to obtain the motion of stars in a star cluster. However, if there are electrical bodies nearby or if the plasma region is a representative part of a larger plasma, calculation of the force from interparticle distances becomes too complicated.

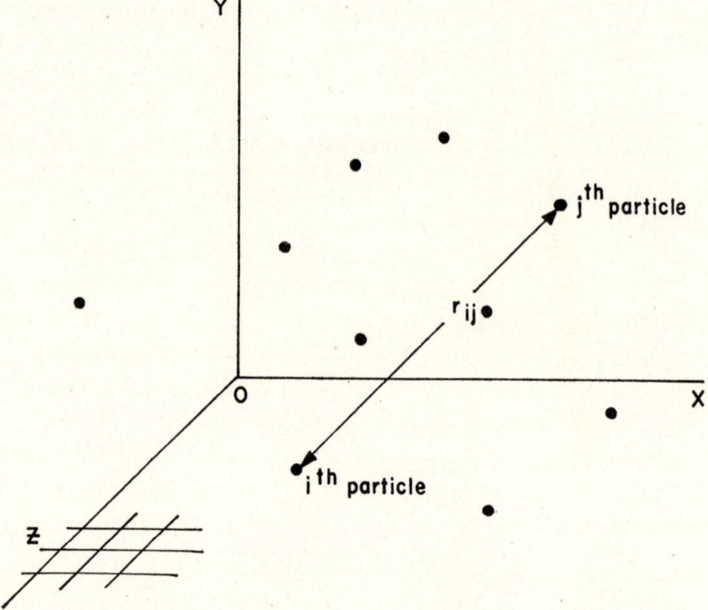

Fig. 1.3 Particle positions; distance r_{ij} from ith to jth particle; portion of a grid (or mesh) in $x - z$ plane

A second method is to replace the vector sum over $(1/r_{ij}^2)$ with the *electric field* which is the force on a unit charge due to all other charges, but is calculated from the *density* of charges, a far easier quantity to obtain than the r_{ij} sum; this step also smoothes over the jaggedness of the force (peaks where particles approach one another, $r_{ij} \to 0$). The charge density and field are calculated, once each time step, with values obtained, not from the exact charges, but from charges placed at *grid points* throughout the plasma. The grid or mesh is a set of lines running parallel to coordinates, spaced finely enough to resolve the detail desired in the problem at hand, but which may well be coarse compared with the average interparticle spacing. (Part of an x–z grid is in Fig. 1.3.) The field and hence force at the particle is then obtained by suitable averaging among the fields at nearby grid points, an approximation. If there are metal or dielectric bodies in or bounding the plasma, the electric fields at these boundaries are usually known and present no major problem. There is no problem with examining a portion of a larger plasma, by assuming that the plasma is periodic. The field approach is the most popular and considerable effort has been put into making field solutions more rapid and accurate.

For typical numbers, in one dimension (all action along x), one might use 100 to 1000 cells with $N = 10{,}000$ particles; in 2D, this might become 32×32 up to 128×128 cells, with about 20,000 to 100,000 particles; in 3D, several workers have set up meshes on the order of $32 \times 32 \times 32$ cells and only a few experiments have been tried, with up to 1,000,000 particles. Note that the progression is from 1000 to 16,384 to 32,768 cells. Present methods convert the charge positions into charge densities associated with each grid point and then solve for the field at each grid point. In the computer memory the field over-writes the density, but there is always the one big field-density array sitting in the memory. This array poses no problem in 1D, a small to medium problem in 2D (depending on memory available) and a large problem in 3D. This memory demand is compounded by requiring the memory to keep each particle position and velocity, only $2N$ quantities in 1D, but $4N$ in 2D (x, y, v_x, v_y) and $6N$ in 3D; of course, as the particles are advanced one by one, this information can be put outside of the main memory without any loss in time, if handled properly.

The computer solves (integrates) the following differential equations:

a) to obtain electric potential, ϕ, from the charge density, ϱ, Poissons equation, done once each time step (say, 1000 times per simulation),

$$\nabla^2\phi = -\frac{\varrho}{\varepsilon_0} \tag{a}$$

$$\left(\nabla^2 \equiv \frac{\partial^2}{\partial x^2} + \frac{\partial^2}{\partial y^2} + \frac{\partial^2}{\partial z^2}, \quad \text{a differential operator}\right)$$

b) to obtain electric field, **E**, from the potential ϕ, done once each time step,

$$\mathbf{E} = -\nabla\phi \tag{b}$$

$$\left(\nabla \equiv \hat{a}_x \frac{\partial}{\partial x} + \hat{a}_y \frac{\partial}{\partial y} + \hat{a}_z \frac{\partial}{\partial z}, \quad \text{a differential operator}\right)$$

c) to obtain particle velocity, **v**, from the force $q\mathbf{E}$, done separately for each particle each time step, (say 10,000,000 times per simulation),

$$m\, d\mathbf{v}/dt = q\mathbf{E} \tag{c}$$

d) to obtain particle displacement, **x**, from the velocity, done separately for each particle each time step,

$$d\mathbf{x}/dt = \mathbf{v} \tag{d}$$

The set (a), (b), (c), (d) is that used to obtain the results of Part II and part of Part III.

The integrations from force to new positions may follow the time stepping shown in Fig. 1.4. The particle acceleration (force/mass) is calculated at time n from the particle positions at time n and velocity at times $n-\frac{1}{2}$, $n+\frac{1}{2}$, which is then used to obtain the velocity at time $n+\frac{1}{2}$; this velocity is used to advance the particle position. The process then repeats. Higher order integration methods requiring v and x at more times (i.e., $n-1$, $n-2$, etc.) are seldom used because they demand larger memories.

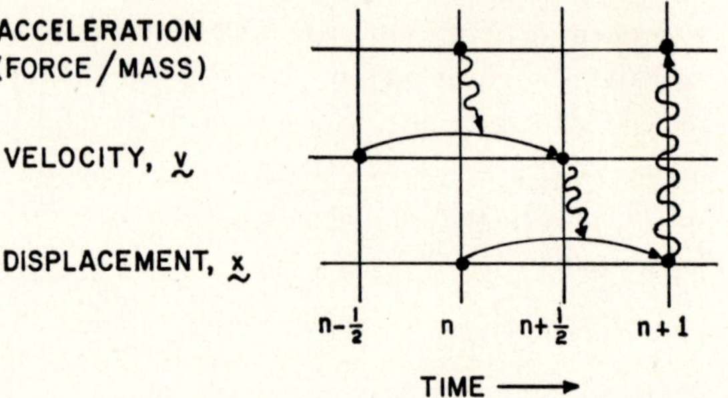

Fig. 1.4 One possible time stepping for integration of equation of motion

If we add a static magnetic field, $\mathbf{B}_0$, then the force changes from $q\mathbf{E}$ to $q(\mathbf{E} + \mathbf{v} \times \mathbf{B}_0)$. If the current densities, $\mathbf{i}$ (or particle fluxes), are large enough to create substantial self magnetic fields, these are calculated from the magnetic vector potential, $\mathbf{A}$, solving a vector form of Poisson's equation,

$$\nabla^2 \mathbf{A} = -\mu_0 \mathbf{i}$$

and $\mathbf{B}$ from $\mathbf{B} = \nabla x \mathbf{A}$. ($\mu_0$ is the permeability of vacuum, a constant).

The above field equations apply to static or slow field variations where the velocity of light, c, may be considered infinite. However, if $\mathbf{E}$ and $\mathbf{B}$ change very rapidly, then the full set of Maxwell's field equations must be used, with c finite. For example, Poisson's equation changes to

$$\left(\nabla^2 - \frac{1}{c^2}\frac{\partial^2}{\partial t^2}\right)\phi = -\frac{\varrho}{\varepsilon_0}$$

Also, if particle velocities become large compared with c, then a relativistic mass change must be allowed and the acceleration changes from

$$m(\partial \mathbf{v}/\partial t) \quad \text{to} \quad \partial(m\mathbf{v})/\partial t$$

All of the formulations given above have been used in plasma simulation, with examples of solutions with applied and self $\mathbf{B}$, electromagnetic waves and relativistic effects given in Part III.

This is a rough view of how it is done; the doing in detail depends on the problem to be solved—and style.

PART II
CONTRIBUTIONS OF COMPUTER SIMULATION TO PLASMA THEORY

This section will be devoted almost exclusively to the work at Princeton on one-dimensional electrostatic plasma models and their applications to the verification of plasma theory. While there are quite a number of other calculations which relate to plasma theory, these cover a wide range of methods and it is difficult to give an adequate discussion of them in a short survey. References to a number of these other efforts will be given where appropriate.

The charge sheet model

The charge sheet model is the one which has been employed for a large number of one-dimensional plasma investigations. There are two such models.

The *first model* consists of negatively charged sheets imbedded in a fixed uniform neutralizing background. (Dawson, 1962a.) The sheets are all parallel and move only in their common normal direction (see Fig. 2.1). The sheets are allowed to pass freely through each other and through the background. Their motion is computed self-consistently from the electric fields.

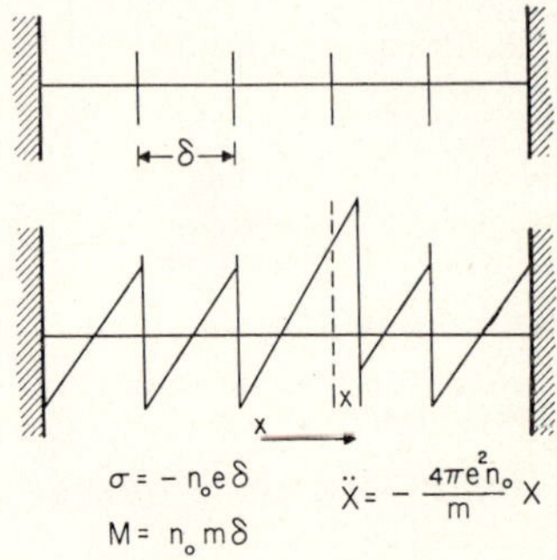

Fig. 2.1 One-dimensional plasma model

This is a reasonable model for motions involving electrons. Due to the light mass of the electrons compared to that of the ions, the ions can be considered infinitely massive and immobile by comparison. Further, for a plasma, the kinetic energy of a particle is much larger than its potential energy due to interactions with other individual particles. It is only at the elevated temperatures where this is so that the electrons can be broken away from the nuclei of the atoms so as to form a plasma. As a consequence, the particles move more or less freely responding primarily to the gross fields of many particles. Thus, as the sheets are to represent the motions of electrons, they are allowed to pass freely through each other. Encounters between particles do produce small deflections which accumulate over long distances so as to randomize a particle motion. The distance required for such a deflection (the mean free path) is generally very large and for many situations these deflections can be ignored.

The *second model* consists of both positive and negative charged sheets. The charges on the two species are equal in magnitude, but opposite in sign

(Smith and Dawson, 1963). The mass of the ion sheets is generally large compared to that of the electron sheets. Generally, mass ratios (m_i/m_e) in the range of 10/1 to 100/1 are used. Realistic mass ratios of 2000/1 or more are rarely used because then the ions move so slowly that the computer spends most of its effort on what the electrons are doing. In this second model, as in the first, the charge sheets are allowed to move freely through each other and move only in the common normal direction.

As both models consist of discrete particles they exhibit rapidly fluctuating fields associated with the individual particles; because of these fields, the models contain collisional type effects. These collisions are, of course, quite different from those between electrons or electrons and ions in a real plasma. Nevertheless, the theoretical methods used to study collisions in real plasmas can be applied to these models. Thus, we can obtain quite a detailed check of these theoretical methods and the approximations that go into them. An understanding of these effects is essential to using the models to check many theoretical predictions.

Checking the theory of a plasma in thermal equilibrium

One of the first uses of the sheet model was to check the statistical mechanics of thermal equilibrium and fluctuations about thermal equilibrium. Statistical mechanics for equilibrium systems is a highly developed theory, which has had a great many successes in predicting the thermodynamic properties of systems of neutral particles. The theory can also be applied to systems of charged particles (Landau and Lifshitz, 1958) and, in particular, to the charged sheet model (Dawson 1962, Smith and Dawson 1963, Lenard 1961, 1962, 1963). Because of the long range nature of the forces between particles, the theory for plasmas is somewhat more subtle than that for gases of neutral particles. Nevertheless, the theory has been developed to a high degree. A comparison of this theory with results from the numerical model primarily provides a check on the model and adds to our confidence in using the model for other problems. However, the results also provide us with a more detailed check on the theory than can be obtained by any other means.

Thermal equilibrium and fluctuations about equilibrium

A number of equilibrium properties have been measured on the single species sheet model (Dawson, 1962). The velocity distribution, the Debye shielding of a particle, the mean electric field energy in a Fourier mode, and the distribution of electric fields, as seen by the sheets, were all measured. Results of such measurements are shown in Figs. 2.2–2.5. Other calculations can be found in the work of Eldridge and Feix (1962, 1963), Feix (1967) and Barnes and Dunn (1967).

Distribution functions

Figure 2.2 shows the time averaged velocity distribution for a system of 1000 sheets, and 5.16 sheets per Debye length ($nv_T/\omega_p = 5.16$, v_T is the thermal velocity, n is the density of sheets per unit length of the system, and ω_p is the plasma frequency). The system was started out with the sheets equally spaced and with the velocity of each sheet chosen randomly so that the over-all distribution (number of sheets in a velocity interval Δv about v) was Gaussian [$f(v) \propto \exp(-v^2/2v_T^2)$]. Theory predicts the Gaussian form shown by the solid curve. Since the system was started out with essentially this distribution it is not surprising that the system maintains this equilibrium. However, this time averaged distribution is much smoother than the

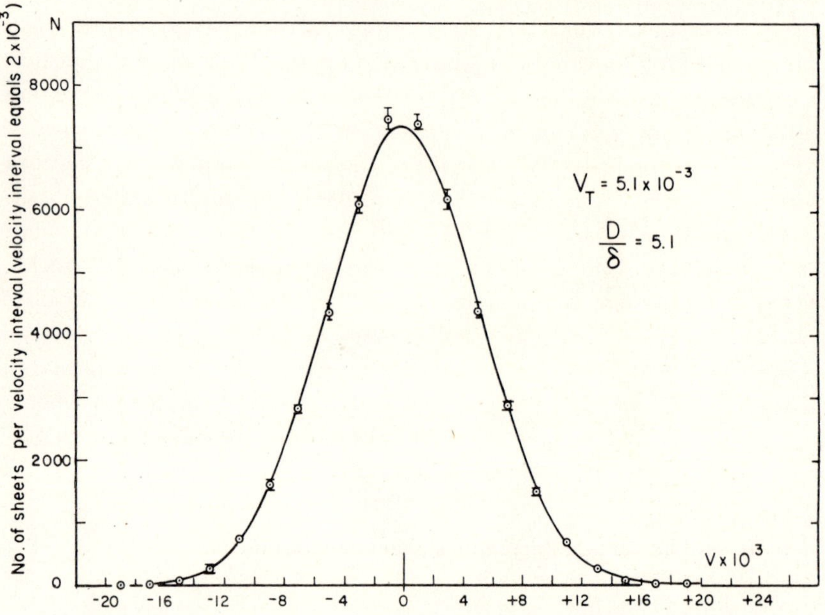

Fig. 2.2 Average velocity distribution function for a 1000-sheet system

initial distribution because the averaging smooths out the fluctuations in the number of particles with a given velocity due to the finite number of particles used. It has also been shown that non-equilibrium distributions relax to a Gaussian. This will be discussed in more detail later.

Debye shielding

A charge embedded in a plasma tends to repel charges of like sign and attracts charges of the opposite sign. Thus, around such a charge, a cloud of charge of the opposite sign forms. This cloud contains on the average an equal, but opposite sign of charge to the embedded one. The dimensions of this charge cloud are the Debye length, $\lambda_D \equiv v_T/\omega_p$. (This idea was introduced by P. Debye (1923) in his theory of electrolytes.) Outside of this cloud, the system of the particle plus cloud looks neutral. This phenomenon is called Debye shielding. The shielding of such an embedded charge was investigated on the single-species charge sheet model. The results are shown in Fig. 2.3. This figure shows the average charge density surrounding the embedded charges. The points are averages obtained from a large number of samples of embedded sheets. One must take such averages to see the shielding because it amounts to only a small bias on the otherwise uniform random distribution of charges in the neighborhood of the test sheet. The shielding amounts to one sheet on the average being absent from a region the size of a Debye

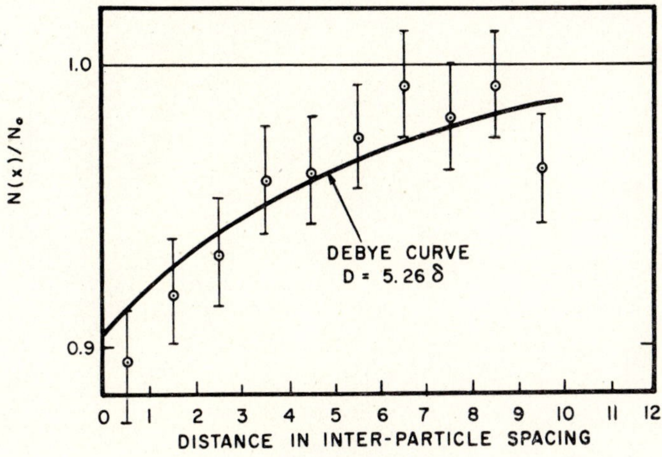

Fig. 2.3 Debye cloud

cloud which contains many sheets (in this case, $10 = 2n\lambda_D$). This small bias can be masked by the fluctuating density in the neighborhood of the test sheet due to random motion of the sheets. Sometimes one will find perhaps 12 sheets in a Debye length, other times maybe 8 or 9. This solid curve in Fig. 2.3 is the theoretically predicted curve; the points are those obtained from the numerical experiment. The error bars are the statistical uncertainties due to the fact that we have used a finite number of test sheets.

Electric field energy

Figure 2.4 shows the electric field energy (mean square electric field) for the k'th Fourier mode [$E(k) = \text{S}\, E(x) \sin kx\, dx, k = 2\pi/\lambda$, λ is the wavelength]. Equilibrium theory predicts that this quantity should be given by

$$\langle E^2(k)\rangle /8\pi = (\tfrac{1}{2}\varkappa T)/(1 + k^2\lambda_D^2)$$

For large wavelengths, long compared to the Debye length, the electric field energy is $\varkappa T/2$ while for short wavelengths (large k) it is much less. The

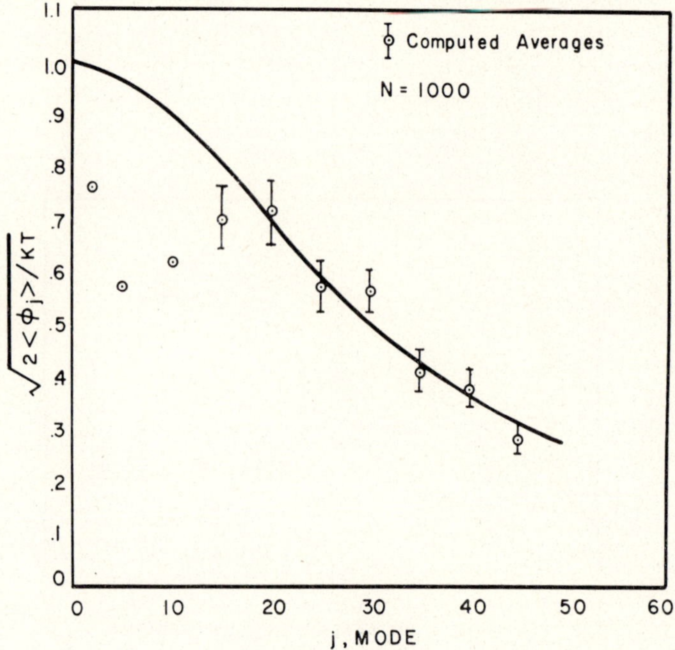

Fig. 2.4 The root mean square amplitude *vs* Fourier-mode number

electric field plays the dominant role in determining the magnitude of the charge density fluctuations for long wavelengths, while at short wavelengths these fluctuations are determined mainly by the random motion of the particles. The solid curve is that predicted theoretically; the points are those obtained from the numerical experiment. The error bars are again the expected statistical uncertainties due to the finite size of the sample. The agreement is within the expected uncertainty except for long wavelengths. The cause of the deviation there is due to the fact that these modes were not started out at their expected equilibrium level and they take a long time to come to equilibrium as will be discussed shortly.

Electric field felt by a particle

Figure 2.5 shows the distribution function for the electric field as seen by the sheets (number of sheets which feel an electric field in a small range about E). The solid curve is that predicted by theory while the points are those obtained from a numerical experiment on a system of 1000 sheets with 10 particles per Debye length.

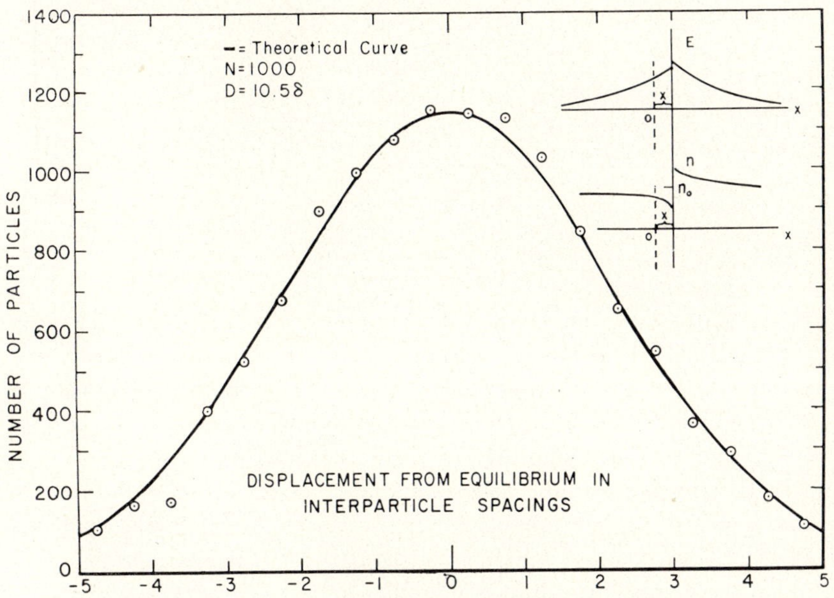

Fig. 2.5 Distribution of displacements from equilibrium position

Drag on a particle

Figures 2.6 and 2.7 show the drag on test sheets which are moving fast and slow compared to the thermal velocities. The points are obtained as follows. Particles are selected with velocities lying close to the desired velocity. Their velocities are then recorded at a series of later times (τ, 2τ, 3τ, ... etc. later) and the average velocity for the group is found.

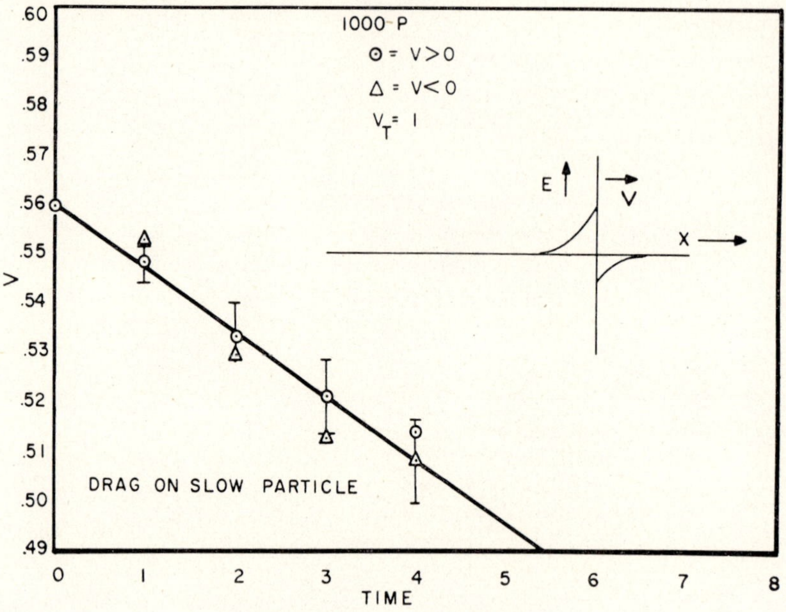

Fig. 2.6 The drag on a slow sheet

A fast particle moving through the plasma excites a plasma oscillation ($kv \approx \omega_p$) and, by this process, is slowed down. It is also accelerated by the random electric fields produced by all the other particles. By averaging over a large number of particles the random accelerations average to zero, but the systematic effects due to the excitation of the wave, remain.

For the slow particles, the drag is not due to the excitation of a wave, but is rather due to the reflection of particles of nearly the same velocity by the repulsive electric field surrounding the test particle. If the particle is moving through the plasma, it overtakes more particles than overtake it and, as a result, it is slowed down. It is also randomly accelerated by the random fields

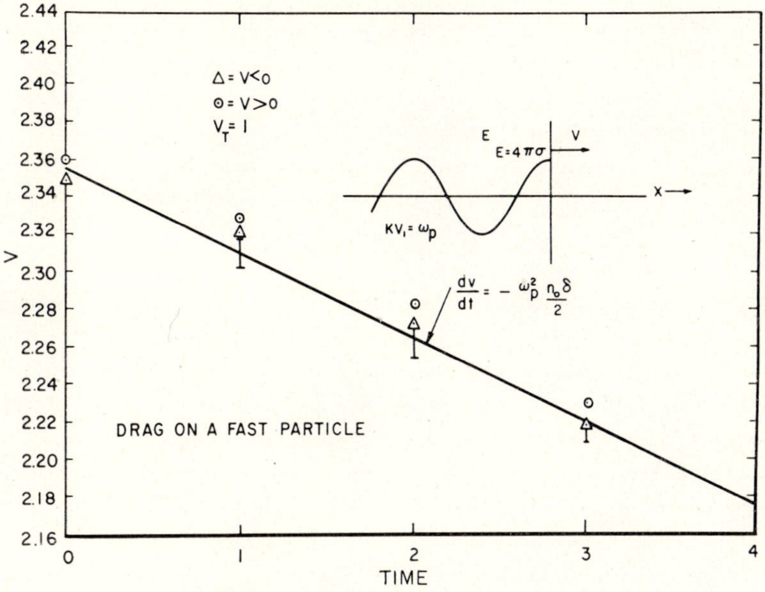

Fig. 2.7 The drag on a fast sheet

associated with the other particles. These random effects again are removed by averaging over many particles.

The solid curves in Figs. 2.6 and 2.7 are the theoretically predicted results. The agreement is again within the expected statistical uncertainty.

Thermalization of a sheet plasma

As we have just seen, the average motion of a sheet moving through a plasma is reduced due to the drag it feels. At the same time it is accelerated by the random fields present in the plasma. The velocity acquired from these random accelerations gives rise to a diffusion or spreading of the velocities of a group of particles. Similar processes take place in real (3 dimensional) plasmas. In 2 and 3 dimensional plasmas, the slowing down and random acceleration compete against each other. Ultimately, they produce a stable distribution, the Maxwellian distribution. For this distribution, and only for this distribution does the diffusion or spreading to higher velocities exactly balance the slowing down of the particles.

One may apply the kinetic theory of plasmas to the one-dimensional sheet plasma. The (present) theory predicts that, for all stable plasmas, the diffusion in velocities will balance the drag and hence there will be no evolution to a Maxwellian. One may give a simple physical argument why this should be so. Two particles colliding in one dimension, will have velocities v_1 and v_2 before the encounter and $\tilde{v}_1$ and $\tilde{v}_2$ after the encounter. Two quantities are conserved for an isolated encounter, the energy ($m\,[v_1^2 + v_2^2)$] and the momentum ($m\,[v_1 + v_2]$). There are only two choices for $\tilde{v}_1$ and $\tilde{v}_2$ which will conserve these quantities; first, no change in velocities, $\tilde{v}_1$ equal v_1 and $\tilde{v}_2$ equal v_2), or, second, an interchange of velocities, $\tilde{v}_1$ equal v_2 and $\tilde{v}_2$ equal v_1. In either case the number of particles with a given velocity is not changed. One might expect that in a plasma this simple two-isolated-particle-collision argument might not apply since many particles are interacting with each other all at the same time due to the long range of the forces. However the (present) theory assumes that all interactions are weak which means that even though there are many simultaneous collisions, they do not interfere with each other so that their effects are simply additive. Thus, the theory predicts no change in the distribution function.

In reality, the simultaneous encounters will affect each other and there will be some change in the distribution function. The change was checked on the sheet model and the actual rate of relaxation to Maxwellian was determined

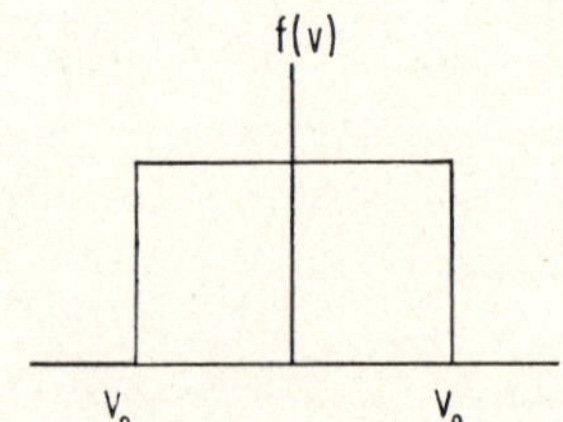

Fig. 2.8 Initial velocity distribution

(Dawson, 1964). The problem which was investigated was that of the time development of a velocity distribution which initially had the square profile shown in Fig. 2.8. The initial velocity of a particle was obtained by computing a random number which had a uniform probability of lying anywhere between -1 and 1 and then multiplying this number by v_0. A number of different v_0's were used to determine the dependence of the time development on the kinetic energy or number of particles per Debye length ($\lambda_D = \langle v^2 \rangle^{1/2}/\omega_p$, $\langle v^2 \rangle = \frac{1}{3} v_0^2$ is the mean square velocity). The initial positions were chosen to be equally spaced. During the first plasma oscillation, Debye shielding clouds develop around each of the particles. The formation of these clouds requires some energy and as a result there is a short period of rapid adjustment that takes place which rounds off the corners of the distribution. After this initial adjustment the distribution evolves very slowly.

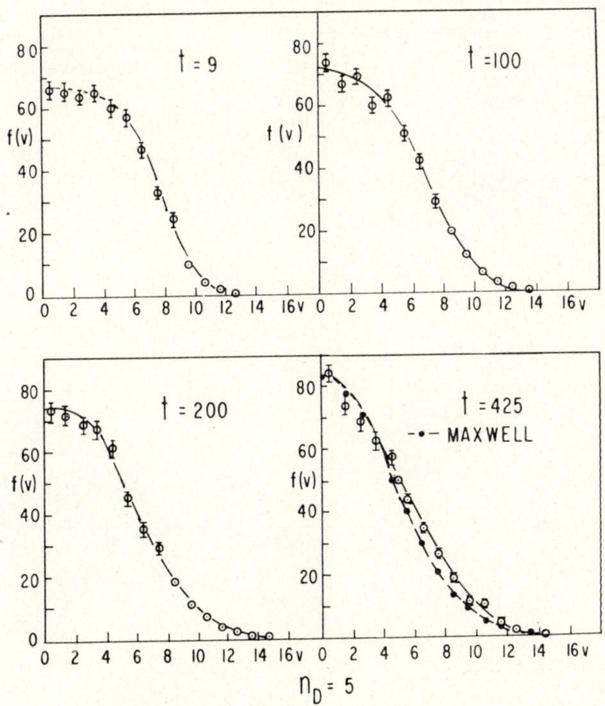

Fig. 2.9 Relaxation of the velocity distribution for $n_D = 5$ to a Maxwellian

Figure 2.9 shows the time development of the velocity distribution for the case of 5 particles per Debye length. The first figure shows the situation just after the transient; the other figures show how the distribution evolves toward a Maxwellian. It is essentially reached by $\omega_p t$ equal 425. This time is much longer than the time required for a group of singled out particles to acquire the background distribution; this latter time is only $10\omega_p^{-1}$.

Figure 2.10 shows a plot of the relaxation time *vs* $(n\lambda_D)^2$ as determined from such calculations. We see that the relaxation time is proportional to $(n\lambda_D)^2$. This indicates that the simultaneous interaction of three particles gives rise to the relaxation since the relaxation time due to two-particle interactions would be proportional to $n\lambda_D$, if it did not cancel out. At the present time there is no theoretical calculation which predicts this relaxation, although, it is possible to estimate a relaxation time which is of the right order of magnitude due to emission and absorption of waves by particle encounters.

One interesting point which was found was that the distribution function undergoes rapid random fluctuations about a mean distribution which gradu-

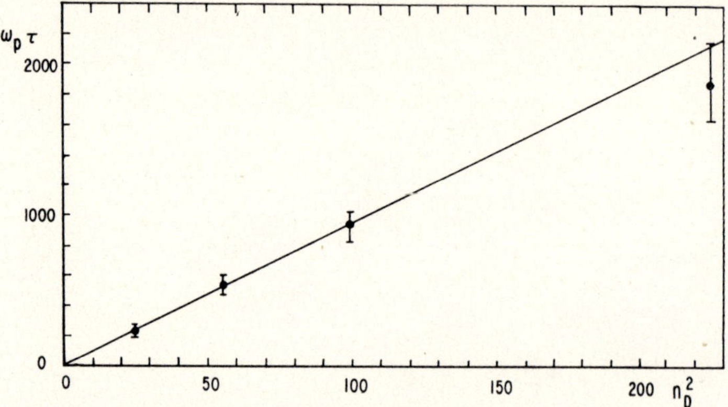

Fig. 2.10 Plot of relaxation time *vs* n_D^2

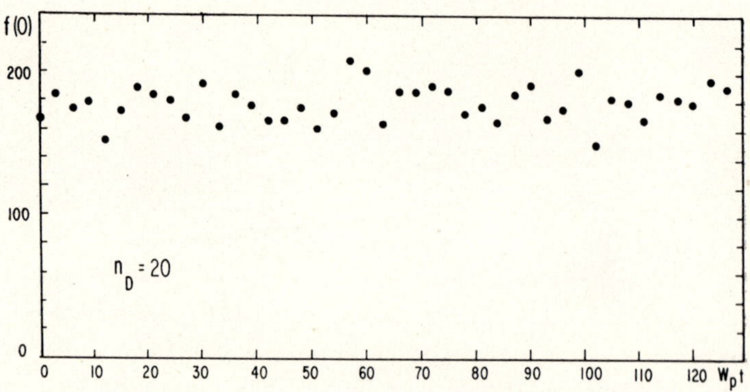

Fig. 2.11 Plot of $f(0)$ *vs* time for $n_D = 20$

ally drifts toward a Maxwellian. The fluctuations in the number of particles with velocities in a small range about zero are shown in Fig. 2.11. The rapid fluctuations are clearly visible. These fluctuations result from the constant exchange of energy between the electric field and the particle kinetic energy. Although this exchange is constantly going on, it is such that it produces very little systematic change. This shows that the fact that the distribution relaxes slowly results from a very subtle balance and thus the calculation provides an important test of the kinetic theory of plasmas.

Longitudinal bremsstrahlung

When an electron encounters an ion it is accelerated and the accelerated electron emits electromagnetic radiation. When an electron encounters another electron, the two electrons suffer equal and opposite accelerations and the radiation fields to a first approximation cancel each other. However, the

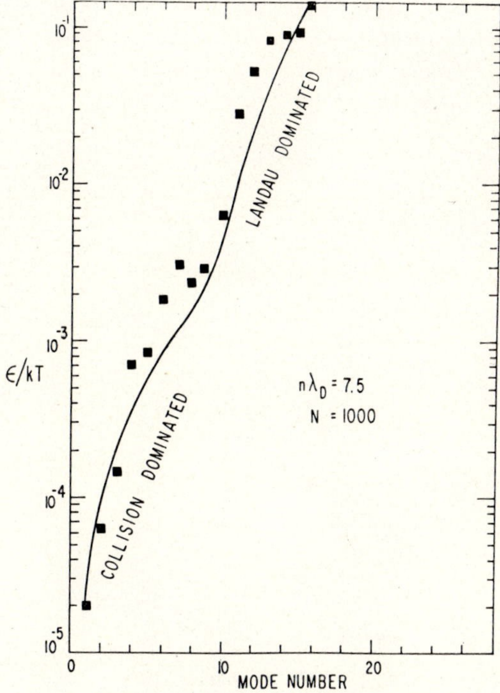

Fig. 2.12 ε *vs* mode number

cancellation is not complete; the amount of radiation is reduced by the factor v^2/c^2, where v is the relative velocity of the particles and c is the speed of light. (For electron–ion encounters the radiation is dipole while for electron–electron encounters, it is quadrupole radiation.)

When two particles encounter each other in a plasma, they can emit not only electromagnetic waves, but they can also emit longitudinal plasma oscillations. A theory for this emission has been developed (Birmingham *et al.*, 1965, 1966) which involves the acceleration of two particles and their shielding clouds when they encounter each other. This theory is esssentially identical to the theory used to describe weak turbulence in a plasma. Thus, a check of this theory of wave emission also provides a check of weak turbulence theory, at least, for the low levels of waves which are encountered here.

The theory for the emission of longitudinal waves was checked on the one species sheet model (Dawson *et al.*, 1969). The emission encountered here is analogous to that from electron–electron collisions (like particle encounters). Figure 2.12 shows a plot of the emission vs wave number (inverse of the wave length). The curve is that predicted by theory. The points are those obtained from a numerical experiment on a system of 1000 particles with 7.5 particles per Debye length. The agreement is quite good, the emission varying over three orders of magnitude.

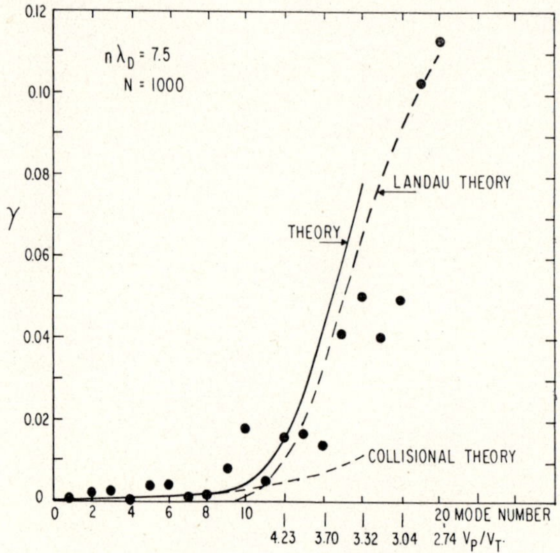

Fig. 2.13 γ *vs* mode number

Closely related to the emission of waves due to particle encounters is the absorption of waves due to particle encounters. The collisional absorption of waves may be found from the emission from the fact that the emission and absorption must lead to thermal equilibrium where each mode of oscillation has an energy $\varkappa T$. Figure 2.13 shows a plot of the damping time for waves against wave number (number of wavelengths which fit in the system). The system contained 1000 sheets and had 7.5 particles per Debye length. The solid curve is the theoretical one and is the sum of the two dashed curves marked collisional damping and Landau damping. The collisional damping curve is that predicted by the collisional theory just discussed. The damping which is shown in the curve labeled Landau damping does not have its origin in collisions, but rather arrives from the absorption of energy by particles moving at the phase velocity of the wave.

Damping of a large amplitude wave

The damping of a large amplitude wave was investigated on a sheet model containing 1000 sheets with 20 sheets per Debye length (Dawson and Shanny, 1968). Figure 2.14 shows the result. The solid curve is that found in the numerical experiment; the dashed curve is that predicted from small amplitude theory. As can be seen the damping is stronger than that predicted by small amplitude theory. This result is easily understood. According to the small amplitude theory, particles moving slower than the wave are on the average accelerated while those moving faster than the wave are on the average decelerated. Thus, particles moving slower than the wave pick up energy from it while those moving faster than it give up energy to the wave.

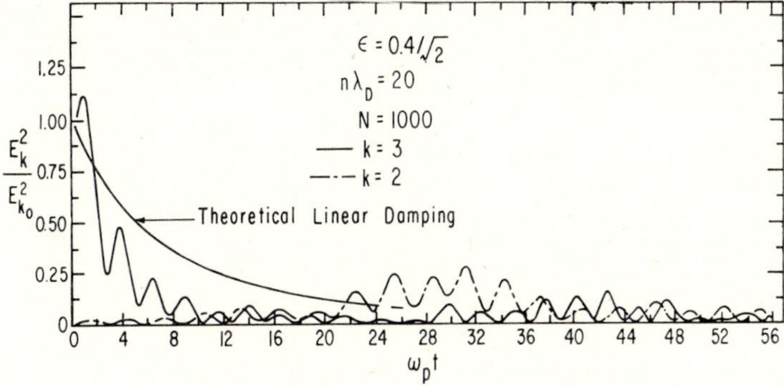

Fig. 2.14 Decay of nonlinear travelling wave

If there are more particles moving slower than the wave than moving faster than it, the wave is damped. For small amplitudes, only the particles moving at nearly the wave velocity interact with it strongly and thus, the damping of the wave is proportional to the slope of the distribution function at the phase velocity as illustrated in Fig. 2.15. When the amplitude of the wave becomes large, the wave can accelerate (or decelerate) particles by a large amount and hence, it can significantly affect particles with much different velocity. Since the number of particles which are moving slower than the wave, and which are strongly accelerated by it, rise rapidly as the amplitude increases (see Fig. 2.15) the damping also rises rapidly. This increase has also been found in other numerical experiments (Armstrong, 1967). The magnitude of this effect can be estimated from the number of particles that are picked up by the wave and the energy they gain and quantitative agreement is found.

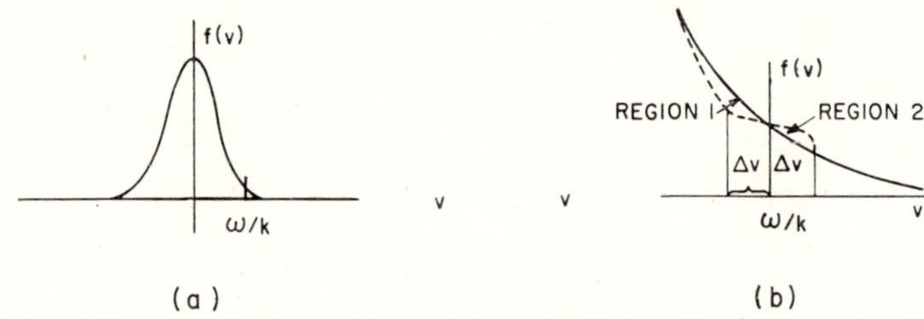

Fig. 2.15 Velocity distribution. Expanded view

Wave–wave coupling

Suppose there are two large amplitude waves propagating in the plasma with frequencies f_1 and f_2 and wavelengths λ_1 and λ_2. The two waves will both have very nearly the same frequency, the plasma frequency. If the wavelengths are large, then the phase velocity of the waves (λf equals the phase velocity) will be large compared with the random thermal velocityof the particles. Thus, there will be no particles moving at the phase velocity of the wave and there will be no damping of the waves by such particles as has already been discussed. Because the waves are large amplitude they will interact nonlinearly. Such nonlinear interaction produces disturbances at the sum frequency $\tilde{f}_1$, ($\tilde{f}_1$ equals $f_1 + f_2$) and wavelength $\tilde{\lambda}_1$ ($1/\tilde{\lambda}_1 = 1/\lambda_1 + 1/\lambda_2$) and at the difference frequency $\tilde{f}_2$, ($\tilde{f}_2$ equals $f_1 - f_2$) and wavelength $\tilde{\lambda}_2$ ($1/\tilde{\lambda}_2 = 1/\lambda_1 - 1/\lambda_2$). Of particular interest is the latter case because its phase velocity ($\tilde{f}_2\tilde{\lambda}_2$) is low since the two frequencies are very nearly equal. Because of this, the phase velocity of this disturbance is comparable to the thermal velocity of the particles and the disturbance can thus interact strongly with those particles moving at its phase velocity. Such particles tend to absorb some energy from the waves in the same way they absorb energy from a single wave. However, the primary effect is an energy transfer from the wave of shorter wavelength to the wave of longer wavelength. The process is a kind of scattering where the energy from the shorter wavelength mode is preferentially scattered from the particles into the longer wavelength wave.

The above conclusion was tested on a one-dimensional sheet model containing 400 sheets with 60 sheets per Debye length (Dawson and Shanny 1968). Two modes were initially strongly excited. Special precaution was taken to see that no particles achieved the phase velocity of either wave. The results are shown in Fig. 2.16. There is an initial transient in which the

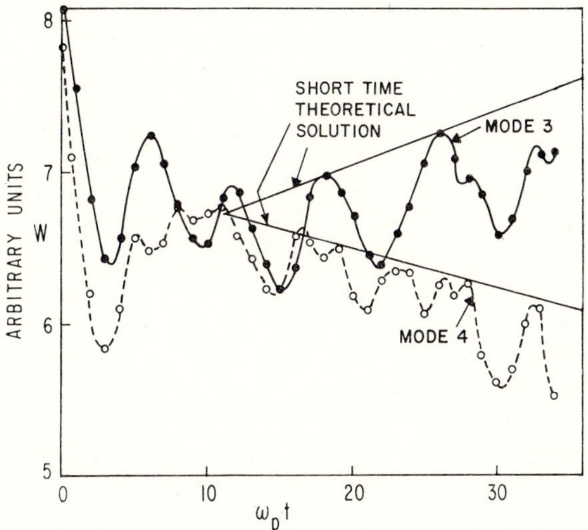

Fig. 2.16 Coupling of modes 3 and 4 by nonlinear Landau damping

energy of both waves decreases. After this the energy of the longer wave length grows while that of the shorter wavelength wave decays in quite good agreement with the theoretical prediction as shown by the straight line. The oscillations seen there are due to beating of the waves against low frequency disturbances associated with the random motions of the particles.

The double-stream instability

A number of numerical experiments have been carried out on the one-species, one-dimensional model on the two-stream instability. The following are brief descriptions of these results.

The first situation which was investigated was the two-beam instability for two cold streams of electrons (Dawson, 1962b). Cold means that the initial velocity of all the electrons in one stream was the same. Initially, the sheets were distributed uniformly in x and were given a velocity of either $+v$ or $-v$, the sign of the velocity being chosen randomly. This results in two streams with equal but opposite velocities and equal mean densities. However, there are local density fluctuations in the density of each stream because there will be regions where more particles of one velocity or the other will be found.

This situation gives rise to the well known two-stream instability. It is somewhat similar to waves produced by wind over water; the winds of the two-electron streams are of course interpenetrating. A number of different wavelengths are unstable and because of the density fluctuations, all unstable modes grow. During the early stages the linearized or small amplitude theory predicts the behavior. Figure 2.17 shows a comparison of the theoretically

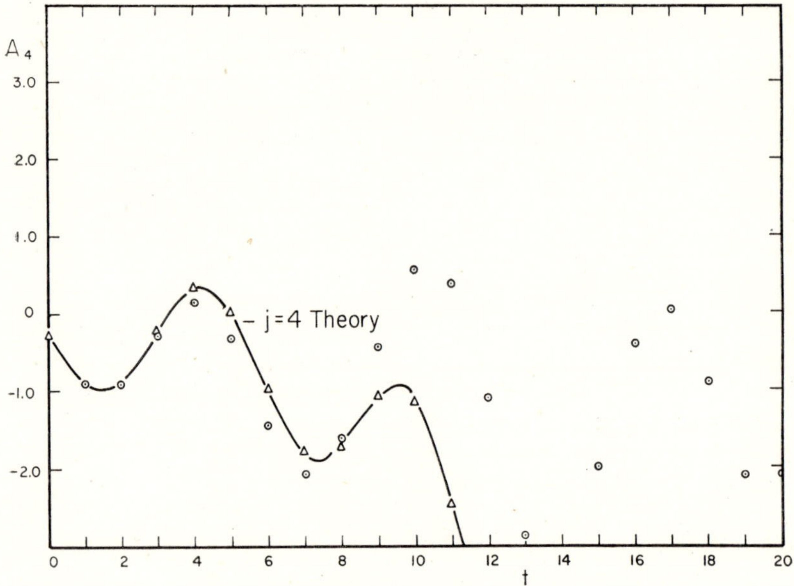

Fig. 2.17 **Amplitude** of mode 4 *vs* **time** for run 1

predicted time development of mode 4 with the results from the numerical experiment. The results are in good agreement to time $\omega_p t = 8$. One may estimate when the small amplitude theory will break down by computing when the perturbed density it predicts becomes equal to the unperturbed density. One finds from such a calculation that the breakdown should come around $\omega_p t = 8$ or 9. It is found that this behavior is typical of the unstable modes, and that the agreement with the theory generally fails at about this time.

Figure 2.18 shows a plot of the per cent of the energy that is stored in the electric field. It has a strong peak at $\omega_p t = 10$. The maximum electric field energy is about 25% of the total energy. After the peak there is a rapid drop in the electric field energy to about 10% of the total energy where it more or less stays for the remainder of the calculation.

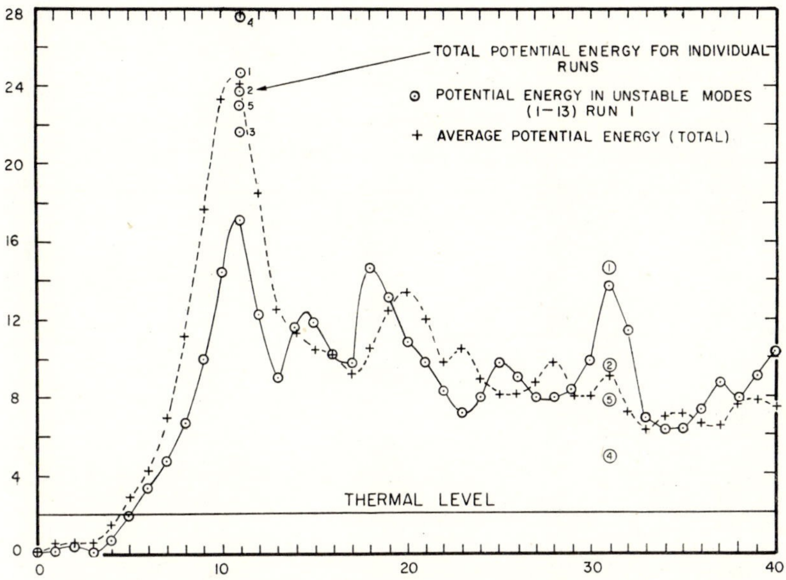

Fig. 2.18 The potential energy *vs* time averaged over all runs and the potential energy of the unstable modes *vs* time for run 1. Potential energy equals $\int E^2/8\pi \, dx$

Roberts and Berk (1967) have carried out numerical experiments on the two-stream instability using a somewhat different one-dimensional model. They showed that if one makes plots of the distribution function in phase

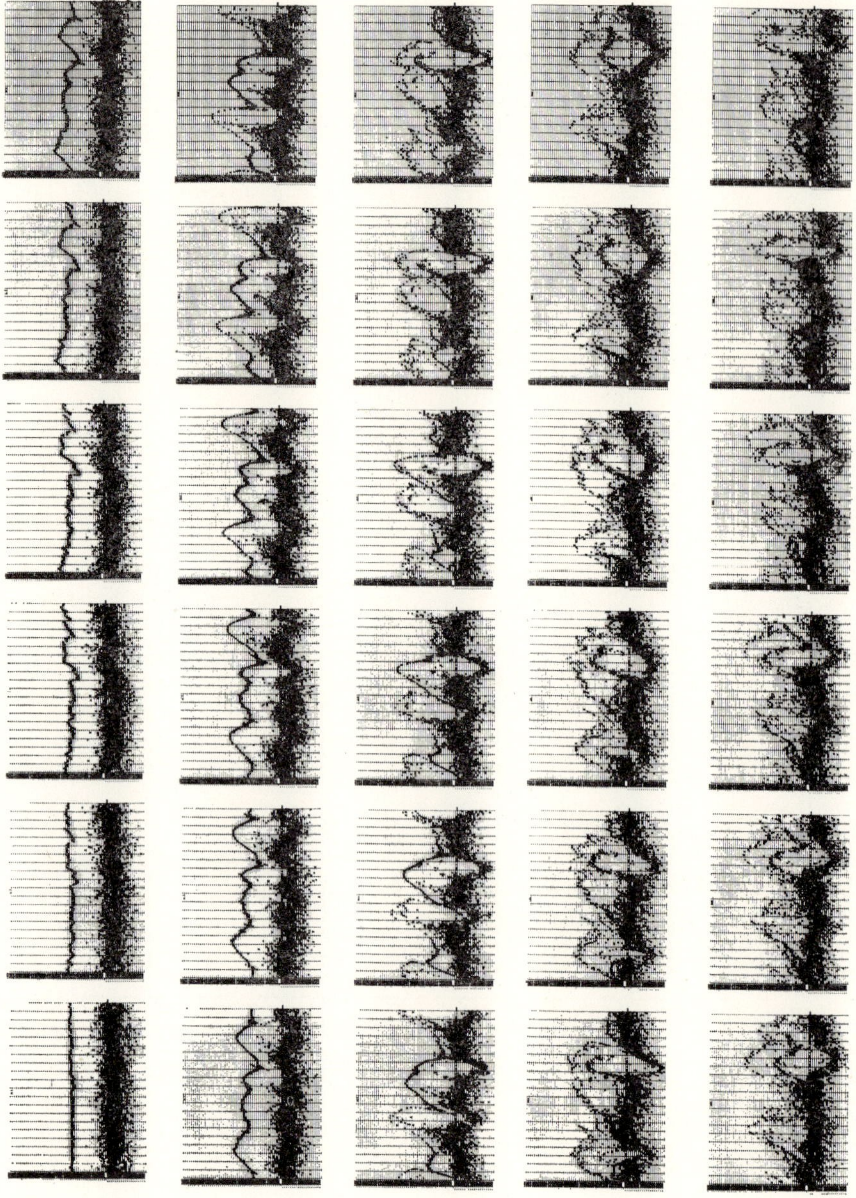

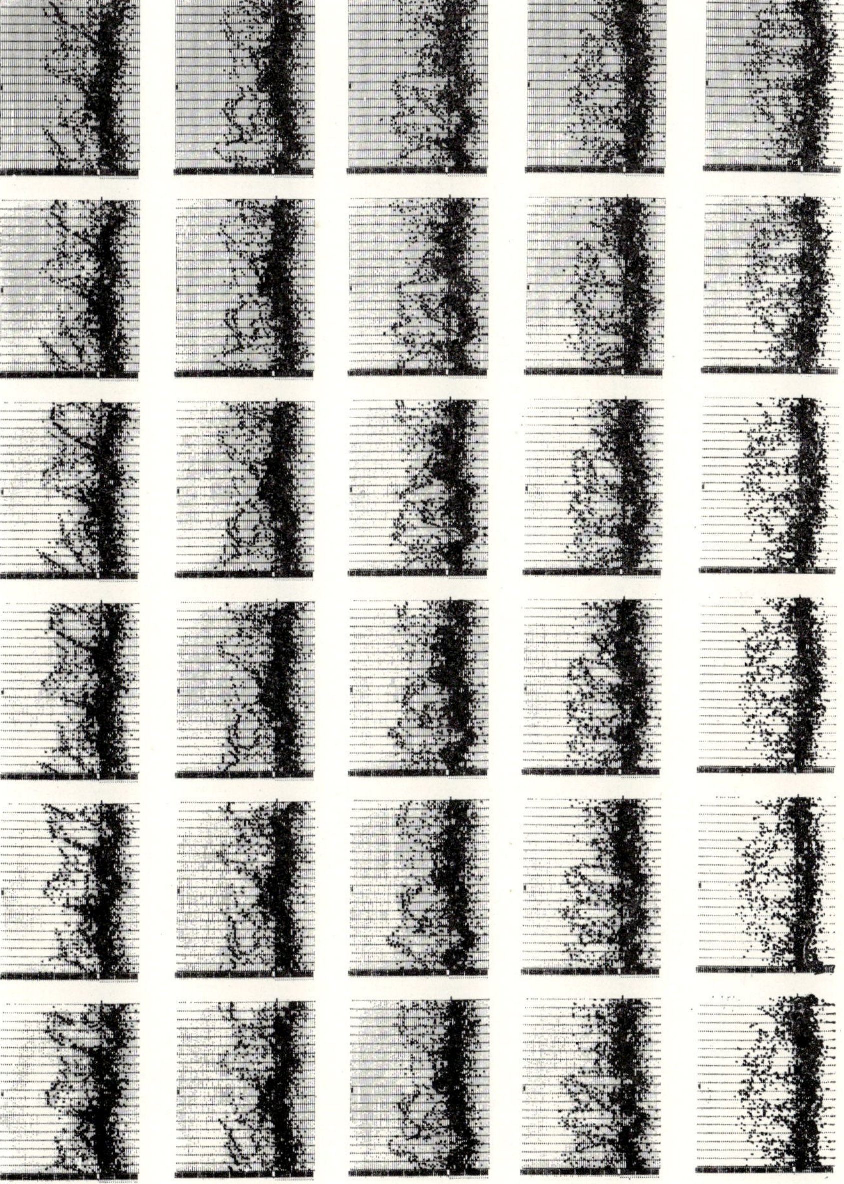

Fig. 2.19 Phase space plots, velocity v versus position x (abscissa), for weak cold beam (line in top left corner) interacting with a warm plasma

space (density of particles with position x and velocity v) that a vortex type structure arrises. It seems likely that a similar structure is formed here and that this is responsible for the more or less constant 10 percent level of the electric field after the peak. Unfortunately such diagrams were not made in this experiment. However, in another experiment in which a weak cold beam was sent through a warm plasma such vortex structure was seen to develop. This is shown in Fig. 2.19. This experiment shows that small amplitude theory predicts the behavior of the two-stream instability up to the time when the streams break up. No theory exists for the plasma behavior after beam break up. However, the qualitative picture for the development of vortices in phase space developed by Roberts and Berk seems to give some understanding of the behavior here. Similar behavior has also been seen by Morse and Nielson (1968).

Current instabilities

The second unstable situation which was investigated was that resulting from the passage of a current through the plasma. For this calculation the second sheet model had to be used since we needed mobile ions as well as electrons.

When one passes a current through a plasma, the mean electron velocity is separated from the mean ion velocity. If the two velocity distributions are Maxwellian and the ions are much colder than the electrons, then the mean velocities must be separated by $\Delta V = (m_e v_{Te}^2/M_i)^{1/2}$ (m_e = electron mass, M_i = ion mass, v_{Te} = thermal velocity of the electrons) before the system goes unstable (Buneman, 1958).

The combined distribution function is $f_- + (m_e/m_i) f_+$; f_- is the electron distribution function and f_+ is the ion distribution function. This combination determines whether the plasma is unstable and has the form shown in Fig. 2.20. The source of the energy for the instability comes from the slowing down of electrons going slightly faster than the wave, as previously mentioned with regard to wave damping. It can be shown that the wave tends to reduce the velocity of particles relative to its velocity. Thus, those particles going slower than the wave are speeded up while those going faster than the wave are slowed down. When there are more particles going faster than the wave than are going slower than it, the wave gains energy and grows.

One might expect that the tendency to slow down fast particles and speed up slow ones would result in a flattening of the distribution in the vicinity of

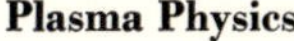

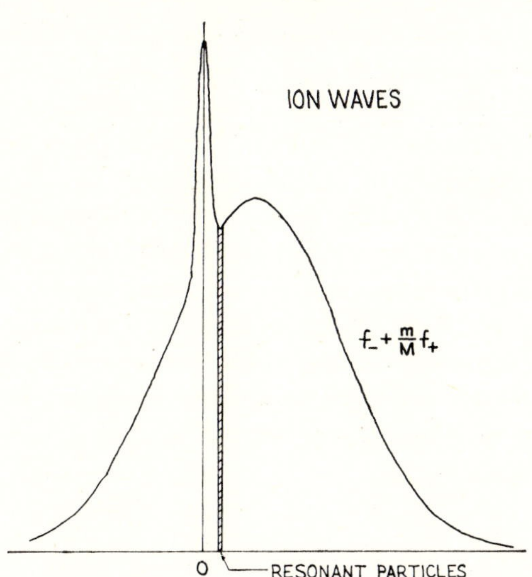

Fig. 2.20 Initial distribution function for low current case

the wave velocity. If this happens, the instability will turn itself off. The numerical experiments show that this indeed happens when there is a small separation between the mean velocities of the ions and electrons.

For large relative drift velocities we might expect that the plasma will not be able to stabilize itself so easily. Therefore, two problems were run, one with a small current and one with a large current. Before describing the results of these calculations we will discuss the situations which were investigated.

Systems containing 1000 electron and 1000 ion sheets were investigated (Smith and Dawson, 1963). Periodic boundary conditions were imposed so that a sheet leaving one end of the system was reintroduced at the other. This is equivalent to putting an infinite set of identical systems end to end. Then the electric field which a particle feels does not change when it leaves one system and enters the next. Thus, the sheet carries the field which it sees with it when it is transferred from one end to the other. This is equivalent to charging the boundaries of a single system. The current passing through the plasma continually polarizes it. The electric field which results soon stops the current unless it is canceled out. We do this by removing charges from the boundary at a constant rate. We may consider the system as being contained between condenser plates which are connected by a constant current generator as shown in Fig. 2.21. The current must equal the current produced by the generator on the average. If the current deviates from this average, an electric field quickly builds up which tends to restore the current to the average.

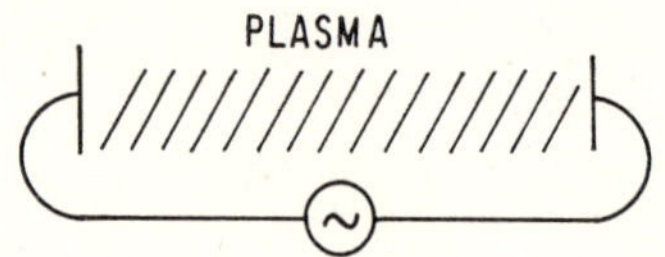

Fig. 2.21 Constant current generator

Results of the calculations

As already mentioned two situations were investigated, one for small drift, $0.6v_{eT}$, and one for large drift, $2.5v_{eT}$ (v_{eT} is the electron thermal velocity). The ion mass was taken to be 25 times the electron mass. The initial conditions on the calculations were the following. The velocities of the sheets were chosen randomly, but in such a way that the distribution functions for both electrons and ions were Maxwellian. The electron distribution was centered about the selected drift velocity and the electron temperature was 10 times the ion temperature, the electron Debye length contained 25 electrons and 25 ions. Both electrons and ions were initially uniformly distributed in space.

The results of the calculations are shown in the figures. Figure 2.20 shows the initial combined distribution function $f_- + (m_e/m_i) f_+$ for the low current case. Figure 2.22 shows the time development of f_-. Zero velocity is the mean velocity of the ions. The distribution flattens out in a very short time, $\omega_p t = 29$, in the region of the unstable waves. After a very long time, $\omega_p t = 273$, the peak even moves over to coincide with the mean velocity of the ions. The distribution, however, has an unsymmetrical form so as to maintain the current. Figure 2.23 shows the potential across the plasma, the current in the plasma, the total energy of the plasma and the total potential energy of the plasma (energy stored in the electric field, $\int E^2/8\pi \, dx$). From this figure one cannot tell whether the plasma was unstable. The potential and current show small oscillations at the plasma frequency. The total energy also shows small oscillations. Due to the fact that the system is driven the total energy rises slightly because of the heating by the current, but the effect is very mild. The potential energy fluctuates about a constant value but

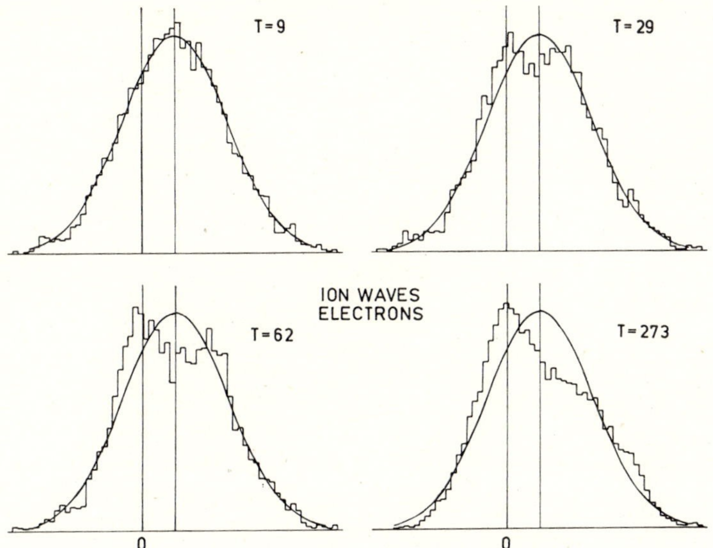

Fig. 2.22 Electron sheet distribution function for a number of times for the low current case

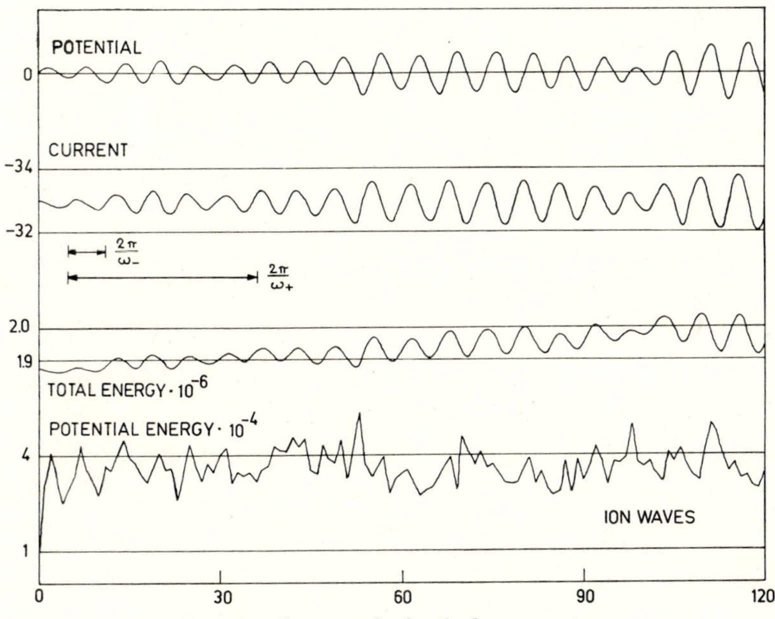

Fig. 2.23 Some results for the low current case

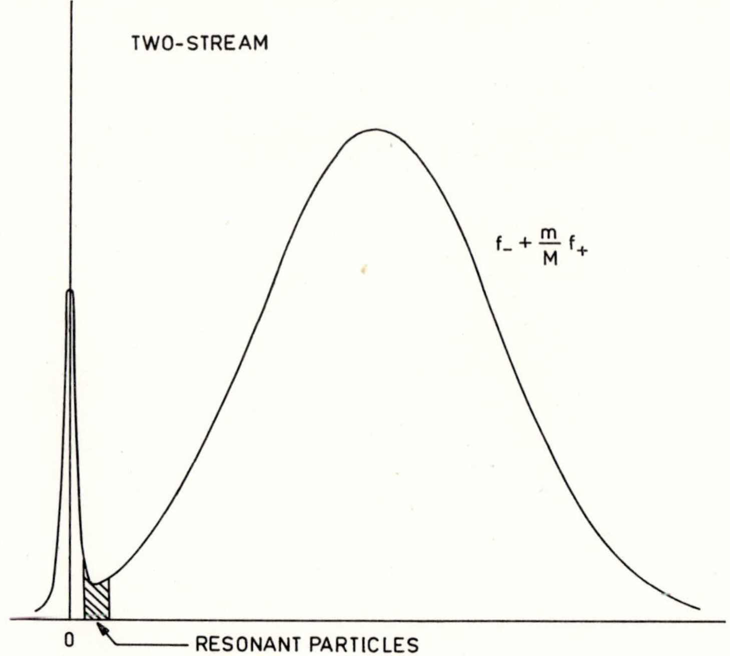

Fig. 2.24 Initial distribution function for high current case

does nothing spectacular. As the velocity distribution shows, the instability shuts itself off very quickly and hence, the instability never really shows up.

The velocity distribution ($f_- + m/M\, f_+$) for the high current case is shown in Fig. 2.24. Figure 2.25 shows the time development of the velocity distribution. As expected the system cannot stabilize itself easily. It does tend to do so, however, and the distribution becomes progressively flatter as time goes on and ultimately stability is regained. The spreading of the distribution and particularly the appearance of particles with large negative velocities (hence particles which have been turned around) show that there are large fluctuating electric fields in this case.

Figure 2.26 shows the time development at the total energy and the potential or electric field energy ($\int E^2/8\pi\, dx$) for the system. We see that the total energy begins to increase rapidly at $\omega_p t = 30$ and by $\omega_p t = 90$ it levels out again. The increase in energy can be attributed to an anomalous resistance through the equation

$$dW/dt = Rj_0^2.$$

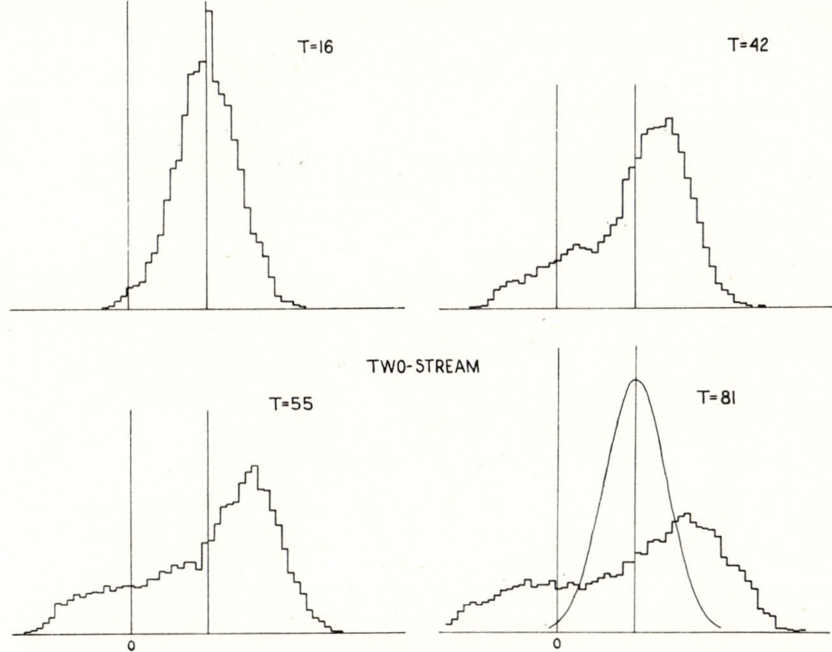

Fig. 2.25 Electron sheet distribution function for a number of times for the high current case

It is hard to make an exact comparison to the resistance of a real plasma since we must decide on what type of plasma it corresponds to; however, it is clear that the resistance is large, doubling the energy in 10 plasma oscillations ($\omega_p t \cong 60$) or essentially stabilizing itself in a few ω_{pi}. It is also clear from the figure that the anomalous resistance is much larger than what existed before the instability got going.

The large increase in the potential energy shows that large electric fields are produced by the instability. Fourier analysis of the electric field shows that the potential energy resides primarily in the unstable or low (long wavelength) modes. The Fourier analysis also shows that the increase in energy is primarily at the particle level; thus, it is randomized. The unstable modes level off at a relatively large amplitude (compared to the thermal level) and then absorb no more energy. At best these modes account for only about 10 percent of the total energy. Figure 2.27 shows the time development of the electric field energy in the third Fourier mode as a function of time. This is

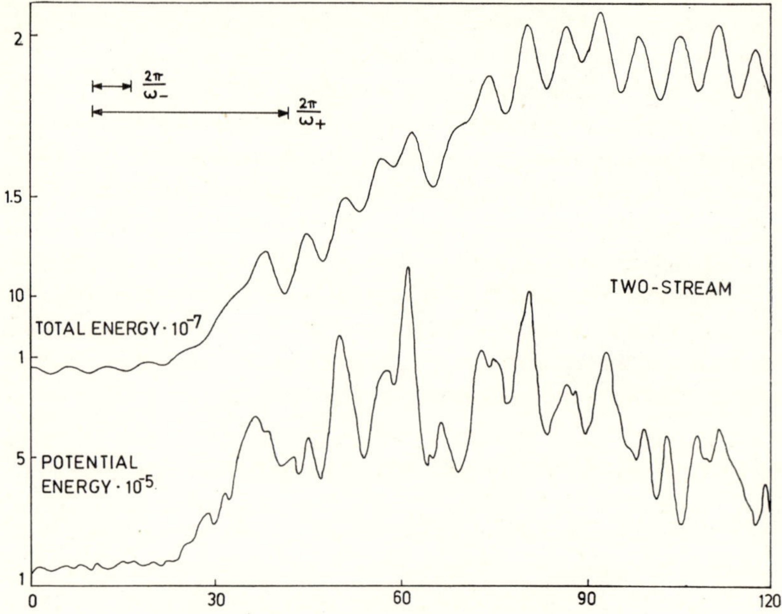

Fig. 2.26 Some results for the high current case

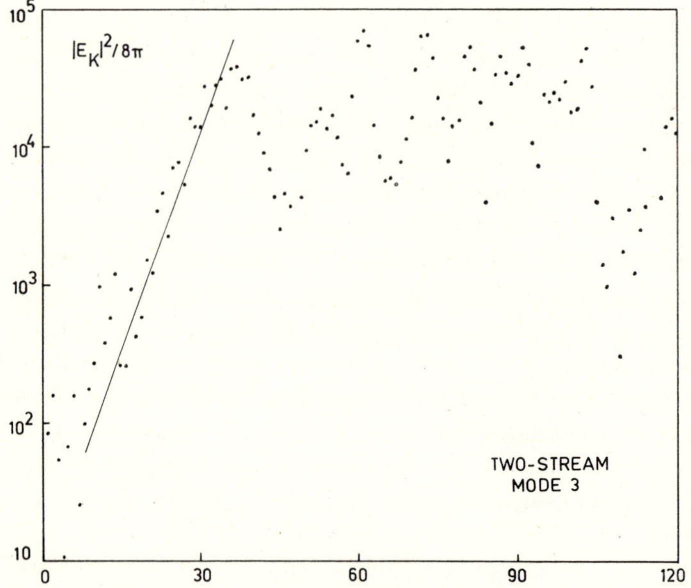

Fig. 2.27 Time development of the electric field energy in the third Fourier mode

the most unstable mode at the start of the calculation. It grows between two and three orders of magnitude and then levels off.

A number of theoretical calculations (Buneman 1958, Sagdeev and Galeev, 1966) have predicted an anomalous resistance for plasma carrying large currents. Such anomalous resistance has been found in some recent experiments. This numerical experiment confirmed the conjectures about the possibility of an anomalous resistance more than 5 years before such an effect was demonstrated convincingly in laboratory experiments (Dimock and Mazzucato, 1968, Bodin, McCartan, Newton, Wolf, 1968). While the detailed mechanism for the anomalous resistance may be different for these simplified models than it is for a real plasma, the mere fact that a mechanism exists lends support to theoretical calculations of such effects.

PART III SOME RESULTS OF TWO-DIMENSIONAL SIMULATION

Two and three-dimensional simulation should be able to reproduce much that exists in laboratory or space plasmas, bounded or unbounded. Some practitioners in $2D$, $3D$ started out with these ends in mind and tried to make their "computer plasmas" as much like real plasmas as practicable; they have achieved success with some models and have had to relax their ambitions with others. The limitations to progress appear primarily in enhanced noise due to use of too few particles; these limitations are being overcome by several steps. One step, brute force, is to use *more particles*; a million particles have been used at several laboratories through efficient handling of external memories. Another step, is to use more sophistication per perticle in order to reduce the noise associated with the grid and that with particle encounters. This step is to change from point to *finite-size particles* (Birdsall and Fuss, 1968a, b, c, 1969; Dawson and Hsi, 1968; Hsi *et al.*, 1968; Landon *et al.*, 1968). Each particle is made large enough to overlap a number of grid points and, hence, is shared with many of the $1D$, $2D$, or $3D$ cells; this sharing reduces grid noise. The particle diameter may also be made the same or larger than a Debye length, greatly reducing particle scattering as well as smoothing out phenomena on a scale less than λ_D, the small scale activity that is not always wanted. By these methods and others being developed, noise appears to be reducible while the desired collective plasma activity is retained.

The present state of $2D$, $3D$ results tends to be phenomenological, advancing into more quantitative descriptions. Thus, what will be shown in this section will be *pictures* of what occurs, some with *time scales* of occurence and a few with *rates* that can be checked with theory. We will show changes in shapes of boundaries, initiation of structure in density, formation of vortices, or rings or "venetian blinds" and other evidences of turbulence. We will also show changes in velocity distributions, with rapid relaxation from highly unstable to roughly Maxwellian distributions. Each of these effects is *nonlinear*, and thus essentially unavailable analytically. One of the objects is to help the theorist devise theories (which may also start out phenomenologically) for more general situations from these particular examples. Another object is to show the experimentalist what to look for in terms of gross plasma changes (boundaries, internal structure, velocities) so that he can set up his diagnostic instruments to look in certain wavelength and frequency ranges, and use certain time scales.

The computer experimentalist has been well occupied with finding the long-time evolution of the many plasma instabilities known from linear theory. These effects start from small perturbations (where the assumptions and approximations made in the theory are still valid) and grow exponentially in time and or space to large amplitudes where they become limited by nonlinear effects. Some instabilities have been found to be disastrous, pumping plasma rapidly across a magnetic field to the vacuum vessel walls; others have been found to be self-quenching (Byers, 1966). One combination of two instabilities has been shown to end up in a stable state (Byers, 1967). All such results as these are also helpful to the experimentalist, giving clues as to why a particular plasma, ripe for a given instability, was or was not contained.

Diocotron instability

This example considers the behavior of a thin ring or cylinder of electrons in a magnetic field, as sketched in Fig. 3.1. There are no ions in this run so it is not strictly a plasma; however, the behavior to be seen may occur in a plasma where there is a charge imbalance (a few more electrons than ions in some localized region) at quite low plasma density. The instability has a classical counterpart in hydrodynamics, called the *Kelvin–Helmholtz* instability, due to shear flow across a fluid. Here, it will be called the *diocotron* instability, a name acquired from magnetron device research of many years ago. The electron cylinder will be shown in the computer experiment to fill in, an effect which also occurs in devices.

The computer experiment was started with the electrons arranged cylin-

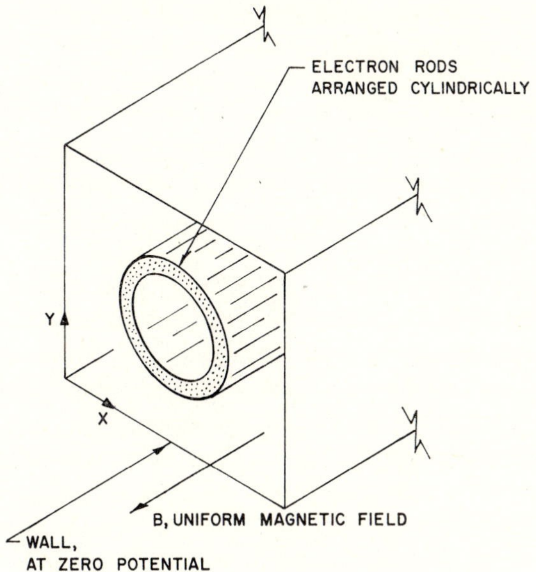

Fig. 3.1 Model for cylinder of electrons

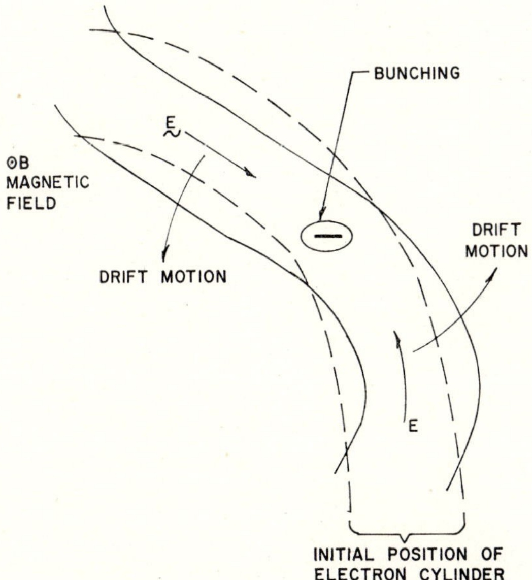

Fig. 3.2 Initiation of the instability due to bunching; the *E* fields of the bunch set up the drift motion, as shown

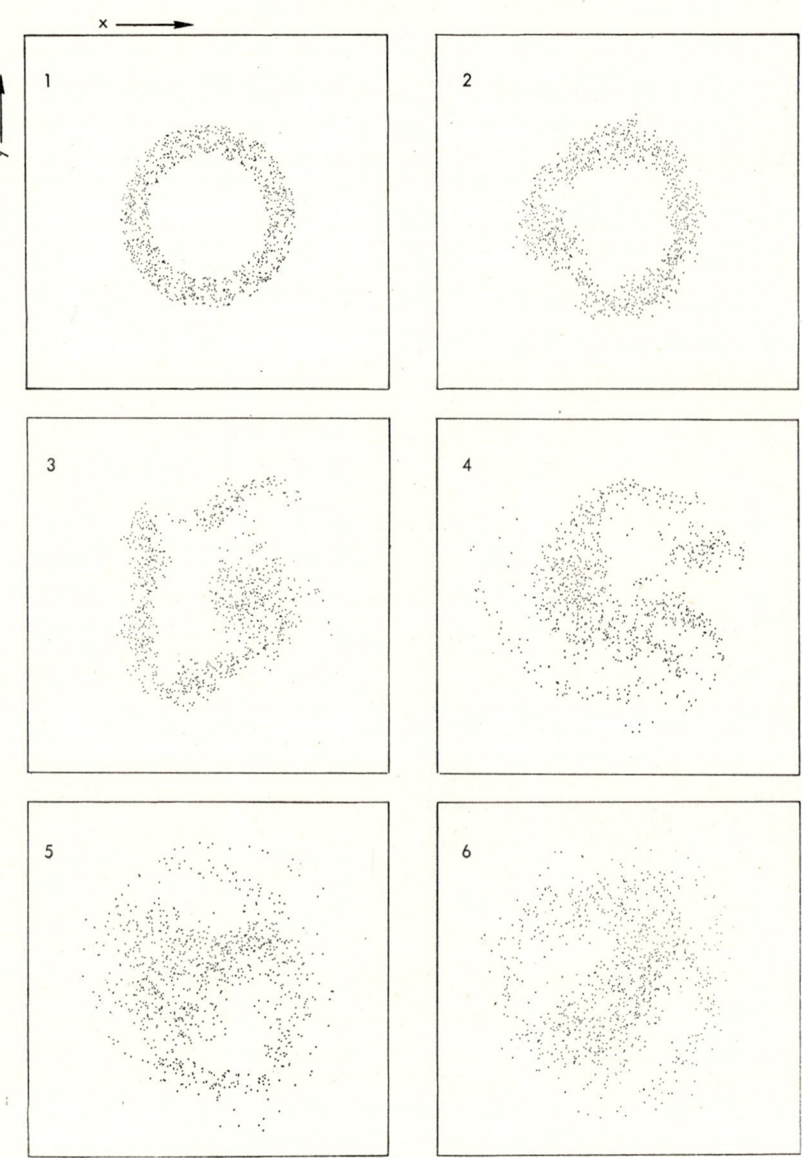

Fig. 3.3 Development of diocotron or Kelvin–Helmholtz instability for a cylinder of electrons in a uniform B_z field. These are snapshots of electron positions roughly evenly spaced in time, about 32 electron–cyclotron periods apart

drically, as shown in Fig. 3.1. This was a $2D$ experiment, so that the electrons move only in x and y and are represented by charged rods, providing uniformity along z. The magnetic field is directed along z. At $t = 0$, there were 500 electron rods randomly uniformly distributed in the cylinder, each with a velocity chosen randomly uniformly (i.e. a square velocity distribution). The density was relatively low, given by $(\omega_{pe}/\omega_{ce})^2 \cong 0.1$. (We have since learned that blind use of random numbers generally enhances noise and is to be avoided.)

There are three stages to the motion. First, some nonuniformities arise in the density (bunching), as sketched in Fig. 3.2; the bunching sets up the

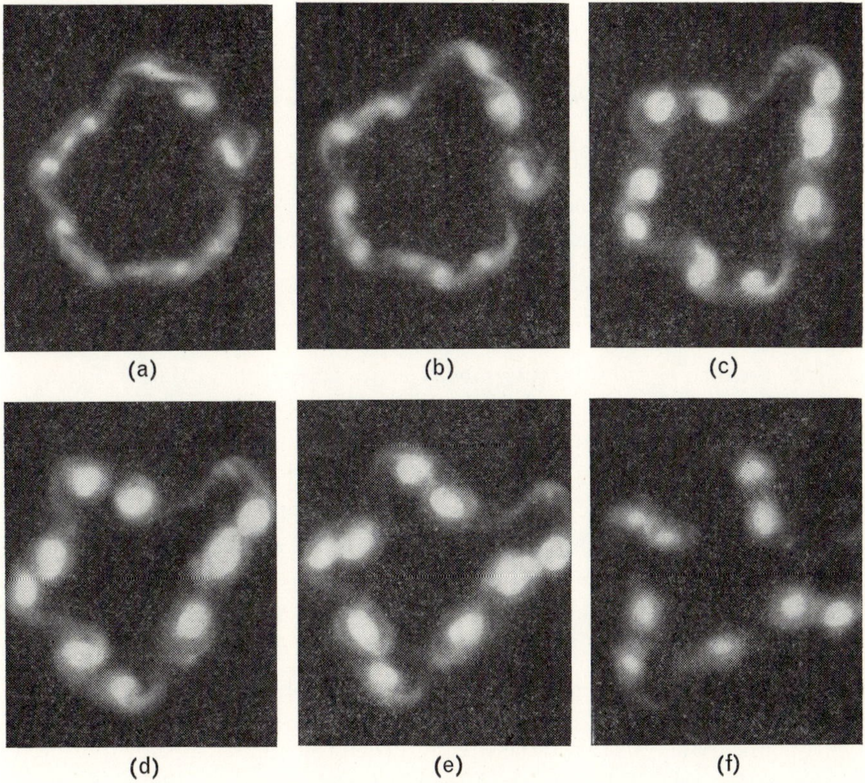

Fig. 3.4 Development of instability of an electron stream in a uniform B_z field. These are snapshots of the stream on a fluorescent screen, supplied by C.C. Cutler (Bell Telephone Labs., unpublished). Viewing the stream at different distances from the cathode is similar to viewing the computer experiment at different times, Fig. 3.3

electric field, **E**, which gives rise to a $\mathbf{E} \times \mathbf{B}$ drift motion, as shown. This initial sinuous distortion is predicted, from linear analysis, to grow exponentially in time. Second, the cylinder distorts into a set of vortices as shown in the early frames of the computer experiment sequence in Fig. 3.3; the vortex state is not predicted by linear analysis. Third, after some relatively long time, during which the vortices interact with each other in a metastable state, the vortices finally coalesce, drift toward the axis and form one large vortex. (There may be other models where the vortices live indefinitely.) The computer experiment ran about 10,000 time steps and was not carried further than that shown (Birdsall and Fuss, 1968a).

Hollow electron beams were shown to have very similar behavior by Webster (1955), and Cutler (1956) as seen from photographs of beams in Fig. 3.4. They did not see final states of just one vortex, possibly because their tubes were long enough to see only tens of electron cyclotron periods; our frames are spaced about 32 electron cyclotron periods apart. The similarity between the laboratory and the computer experiments is sufficiently good to venture that the causes and effects were the same, even though the charge densities, magnetic fields, and initial velocities were not identical.

Negative mass instability

The so-called *negative mass instability* arises when a thin cylinder or ring of electrons rotates at relativistic speeds in a uniform applied magnetic field. The interaction now is made more complicated in that the electron beam also generates a self magnetic field and electromagnetic interactions must be included (as against only electrostatic interaction in the diocotron instability). The basic mechanism for this instability was first investigated by J. C. Maxwell in 1890 in his essay on the stability of Saturn's rings. It can be shown that wherever the frequency of rotation of particles in a ring is a decreasing function of their energy, interaction through repulsive (e.g. electron–electron) forces makes the ring unstable. Grewal and Byers (1969) have done computer experiments for this configuration on the model shown in Fig. 3.5. It is known from linear analysis that if the stream is perturbed (much as shown earlier in Fig. 3.2), there will be growth of the perturbation. Grewal and Byers found that thr growth ended in formation of vortices, as shown in Fig. 3.6, very much the same result as with electrostatic interaction. But, there are important differences. In the relativistic negative mass case, the *full* set of Maxwells equations must be solved numerically, thus, changing the requirements from solution of Poisson's equation, which is time independent, to a solution for the wave equation, a partial differential equation involving time. Furthermore, the relativistic negative mass instability is greatly enhanced due to resonances with the natural frequencies of the surrounding structure (the cavity modes). When the frequency of rotation of particles is close to a cavity frequency, the vortices resulting from the instability will generally have field energy increased many fold over the field energy of the unperturbed beam.

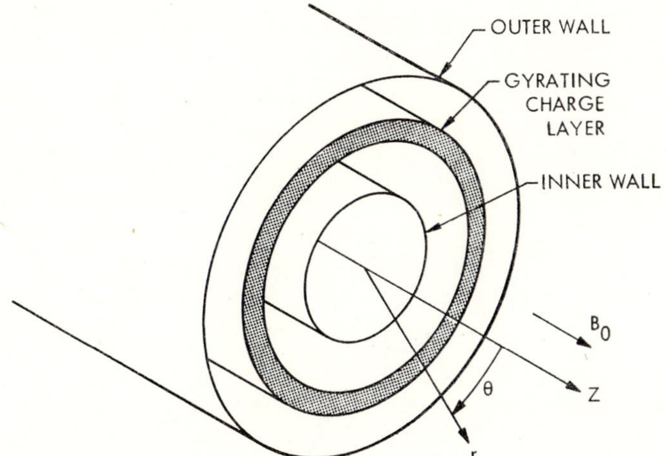

Fig. 3.5 Model of charge layer for negative mass instability

These results relate to several important devices, among them, the Astron *E*-layer (proposed fusion plasma containment mechanism), the ERA (electron ring accelerator) and the electron-potential-well ion accelerator (HIPAC, Janes *et al.*, 1966). Vortices such as those found here have been seen in laboratory experiments.

Plasma buildup in a fusion plasma experiment with beam injection (Alice experiment)

This simulation experiment has not yet been made close enough to the laboratory experiment to be labelled successful, so that the reader interested only in successes may skip on. However, the Alice modelling attempted has illustrated the ease with which different experimental configurations may be dry run on the computer with low cost relative to the Alice (or any other fusion) laboratory experiments.

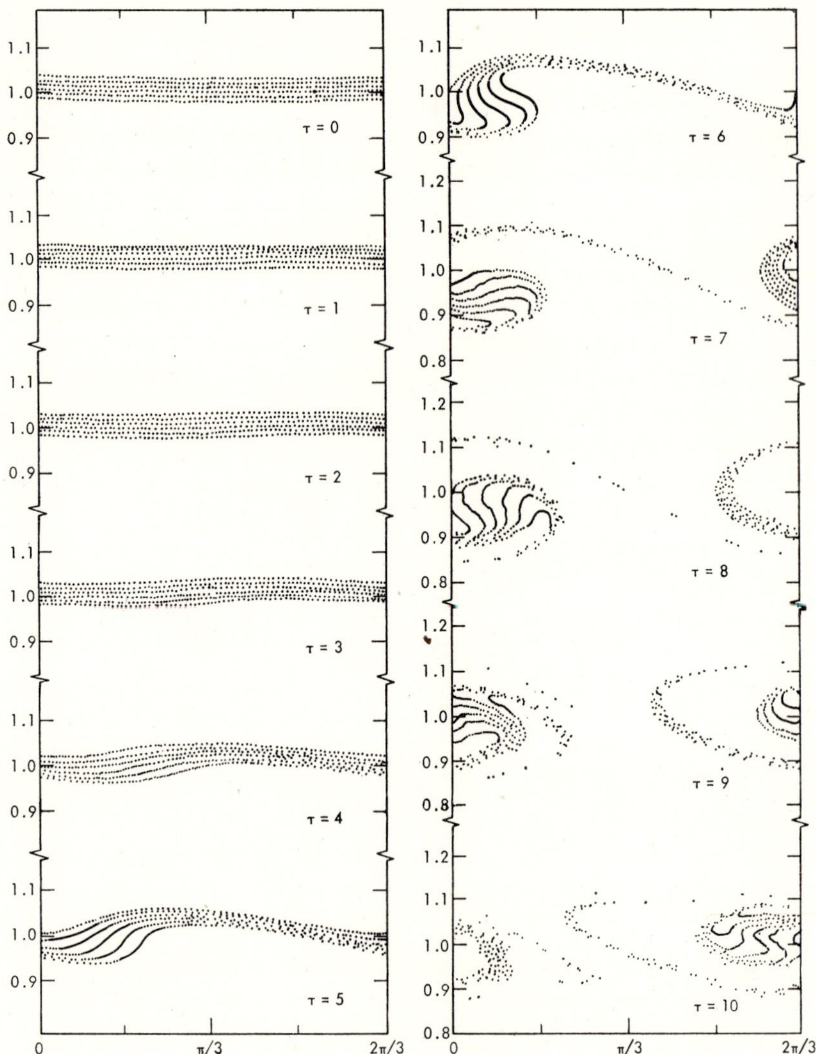

Fig. 3.6 Shape of the charge layer at various stages during the growth of the instability. The mode has three azimuthal variations so that only $\frac{1}{3}$ of the charge layer is shown azimuthally; the horizontal coordinate is θ. The vertical coordinate is the radius, r, with the inner and outer walls at $r = 0.6$ and 1.4, respectively. The time, τ, is given in terms of gyro-periods

The model is shown in Fig. 3.7. An atomic beam is injected from the left, into a magnetic field, **B**, (out of the paper) with high velocity, **v**, so that the Lorentz electric field, $\mathbf{v} \times \mathbf{B}$, can strip an electron from the atom; both electron and ion can then be trapped in the box by the magnetic field. **B** is not uniform, but may be stronger in the center (as in mirror containment fields, called a maximum-B configuration) or weaker (as with axial current carrying bars, Ioffe bars, added to mirror coils, called a minimum-B field). The laboratory ionization is not uniform along the atomic beam path and in the simulation can be programmed as desired, in time and space. The ions and electrons, once formed, will precess about the axis (the line of minimum or maximum B), in opposite directions. In addition to the precession, the ions will have a rapid circular (cyclotron) motion with relatively large (Larmor) radius, as shown in Fig. 3.7 (for max-B field); the electrons will have much smaller precession velocity and virtually zero Larmor orbit size. As a result

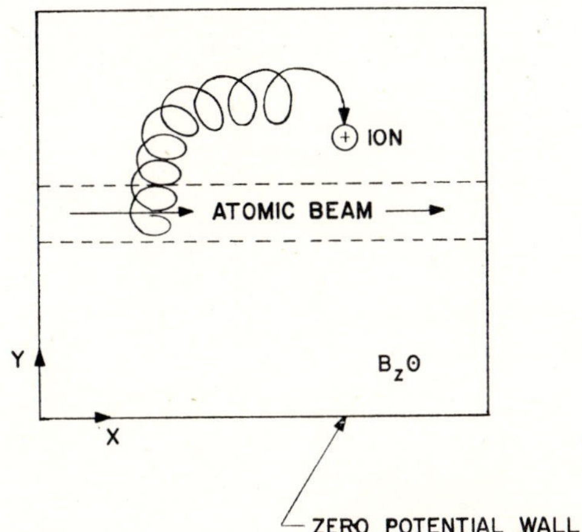

Fig. 3.7 Model for atomic beam injection. The path of the beam is shown, along with a typical ion path, with fast gyro-motion and slow precession due to the magnetic field. $B_z\,(x, y)$ being stronger at the center than at the wall

of the differences in the electron and ion motion, there can be substantial charge separations and large electric fields arise which force the ions to the walls as shown in Fig. 3.8.

In beam injection machines, the densities reached so far are relatively low; (about $10^9\ \mathrm{cm}^{-3}$, about $(\omega_{pi}/\omega_{ci})^2 \cong 0.3$, limited by several factors, in-

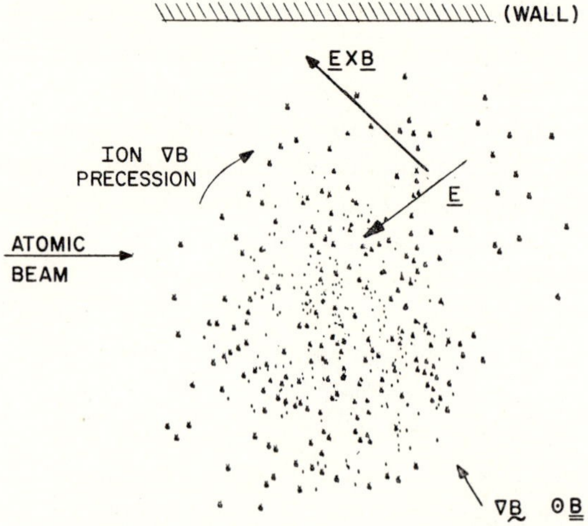

Fig. 3.8 Ion and electron motion in a maximum-B field. The electrons and ions separate, setting up large electric fields, E, which then drive the ions to the wall, via $\mathbf{E}\times\mathbf{B}$ drift Ions are $*$; electrons are •.

cluding instabilities); as the energy is high, the Debye length is large, on the order of the machine size, so that in one sense, a plasma is barely formed (plasmas, by definition, are many Debye lengths across).

However, at present there are two unreal parts to the simulation. One is that the laboratory buildup time is about 50,000 ion precession periods and many times more cyclotron periods; to depict this behavior would mean days of computer time. Our choice was to use amperes of atomic beam, rather than the experimental value of milliamperes. This rapid injection leads to large electric fields and the interactions in the first few precessions cause rapid loss of plasma. The second problem is that the low energy electrons tend to stay in the beam path setting up large **E** fields which drive out the ions, as was seen in Fig. 3.8. In the laboratory model, these electrons would set up axial (along B, in and out of the paper) electric fields and blow themselves out the ends (along the third dimension); this motion is not allowed in the simulation although an ad hoc axial loss could be programmed.

Nonetheless, we found a number of effects of interest, as follows.

The buildup of plasma with max-B and min-B fields was compared. The latter was clearly better for reasons easily seen. Here the ion precession ($\sim\nabla B$) brought the ions around the electrons in such a way that the charge separation electric field tended to be radial (see Fig. 3.9), setting up a more or less solid body rotation of electrons and ions together, with some desirable charge mixing. Beyond a well depth (maximum to minimum B ratio) of 3, the plasma was contained, up to density of $(\omega_{pi}/\omega_{ci})^2 \cong 0.24$. To try different magnetic field shapes or well-depths was simply a data card charge, corresponding, in the laboratory, to new (and expensive!) coils on the machine.

Slower injection rates were found to lead to larger densities at saturation or to no saturation at the end of computation; this confirms the need for more time to build up a more uniform plasma.

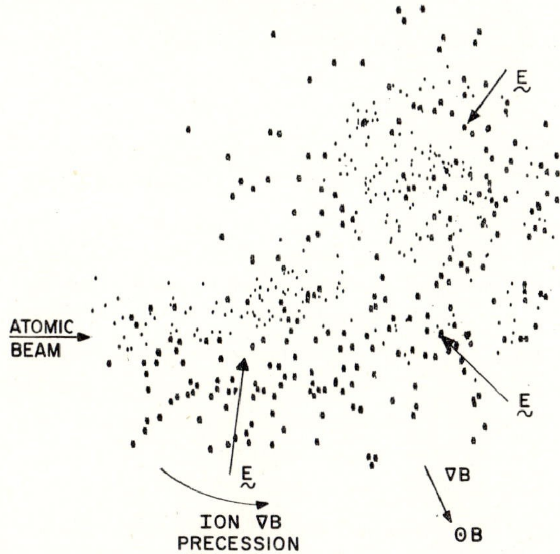

Fig. 3.9 Same as Fig. 3.8, but for minimum-B field. The difference is that the ion precession is reversed, the charge separation is altered, so that the $\mathbf{E}\times\mathbf{B}$ drift is now azimuthal rather than to the wall

A second beam was injected from the right with the total current kept the same. Smaller E fields were generated (quadripolar in shape, rather than dipolar) and higher saturation density was obtained. As beam injectors (beam lines) cost about $250,000, it may be comforting to the experimentalists to have encouragement from preliminary computer runs.

Further work on injection machine simulation should be done in 3D.

Instability due to non-equilibrium velocity distribution

Gases in equilibrium, we are taught, have a Maxwellian distribution of velocities, and gases not in equilibrium tend to seek equilibrium. We expect the same for plasmas. However, the most interesting part may be the rate of relaxation, the possible intermediate states, and the production of higher velocity particles than one might expect.

For example, it is of interest to follow the relaxation of velocities of plasmas with all particles having the same speed perpendicular to a magnetic field (so-called delta-function distributions, $f(v_\perp) \sim \delta(v_\perp - v_0)$, as occurs in beams or Alice injection) where classical collisions, large angle scattering, play little or no role. In this example, collective electrostatic interactions (instabilities) do the velocity spreading. (In neutral gases, collisions are the main cause in setting up and maintaining Maxwellians.) The reason for interest here is that there is some evidence from fusion plasma experiments that these instabilities play an essential role in the production of bursts of particles and radiation of certain frequencies. These instabilities may even play a desirable role in converting of beam-like to Maxwellian distributions. Computer simulation is being used to study the relaxation in detail, *and* to see what can be done to reduce the undesirable effects (particle loss).

The model here is one of as uniform a plasma as possible, in a uniform magnetic field, with no walls, simulating a small region of an infinite homogeneous, isotropic plasma. Effects along the magnetic field are ignored. The

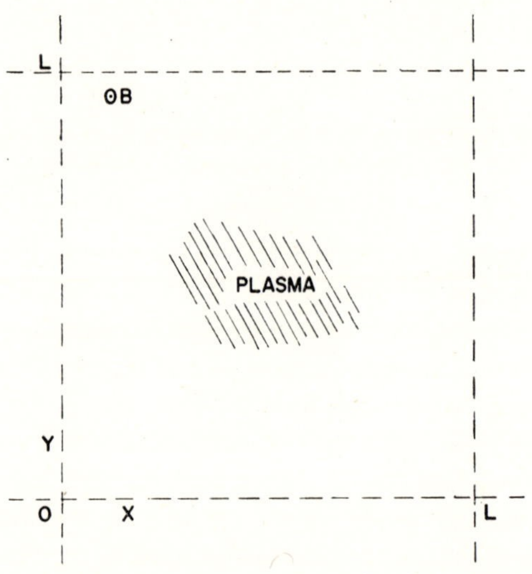

Fig. 3.10 Coordinates for periodic model. The "boundary lines" are fictitious, simply grid lines used for reference. Plasma fills all space and this region is but a small sample of space

model is shown in Fig. 3.10. The "boundary lines" from $x = 0$ to $x = L$ and $y = 0$ to $y = L$ are fictitious, delimiting the region under study. Space is considered to be made up of rectangles, L by L, the same as this one. We call the model *periodic*, with longest wavelength, λ, given by L; we can measure fields, densities and so on to a much finer scale, roughly down to a

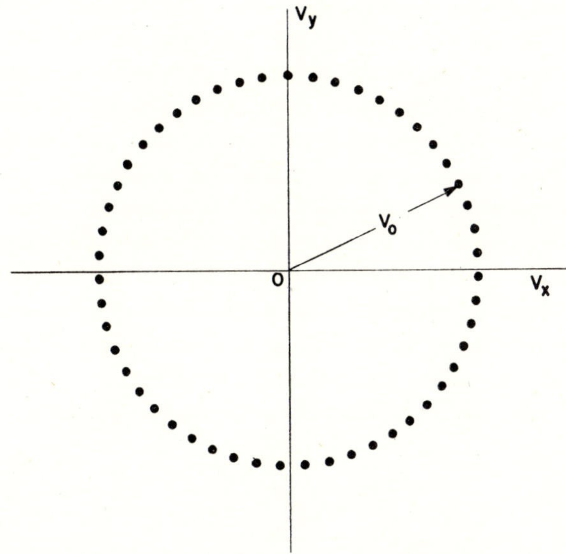

Fig. 3.11 Velocity space, showing initial velocities of ions. Each dot is an ion, located on a circle of radius (speed) v_0

minimum wavelength given by $\lambda = 2L/G$, where G is the number of grid lines (or cells) used in the interval $x = 0$ to $x = L$. Typically G may be 32 or 64 or 128, depending on how fine grained a description is needed.

We were interested in what the ions do and ignored the electron motion, simply by making the electrons have uniform density for all time. (This is a physically justifiable trick, but not open to laboratory experimentalists.) The action obtained is all due to the ion motion, essentially due to groups of ions streaming through other groups, creating electric fields which in turn enhance or destroy the groups, altering the cyclotron orbits. Starting from a condition that all ions have the same speed, $v = v_0$ (but have a uniform distribution of directions, shown in Fig. 3.11), this interaction rapidly spreads out the speeds, ultimately leading more or less to a Maxwellian distribution, which has all speeds, with the "dots" in Fig. 3.11 becoming densest near the origin, $v = 0$.

The computer experiment appears, from the above, to be almost child's play, with uniform density and speed, no boundaries, and no apparent problem about setting up initial near-equilibrium conditions as there would be with non-uniform plasmas. In our first runs, we observed the expected interchange between kinetic and potential energy, the production of near cyclotron harmonic frequencies and rapid broadening of the velocity distribution. However, looking closer, we found that we were essentially following a noise dominated interaction (not an exponentially growing instability), as our system initially had too large velocity and field fluctuations. The unanticipated difficulty was that the instability of interest saturates (stops growing) when the energy in the fields (that due to the interactions of the groups) reaches about one per cent of the total energy (essentially the kinetic energy). For the interaction to be seen, the initial field energy must be much smaller yet, say, $\frac{1}{100}$ per cent of the total. Because we must use far fewer than the actual number of ions, each of the computer plasma ions will have much more of an effect; that is, for a real plasma, the expected nonuniformity in space or velocity will contribute far less to collisions or noise than for a computer plasma (one computer ion stands for about 10^{12} plasma ions). Hence, we had to improve on our initial uniformity and we did this in two ways.

First, more computer ions were used, increased from 70 by 70 in our x, y grid, to 128 by 128. We estimated that this increase from 4900 to 16,384 particles would be sufficient to reduce the random velocity spreading rate. [Note that experiments in $1D$ typically use 1000 to 10,000 particles in one period L; in $2D$, doubly periodic, we cannot hope (in the near future) to use 10,000 by 10,000 particles.] Initial runs still used a random number generator to choose the initial velocity directions; the results did start at smaller initial total fields and the instability expected was clearly seen growing from small amplitude through to saturation (Birdsall and Fuss, 1968b).

Second, the velocity direction choices were made more nearly uniform, even in small regions (as suggested by J. A. Byers), by programming a uniform spacing of velocity phase, without the random numbers. This step improved the start sufficiently that the initial potential energy was much less than 10^{-4} of the total energy; by adding spatial spectral smoothing, this was reduced to 10^{-24}. The instability growth was then observed much more clearly; indeed, from the magnitudes of the potentials of various modes, as shown in Fig. 3.12, we observed the longest wavelength modes to grow from

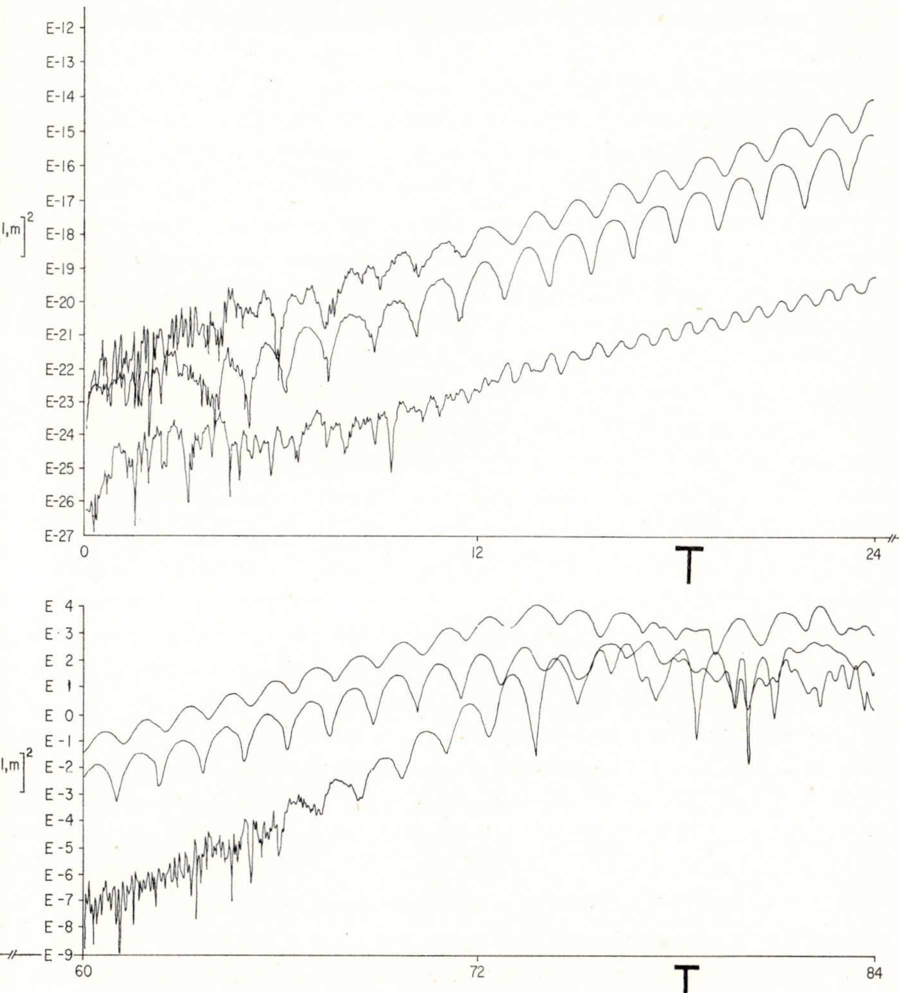

Fig. 3.12 Growth of the three longest wavelength modes in time, in terms of their potentials squared (~ energy), starting from the beginning of the experiment through to saturation (at $t \cong 74$) and beyond. Each vertical division is a factor of 10, so that growth over about 10^{25} or 250 decibels is shown

very small amplitudes *over a range of* 10^{25} before the growth stopped at saturation. This is growth of *250 decibels*, a far greater range than could be measured in the laboratory. The growth rates and frequencies agreed within a few percent with those of linear analysis. Typical electric field energy growth is shown in Fig. 3.13.

One of the most striking physical results is the rapid change from nearly single speed ions, $v \cong v_0$, to a distribution of speeds from $v = 0$ to $v \cong v_0$ in about one ion cyclotron period, a laboratory time of typically 10^{-7} to 10^{-8} seconds. The most rapid part of this evolution of ion speeds is shown in the sequence given in Fig. 3.14; $T = 54$ to $T = 57$ is about one half a cyclotron period. In a $3D$ laboratory experiment, say, with a mirror magnetic

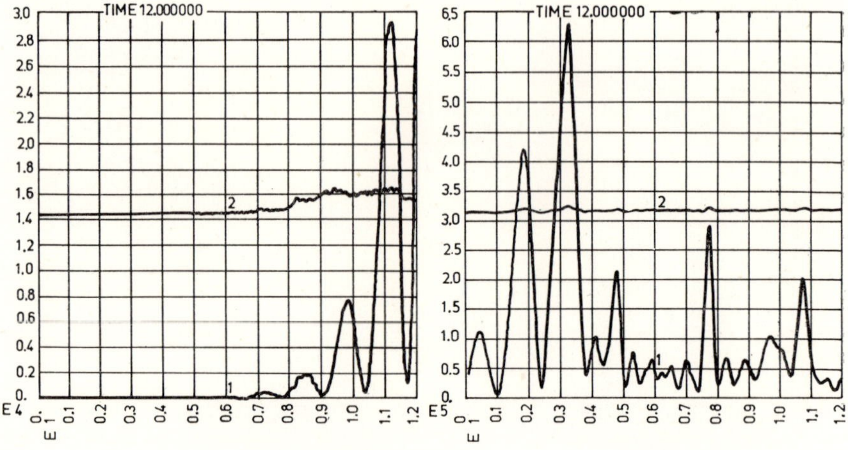

Fig. 3.13 Growth of electric field energy with time from $t = 42.08$ to 54.08 (middle) to 66.08; saturation is reached at $t \cong 59$. The vertical scale is linear, with peak of 0.2 percent (at left) and 4.4 percent (at right) of total energy. The almost straight line in the middle is the total energy on the same scale—zero suppressed—which is seen to have fluctuations far smaller than the potential energy variations—indeed, less than 1 part in 10^4—; this is a good test on the accuracy of the numerical methods, suggested by J. A. Byers. (Not the same run as Fig. 3.12)

field shape, some of the ions driven below a given value of $v_\perp/v_\parallel$ would tend to be lost out of the open-ended magnetic field, observed as a burst of particles, preceded by a burst of radiation. As such bursts have been seen in fusion plasma machines and yet remained unexplained, the above simulation behavior may be part of their evolution.

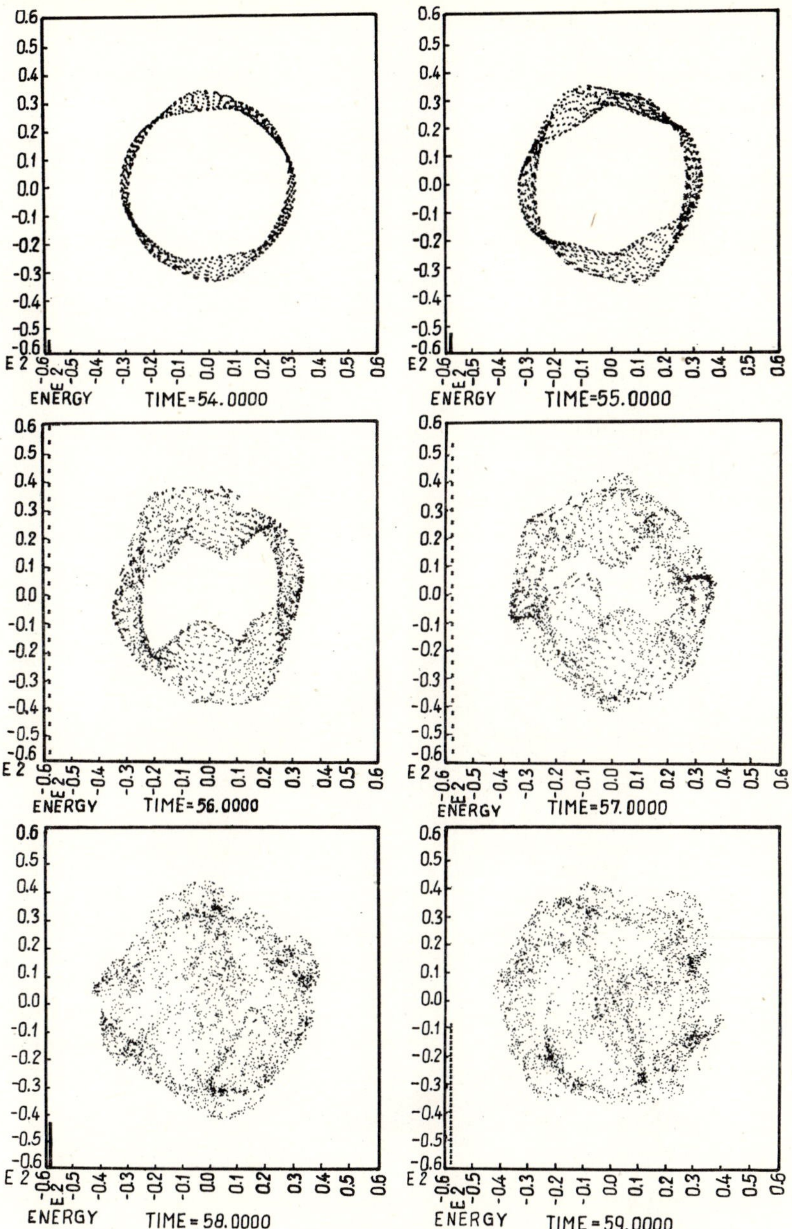

Fig. 3.14 Evolution of single speed velocity distribution, $v_\perp = v_0$, near the saturation of the instability, as viewed in $v_\perp$ velocity space, v_x versus v_y. The interaction causes rapid acceleration and deceleration of the ions, with the most noticeable feature being the diffusion of velocities toward zero, which thus wipes out the mechanism for growth. The main part of this action takes place in about half of a gyration (cyclotron) period. (Same run as Fig. 3.13)

Shock waves in collisionless plasmas

The model here resembles that discussed earlier, Fig. 3.1, with the cylinder of electrons replaced by a cylinder of plasma and the vacuum space replaced with a plasma of lower density; the periodic boundary conditions of Fig. 3.10 are used rather than zero potential walls. However, any resemblance stops here, as the density is much higher such that the force of the self magnetic (induction) field of the ions is included, with only a small charge-separation electric field present. The electrons are treated as making up a massless background (a good assumption, as they have $\frac{1}{4000}$ the mass of the deuterium ions used in the simulation) moving only to maintain charge neutrality to lowest order. The ion motion is treated exactly. Shonk and Morse (1968) have used 64,000 particles in a 50 by 50 mesh to observe various shock wave nonlinearities, starting from the configuration shown in Fig. 3.15a. The dark ring of piston particles is six times the denser than the rest of the plasma. One interesting

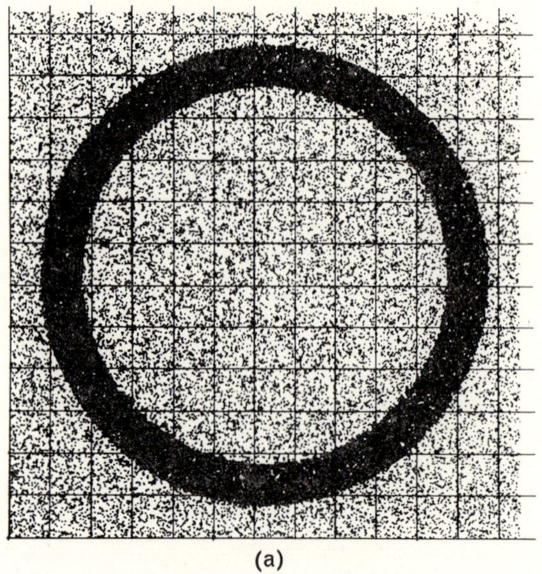

(a)

Fig. 3.15 Evolution of a cylindrical shock wave as driven radially inward by a magnetic field (out of the paper) which is larger outside of the dense plasma cylinder (six times denser than the rest), in x–y space. The interesting part in the fin-like structure in *d*, *e* which has also been seen in photographs taken of θ pinches *(Continued)*

experiment was to start with a magnetic field outside the piston particles larger than that inside; this field is a piston which drives the plasma radially inward. The particle positions at later times are shown in Fig. 3.15b–e. This

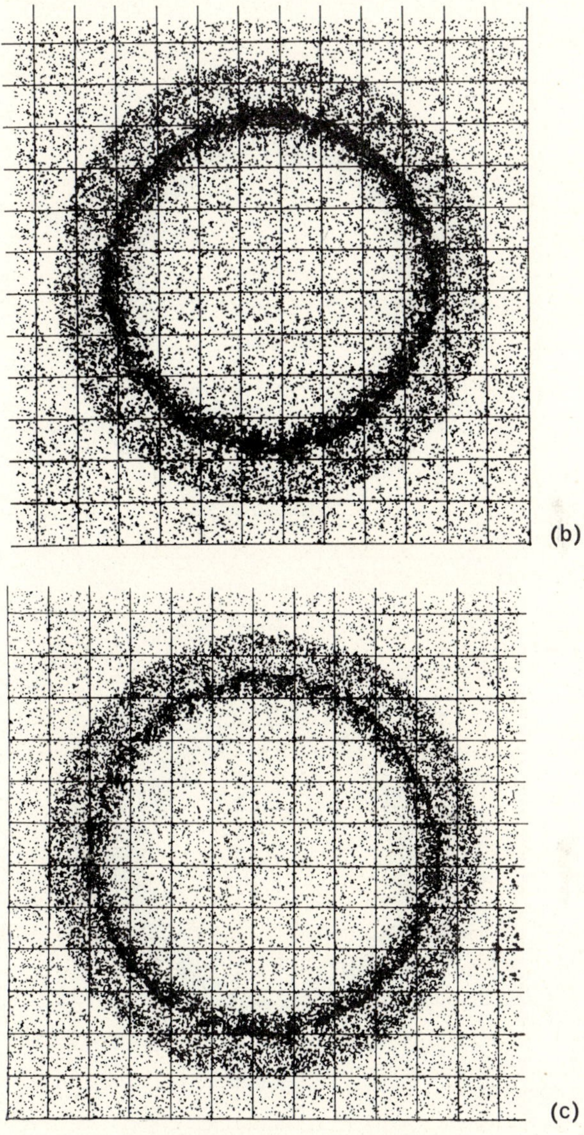

Fig. 3.15 (*Continued*) top—b; bottom—c

sequence shows the development of instabilities at the interface of the cylinder and the driving field. Similar instabilities have been reported in photographs taken looking axially (along the magnetic field, **B**) in a theta pinch (meaning that the current in the ring is in the azimuthal direction) shock devices.

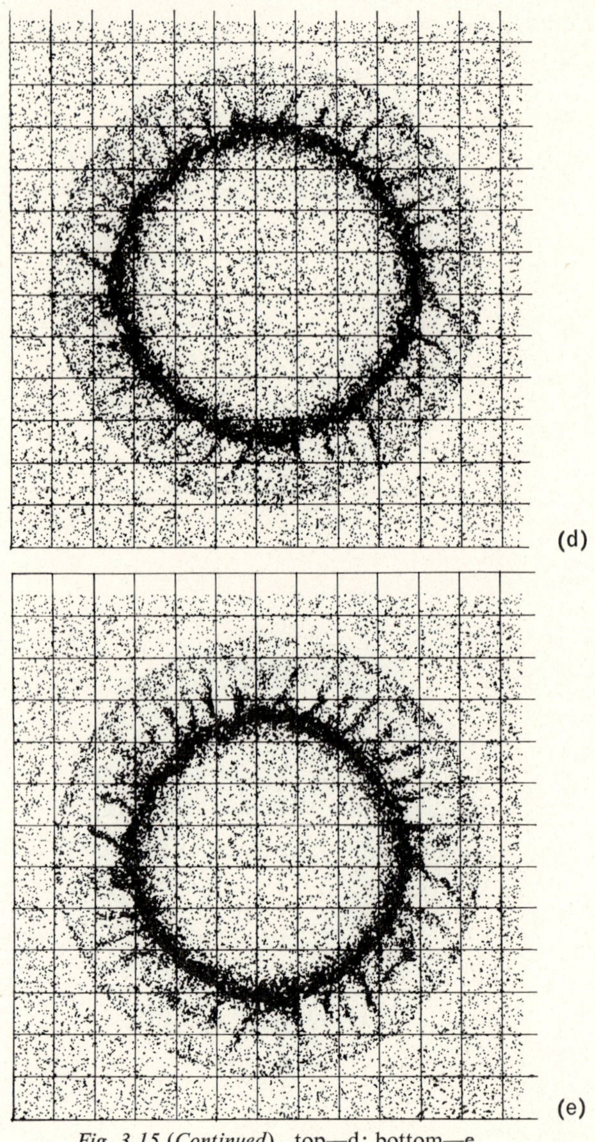

Fig. 3.15 (*Continued*) top—d; bottom—e

The valuable point here is that the usual theories, MHD, Vlasov, or weak turbulence either have flaws or are analytically formidable. The particle motion simulation shown here has none of these problems, starting from the basic equations of motion and the equations of the fields produced by the particles. The results are also easy to interpret, and have bearing on those observed in the laboratory.

Axial break-up of a plasma cylinder

A hollow plasma cylinder in which the electrons circle the axis (forming a current which provides a magnetic field opposing that applied) can have the magnetic field on the cylinder axis reversed from the applied field, such that the plasma sits in a magnetic well, as sketched in Fig. 3.16. This is called an Astron-like configuration. Dickman, Morse and Nielson (1969) have simulated an approximation to this configuration in which the diamagnetic current, J_θ, setting up the reversed field dominates over the particle flow that tends to neutralize any charge imbalance; they neglected electrostatic fields. Starting from an initial condition of equal energies along and normal to the magnetic field which linear analysis predicts is unstable, using 20,000 particles, they found that the ring tears apart axially as shown in the sequence

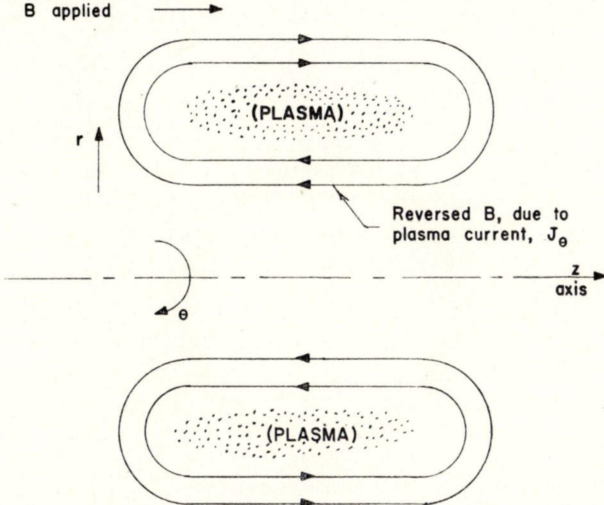

Fig. 3.16 *E*-layer or Astron-like configuration made from having large electron current in the θ, azimuthal, direction in an applied B_z, axial field. For sufficient J_θ, a minimum-B region occurs in the region labelled plasma

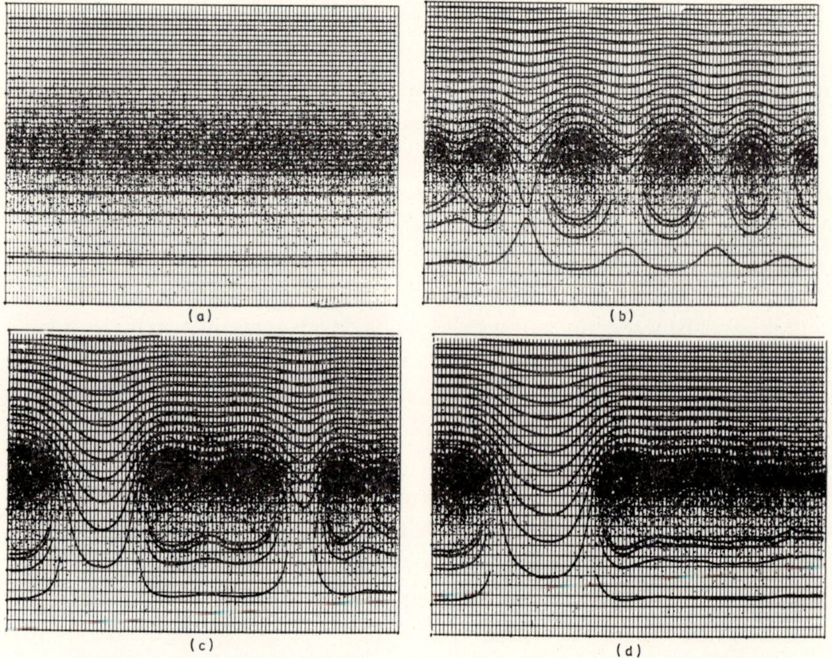

Fig. 3.17 Break-up of the E-layer into rings, then coalescence of the rings back into an E-layer with one gap. This is two-dimensional, r–z, with no azimuthal variations allowed or accounted for. Solid lines are B lines

in Fig. 3.17. The solid lines are the magnetic field lines. The plasma forms a number of rings about the axis, which then coalesce, ending up with only one gap. In runs not shown, by making the axial configuration more like that of the Astron machine, a single long layer is observed at long time, as desired for Astron. Physically, the tearing apart occurs along with an increase in the axial temperature, $T_{||}$, which in turn decreases the tendency for this instability to occur. The short wavelength modes grow first in time as they have less inertia, but the long wavelength modes have larger energy available to drive the instability.

EPILOGUE

We have reviewed a limited number of results in one and two dimensional plasma computer experiments, far fewer than have been done.

In $1D$, the results have added substantially to plasma theory either in small amplitude simulation, or experiments beyond laboratory capability, or in large amplitude nonlinear problems beyond analytical ability. A large amount of theory has been verified and some new theory has been developed in response to simulation results; still much remains to be done in $1D$.

In $2D$ models, problems have been solved with motion in a plane normal to the magnetic field and in the plane of the magnetic field. In both planes, the growths of a number of instabilities have been followed through to nonlinear limits, with strong relevance to laboratory experiments. The $2D$ results so far are really just a beginning with considerable room left for fruitful exploration. The next step in computer simulation is the obvious one, to *three dimensional models*, with no restrictions as to observations in just a normal or parallel plane, and including the boundaries normally used in the laboratory or in space. This step presents difficulties of many kinds, benefiting from the experiences in 1 and $2D$, but with much hard work ahead.

The objects are to observe, to perform experiments much as would a laboratory experimentalist, to try experiments on the computer *prior* to laboratory attempts, to guide what to build and what to look for in the laboratory, to obviate as many trial-and-error experiments as possible. Such a record is already a reality in some laboratories in solid state research. If results in just one area of plasmas, controlled thermonuclear fusion research, are accelerated by any appreciable amount, the computer simulation effort will have paid its way many fold.

References

The reader who wishes to broaden his background in this growing area of plasma computer simulation would find helpful reports on what is becoming an annual symposium: Symposium on Computer Simulation of Plasma and Many-Body Problems, Williamsburg, Virginia, April 19–21, 1967, available as NASA report SP–153; for prior work, this contains a bibliography and classification of related research in the period 1950–1966.

Conference on Numerical Simulation of Plasma, Los Alamos Scientific Laboratory, September 18–20, 1968, available as LASL report LA–3990. (Abstracts appeared in *Bulletin of the American Physical Society*, Vol. ***13***, (12) pages 1744–1748, 1968.)

Third Annual Numerical Plasma Simulation Conference, Stanford University, Stanford Calif. Sept. 2–5, 1969 (abstracts only).

References in text:

Armstrong, T.P., *Phys. Fluids* ***10***, 1269 (1967).

Barnes, C., and Dunn, D.A., NASA report SP–153, p. 31 (1967).

Birdsall, C.K., and Fuss, D., *Bull. APS*, ***13***, No. 2, 283 (1968a).

Birdsall, C.K., and Fuss, D., *Bull. APS*, ***13***, No. 11, 1556 (1968b).

Birdsall, C.K., and Fuss, D., Conference on Numerical Simulation of Plasma, Paper D1, September, 1968 (1968c).

Birdsall, C.K., and Fuss, D., *Jour. Comp. Phys.*, *3*, No.4, 494 (1969).

Birmingham, T., Dawson, J., and Oberman, C., *Phys. Fluids* ***8***, 297 (1965).

Birmingham, T., Dawson, J., and Kulsrud, R., *Phys. Fluids* ***9***, 2013 (1966).

Bodin, H.A.B., McCartan, J., Newton, A.A., Wolf, G.H., Proceedings of the Third Conference on Plasma Physics and Controlled Nuclear Fusion Research, Novosibirsk, USSR, No.CN–24/Kl, 1968.

Buneman, O., *Phys. Rev. Letters* ***1***, 8 (1958).

Byers, J.A., *Phys. Fluids* ***9***, 1038 (1966).

Byers, J.A., *Phys. Fluids* ***10***, 2235 (1967).

Cutler, C.C., *J. Appl. Phys.* ***27***, No.9, 1028 (1956).

Dawson, J., *Phys. Fluids* ***5***, 445 (1962a).

Dawson, J., (1962b) *Nuclear Fusion:* 1962 Supplement, Part 3, 1033.

Dawson, J., *Phys. Fluids* ***7***, 419 (1964).

Dawson, J.M., Hsi, C.G., Shanny, R., Conference on Numerical Simulation of Plasma, Paper A1, September, 1968.

Dawson, J., Shanny, R., *Phys. Fluids* ***11***, 1506 (1968).

Dawson, J., Shanny, R., and Birmingham, T., *Phys. Fluids* ***12*** (1969).

Debye, P.J., and Huckel, E., *Phys. Zeits.* ***24***, 185, 305 (1923).

Dickman, D.O., Morse, R.L., and Nielson, C.W., *Phys. Fluids* ***12***, 1708 (1969).

Dimock, D., and Mazzucato, E., *Phys. Rev. Letters* ***20***, 713 (1968).

Eldridge, O.C., and Feix, M.R., *Phys. Fluids* ***5***, 1076 (1962).

Eldridge, O.C., and Feix, M.R., *Phys. Fluids* ***6***, 398 (1963).

Feix, M.R., NASA report SP–153, p.3 (1967).

Grewal, M.S., and Byers, J.A., *Plasma Phys.* ***11***, 727 (1969).

Hsi, C.G., Dawson, J.M., Boris, J.P., Kruer, W., and Bernstein, I.B., *Bull. APS*, ***13***, No.11, 1555 (1968).

Janes, G.S., Levy, R.H., Bethe, H.A., Feld, B.T., *Phys. Rev.* ***145***, No.3, p.925 (1966).

Landau, L., Lifshitz, E., *Statistical Physics*, Addison Wesley (1958), p.229 (also see Chap. XII for the theory of fluctuations).

Langdon, A.B., Birdsall, C.K., McKee, C.F., Okuda, H., and Wong, D., Conference on Numerical Simulation of Plasma, Paper D2, September, 1968.

Lenard, A., *J. Math. Phys.* ***2***, 682 (1961).

Lenard, A., *J. Math. Phys.* ***3***, 778 (1962).

Lenard, A., *J. Math. Phys.* ***4***, 533 (1963).

Morse, R.L., Nielson, C.W., Conference on Numerical Simulation of Plasma, Paper A4, September, 1968.

Roberts, K.V., Berk, H.L., *Phys. Rev., Letters* ***19***, 297 (1967).

Sagdeev, R.Z., and Galeev, A.A., Lectures on the Non-Linear Theory of Plasma, International Atomic Agency Report, IC/66164 (1966).

Shonk, C.R., and Morse, R.L., Conference on Numerical Simulation of Plasma, Paper C3 September, 1968.

Smith, C., and Dawson, J., Princeton University Plasma Physics Laboratory Report, MATT–151 (1963).

Webster, H.F., *J. Appl. Phys.* ***26***, No.11, 1386 (1955).

CHAPTER 14

Computer Solutions in Continuum Mechanics

FRANCIS H. HARLOW

1 INTRODUCTION

CONTINUUM MECHANICS is the study of the motions of bulk materials. Fluid flow is an example; the elastic oscillation of a block of rubber is another. In all cases, a characteristic feature is the continuous variation of properties from point to point within the material. In the effluence of smoke from a stack, for example, there is a smooth change of velocity from the center of the jet to the stagnant air nearby. Likewise there are variations in temperature and density, such that the hot smoke is less dense, and thus more buoyant than the adjacent air.

To "solve" a problem in continuum mechanics means to predict how a material will change its configuration as a result of internal and external forces. The great difficulties encountered in obtaining such solutions arise especially because the forces depend upon the progress of the dynamics itself, and cannot be specified ahead of time. Mathematically, this fact is expressed by the nonlinearity of the governing equations, and by the coupling together of equations into sets involving a host of simultaneously changing field variables.

Of all the classes of continuum mechanics, perhaps the most diverse is the one called fluid dynamics. A century ago, water was the principal fluid of interest; its distinction from an elastic body was clear and unambiguous. Even today, we consider a fluid to be a material without elastic properties, that is, one which can be adapted to any new static configuration without residual stress. In many circumstances of interest, however, we now know that the definition of a fluid is considerably more complicated, in that materials may exhibit a continuous transition from elastic to fluid-like, with a "plastic" state between the two to further confuse the issue. Furthermore, a fluid may be capable of action at a distance, as for example when it is in a plasma state; and several other possibilities could be cited, in which the strict definition becomes clouded.

Thus, to demonstrate the power of computers for solving problems in continuum mechanics, the subject of fluid dynamics in its broadest sense is especially appropriate, and this chapter will concentrate on several aspects for its examples. Water in its usual types of flow circumstances still serves to illustrate many of the most interesting fluid flow phenomena. In addition, however, the high speed flows, in which "solids" behave as fluids, serve well to show the intriguing properties of shocks and rarefactions that play such crucial parts in modern engineering and science. These, too, will receive much emphasis in this chapter.

Our principal purpose is to show how a fluid state can be represented by the data stored into a computer, and how the computer can, in turn, process the initial-state data into a series of results that express the evolution through time of the solution.

2 THE REPRESENTATION OF A FLUID

Since the memory of a computer can hold only a finite number of data, it is apparent that the exact details of a continuous fluid cannot be stored. The first requirement for fluid dynamics calculations, therefore, is a method for approximating the configuration by means of a restricted set of numbers. The usual way of accomplishing this is to imagine a grid, or mesh of cells, through which the fluid moves, or which may move with the fluid. This imaginary grid serves only to divide the region of interest into a finite number of subregions (the cells), but has no effect on the flow itself. The fluid properties in each cell are then assumed to be represented by *averages* over the cell. Thus, for example, the "temperature of the fluid in cell number j" is actually an average temperature of the fluid throughout the cell. At any instant in time, only one number is needed to describe that temperature, which, accordingly needs only one storage space in the computer.

Consider the example of a cubical can of water. Suppose the space within the can were divided into one thousand cubical cells, ten by ten by ten. (These are purely imaginary cells, in that the water can circulate through the can without impedance, with no effect from the cells.) If, for every cell, the average fluid velocity in it is known as a function of time, then this information gives a fairly accurate description of what is happening in the can. The finest turbulence eddies would not be detected, but the gross features of the swirling motion would be quite clear.

If the can were divided into fewer cells, the details would be less well recorded; eight cells (two by two by two), for example, would enable only the crudest description to be recorded, while one cell (the whole interior) would allow virtually nothing of the motion to be described. At the other extreme, one million cells (one hundred in each direction) would begin to resolve some of the more detailed turbulence structure, but this begins to tax significantly the capabilities of available computers, both for storage and for the time required to calculate so much data. This limitation becomes especially noticeable when one realizes that up to *ten* quantities or more may be required for each cell, and the complete computer solution may require up to several *billion* operations to process all the data.

Numerical fluid dynamicists commonly use either of two types of cells. If the fluid moves from cell to cell, as implied for the water-in-the-can example, then the mesh is called Eulerian. In contrast, the mesh may actually move with the fluid in the calculations, like wispy, massless sheets of cellophane that drift with the currents and give no impedance; this second type of mesh is called Lagrangian. It is apparent that the occurrence of a vortex or other strongly deforming motion will quickly twist a Lagrangian mesh into a badly distorted mess. Indeed, under such circumstances, the computer logic becomes meaningless without the use of special (very complicated) techniques, so that Lagrangian meshes are seldom used to study highly distorting flows.

"Particle" meshes are also used to show the fluid configuration, usually in conjunction with an Eulerian mesh through which they move. Each particle represents the center-of-mass position of a small element of fluid. It is quite useful to examine a plot of the coordinates of all the particles; several of the examples illustrating this chapter show how well the changing fluid configuration thereby can be visualized. Strictly speaking, the particles form a Lagrangian mesh, but the main field variables of the calculation are tied to the coexistent Eulerian cells, so that distortions of the particle array have no effect on the solution logic.

There are various other ways that have been proposed for the representation of a continuous field of variation by a finite set of numbers. One that has often been tried, with but moderate success, is the use of truncated series expansions. Fourier series, for example, are capable of describing a variable field to a degree of accuracy that depends upon how many terms of the expansion are retained. The computer stores into its memory the Fourier coefficients, and calculates the changing configuration through adjustments of this finite set of numbers. The principal difficulties with such expansion techniques are (a) lack of flexibility in boundary conditions, (b) difficulty in expressing the physical principles of fluid motion in rigorous form when the series are truncated, (c) slowness in the computer manipulation of long series. Accordingly, this chapter does not consider further the discussion of this or other types of unusual representations.

3 CALCULATION OF THE DYNAMICS

The mathematical equations that describe fluid flows are precise statements of a very simple physical principle. This principle can be stated in a form that is particularly appropriate for our imaginary mesh of cells. It relates to the three basic entities, mass, momentum and energy. Specifically,

"The change in any of these quantities in any subregion of space is exactly balanced by the net amount of that quantity passing through the boundaries of the subregion."

This is a conservation principle, stating that mass, momentum and energy cannot be created or destroyed. Instead they are only moved about, or in some cases transformed from one state to another. The principle is based on the concept of "flux", which is the amount of a quantity passing through unit area per unit time. In fluid dynamics, there are three types of flux that enter into most types of problems. Foremost among these is the convective flux of mass, momentum or energy that results from the fluid motion itself. Like wisps of dye that drift with the current, any imbedded quantity is convected, and may thereby have much influence on the changing configuration.

For the mass of the fluid, convection is indeed the only way in which material can move through space (neglecting the slow process of diffusion, which is of importance in only a restricted class of problems). For momentum and energy, convective flux is also important, but the effects of pressure also contribute to the transport of those quantities, *relative to the fluid.* Thus, for example, through the exertion of a pressure force, one element of fluid can accelerate another, thereby producing a flux of momentum. Likewise one can do work on the other, thereby transporting energy through the fluid in what is called a work flux, or energy flux.

It is customary to express all of these concepts in the form of partial differential equations, but these formulations need not concern us here. Of more importance are the physical principles themselves. Indeed, when the numerical analyst adapts the differential equations to a description of what happens in his imaginary set of cells, he must do so in such a way as to preserve exactly the underlying physical principles that formed a basis for the equations in the first place. Failure to do so (a common oversight) has often resulted in disappointing results that do not represent accurately the true fluid-flow processes of interest.

From this point of view, then, it is easy to visualize quite realistically the manner in which a fluid dynamics calculation is carried out. Into the computer memory are stored the necessary data for each cell, describing the entire fluid configuration at some initial time. The object of the calculation is to process all the data, using the flux principle described above, in such a way that the cell data describe accurately the fluid configuration at a short time later. If this can be accomplished, then a repetition of the process again and again will result in a continuing description of how the flow pattern develops from the initial state into its subsequent changing configuration.

The result is closely analogous to the frames of a motion picture. If the frames are taken close enough together in time, then the result is essentially like the smooth and continuous evolution of what is happening. If the frames are taken too far apart, then too much detail is missed and the result is not useful. So it is with the computational cycles; the interval of time (which is controllable by the programmer) must not be too great or the results will be meaningless. On the other hand, it is wasteful of computer time to have the interval of time be too small. Choice of the proper interval depends upon several rather interesting and unexpected properties of the numerical analysis, and will be discussed further in Section V.

It is worth noting, incidentally, the close resemblance of a computer calculation with the performance of an actual physical experiment. The computer, with its program of instructions, is like the hardware for the experiment. Initial and boundary conditions in the computer correspond to the same features in the laboratory. In both cases, these are all assembled and made ready, and the process started. In both cases, the results are then produced "automatically", with the operator recording the results periodically as they are evolved. The operator is simply an observer, with minimal participation in the proceedings as they take place. When the "experiment" (a word applicable in both cases) is complete, he then can analyze his recorded results at leisure, testing their accuracy and usefulness, and deriving whatever conclusions are appropriate.

In both the physical experiment and the calculation, the questions of accuracy are related to analogous features. In the laboratory, there are motor vibrations, lens distortions, and a host of similar sources of error. Calculations may suffer from lack of resolution (too few cells), numerical fluctuations, programming errors, and similar sources of error. Thus, whenever the results of physical and numerical experiments correspond closely (as they, in fact, often do) there is cause for much delight; and when they disagree, the intriguing challenge to find out why, often results in the improvement of both kinds of investigation.

4 TWO TYPES OF FLUID FLOW

Numerical fluid dynamics is generally concerned with two quite different types of fluid flow problems. One of these is compressible (high-speed) flow while the other is incompressible (low-speed) flow. The accurate definition of these two types of dynamics is given in terms of the sound speed in the moving material. For air under normal conditions, this speed is 1100 feet per second, or roughly one mile in every five seconds, or 700 miles per hour. In water it is 5000 feet per second, while in iron the speed is 16,000 feet per second.

Thus, we find that low-speed flow, in which the fluid is virtually incompressible (unless confined), is that type of motion in which the material moves slowly compared with its sound speed. Water, in all ordinary circumstances, behaves as though it were completely incompressible. The air about an airplane moving at 200 miles per hour is likewise an incompressible fluid, except perhaps immediately adjacent to the propeller. A jet aircraft, however, may travel faster than sound, in which case the air dynamics is that of a highly compressible fluid, and such phenomena as the "sonic boom" can result.

Explosive chemicals can accelerate adjacent pieces of rock or metal to speeds of about 30,000 feet per second, much higher than the speed of sound in iron, with the result that almost any material immediately nearby will behave as a compressible fluid, and can be studied with the same computer programs that examine the dynamics of such fluids as air.

From a computational point of view, there is a fundamental difference between the two types of flow. For high-speed flows, sound signals propagate through the material no faster than the fluid itself is moving. If, for example, the motion is compressive (as when a burst of explosive drives a chunk of iron into a block of aluminum), the fluid simply piles up into a dense layer just ahead of the "piston", and the fluid just in front of this remains completely unchanged until the front of the layer (which is called the shock front) arrives. What this means, then, is that the behavior of each element of fluid affects only those that are immediately adjacent. Thus, the changes that occur in a particular computational cell can be calculated each cycle solely in terms of influences from the closest cells around it.

In contrast, low-speed flows are characterized by motions much slower than the speed of sound, so that an occurrence in one part of the fluid can almost instantaneously influence the dynamics everywhere else. Consider, for example, the motion of water in a very long pipe initially filled with water at rest. Turning on the faucet at one end results in a flow out the other that occurs with virtually no observable delay. In numerical calculations, this means that the change occurring in any one cell must be calculated in terms of influences from all the other cells in the system.

Mathematically, the high-speed flows are termed hyperbolic, and are governed by the theory of characteristics. Low-speed flows are elliptic, and require the solution of a finite-difference Poisson's equation, through matrix inversion, iteration, or some other complicated process. It is not appropriate here, however, to discuss further these mathematical aspects, as they have been presented in numerous reports, some of which are listed in the bibliography.

Sometimes the flow of fluids is divided into two somewhat different classifications, viscous and nonviscous. For numerical calculations, the distinction is not so clear-cut, in that even if the viscosity is not explicitly included, the inaccuracies of cell-wise fluid representation often manifest themselves in the form of an "effective" viscosity. This modifies the flow in a manner quite similar to the effects of real viscosity, and usually limits the degree to which those effects can be calculated.

Another reason that the viscous, nonviscous distinction is not very important in numerical fluid dynamics is that the incorporation of the full stress tensor is not a very difficult task, even for quite general types of high-speed flow. Although the necessary equations are quite long, there is not much added complexity to their solution. Indeed, one of the most powerful aspects of numerical methods lies in their ability to incorporate the complex properties of materials. The advantages extend to numerous features besides viscosity. Examples are heat flow, exotic equations of state, chemical or nuclear reactions, and interactions with electromagnetic fields. Each of these lengthens the equations, or may even introduce additional equations, but the basic conservation principles still apply, and the basic ideas for their application to cell-wise calculations remain appropriate.

5 NUMERICAL STABILITY AND ACCURACY

One of the most intriguing parts of the development of new methods for numerical fluid dynamics is the elimination of numerical instability and the enhancement of accuracy. We already have seen that a necessary condition for accuracy is smallness of the cells and shortness of the time interval per cycle. In both cases the specific criterion is that in the space or time interval, the magnitude of change in any field variable must be small compared with the magnitude of the variable itself. This condition is far from sufficient, however, in that a meaningful accuracy criterion must assure that the *cumulative* errors are small. Thus, if the error per cycle decreases in proportion to the time interval per cycle, then nothing is gained in the cumulative effect by decreasing that interval. In some techniques, the accumulated error actually increases without bound as the time interval per cycle is cut. For such a technique there exists an optimum interval, but this may vary in an unpredictable manner through the course of the calculation. These insidious matters are often difficult to detect, and any assessment of their effect may be impossible to anticipate for new problem circumstances.

More spectacular than the manifestations of inaccuracy are the results of numerical instability. (*Hydrodynamic* instability is a physical process of great interest; it should not be confused with *numerical* instability, which is a computational difficulty that can rapidly reduce the calculated answers to nonsense.) Numerical instability can occur in many ways. It is always detectable with ease, and therefore is seldom misleading. (Recent exceptions, however, can be found in the investigations of turbulence onset, in which the output of unstable computer runs have been interpreted as bearing significantly on fluid turbulence phenomena.)

To understand numerical instability, consider the following analogy. Suppose that on each swing of a pendulum a large accelerating force, F, is applied for a very short time interval, T. The product, FT, is the amount of momentum imparted each swing. If this is exactly balanced by the retarding air friction, then the amplitude of swing will remain constant. For a fixed value of F, a decrease in T means that the imparted momentum is insufficient to overcome the friction, and the amplitude will decrease. Conversely, an increase in T will result in an increase in amplitude, and eventually a value of T would be found in which the swinging would be so wild as to turn the pendulum upside down, or even tear it from its mounting.

Numerical fluid dynamics instability is like the pendulum that is accelerated for too long each swing. For a given set of internal fluid forces (that depend upon the nature of the problem and not on the cell size or time interval per cycle), the momentum imparted each cycle to the fluid in any particular cell depends upon the length of the time interval per cycle. When that interval is small, then the viscous (frictional) forces can absorb the tendency to overshoot in value, and the pressures can readjust themselves appropriately. But when the interval becomes too large, the amplitude of cell-wise oscillation can increase, even to the point of ridiculously large motions that are completely unrealistic.

This analogy demonstrates but one type of the many possibilities. Others are less easily pictured in homely terms, but are nevertheless just as real in their disastrous effects. Fortunately, many of them can be predicted by relatively simple analytical techniques, and thereby avoided by one or another means. (In many cases, an instability is, strictly speaking, tolerated, but its effect is made negligible by techniques that are effective in preventing further growth before the amplitude has increased to disastrous size. These are called "bounded instabilities", and the art of knowing to what extent they can be tolerated in any circumstance is a part of numerical fluid dynamics "lore" that still defies rigorous mathematical analysis.)

One very practical aspect of the matter of accuracy and stability is directly related to the capabilities of the available computers. In general, the smaller the mesh spacing, the more accurate is the computational result. But decreases in mesh spacing require not only much more computer storage, but they also require decreases in the time interval per calculation cycle. The question thus arises: Which is generally the more limiting, the memory available to store the cell-wise data, or the computer time available to process the solution through the desired elapsed time?

To put this question in its proper perspective, consider the following fairly realistic example of a calculation that taxes the capability of available computers. The study of meteor impacts onto a sandwiched multilayer space vehicle case would require about 100,000 "words" of computer memory (which is close to the maximum "fast access" memory of contemporary computers) and would typically consume roughly five hours of computing time for completion. A problem with twice as many computational cells will require approximately twice as much computing time *per cycle*. But the time interval per cycle is forced (by requirements of accuracy and stability) to be somewhat less than in the first study, so that the computing time for the problem would be increased by significantly *more* than a factor of two. A 200,000-word problem would require from fifteen to twenty hours of time to complete! Thus it becomes apparent that future computer requirements are not just for enormous increases in the memory size, but also for significant enhancement of speed, with the emphasis more on the latter than on the former.

6 THE USES FOR COMPUTER SOLUTIONS

Since large amounts of effort and money are expended to obtain computer solutions of problems in fluid dynamics, the question arises as to the usefulness of the results. In particular, is the investment adequately repaid? To answer this, consider the ways in which computer solutions are utilized. In the next section this is illustrated by a variety of examples; here we show how a typical example fits in with the other types of research techniques available to the scientist or engineer.

The classical methods for investigating fluid flow phenomena have been of two types, experimental and theoretical. The experimental method means the direct or indirect observation of fluid flows themselves. In a laboratory wave tank, observations of the water motion may be quite direct, although the subsurface behavior may require special tracer techniques to enhance visibility. On the other hand, the observation of gaseous motions within the sun are necessarily quite indirect, utilizing the interpretation of spectral measurements and drawing inferences of subsurface behavior from observable sunspot erruptions.

Theory, in contrast, is the examination of relationships among experimental observations, with the principal goal of being able to predict ahead of time the behavior of a natural process. Thus, for example, the repeated observation of water splashing from a rock thrown into a pond leads to an empirical prediction formula for splash height as a function of size, weight, speed and angle of rock impingement. To test his hand at the creation of theory, the investigator might then try to modify the splash formula to apply to the effect of a rock thrown into tar, or some other viscous fluid. If he is clever, he will utilize reasoning based on energy and momentum considerations, taking into account the nature of the forces to be expected from viscosity. He then will test his theory with a pot of tar and some appropriate rocks, and if there is agreement, then he will conclude that the theory is valid. Indeed, his confidence might be so high that he would design the casing of a space vehicle in such a way that the splash predictions as applied to meteor impact (not testable in the laboratory) would indicate complete safety for the occupants inside. Hopefully, the prediction would be correct, but whether it is or not, his theory would certainly be a better basis for designing the case than if there were no prediction method at all for anticipating the splash.

Strictly speaking, computer calculations are purely theoretical. We refer to them as experimental, but only in the sense that the results cannot, in general, be completely anticipated, even qualitatively. Nevertheless, the basis for computer runs is purely a theoretical expression of the basic principals of mechanics, the conservation laws described in Section III. These laws are accepted as being virtually incontrovertible; the manner in which they are applied to fluids is likewise accepted as correct for large classes of circumstances. Their meaning for specific complicated problems, however, may be extremely difficult to discern without the aid of computers. Thus, the computers have not accomplished much that is new in principle; their advantage is that they can do problems with speeds that are fantastically fast. This means that problems formerly requiring hundreds of years to solve can now be done in a few minutes or hours.

This, then, enables us to see quite clearly how the computer results can be used in our splashing-rock example. For the experimenter, the output can be observed almost exactly as in his laboratory, and he can use the data for his correlation formulas in the same way as if he had measured the results directly. More important, however, is the fact that he can get results for examples that lie outside the range of his experimental equipment, thereby extending considerably the confidence of his conclusions. Conversely, he also can use the results of presently-computable circumstances to design experiments that test the many types of fluid-flow problems that cannot yet be computed. Indeed, it is this last, the use of idealized solutions for complicated experimental design, that furnishes one of the most valuable applications of the calculations. Innumerable examples could be cited in

21 Fernbach/Taub (1403)

which computer solutions have enabled the avoidance of engineering design errors that would have cost many times more (in time and money) than the price of the calculations.

Theoreticians also find tremendous value to computer solutions in their search for valid relationships among physical processes. Through close examination of the detailed results, it has often been possible to distinguish which aspects of the flow are crucial to the results, and which are of less importance. Such insights provide numerous new ideas regarding the analytical simplifications that are reasonable, and the formulation of prediction techniques which can be applied with confidence.

7 EXAMPLES OF FLUID FLOW CALCULATIONS

The examples chosen here utilize some of the numerical techniques in current use at the Los Alamos Scientific Laboratory. In each case, the computer output is shown exactly as it is recorded during the calculations, and processed by the Stromberg–Carlson–4020 Microfilm Recorder.

Several computer techniques have been employed, as appropriate for different circumstances. No universal computing method has yet been developed that can handle all types of problems. The various methods have come to be known by a variety of shorthand names. Some examples of these are MAC (Marker-and-cell), PIC (Particle-in-Cell) and FLIC (Fluid-in-Cell). These and others are listed in the bibliography, along with some of the more important references.

Each of these basic *methods*, incidentally, usually has a number of different computer *programs* written to utilize the technique. Each program is arranged to give maximum versatility and efficiency for its particular class of problems, and various versions may be in use at one or several different computer installations. Thus, for example, some well-known programs utilizing the PIC method are DUX, VALLE, PICWICK, SHELL, and the newly-written, highly-versatile MESA.

Several different types of output display are illustrated in the figures. In no case is the mesh of computational cells included; Fig. 1, however, shows a gridwork somewhat coarser than the computer grid, superimposed over the particle configuration in order to facilitate detailed measurements of the results.

The most useful type of representation is usually in the form of the particle plots, as in Figs. 3, 4 and 6. Almost as valuable are various types of contour

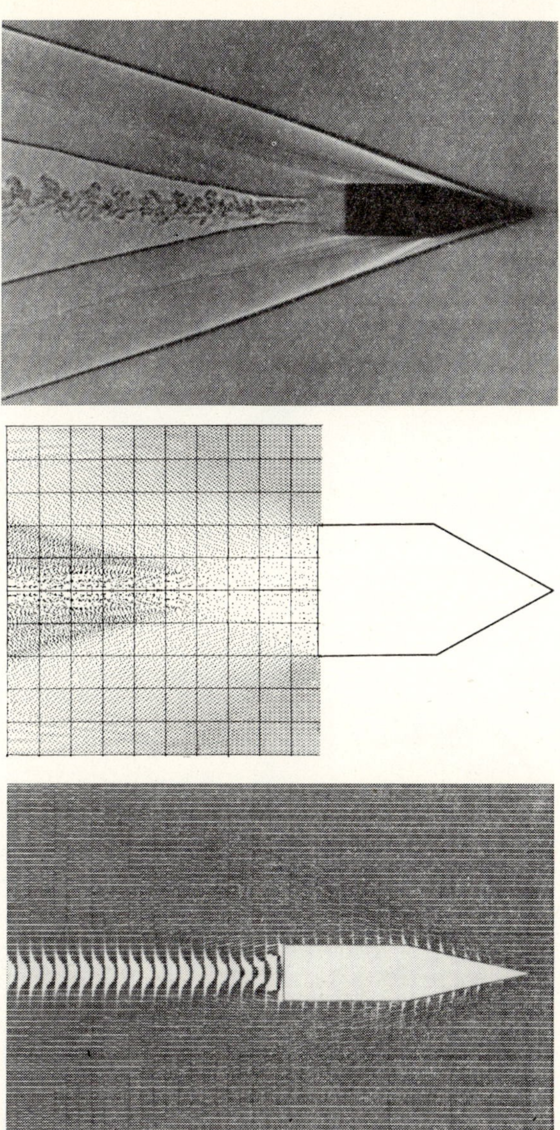

Fig. 1 The high-speed flow of air past a bullet is shown both experimentally (at Mach number 3.98) and calculationally. The top view shows the shock waves sent off to each side, contributing to the "bang", or to the "sonic boom" if this were an aircraft. In the wake behind the bullet there is a central turbulent region with a wake shock on each side. The middle view shows a detailed PIC-method calculation of the wake region for a similar type of bullet, while the bottom view is a FLIC-method calculation showing velocity vectors. In this last, notice how the fluid speed actually reverses immediately behind the base. Top, courtesy of David L. Merritt, U.S. Naval Ordnance Laboratory. Middle, calculation by A. A. Amsden. Bottom, calculation by R. A. Gentry and R. E. Martin

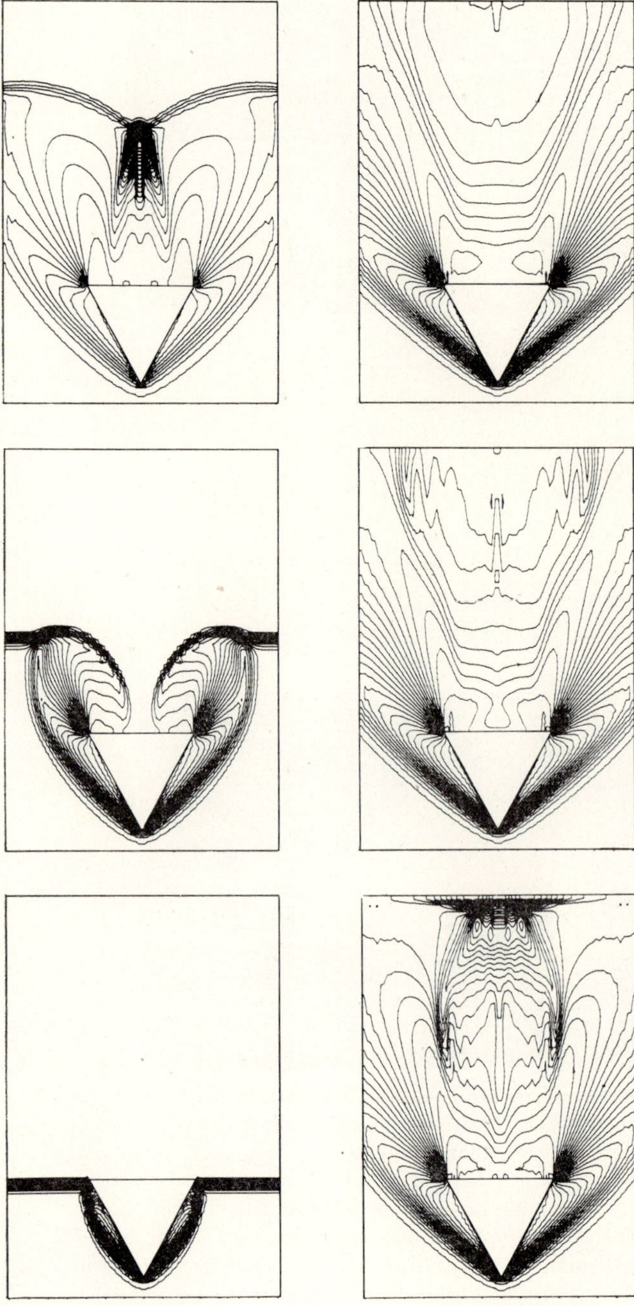

Fig. 2 These views show a selected sequence of frames from a computer motion picture of the interaction of a shock wave with a cone. Each line is an isopycnic, a line of constant density. In the upper-left frame, the shock, moving from left to right, has just arrived at the base of the cone. The top-middle frame shows the diffraction around the base and the formation of the bow shock. Subsequent frames show further stages in the interaction, and the ultimate achievement of a steady state flow pattern like that of Fig. 1. Calculation by R. A. Gentry and R. E. Martin

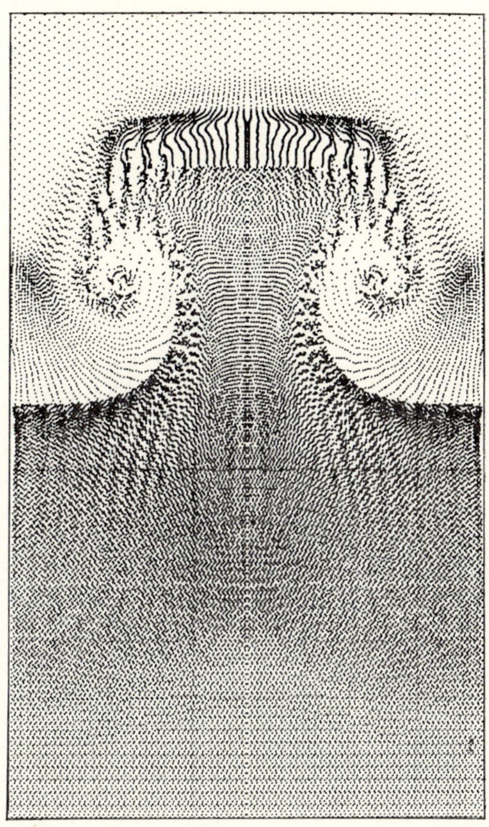

Fig. 3 This beautiful pattern arises in the PIC-method computer solution of the problem of shock interaction with a notch. In this case, the shock was moving upwards through a low-density plastic material when it encountered a dense plastic with a rectangular notch in it. The spiral vortices on the two sides are particularly noticeable features. This study was part of the investigation of what happens to pieces of equipment stationed in the immediate vicinity of a detonating chunk of explosive. Calculation by T. D. Butler and A. A. Amsden

plots, of which two different kinds are shown in Figs. 2 and 8. These contour plots show lines along which some quantity has a constant value; in Fig. 2, for example, the curves show lines of constant density. Such displays are generated by the computer itself, by interpolating among the cell-wise data to find the desired lines. Another type of display, less commonly employed, is the velocity-vector plot in Fig. 1. Each little line shows the local direction and speed of the air flow.

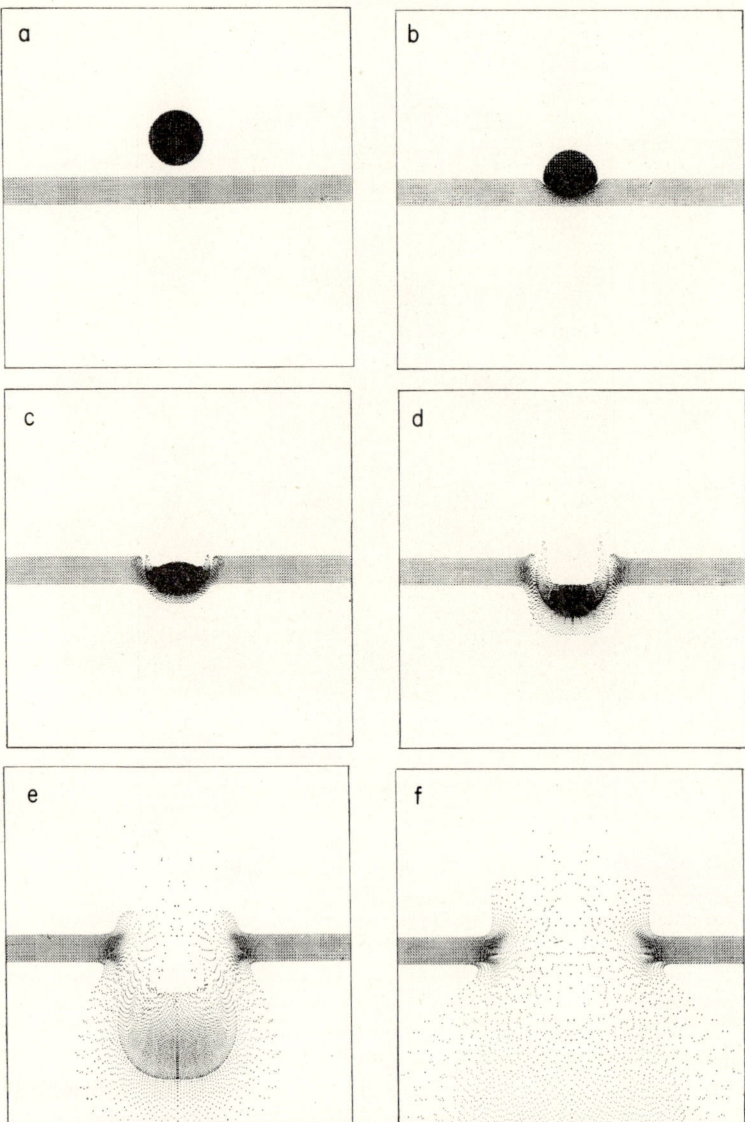

Fig. 4 The collision of a spherical meteor with a metallic plate is illustrated by these PIC-method calculations, selected from the frames of a computer-generated motion picture. The first frame is just before impact, while the next three illustrate the process of collision. The fifth frame, at a time somewhat after the collision is over, shows the tremendous expansion of the meteor that results from shock-heat vaporization. The last frame shows further vapor expansion, which eventually will leave a hole in the plate not much larger than shown at this time. Calculation by J. L. Cook

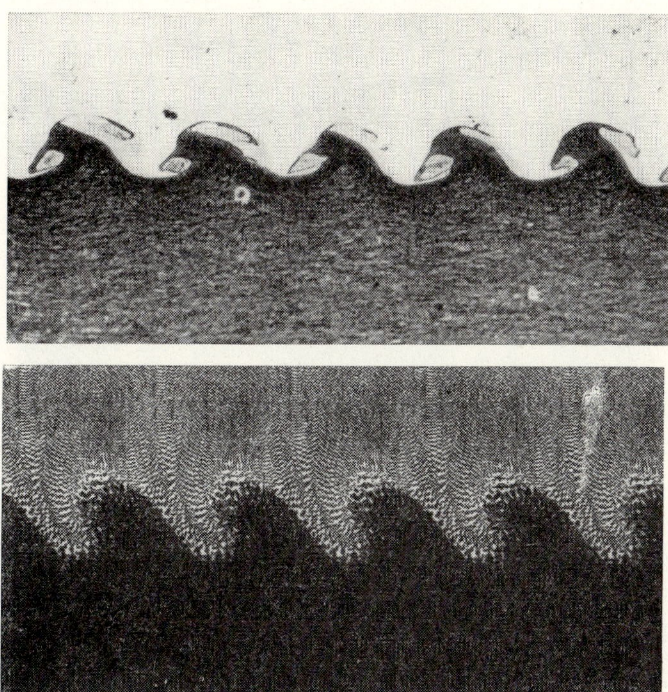

Fig. 5 When two plates of metal are driven together by high explosive charges, they may bond along the interface in a manner shown in the upper picture. In this case, the lighter metal is copper while the darker is brass. The surface waves that arose while the metals were melted have frozen into a pattern of tiny hooks (here much magnified) that strengthen the bond considerably. The lower picture shows a PIC-method calculation of a similar process, by which wave-formation mechanisms can be studied in detail. Experimental results courtesy of John Pearson, U.S. Naval Ordnance Test Station. Calculation by A. A. Amsden

In many cases, it is very effective for display to assemble the output pictures into a motion picture sequence. Since the time interval per cycle is such that only a small amount occurs in each computational cell, the projection at standard motion-picture speeds results in visualization at an appropriate rate for the study of the flow development in detail. The appearance for the meteor penetration problem, for example (selected frames are in Fig. 4), is like that of a slow-motion movie that illustrates the dynamics of one-millionth second in a sequence lasting about thirty seconds. A group of such motion-picture sequences is available on loan—see Bibliography.

Fig. 6 Until recently, the crashing wave on a beach has defied theoretical analysis. The calculations shown here were performed by the MAC method, a computing technique for the study of confined or free-surface incompressible flows. The example shown here illustrates the effect of a wave breaking into the run-back water from the preceding wave. The result is formation of a second smaller breaker which, in the last frame, is falling onto the dry beach. Calculation by W.E. Pracht

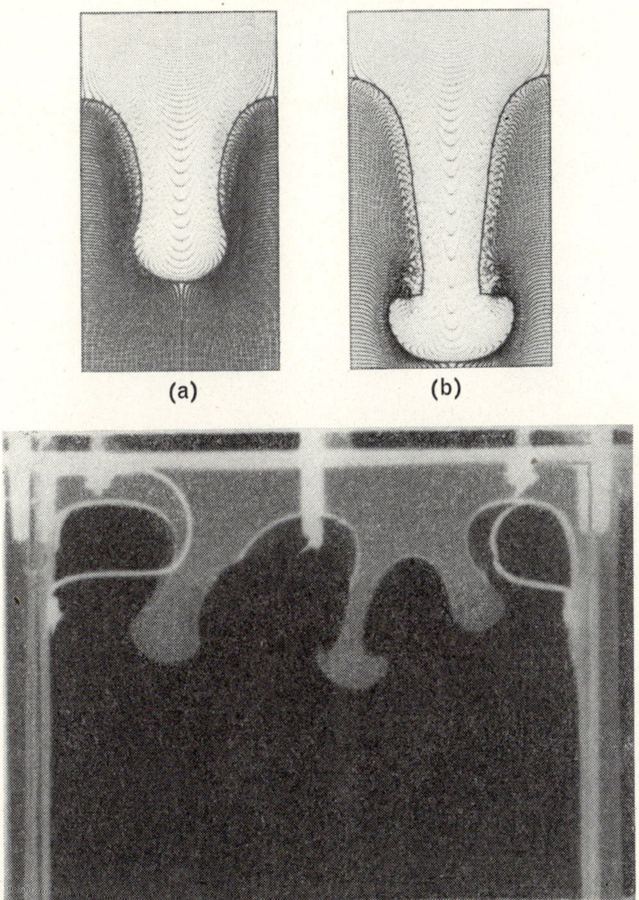

Fig. 7 In this investigation, the initial state has a heavy fluid supported by a diaphragm over a lighter one. When the diaphragm is quickly withdrawn, then the heavier liquid commences to fall in the manner shown in the bottom picture. The upper fluid descends in a series of spikes, or "fingers", while the lower fluid rises in somewhat broader bubbles. The curious part of the process is the formation of "nail-head" flattening on the tips of the spikes, a phenomenon that has been studied numerically only recently. The upper two pictures show MAC-method calculation results at two stages, matching quite closely the stages of two of the spikes in the experimental photograph. Like the experiments, the calculations were started at the initial diaphragm removal time, and progressed through the evolution of results automatically. Both the experiment and the calculation have been assembled into motion-picture form, and the two shown together serves to enhance considerably the demonstration of their agreement. Experimental results courtesy of J.R.Melcher and M.Hurwitz of Dynatech Corporation. Calculation by B.J.Daly

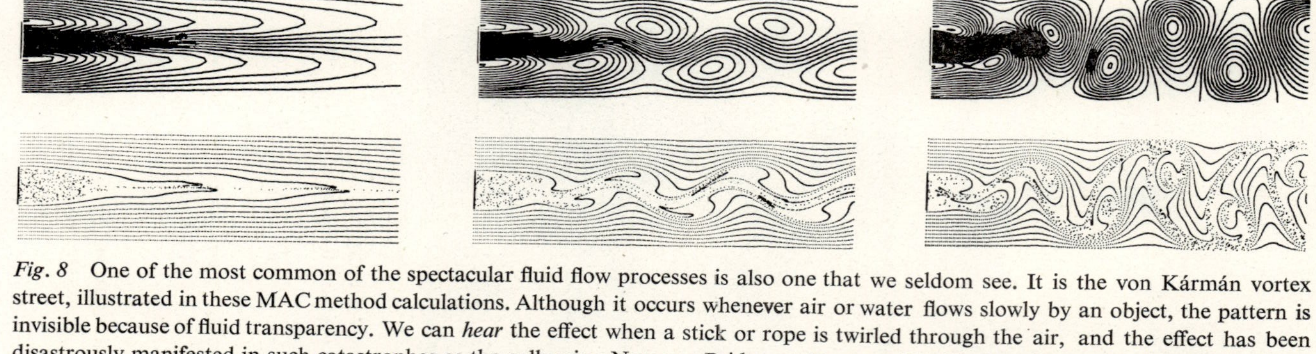

Fig. 8 One of the most common of the spectacular fluid flow processes is also one that we seldom see. It is the von Kármán vortex street, illustrated in these MAC method calculations. Although it occurs whenever air or water flows slowly by an object, the pattern is invisible because of fluid transparency. We can *hear* the effect when a stick or rope is twirled through the air, and the effect has been disastrously manifested in such catastrophes as the collapsing Narrows Bridge at Tacoma, Washington. Here, the calculation shows development of the pattern behind the back end of the object. Each view is shown as a pair, with the streamlines above and streakline patterns below. Streaklines are patterns formed experimentally by the presence of filaments of smoke that contort with the motion of the fluid. Calculationally they are formed by the motion of computational particles introduced into the stream at the left side, adjacent to the back end of the obstacle. The streamlines show fluid-flow direction in a coordinate system corresponding to the object being towed through fluid otherwise at rest. When all of the frames from the calculation are projected in sequence (as in the motion picture described in the Bibliography) the result is a smooth visual description of the process from beginning to end. Calculation by A. A. Amsden

In addition to the examples shown here, numerical calculations of fluid dynamics problems have been performed for other applications in a diverse set of engineering and scientific fields. The initial impetus for generation of these techniques was closely related to the development and effects-testing of nuclear weapons, and this field of investigation still remains a primary user of the high-speed flow calculations. Some additional examples of fluid dynamics calculations will show the wide scope of their applications:

1 In geology and oceanography, the study of sediments carried by underwater currents.

2 In agricultural engineering, the flow of water from reservoirs, under sluice gates and from broken dams.

3 In chemical engineering, the mixing of a jet of one fluid with another surrounding it; also, the dynamics of bubbles and droplets of various chemicals interacting with each other.

4 In meteorology, air dynamics on both local and global scales; also, raindrop dynamics during collision.

5 In space vehicle engineering, fuel sloshing problems in the presence of very small gravity.

6 In engineering safety, the motions of poisonous liquids or gases from a broken storage tank.

7 In naval design, the interaction of smooth or turbulent water with a moving object.

8 In heat engineering, the convection currents set up in heated fluids or gases, and the rates of heat transport by these currents.

9 In medicine, the flow of blood and other fluids.

These specific examples describe some of the work just recently completed or now in progress. Many others could be cited, in this rapidly expanding field of investigation, pertaining to almost every aspect of technology.

Looking to the future, it is clear that much remains to be done. Many interesting problems cannot yet be meaningfully solved. Some of the principal areas of method development for new numerical techniques are the following:

1 Effective methods for the calculation of turbulence. This elusive but extremely important phenomenon is of such a fine scale that computational cells cannot resolve it.

2 Better ways to incorporate more exotic fluid properties, such as elasticity and plasticity, surface tension, phase transitions, electromagnetic properties, and various "non-Newtonian" effects.

3 Improved accuracy in the fine resolution of strongly distorting details, such as a thin film of material imbedded in the flow.

4 New ways to calculate the movement of dust or other particles through fluids, and their effects on the fluid flows themselves.

5 Generalizations of techniques to unusual types of coordinates systems, in order to study flows near peculiarly-shaped walls.

6 Creation of efficient methods for studying low-speed flows in which the effects of compressibility are small but not negligible.

Already there has been much progress in these and related fields, and new, powerful methods can be expected soon to undertake an ever widening range of investigations.

Acknowledgments

I am very grateful to my associates at the Los Alamos Scientific Laboratory who furnished the examples illustrating this chapter. This work was performed under the auspices of the United States Atomic Energy Commission.

Bibliography

This compilation is based upon investigations at the Los Alamos Scientific Laboratory, and omits much excellent work performed at other institutions. Publications with an LA number are available from the Laboratory's Report Library.

A *The Particle-in-Cell (PIC) Method:* A combined Eulerian–Lagrangian computing method, which is suitable for solving multi-material compressible-fluid problems involving large fluid distortions.

1963 Harlow, F.H., "The Particle-in-Cell Method for Numerical Solution of Problems in Fluid Dynamics", *Proceedings of Symposia in Applied Mathematics* ***15***, 269.

1964 Harlow, F.H., "The Particle-in-Cell Computing Method for Fluid Dynamics", in *Methods in Computational Physics*, Vol. 3, B. Alder, S. Fernbach and M. Rotenberg, Eds.; Academic Press, New York.

1965 Amsden, A.A., and Harlow, F.H., "Numerical Calculation of Supersonic Wake Flow", *AIAA Journal* ***3***, 2081. Also republished in the AIAA Series of Selected Reprints Vol. 4, *Computational Fluid Dynamics*, C.K. Chu, Editor.

1965 Gage, W.R., and Mader, C.L., "Three Dimensional Cartesian Particle-in-Cell Calculations", LA–3422.

1966 Amsden, A.A., "The Particle-in-Cell Method for Calculation of the Dynamics of Compressible Fluids", LA–3466.

1966 Harlow, F.H., and Pracht, W.E., "Formation and Penetration of High Speed Collapse Jets", *The Physics of Fluids* **9**, 1951.

1966 Mader, C.L., Taylor, R.W., Venable, D., and Travis, J.R., "Theoretical and Experimental Two-Dimensional Interactions of Shocks with Density Discontinuities", LA–3614.

B *The Fluid-in-Cell (FLIC) Method:* An Eulerian computing method, which can be used for one-material compressible-fluid problems involving large fluid distortions.

1965 Butler, T.D., "Numerical Calculation of the Transient Loading of Blunt Obstacles by Shocks in Air", *AIAA Journal* **4**, 460.

1966 Gentry, R.A., Martin, R.E., and Daly, B.J., "An Eulerian Differencing Method for Unsteady Compressible Flow Problems", *Journal of Computational Physics* **1**, 87.

1966 Mader, C.L., and Gage, W.R., "2DE—A Two-Dimensional Eulerian Hydrodynamic Code for Computing One Component Reactive Hydrodynamic Problems". LA–3629–MS.

1967 Butler, T.D., "Numerical Solutions of the Hypersonic Sharp–Leading–Edge Problem", *Physics of Fluids* **10**, 1205.

C *Lagrangian Method:* A technique for calculating compressible fluid flows.

1955 Kolsky, H.G., "A Method for the Numerical Solution of Transient Hydrodynamic Shock Problems in Two Space Dimensions". LA–1867.

1963 Mader, C.L., "Stretch Sin—A Code for Computing One-Dimensional Reactive Hydrodynamic Problems". LA–DC 5795.

1965 Browne, P.L. and Hoyt, Martha S., "HASTI—A Numerical Calculation of Two-Dimensional Lagrangian Hydrodynamics Utilizing the Concept of Space-Dependent Time Steps". LA–3324–MS.

1965 Browne, P.L., "Rezone, A Proposal for Accomplishing Rezoning in Two-Dimensional Lagrangian Hydrodynamics Problems". LA–3455–MS.

1966 Mader, C.L., "The Two-Dimensional Hydrodynamic Hot Spot". LA–3450.

1968 Browne, P.L., and Wallick, K.B., "A Brief Discussion of a Method for Automatic Rezoning in the Numerical Calculation of Two-Dimensional Lagrangian Hydrodynamics", *Proceedings of the IFIP (International Federation of Information Processing) Congress*, 68, Edinburgh, Scotland. Booklet I, p. 131.

D *Stream-Function and Vorticity Method:* An Eulerian method for one material problems involving incompressible fluid flow with confining boundaries.

1963 Fromm, J.E., "A Method for Computing Nonsteady, Incompressible, Viscous Fluid Flows", LA–2910.

1963 Fromm, J.E., "The Time Dependent Flow of an Incompressible Viscous Fluid", in *Methods of Computational Physics*, B. Alder, S. Fernbach, and M. Rotenberg, Eds., Academic Press, Inc., New York.

1963 Fromm, J.E., and Harlow, F.H., "Numerical Solution of the Problem of Vortex Street Development", *Physics of Fluids 6*, 975. Also republished in the AIAA Series of Selected Reprints, Vol. 4, *Computational Fluid Dynamics*, C.K. Chu, Editor.

1964 Harlow, F.H. and Fromm, J.E., "Dynamics and Heat Transfer in the von Kármán Wake of a Rectangular Cylinder", *Physics of Fluids* **7**, 1147.

1965 Fromm, J.E., "Numerical Solutions of the Nonlinear Equations for a Heated Fluid Layer", *Physics of Fluids* **8**, 1757.

1965 Harlow, F.H., and Fromm, J.E., "Computer Experiments in Fluid Dynamics", *Scientific American*, Vol. 212, No. 3, 104 (March 1965).

E *The Marker-and-Cell (MAC) Method:* An Eulerian method, which can be used for multi-material incompressible fluid flow problems. The method permits the treatment of flow with free surfaces, as well as confined flow.

1965 Harlow, F.H., Shannon, J.P., and Welch, J.E., "Liquid Waves by Computer", *Science* ***149***, 1092.

1965 Harlow, F.H., Welch, J.E., Shannon, J.P., and Daly, B.J., "The MAC Method", LA–3425.

1965 Harlow, F.H., and Welch, J.E., "Numerical Calculation of Time-Dependent Viscous Incompressible Flow", *The Physics of Fluids* ***8***, 2182.

1965 Harlow, F.H. and Welch, J.E., "Numerical Study of Large Amplitude Free Surface Motions", *The Physics of Fluids* ***9***, 842.

1967 Daly, B.J., "A Numerical Study of Two-fluid Rayleigh-Taylor Instability", *The Physics of Fluids* ***10***, 297.

1967 Harlow, F.H., and Shannon, J.P., "The Splash of a Liquid Drop", *Journal of Applied Physics* ***38***, 3855.

1967 Harlow, F.H., and Shannon, J.P., "Distortion of a Splashing Drop", *Science* ***157***, No. 3788, 547.

1967 Daly, B.J., and Pracht, W.E., "A Numerical Study of Density Current Surges", *The Physics of Fluids* ***11***, 15.

1967 Hirt, C.W., and Shannon, J.P., "Free Surface Stress Conditions for Incompressible Flow Calculations", *Journal of Computational Physics* **2**, 403.

F *Turbulence.*

1967 Harlow, F.H., and Nakayama, P.I., "Turbulence Transport Equations", *Physics of Fluids* ***10***, 2323.

1968 Harlow, F.H., and Nakayama, P.I., "Transport of Turbulence Decay Rate", LA–3854.

1968 Amsden, A.A., and Harlow, F.H., "Transport of Turbulence in Numerical Fluid Dynamics", *Jour. Computational Physics 3*.

1968 Harlow, F.H., "Transport of Anisotropic or Low-Intensity Turbulence—LA–3947.

1968 Hirt, C.W., "Computer Studies of Time-Dependent Turbulent Flows", *Proceedings of the IUTAM International Symposium on High-Speed Computing in Fluid Dynamics*, Monterey, Calif. August 1968.

G *Miscellaneous.*

1965 Welch, J. Eddie, "Moving Picture Computer Output", *Computing Reviews* ***7***, 355.

1966 Welch, J.E. "Computer Simulation of Water Waves", *Datamation* ***12***, No. 11, 41.

1967 Hirt, C.W., and Harlow, F.H. "A, General Corrective Procedure for the Numerical Solution of Initial-Value Problems", *Journal of Computational Physics* **2**, 114.

1968 Hirt, C.W., "Heuristic Stability Theory for Finite Difference Equations", *Journal of Computational Physics* **2**, 339.

1968 Harlow, F.H., and Amsden, A.A.,"Numerical Calculation of Almost Incompressible Flow", *Jour. Computational Physics*, Vol. ***3***.

1968 "Computer Fluid Dynamics", Los Alamos Scientific Laboratory Motion Picture Number Y–204.

CHAPTER 15

Computers and Reactor Design*

E. H. BAREISS

1 INTRODUCTION

ON DECEMBER 2, 1942, beneath the West Stands of Stagg Field (The University of Chicago athletic stadium) a small group of scientists under the leadership of Nobel prize winner Enrico Fermi witnessed the advent of a new era in science. At 3 : 36 P.M., Chicago time, George Weil withdrew a cadmium-plated control rod from the first nuclear reactor. The energy of the atom was unleashed and controlled![1]

Creation of the first reactor made it possible to release and use the huge forces locked in the atom. This energy was initially employed in wartime, for atomic bombs. Then, over the years, other reactors—more technologically sophisticated, more ingenious, more powerful—were built to channel the energy of the atomic nucleus into peaceful pursuits: the generation of electricity, the conquest of disease, the pursuit of knowledge, the identification, measurement and testing of materials, the propulsion of ships and rockets, and many other uses.[1]

Fermi, it has been said, created the *atomic age* with a slide rule in his hand. Yet, a parallel development emerged: the invention of the sequence controlled high speed digital computer. The atomic energy research and atomic energy weapon studies carried on in the decade after World War II provided a great deal of the necessary impetus for the rapid development of the electronic digital computers; such machines were being designed and built during this war as aides to various research efforts.

---

* Work performed under the auspices of the United States Atomic Energy Commission.

Fig. 1 Arrangement of fuel in the core of the Experimental Boiling Water Reactor at Argonne National Laboratory. This is a view looking down into the reactor vessel with the vessel head removed (Argonne National Laboratory)

The Nuclear Sciences have provided motivation and direction for the Computer Sciences, and the Computer Sciences have provided the means for solving problems in the Nuclear Sciences which could not have been attacked in any other way. The need for scientific computation was felt so strongly by the National Laboratories and other laboratories of the AEC contractors that in the early 1950s they began active cooperation with the Institute for Advanced Study group, organized by John von Neumann, and started development of stored-program computers closely modeled after the machines developed by this group. The machines that emerged from this

Fig. 2 Enrico Fermi (1901–1954) (Drawing from ANL Photo)

Fig. 3 John von Neumann (1903–1957) (Drawing from ANL Photo)

effort were the ORACLE at Oak Ridge National Laboratory, the AVIDAC at Argonne National Laboratory, and the MANIAC I at Los Alamos Scientific Laboratory. Also, at the end of 1952, the AEC purchased model No. 4 of the first commercially available electronic computer, UNIVAC I and installed it at the AEC computer facilities at New York University in 1953. The reliability of this machine together with its large-capacity storage, made it an attractive and much-used machine for reactor design problems.[4,10,45]

While the AEC-sponsored Laboratories continued to produce a second series of general-purpose machines (MANIAC II, GEORGE, ILLIAC II, and even later, MANIAC III), the trend changed toward acquisition of commercial, general-purpose computers. AEC computer research became restricted to the special-purpose applications or the far-out beyond the state-of-the-art efforts, such as the cooperative efforts between Remington Rand and the Livermore Laboratory that produced the LARC and those between Los Alamos and IBM in designing STRETCH. However, the bulk of the computing in the nuclear industry and the AEC laboratories from the late 1950's to date has been done with workhorse commercial machines—IBM 650, 704, 7090, and now the System 360; the Control Data Corporation's 1604, 3000 and 6000 series; the UNIVAC 1107 and 1108; General Electric's 625 and 635; and the Philco 2000 series computers.[10]

Because of the urgency of large scale computations in nuclear reactor design, an extensive exchange of computer codes established itself soon. In 1955, the first bibliography of digital computer programs for use in reactor research was published. In 1956, a nuclear codes group for nation-wide exchange of codes was formed. This consisted of representatives of approximately forty installations. Meetings were held in conjunction with meetings of the American Nuclear Society, and an AEC-supported Newsletter covering new codes was published at New York University. In 1959, the nuclear codes group petitioned the American Nuclear Society for the establishment of a division within the Society to represent mathematical and computing interests. The division was officially set up as the Division of Reactor Mathematics and Computation, with a codes center headquartered at Argonne National Laboratory and funded in part by the Atomic Energy Commission. Although designing and debugging a large reactor code may take several years, the collection of codes at the Nuclear Codes Center has increased quite rapidly. One reason is that a good old code for basic problems tends to stay in demand and that in the normal course of events a code may be converted, without major modifications, for use on newer computers. In the same spirit, the European Nuclear Energy Agency (ENEA) set up in 1964 its Computer Program Library at the Joint Research Center of Euratom at Ispra, Italy. Thirteen countries support this library. As of today, some one thousand codes have been offered to the library.[18]

As further illustration of the growing use of computers in reactor development, the percentage of computer-oriented papers published in "Nuclear Science and Engineering" (The American Nuclear Society Journal) has grown from eight percent in 1956 to about sixty percent in 1969.[18]

Nuclear reactor development was supported primarily from government funds in the past. Now that costs of power generated from nuclear reactors are competitive with costs for other sources of power, a growing interest and activity in commercial enterprise in this field has developed. Computers became important to provide better design: to establish better limits on the power that a given system can produce with safety; to study the cost of power, including analyses of the fuel-cycle and the ecologic effects besides the conventional factors of cost analysis.

The role of the computer in reactor design is the topic of this chapter. It is impossible to cover all aspects of computer–nuclear energy interaction; instead, we have chosen to touch on those topics we consider to be of major interest, in particular nuclear power reactors and digital computers. The slide rule and desk calculator have been replaced by ultra-high speed electronic computers costing millions of dollars. We shall attempt to explain why this is so.

2 TYPES OF NUCLEAR REACTORS

Today's nuclear power plants are similar to conventional thermal power plants insofar as each type uses steam to drive a turbine generator that produces electricity. The method of production of the steam, however, makes an essential difference between a nuclear and conventional plant. In nuclear plants, fission reactions generate heat, and this heat is transferred, sometimes indirectly, to the water that produces the steam.[29]

The fission process requires a particular kind of heavy element, such as uranium (^{233}U, ^{235}U, ^{238}U), plutonium (^{239}Pu, ^{241}Pu) or thorium (^{232}Th) as a basic material. Let us consider uranium. Natural uranium is a mixture of three isotopes: ^{238}U representing 99.27 percent, ^{235}U representing 0.72 percent and ^{234}U representing only 0.0056 percent of the total. An atom of one of these isotopes, uranium-235, can most readily undergo fission when a free neutron strikes its heavy central nucleus. The nucleus breaks into two pieces that fly apart at high speed; in addition, two or three new neutrons are released, as illustrated in Fig. 4. The released neutrons cause a chain reaction by initiating new fissions in other ^{235}U atoms. The kinetic energy of the flying fission fragments is converted to heat when they collide with surrounding atoms. Some other contributions to heat generation other than from stopping of fission fragments are: slowing down of the fission neutrons by collision with other atoms; capture of radiation caused by neutron inelastic scattering; capture of γ-rays emitted in the fission process or when neutrons are captured.[29,41]

Sustaining the chain reaction is important because approximately 30 billion fissions must occur to release one Watt-sec of energy. If the chain reaction

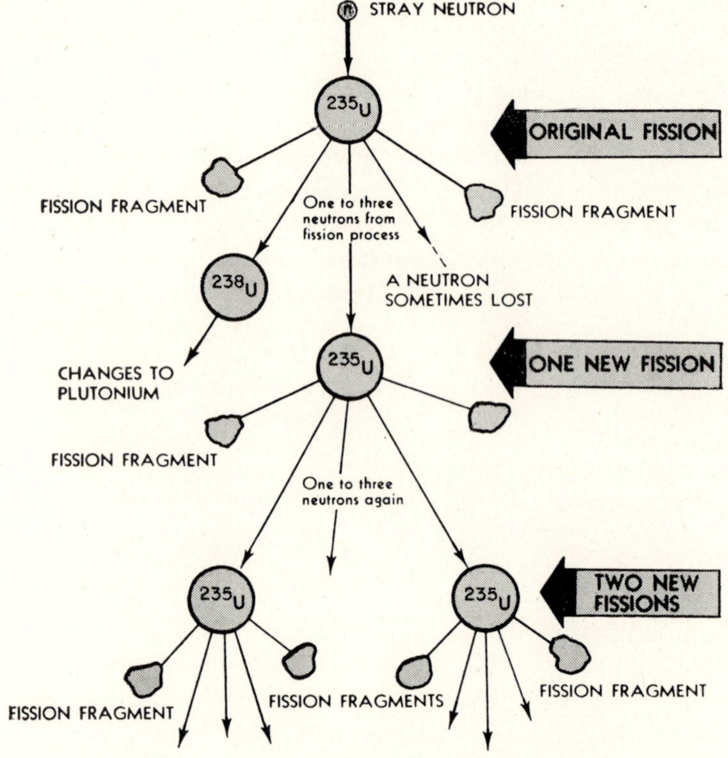

Fig. 4 Schematic of a chain reaction resulting from fission of uranium-235 atoms. Radiation effects are not shown (U.S.A.E.C., Reference 15)

is to be useful, the fissions must occur at a desired rate, and the heat that is generated by the process must be removed. The job of the nuclear power reactor is to provide an environment in which fission reactions can be initiated, sustained, and controlled, and to make possible recovery of the resultant heat. All reactors—whether used for research, power production, or other purposes—have in common this basic function: they allow a neutron chain reaction to take place safely under controlled conditions.[29,30]

Reactors are conveniently classified as *thermal*, *intermediate* and *fast reactors*. In thermal reactors, fission is most likely induced by slow neutrons with a speed of about 7000 ft. per sec ($\approx$5000 mph) in fast reactors by fast neutrons with a speed of roughly 30,000,000 ft. per sec ($\approx$20,000,000 mph). The desirable neutron speed for intermediate reactors lies between the given values. Certain components are found in almost all thermal power reactors (see Fig. 5):

1 The fuel, e.g., slightly enriched uranium or plutonium, which undergoes fission to produce neutrons and to release energy.

2 The moderator, which slows down the neutrons so as to cause fission more readily. The moderator is usually water (because of its hydrogen atoms), "heavy" water, beryllium, or carbon; the lighter the element the better it moderates. (It is undesirable to have a moderator in a fast reactor.)

3 The coolant, which flows through the core to remove the heat released by the fission process. The fuel, moderator, and coolant surrounding the fuel are usually referred to as core (Fig. 1). (The core includes also structural material, control rods, etc.)

4 The blanket, a layer of fertile material, such as uranium-238 (or thorium-232) placed around the fissionable material in a reactor. It captures neutrons, otherwise lost, to produce fissile material such as ^{239}Pu (or ^{233}U) that can be processed for later use as fuel.

5 The reflector, which surrounds the core. (Some neutrons escape from the core, collide with atoms in the reflector, and are returned to the core.) The same substances used for moderators are useful as reflectors.

6 The control system, which safely regulates and shapes the distribution of power generation during the fuel cycle and permits shutdown of the reactor.

7 The shield, which prevents radiation hazards from the intense radiation produced in the core.

Frequently some of these items have more than one basic function; for example, in some reactors, the moderator may also be the coolant, reflector, and shield.[30]

Before we consider the differences in reactors, let us consider something that has a bearing on the development of nuclear power plants in our country: the nuclear fuel resources are not unlimited. It is obvious that nuclear power plants would not have much appeal if they used up the available fuel in a relatively short time. Among the different types of nuclear reactors there are wide differences in the net consumption of nuclear fuel. First, there are

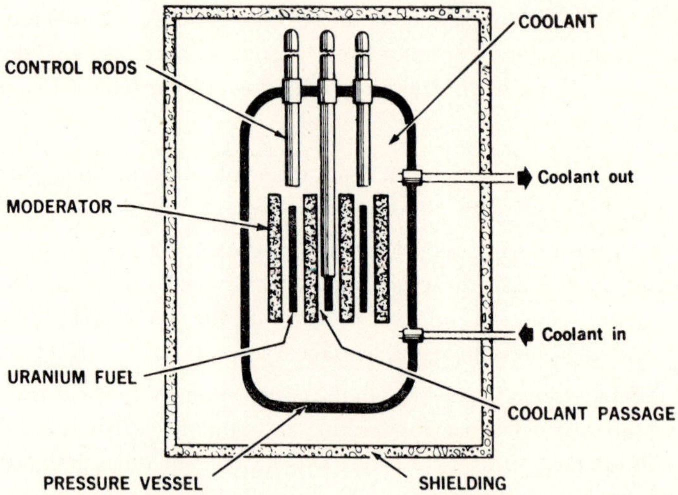

Fig. 5 Schematic of a Thermal Reactor. A power reactor is basically a nuclear heat source in which the fissioning of atoms of nuclear fuel is controlled and the heat of fission put to useful work. This simplified diagram shows the basic components of a thermal reactor (Adapted from Reference 30)

reactors that have a high net fuel consumption; these are used in most of the nuclear power plants operating in the United States today and are usually *Boiling Water Reactors* and *Pressurized Water Reactors*. Next, come the reactors with a low, but positive, net fuel consumption; these have been developed to a considerable extent such as Gas Cooled Reactors and Heavy Water Reactors. The ultimate reactors, insofar as fuel conservation is concerned, are those that have a negative net fuel consumption, which means they produce more fuel than they use. These are known as *Breeder Reactors* and may be popular for central station nuclear power plants that will be in operation in the future.[29]

Because many scientists and engineers concerned with the development of efficient and safe nuclear reactors consider the breeder, at this time, a promise for abundant energy in the future, we describe briefly how a breeder works. This may also serve as a first introduction to some of the many problems in reactor design, all of which demand a quantitative, i.e., mathematical and numerical solution to assure dependable design work.

A uranium-235 or plutonium-239 atom can fission when its nucleus absorbs a neutron. The fission reaction releases free neutrons (as illustrated in Fig. 4 for ^{235}U) that may, in turn, initiate other fissions. Some of the released neutrons, however, are wasted by being captured by the structural material of the reactor, the control elements, or the coolant, or may actually leak from the reactor. The breeder concept puts the wasted neutrons to work and exploits the characteristics of certain fertile materials. A fertile material is a material, not itself fissionable by thermal neutrons, which can be converted into a fissile material. There are two basic fertile materials, uranium-238 and thorium-232. When these fertile materials capture neutrons, they will be partially converted into fissile plutonium-239 and uranium-233, respectively. By careful selection and arrangement of materials in the reactor—including, of course, fissionable and fertile isotopes—the neutrons which are not needed to sustain the fission chain reaction can fairly effectively convert fertile material into fissionable material. If, in the mean, for each atom that fissions, more than one atom of fertile material becomes fissionable material, the reactor is said to be breeding. (If a reactor produces some fissionable material, but less than it consumes, it is called a *converter reactor*.)[20, 29]

One type of breeder reactor that has already operated successfully (EBR-2 of Argonne National Laboratory) is the liquid-metal-cooled breeder reactor; illustrated by Fig. 6. Obviously this system has more components than other types of reactors. One unit which is not necessary for the previously mentioned reactor types is an intermediate heat-transfer loop between the reactor coolant system and the turbine water–steam system. The intermediate loop also uses liquid metal (as does the reactor coolant system), because it has excellent heat-transfer characteristics. The metal may be sodium or a combination of sodium and potassium.

The power reactors which do affect or will soon affect our daily lives are summarized on Fig. 7, giving their geographical distribution. There are other types of power reactors which we have not mentioned because they do not

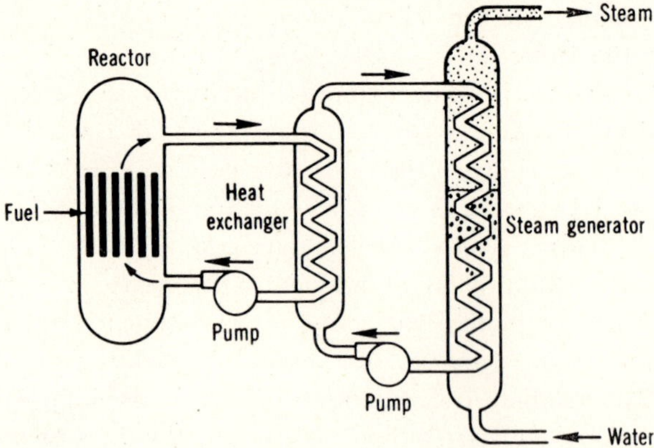

Fig. 6 Steam-supply components in a liquid-metal-cooled breeder reactor (U.S.A.E.C., Reference 29)

look promising at the moment. The problem of conversion of heat is now solved mechanically. However, Los Alamos scientists built and tested a nuclear thermionic conversion unit, called a plasma thermocouple. The goal of thermionic conversion is to convert nuclear power directly into electrical power. In the future, depending upon net efficiency and cost, it may become possible to build reactors out of fuel elements made up of plasma thermocouples at which time thermionic conversion would be a reality.[51]

Besides power reactors, there are hundreds of research reactors all over the world.

Although we have considered in this section only a few physical aspects of nuclear reactors, it is obvious that for any efficient reactor design, besides considerable ingenuity, mathematics and computations are necessary. The next section deals with basic nuclear reactor mathematics in simple terms.

3 ELEMENTARY REACTOR MATHEMATICS

We attempt here to give the main principles of reactor mathematics in a manner that will be comprehensible to an enlightened layman. For the mathematically inclined reader, we refer to the reference list at the end of this chapter, where titles can be found which treat the mathematical aspects rigorously and in detail.

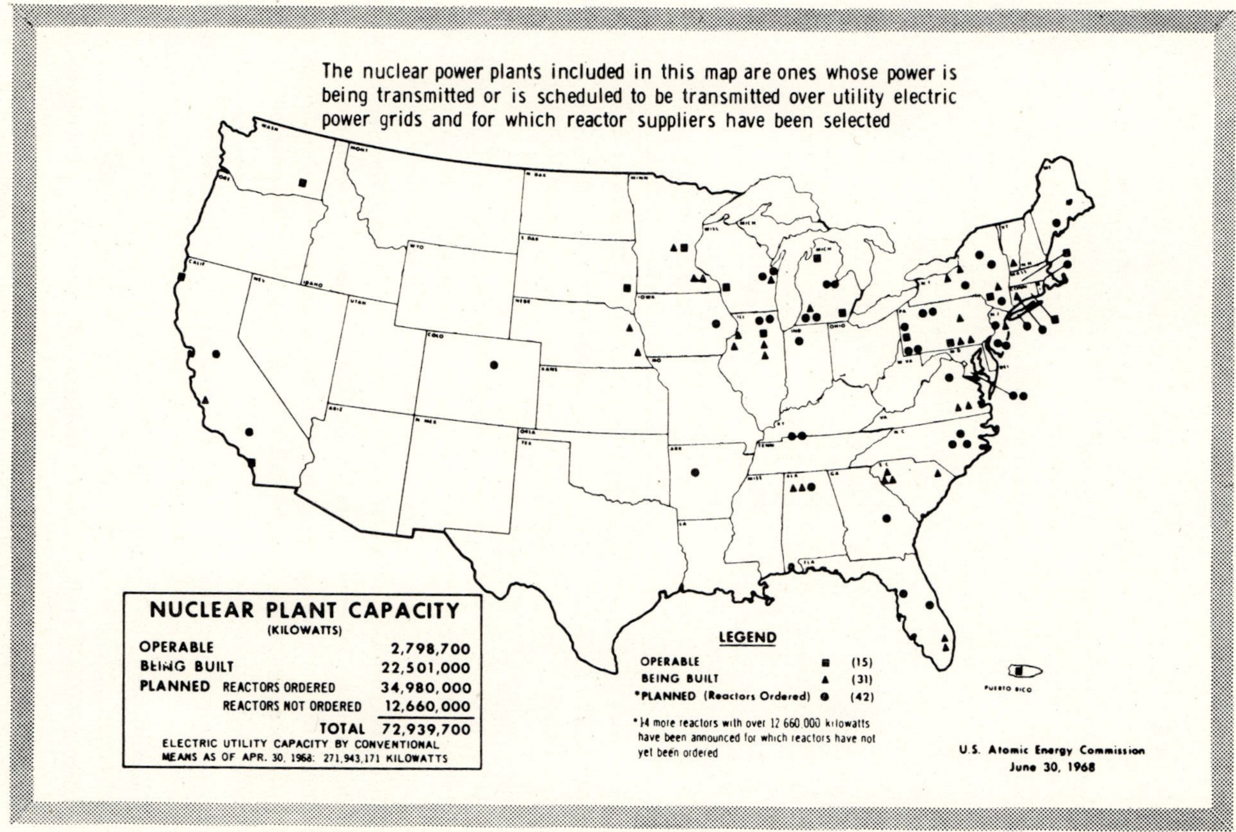

Fig. 7 Nuclear Power Plants in the United States (U.S.A.E.C., Reference 49)

Classification of reactor calculations

Figure 8 lists the four major areas of reactor calculations and the important areas in each. The classification can be carried further to show code types and methods of solutions, but soon loses its conceptual value as it gains completeness. The model is intended to provide a general insight and is an intentional oversimplification of a very complex and overlapping field.[4, 37, 45]

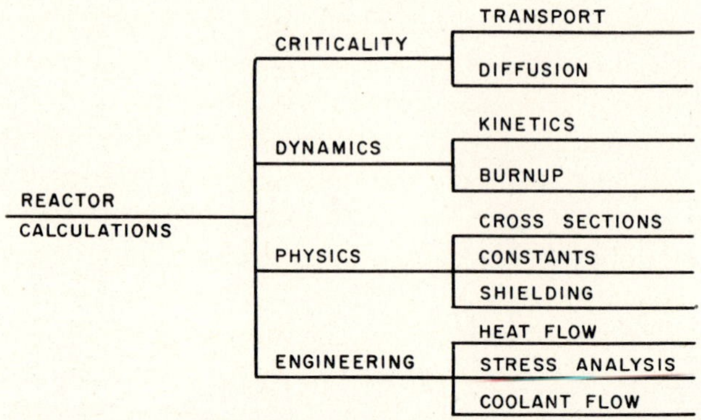

Fig. 8 Classification of Reactor Calculations (U.S.A.E.C., References 4 and 45)

At the heart of all nuclear reactor design is the knowledge of the behavior and distribution of the neutrons in the reactor core. A mathematical theory that predicts this behavior is called neutron *transport theory.* Transport theory takes into account the speed or energy, direction and time-dependence of the neutrons. To simplify the theoretical investigations, one often assumes that neutrons behave isotropically. This means that the neutron flux at any particular point in space is the same in all directions.* The resulting equations are then derived similarly to the equations obtained when considering the diffusion of heat in a body in the presence of heat sources. This approximation is therefore called *diffusion theory.* In certain aspects of reactor design diffusion theory approximations provide, at this time, reasonable results.[6]

There are uses in a design study to investigate a reactor core without con-

* The neutron flux is defined as the number of neutrons of a given direction passing through 1 square centimeter in 1 second.

sidering short term time effects. The time independent transport theory furnishes the mathematical model. Such calculations lead to criticality calculations, i.e., the predictions for the conditions under which the chain reaction can be sustained at a given power level. However, every reactor must be started, when in operation it consumes ("burns") fuel and after a year or so must be shut down for refueling. Therefore, a time-independent approach is a theoretical idealization, and the more difficult problems of reactor dynamics must be solved. Also the possibility of accidents and malfunctioning must be considered. Basic to these calculations is again the transport equation, but now time-dependence, and often delayed neutrons, are important factors to assure that the reactor will not exceed its criticality level by too much, and run out of control. Before such calculations can be performed, the nuclear physicist must supply the data for the coefficients in the equations, such as absorption, scattering and fission cross sections for the neutrons and the alteration of atoms; and constants which characterize the material in the core. Next, reactor shielding is important. The best materials must be found which reflect or absorb neutrons so that the reactor presents no radiation hazard. A thick shield of layers of concrete, iron, or water, or a combination of these, is often used. The basic equations for calculating the shielding thickness is again given by transport theory. However, while the number of neutrons that pass one square centimeter per second may be of the order of millions or billions in the reactor core, at the outer boundary of a shield there should appear only occasional neutrons, commensurate with the ever present neutron dose of the natural atmosphere. In addition, γ-rays which appear as a by-product in the process of neutron capture and in fission, must be absorbed. Therefore, although the transport theory again represents the mathematical model, the numerical techniques to solve shielding problems are expected to be different from calculations of the power distribution in the core. Finally, there still remain the classic engineering problems of heat flow, stress analysis, and safety.

Monte Carlo method

Conceptually, the simplest way to solve the transport problem is to simulate the behavior of the neutrons on a computer. This idea led to the development of the Monte Carlo method. The Monte Carlo method of solving the transport equation may also be considered as a stochastic solution of a multidimensional integral equation. Both concepts are useful. Let us consider the direct simulation of the transport process. A particle, either a neutron or gamma ray, leaves its source, travels through the material, and may react with a nucleus, may be absorbed or may be scattered with a change in speed and direction and continue on through the system. The particle is tracked until it is no longer of interest—e.g., until it escapes, is absorbed, or has a velocity out of range. The simulation of this transport process requires a mathematical description of the geometrical arrangement of the materials; the "cross sections" must be given as functions of location, speed and angular distribution of scattering. The different cross sections define the probabilities with which the straight line path of a neutron is interrupted at a particular point. Thus, a scattering cross section indicates the probability of scattering, a fission cross section the probability that, e.g., a neutron hits a ^{235}U-atom and causes fission, and so on. We also require a mathematical analysis for estimating the quantites of interest; and a well defined program to coordinate the retrieval and storage of information about the past history of the neutrons.[11]

In principle then, almost any transport problem involving processes for which cross sections are known can be solved by Monte Carlo. The practical application of the technique, however, requires the development of methods to reduce the huge number of calculations to a feasible maximum. Because the solution is stochastic, which means it is based on probability assumptions, the sampling of histories that contribute to the answer must be large enough to indicate the typical behavior. For example, if the transmission of a slab is desired, a reasonable sample of those particles that get through the slab must be obtained; the tracking of particles not transmitted is of no use for the end result. Of course, often intermediate results are also required, and then an account for all particles must be made.[11]

Monte Carlo was probably first applied to the solution of transport problems by Ulam and Metropolis in the mid-1940's at Los Alamos. In the following few years, many of the present techniques and ideas were developed by Kahn at the Rand Corp. At first the method was limited to very simple transport problems that used little cross section detail and could be solved on desk calculators. It soon became apparent that precious time could not be wasted tracking unimportant histories, and over the years several techniques for reducing the variance of the statistical behavior of the particle history have been introduced. Some of these techniques use importance sampling (the selection of events is based on a transformed probability distribution so that important events are sampled more frequently); some use the expected value (a part of the random walk is replaced by its average value), and others use Russian roulette and splitting (unimportant histories are terminated with some probability or an important particle is multiplied to give several semi-independent estimates). Nearly all of these variance reduction schemes depend on the user's knowing what is important. Thus much of the development work has consisted in estimating the importance of a particle with respect to its location, its speed, and its direction.[11]

A review paper by Goertzel and Kalos in 1958 already illustrates many of these ideas. It was even shown that for particular problems the solution of the adjoint equation (if known) may be used as an importance function to yield zero variance in the solution of the forward problem; i.e., in theory, if an estimate is obtained only when the random walk stops, then every walk, whatever its path, gives the same result. Since extensive Monte Carlo calculations are in general still expensive, analytic and numerical solutions of the transport equation (to be discussed next) have obtained intensive study.[6,11,22,23]

The neutron transport equation

The equation which describes the time development of the average neutron population in phase space, that is as a function depending on time, position in space, and velocity, is called the neutron transport equation or the Boltzmann transport equation. The name reflects the similarity to the equation derived in "Vorlesungen über die kinetische Theory der Gase" (1893) by the Austrian physicist Ludwig Boltzmann (1844–1906).

Let us define a very small volume element at a point $\mathbf{r}$ in a nuclear reactor such that this volume is unity with respect to an arbitrarily small measure unit. The position and velocity of the neutrons in this volume element change for two reasons. (1) The position changes as a result of the uniform straight line motion of the neutrons. Neutrons enter and leave the volume element (our volume element leaks!). (2) The speed and direction of the motion change because the neutrons suffer collisions. It is also possible for a collision to result in pure capture, in which case the neutron is effectively lost. Furthermore, if collision occurs with a fissionable atom, the neutron might be captured, and often one or more neutrons will be emitted (produced!) either immediately (prompt neutrons) or after some delay in time (delayed neutrons) with different speeds and directions. If we consider neutrons within a narrow range of speed v and direction $\mathbf{\Omega}$ in the volume element, then the rate by which the number of neutrons of speed v and direction $\mathbf{\Omega}$ in this volume element change is given by the equation

rate of change + leakage + collisions
= production of neutrons of speed v and direction $\mathbf{\Omega}$.

In mathematical terms, this leads to the following integrodifferential equation, the *neutron transport* or the *Boltzmann equation*

$$\begin{aligned}\partial N/\partial t + v\mathbf{\Omega} \cdot \operatorname{grad} N + \sigma(\mathbf{r}, v)\, vN \\ = \int \sigma(\mathbf{r}, v')\, v'\, dv' \iiint N(\mathbf{r}, v'\mathbf{\Omega}', t - t') \times \\ \times c_{t'} f(\mathbf{r}, v'\mathbf{\Omega}' \to v\mathbf{\Omega}, t')\, dt'\, d_2\mathbf{\Omega}' + S(\mathbf{r}, v\mathbf{\Omega}, t).\end{aligned}$$

The symbols in this equation have the following meaning:

$c_{t'} f(\mathbf{r}, v'\mathbf{\Omega}' \to v\mathbf{\Omega}, t')$	the probable number of neutrons of speed v and direction $\mathbf{\Omega}$ in the volume element which result from a collision by a neutron with speed v' and direction $\mathbf{\Omega}'$ and which is "produced" with a time delay t' after the collision
$\sigma(\mathbf{r}, v)$	total collision cross section
$N(\mathbf{r}, v\mathbf{\Omega}, t)$	distribution of neutrons in space position, speed, direction of motion, and time
$\mathbf{r}$	position vector
$S(\mathbf{r}, v\mathbf{\Omega}, t)$	independent source strength
t	time
v	speed of neutron
$\mathbf{\Omega}$	direction of motion of neutron.

The solution of this equation for every space point of a reactor and every possible neutron velocity is a formidable task. Analytic expressions have been obtained only for the most simplified versions of this equation, and no computer has ever solved the equation in its general form.[5,6,7,8,21,44]

A great number of different methods have been tried to solve the Boltzmann equation numerically. An often pursued path is to simplify the equation such that the solution to be expected will be just good enough to give a meaningful answer. This simplified equation is then transformed into a form which is more amenable for computational work. Let us consider a reactor core with fuel elements consisting of a great number of parallel, equally spaced, thin and plane plates and operating in a stationary state, that is a state such that time dependence can be neglected. It is then reasonable to assume that the neutron flux will not vary appreciably for neighboring points of a plane parallel to the surface of the fuel plates. However, the flux will change as we pass through the fuel plates and the surrounding moderator in a direction perpendicular to this plane. Therefore, in a first approximation, we retain only this direction as a space variable, and assume the flux to be of constant shape with respect to the other space dimensions.

The neutron flux at any fixed point of the space coordinate is now a function of the direction and speed. Consider only neutrons of a given constant speed and assume the axis for the space coordinate is chosen to go through the centers of the fuel plates. Then the neutron flux can be assumed to be symmetric with respect to the space axis as axis of rotation. Therefore, the directional flux to be calculated depends only on the angle θ between the space axis and the direction of the neutron motion. The range of this angle is between $\theta = 0$ (when the neutrons move in the positive direction of the space axis) and $\theta = 180°$ (when the neutrons move in the negative direction of the space axis). It is for computational reasons advantageous to measure the

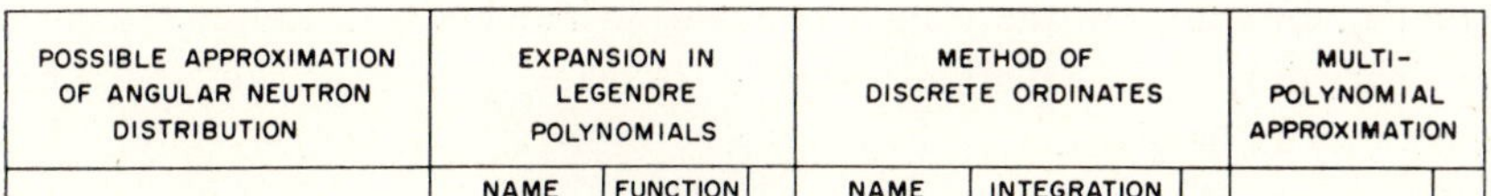

POSSIBLE APPROXIMATION OF ANGULAR NEUTRON DISTRIBUTION	EXPANSION IN LEGENDRE POLYNOMIALS			METHOD OF DISCRETE ORDINATES			MULTI-POLYNOMIAL APPROXIMATION		
	NAME	FUNCTION		NAME	INTEGRATION				
ψ; -1, μ, $+1$	J. Yvon (1951) M. Krook (1955)	$P_n(2\mu+1)$ $P_n(2\mu-1)$	General Electric P_3–Codes	J.B. Sykes (1951)	Double-Gauss Quadrature	David Taylor Model Basin Routines, Westinghouse	Lagrange Interpolation Formulas and Algebraic	Integration	Argonne National Laboratory (1958)
ψ; -1, μ, $+1$	B. Davison J.C. Mark (1944)	$P_n(\mu)$		S. Chandra-sekhar, G.C. Wick (1943)	Mechanical Quadrature				
ψ; -1, μ, $+1$				V. Kourganoff (1950)	Newton-Cotes				
ψ; -1, μ, $+1$								B. Carlson S_n–Codes	LASL (1953)

Fig. 9 Classification of Existing One-Dimensional Transport Theory Routines in Slab Geometry (U.S.A.E.C., Reference 6)

angle θ by its cosine $\mu = \cos\theta$. The range of μ is then between -1 (for $\mu = \cos 180°$) and $+1$ (for $\mu = \cos 0$). We note that for $\theta = 90°$ or $\mu = 0$, the neutrons move in planes perpendicular to the axis and parallel to the surface of the fuel plate. For this most simple example, two mathematically equivalent methods have been used most widely to obtain numerical solutions (Fig. 9):

The first method assumes that the solution for the neutron flux ψ of our simplified transport equation can be approximated by a finite sum of Legendre polynomials $P_k(\mu)$, i.e.,

$$\psi = a_0P_0(\mu) + a_1P_1(\mu) + a_2P_2(\mu) + a_3P_3(\mu) + \cdots + a_nP_n(\mu)$$

where the coefficients a_k are dependent on the space variable. The Legendre Polynomials are polynomials such as $P_0 = 1$, $P_1 = \mu$, $P_2 = \frac{1}{2}(3\mu^2 - 1)$, $P_3 = \frac{1}{2}(5\mu^3 - 3\mu)$, and with fast increasing complexity for the following $P_k(\mu)$, and have the important property to yield the best possible polynomial approximation (in the least square sense) if the term $c_{t'} f(\mathbf{r}, v'\mathbf{\Omega}' \to v\mathbf{\Omega}, t')$ in the Boltzmann equation can be approximated by a constant. These approximations are known as P_n-approximations and were used as early as 1944 (Fig. 9). In addition, the transport equation reduces to a finite system of ordinary differential equations for which in many cases an elegant analytical solution can be given.

The second method determines the solution for the neutron flux ψ of our simplified transport equation for predetermined discrete angular directions θ_k or direction cosines $\mu_k = \cos\theta_k$ $(k = 0, 1, 2, \ldots, n)$. Under the same assumptions as before, and spacing the discrete ordinates according to a procedure by Gauss (1777–1855) for the numerical integration of a function over the interval $[-1, 1]$, the two methods can be shown to be mathematically equivalent. This means, the two methods yield theoretically indentical mathematical errors for the solutions ψ of the same original problem. The method of *Discrete Ordinates* was already used in 1943 to solve transport problems in astronomy (Fig. 9).

In practice, the approximation of the directional flux by a single polynomial equation leads very often to unsatisfactory results for a variety of mathematical reasons which we will not enumerate here. Furthermore, in many problems the directional flux can have a discontinuity at $\mu = 0$, e.g., at the points where fuel and moderator (say uranium and water respectively) are in contact (see the uppermost diagram in Fig. 9). In order to account for this discontinuity, the flux ψ over the two intervals from $\mu = -1$ to $\mu = 0$, and from $\mu = 0$ to $\mu = 1$ are often approximated independently. In the case of expansion in Legendre polynomials, the function ψ takes then the form, $\psi = \psi^-$ when $\mu = 0$, $\psi = \psi^+$ when $\mu > 0$ where

$$\psi^- = a_0^- + a_1^- P_1 (2\mu + 1) + a_2^- P_2 (2\mu + 1) + \cdots$$

and

$$\psi^+ = a_0^+ + a_1^+ P_1 (2\mu - 1) + a_2^- P_2 (2\mu - 1) + \cdots$$

In the method of discrete ordinates one treats similarly the intervals $[-1, 0]$ and $[0, 1]$ independently by assigning to each the most desirable "Gaussian abscissas" μ_k.

The reactor engineer's main interest is not in the directional neutron flux but in the scalar neutron flux. The scalar neutron flux is determined by the area under the directional flux ψ for the interval $[-1, 1]$, multiplied by 2π, and its distribution in the reactor core is a principal factor in determining the efficiency and power of a nuclear reactor. Hence, for engineering applications, it is not necessary to use the best theoretical method to obtain the directional flux, but the most practical methods that yield "good enough" answers for the scalar flux. As illustrated in the third diagram of Fig. 9, equally spaced coordinates are often used. One can simplify even further and approximate the directional flux between the $n + 1$ equally spaced discrete coordinates by n segments of straight lines as illustrated in the last diagram of Fig. 9. A set of computer codes, based on this approximation, are known as S_n-codes and, with appropriate precautions and built in checks, have proven very useful.

What all this means is this: For different special cases of investigation, individually designed approximations will yield in general the best results. If one restricts the approximation functions for the directional flux to polynomials over any number of subintervals of $[-1, 1]$ we have an infinite variety of possibilities to solve the problems. Fortunately, all the methods can be summarized by the general method of "*multipolynomial approximation*", for which a general computer code can be written (see last column of Fig. 9). Therefore, the two different methods described above represent only specific, although important, examples for a general approach.

In a similar fashion, one can make approximations in the space variable. It is then the task of the numerical analysts to combine these approximations and derive a set of reliable formulas for each point of interest in the reactor core that can be evaluated step by step by a high speed computer. What we have considered so far, and this should be borne in mind, is only one phase of a simple problem in neutron transport in one space dimension. Let us consider next, without taking resort to mathematics, the mechanics of treating a problem in two space dimensions.

In calculating the space dependence, one often covers the entire reactor with a hypothetical grid, but suitable for mathematical treatment, and then selects a representative part, called a cell, for which calculations can be performed within a reasonable time. If one is fortunate, the configuration of a reactor core may look like the Yankee Core I, as illustrated in the left side of Fig. 10, where the grid and the cells are obvious. The left side of Fig. 11 shows a cell for which power calculations have been made. This cell has been further subdivided to obtain a network of small enough mesh width to guarantee sufficiently accurate calculations. The accuracy that can be expected from such theoretical predictions has been proven quite satisfactorily by comparison with experimental results.

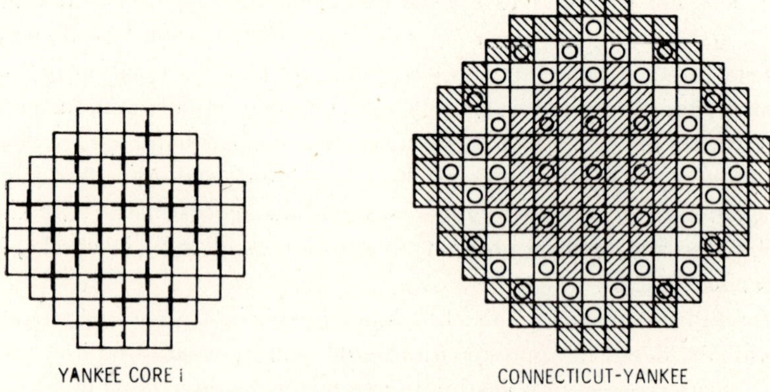

Fig. 10 First and Second Generation Reactor Core Configurations. Solid cruciforms in Yankee Core I are control rods. Circles in Connecticut-Yankee Core represent control rod clusters. Shaded and blank areas designate enriched fuel zones (Westinghouse Corporation. Adapted from References 31 and 33)

However, the high performance requirements for the current reactors demand an even more accurate knowledge of the power distribution and thermal conditions within the core. A major problem due to the large size of these reactors is the increased sensitivity of the power distribution to small changes in reaction rates in the various regions of the core.[31, 33]

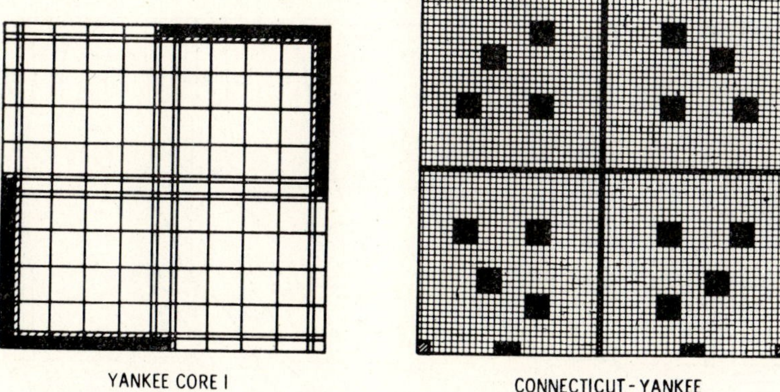

Fig. 11 First and second Generation Detailed Assembly Calculation Mesh. Solid areas designate possible control rod positions; other areas designate fuel. A collection of four quarter-assembly representations with symmetry at outer boundaries are given for Connecticut-Yankee (Westinghouse Corporation. Adapted from References 31 and 33)

The larger and more complex reactors of today, with about 40,000 fuel pins in the reactor core, with partially inserted control rods, and absorption which is depleted in time, also require a refined spatial representation. This is illustrated again by Fig. 10, which gives a comparison of a typical two-dimensional representation for a first and the later generation Pressurized Water Reactor. Figure 11 shows the corresponding detailed calculational mesh. The simplicity of the first generation reactors made it reasonably amenable to a rather gross representation, whereas the complexity of the recent cores requires a detailed representation that is practical only on advanced computers.[31, 33]

As an aid to the design engineer, computer generated graphical displays have become available. These range from simple curve plotting of power and flux distribution to isometric projections and perspective stereo projections. For example, a contour plot of power in a quarter section of a core of the Yankee type is presented in Fig. 12. More detailed contour plots of the type displayed are employed in power distribution and fuel management studies.[31, 33]

Space–time reactor kinetics is another area of interest in reactor analysis that requires a high degree of mathematical sophistication and computational

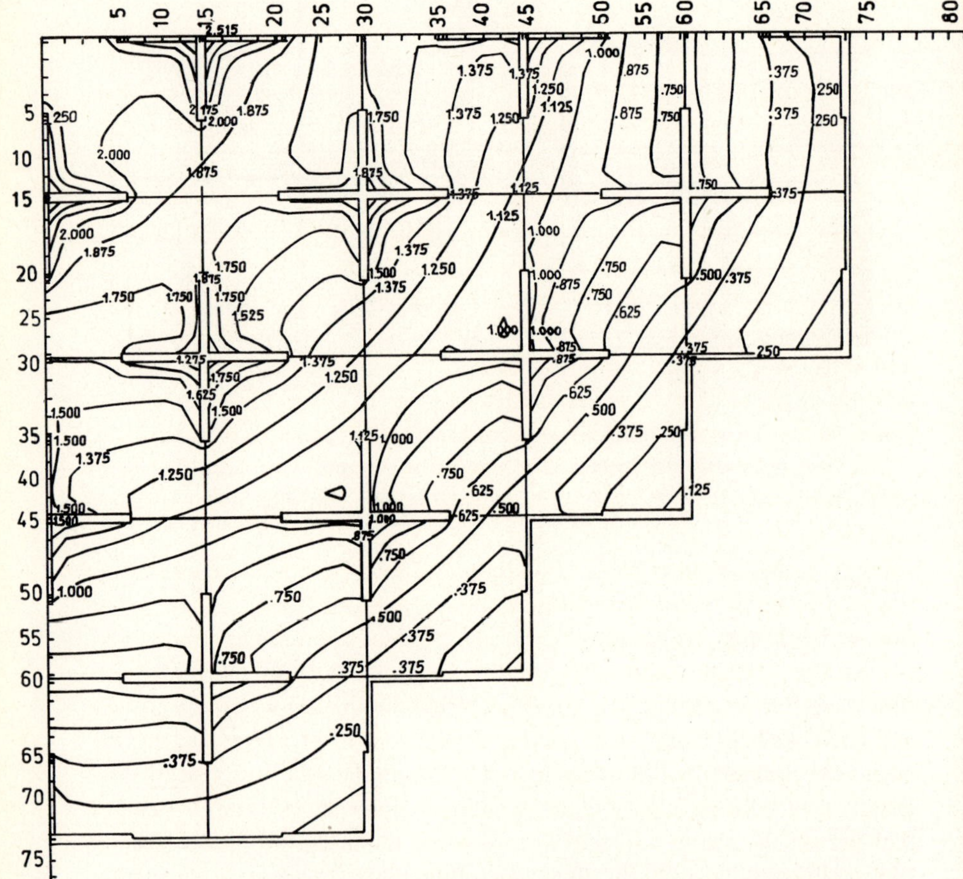

Fig. 12 Relative Power Contour Plot for a Yankee-type Reactor. Figure was generated on automatic plotter coupled with computer. Contours are given for relative power in core (Nuclear News, Reference 33)

power. Also, radiation shielding, in particular analyzing neutron- and gamma-shielding problems, belong to the same category. Because they are also based on the Boltzmann equation, the development of shielding methods has been closely related to nuclear core analysis, and computer programs have been shared between the two disciplines, with appropriate modifications for the particular application.

Engineering calculations

We have explained so far only the neutron transport problem. This problem, centrally important as it is, represents only one part where intensive mathematical and computational research is performed. Neutron physics, basic to nuclear reactor design, also poses a steady demand on mathematical and computational research. But, side by side, the classic problems of engineering are always present, and stress analysis, as an example, remains as important as ever. It is interesting that similar types of equations occur in neutron diffusion and in stress analysis. Thus similar mathematical techniques are applied, and with the increase in power of the electronic computers, a more reliable analysis of the reactor vessels is possible. Figure 13 shows that three-dimensional stress analysis calculations for reactor vessels are now a reality.

The usefulness of mathematical solutions by computer in still other fields of reactor power plant design has been growing very rapidly: calculations of coolant flow, void distributions, heat transfer, and vibrations have become routine on the computer, with virtually every design engineer either modifying or utilizing standard codes from the design program library.* Data reduction, curve fitting, and automatic graphic reproduction are common practice among experimentalists and reactor operations analysts. Fuel and structural material properties as a function of irradiations have been analyzed, and correlated on computers. Analog simulation on digital computers has been used successfully in systems analysis. Comprehensive cost analysis programs have been developed and preliminary optimization codes prepared that permit the sizing and selection of plant components on the basis of specified performance and operational constraints. Management guidance for project scheduling and control has also been made more effective by the use of computers.[33]

So far, we have talked about a wide array of mathematical programs that solve individual problems of reactor design. The real problem for the reactor designer is now to make efficient use of the available computer codes. These aspects are discussed in the following section. However, before we go on, it seems worthwhile to discuss briefly just what it is that makes a computation

---

* A computer code (or program) is a detailed sequence of precise instructions in such terms that a computer can execute them to solve a given numerical (or algebraic or logic) problem. Usually, codes are first written on paper, then "transcribed" on punched cards or magnetic tape, which serve as input for the computer.

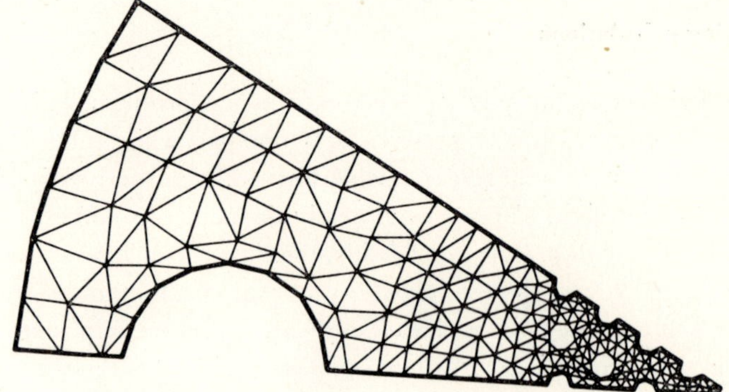

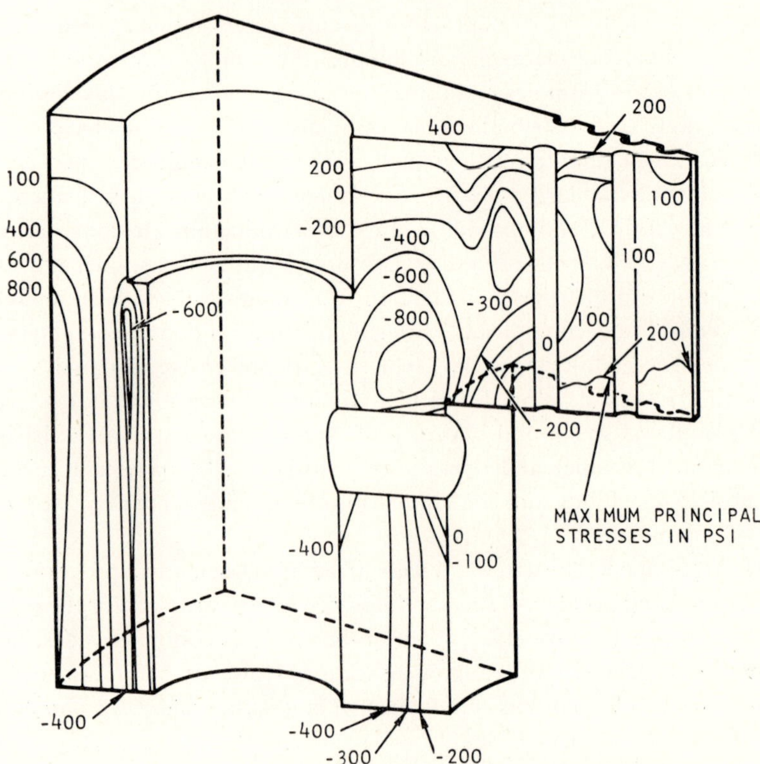

Fig. 13 Example of Three-dimensional Analysis: Prestressed Concrete Reactor Vessel (Gulf General Atomic, Inc., Reference 36)

problem difficult or complex. One way of looking at a numerical problem is to count the number of variables on which the function to be evaluated depends. A function of a single variable, as defined by a differential equation, is often easy to compute, even by hand, whereas a function of two or three variables almost always requires machine computation. The reason for this is straightforward. Suppose for the sake of argument that we know a distribution well enough if we know its value at 100 points. Then, in order to have the corresponding amount of information for a distribution depending on n variables we need to know the function at 100 points. As an example, assume we want certain neutron transport calculations. Here we have three variables for the space position, one variable for the neutron speed, two variables for the direction of the neutron velocity, and one time variable, i.e., a total of seven variables. Thus, for this particular case, the number of points to be calculated would be 100^7 = 100,000,000,000,000 = 100 trillion points, an impossible task![39]

Now, whatever the complexity of the analytical problem, ultimately the calculation is going to come down to dealing with each value of the answer function and performing some arithmetical operations on it, and on other related values of the function. Evidently the amount of work that has to be done will be in some manner proportional to the amount of information that has to be handled, and the number of calculations that have to be carried out with each piece of information. If the problem involves a flux distribution in space and time, the preceding argument indicates that we are going to have to deal with 100 trillion function values and carry out arithmetical operations with each one of them. This is the reason why general transport calculations in space and time are not carried out, even though the basic equations describing what goes on in a reactor are known well enough to make some of these calculations possible in principle.[39]

Since quite useful design calculations are carried out on a routine basis, however, there must be a way around this difficulty. From the standpoint taken here, there is only one way to solve a complicated problem which in itself is beyond one's capacity: that is to break it down into a series of simple problems which, as a whole, take less time to solve than the original complex problem. This can be done by assigning to all functions needed to calculate the required solution of the problem an average value with respect to all the variables but one, and then calculating the dependence of the solution as a function of the single remaining variable. The approximation can only be good if there is no strong interaction between the various sections of the problem. The resulting advantage is that the problem is reduced to a set of one-dimensional problems which, on the same basis as before, would require handling something like $100 \cdot n$ pieces of information—a very great reduction over the direct calculation. In our example, the calculations would reduce from evaluating 100 trillion points to evaluating only $100 \cdot 7 = 700$ points.[39]

This approach applies not only when the complexity is due to many independent variables but also when the complexity is due to the interaction among various physical boundaries in the system such that the problem could be solved if either boundary were absent or replaced by some average effect. This general approach, of course, is exactly what leads to the separation of flux calculations into a local cell and a gross core flux calculation; and to other procedures for synthesizing two- or three-dimensional flux distributions from one-dimensional ones. Unfortunately, the general method* of "separation of variables" is not always possible. Then, only mathematical ingenuity or engineering intuition can help overcome the computational difficulties.

4 THE PROBLEM OF COMPUTATION IN NUCLEAR DESIGN

Design philosophy

The use of digital computers for nuclear design has grown to such an extent in the past twenty years that their importance is thoroughly recognized today. The need for a computer in designing a high-performance reactor is quite comparable to the need for critical experiments.† All national laboratories have large computers and mathematics groups. So do all large commercial companies.

In this section we are concerned with the general philosophy of design and why a computer is useful, how a modern computer affects the organization of the reactor codes that are to be used, and the influence of the computer on reactor physics and analysis needed in developing codes. In order to draw some general conclusions, we find it necessary to bypass some detailed facts.

---

* Maybe we should better call this method "stripping off variables".

† A critical experiment verifies or supplements calculations of the critical size and other physical data affecting a reactor design. The power is kept so low that no heat removal system is required.

Perhaps the main one is that almost any statement about digital computers should be dated because of the rate at which their capacity has been changing (and also the ability of the user to utilize that capacity). The capability in terms of operating speed of the largest available computer has more than doubled every two years. While this tendency obviously will not continue indefinitely, it has certainly been true for the past twenty years. A second fad is that the discussion will be in regard to the design of complex, high-performance reactors. In the case of a simple reactor, where the essential physics calculations can be done with pencil, slide rule and desk calculator, there is little need or justification in going to a computer to carry out the design work. It is primarily in the case of a high-performance system where optimization and refinement have a large economic payoff, that it is feasiable, and indeed essential, to make use of mechanized computations.[39]

We start by asking what engineering design is. In general, and often in vague terms, the designer is given a class of physical systems and a description of expected behavior, or performance. The system is undetermined to the extent that many of the parameters which determine the performance are design variables. The designer's problem is to choose those values which will give the best possible performance consistent with engineering and economic limitations and requirements of safety. Unfortunately, there is no direct method of solving such a problem. For this reason, experience, common sense, intuition, and luck play a large part in the design process. If the engineering system is a simple one (and simple in this context means having few design variables), if the interactions between the design variables are not too strong, and, most important, if similar systems have been designed in the past, the designer can do fairly well with a handbook approach. If the above conditions are not met (and they often are not met in the design of a new type of reactor), the designer must choose many combinations of the design variables, analyze the performance of each resulting system and choose the one which gives him optimum performance. The process is far from straightforward since in carrying out the analyses he must gain the intuition which is necessary to keep the number of cases from growing out of all bounds.[39]

Conversely, the analysis of a final design configuration, e.g., as submitted by competitive bidding for a contract, would be a relatively straightforward problem. The extent to which it can be done is solely a question of the state of the art at the time the design is in process, depending mainly on how much physical data is available and what analytical techniques are practical at the time. The designer's prime interest is in the relation between the possible design configurations and their performance, and not in the mathematical sophistication and techniques of computing this performance. To do his job efficiently he will spend most of his time in establishing configurations that meet the design constraints, trying to integrate the various conflicting requirements of neutronics, heat transfer, materials and structural limitations on the design in order to get a system that will work. He seeks to understand the relationship of the design variables so that he can decide what changes will make the performance better. What the designer may dream about and desires, is a complete package or black box into which he can put a description of a particular design, and which will present him with a prediction of the system performance. (The prediction would not generally be as accurate as the designer would like, but at least he should obtain some estimate of its error.) In partially fulfilling this basic requirement the high speed computer finds its natural place. Computer codes such as described in the previous section can effectively carry out the repetitive calculations of system performance, leaving the designer free to think about his primary responsibility—designing the best system he can.[39]

General organization of computer codes for reactor design

While the preceding remarks apply to design in general, they are especially true for the case of nuclear reactor design. The reasons for this are clear: A power reactor is a complex system with respect to the number of design parameters that must be determined, and the strong coupling between the design variables. The system cannot be designed by optimizing for a single variable at a time but must be designed as a whole. In addition, because of the competition from non-nuclear systems for both power and propulsion, it is necessary to carry on a sophisticated design analysis. Finally, the high cost of power reactors and the fact that often experimental data are not available until after major design decisions have been made, require great reliance on advanced analytical and computational work.[39]

Although it is not desirable to separate nuclear design (including core, blanket, and reflector) from the design of the system as a whole, it is convenient to do so. One part of nuclear design is the calculations which insure that the reactor will go critical and that it can be controlled. If the reactor is a power producer, the calculations must insure that the power is generated where it can be used effectively; they must insure that the cost of power is low; they must insure that the reactor can be operated safely. If it is a breeder reactor, the calculations must insure that the production of fissionable material is efficient. It is important to emphasize the difference between this basic design work as discussed above and the work in reactor physics and mathematics which creates methods for the analysis of specific configurations.

The kind of reactor codes which the designer will need depends to a very large extent on economic considerations.

To begin with, the largest available machines are usually the best ones because of the lower unit cost per computation and because of the more advanced software systems available which greatly reduce the programming time. Besides the speed advantage, the larger fast-access memory reduces the premium on cleverness in getting an efficient code. Furthermore, simple as well as complex problems should be done on a computer whenever they are performed very often. Calculations that might take only a day by hand may take less than a second on a modern computer. Thus it is not a good idea to restrict the use of a large machine to lengthy transport theory or diffusion theory codes.[39]

The input and output form a very important part of the code and must be given as much attention as the physics involved. The general principle is that the input, wherever possible, should require a minimum of work by the designer to describe the system. It should contain primarily engineering quantities, and not mathematical or theoretical decisions that can be made by the computer when properly coded. Of course, the requirement of a minimum input complicates the internal logic of the whole code. However, the programming has to be done only once, whereas data input sheets have to be filled out every time the program is used. Another important point, which underlines the advantage of a large computer, is that all data transfers should be mechanized as far as possible. There should be no transcribing of data by hand or intermediate hand calculations if this can be eliminated; people often constitute the weak links in any data processing system.[39]

The last point concerns the organization of a computation system for reactor design in blocks or modules. As soon as a code is developed and in operation for a short time, both the mathematician who devised it and the users will have thought of all sorts of improvements which should be incorporated. Since revising a code which has been completed is not a simple matter, especially if the original programmer is not available, it is very desirable for the code to be organized in independent blocks so that changes in one part of the computation will not mean revising the entire code. The second advantage of this block organization stems from the economic requirement of generality. It would be convenient for the designer to have just one reactor code, that would analyze any reactor no matter how complicated. In practice this is an impossible task because of the tremendous range of designs that are considered. It would be too costly if only detail studies are needed. The alternative is to organize the computation into what might be called basic or unit operations which apply to any reactor, and meet the requirement of generality by various combinations of these basic blocks, or modules. An attempt to satisfy this demand is the creation of the ARC system discussed below.

Because of the huge amount of computations required, the following question arises: What "kind" of accuracy is expected by the designer to make these calculations worthwhile? We have seen in the last section that by the general method of "separation of variables" the amount of calculation can often be reduced drastically, but with a price in loss of accuracy.

While the accuracy of this kind of approximate calculation is less than that of a direct approach, its usefulness is illustrated by scoping analyses and scoping calculations. Since the designer is interested in comparing a wide variety of possibilities on a common basis and drawing conclusions on their relative performance, and since a critical experiment will almost certainly be done, he has much more interest in consistent results which show the right qualitative trends than in accurate and detailed results limited to a few specific cases. An example of this is the war-time work on graphite production reactors, in which diffusion theory was used successfully for lattice cell calculations even though it was known to be inexact.

Also the lack of sufficient nuclear data may very often fail to justify anything more than a simple approximate calculation. The design computation system, to some extent, should be regarded as a curve-fitting system in which gaps in the input data and ambiguities or uncertainties in the physics are resolved by fitting whatever experimental data is available. The designer is interested in a practical design system that can be used within the deadlines which govern the reactor project, rather than in questions of theoretical accuracy or extreme refinement for its own sake.[39]

The Argonne reactor computation (ARC) system

While the reactor physicist is primarily concerned with solving the Boltzmann equation for a particular physical configuration, we have seen that the reactor designer is involved with numerous adjunct problems to further define the interaction of the reactor with its environment. He has to cope with technological and economic constraints.

Table I summarizes the problems confronting the reactor designer as discussed in the previous paragraph. Three broad categories are indicated, and several examples are listed in each of these.[42]

Table I Scope of reactor problems

Problem category	Examples
Cross-section Manipulations	Unresolved resonances, resolved resonances, elastic scattering, inelastic scattering, n, $2n$ considerations, thermal-region treatment, ...
Solution of the Boltzmann Equation	1D, 2D, 3D diffusion theory, Pn-transport theory, Sn-transport theory, Monte Carlo treatment, ...
Adjunct Calculations	Fuel cycle, neutron inventory, perturbation analysis, kinetics, thermodynamics, hydrodynamics, engineering considerations, spatial synthesis, economics, ...

In the past, individual reactor problems were solved with the help of computers by using a number of self-contained and unrelated codes, each having unique input requirements and furnishing results on a variety of output media such as printed page, tape, and punched cards. Problems based on the application of several codes, where the results of one code are (partially) needed as input for another, normally required several unrelated computer runs with intervening human manipulation of data. The reason for this incoherency is, of course, historical. At the beginning of nuclear technology, when the direction of reactor development was quite uncertain, when the computer industry was in its infancy, and when the programming of even a small code was both a mathematically laborious and artistic enterprise, each individual group tried to optimize the achievement of its own research goals. Machine time was still expensive and not easily available. But soon the scientists started to exchange codes among different laboratories. Though input and output were not standardized, the exchange resulted in great savings of time and manpower for the immediate needs.

However, there have been early attempts to automate the computer usage with regard to a sequence of inter-related calculations. In 1956, General

Electric's Knolls Atomic Power Laboratory (KAPL) started with the KARE system. The Argonne National Laboratory conceived the idea of the MACH 1 system in the early sixties. Then, with the appearance of the third generation computers and their greatly improved speed and direct access-storage facilities, the idea of "control of linked calculations" became popular because of the anticipated economic advantages. In 1964 KAPL started programming the NOVA system, in 1965 Argonne began its first version of the ARC system, and by 1968, the Savannah River Laboratory had announced its JOSHUA system, the French AEC its CODNUC system, the Japanese Hitachi, Ltd. its NCCS (Nuclear Codes Control System) and EURATOM its CARONTE.[46]

We shall describe only the Argonne Reactor Computation (ARC) system because it is representative of the current trend in linked calculations.

The ARC system, which has been implemented for an IBM System 360 Model 50–75 computer complex, represents a break with past computer methods. It enjoys two major advantages as compared with conventional computer codes. These are:

1 The ability to automate the linkage of computational modules in any complex manner, without human intervention for data handling, in order to attack sophisticated reactor problems.

2 The ability to store away, and, at a later time, retrieve data generated by an ARC system run for recomputation of previous problems with data changes, or for extension of previous problems.

The ARC system consists of

1 *The Datapool:* The collection of data and codes that can be used by the ARC system.

2 *Computational Modules:* Collection of conventional subroutines.*

3 *Standard Paths:* Control programs similar to the driver or executive portions of large conventional codes containing numerous subroutines or overlays.

---

* Subroutines are codes that can be called upon by other codes to solve a given problem. E.g., codes to calculate trigonometric or exponential functions, etc. are often in the form of subroutines.

4 *Operating System:* Collection of subroutines which form the inter-relation between the ARC system and the computer hardware.

The computational modules and standard paths are directed by the operating system in an automated and unified manner.

The route to be taken through the ARC system may be determined by selecting a previously cataloged FORTRAN IV control program known as a standard path, or by programming the required path at run time. The only limit to the complexity of the route to be taken through the ARC system is that imposed by the ingenuity of the user.[42]

The individual computational modules are restricted in scope. For example, neutronics calculation modules such as for one-dimensional diffusion, or two-dimensional transport problems, are concerned only with the algorithms involved in computing neutron fluxes, multiplication factors, or whatever intermediate result is needed. They are, however, not involved in manipulating any input or output.[42]

Furthermore, the computational modules are standradized with regard to their interface quantities. In particular, the adjunct calculation modules, which depend to a large extent on data generated by the neutronics-calculation modules, will be compatible with any one of these modules. Thus, for example, only one Neutron Inventory module is required for the ARC system, rather than one for each different type of neutronics-calculation module.

Once the interfaces are defined for any of the computation modules in the ARC system, these modules can be programmed with limited knowledge of other modules of the system. This aspect of the system is extremely important in that it permits the inclusion of additional computational modules without any disruption of the systems. It also encourages module exchanges with other laboratories since the generally tedious input–output efforts are handled separately. The modules are programmed using FORTRAN IV, a widely used computer language, in order to facilitate their use by other installations.

The independence of the various computational modules is actually made possible by use of the datapool concept.

The ARC system uses a datapool in which all external inputs, the computational modules, the standard paths, the utility programs, and the results generated by all the computational modules are stored under the direction of the operating system. Similarly all the data needed by a module are retrieved from the datapool by the operating system.

In its most elementary form, the datapool consists of files on one or more magnetic tapes. However, since the disk and data cell storage capacity of modern computers is measured in the tens of millions, modern random-access memory devices are more appropriate and are utilized for the datapool of the ARC system, supplemented by tape storage for low-activity files.[42]

Figure 14 is a schematic representation of the ARC system. The configuration represented corresponds to the situation in which the operating system has transferred control to a standard path which has in turn called a particular computation module into the fast core from the datapool.

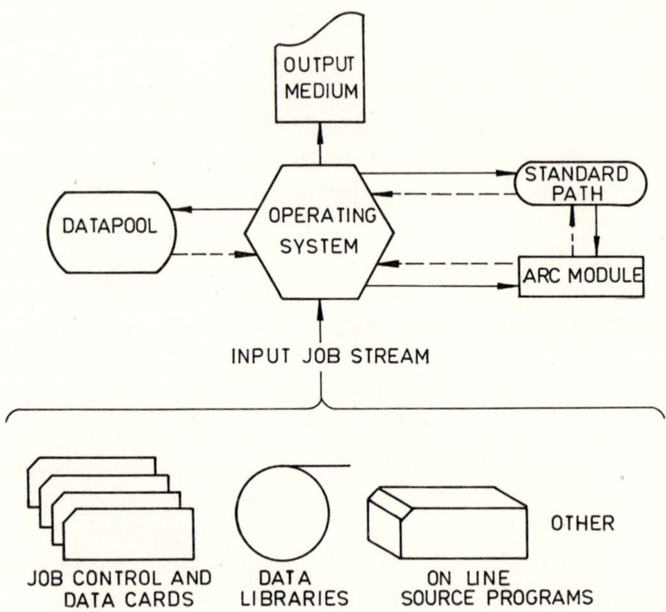

Fig. 14 Schematic Representation of the ARC System (Argonne National Laboratory)

The ARC system is open-ended and can be extended indefinitely. In particular, use of the control FORTRAN compiler is available so that the user may generate at run time a module or path not yet available in the system. In the latter case, the program generated is designated as a nonstandard path. Clearly, all standard paths were originally nonstandard paths.[46]

From the user's point of view, the ARC system facilitates reactor-design calculations that involve many independent computations. These computations may be processed serially. They may also perform logical decisions.

A standard path may be used to link ARC system program modules in such a manner as to effectively generate the equivalent of a present-day self-standing code. On the other hand, because of the flexibility of the system, one may easily use standard paths to perform comparative calculations, either with regard to variations in input data, or with regard to computational algorithmic methods. Indeed, since logical decisions can be made during execution of a path, the particular algorithms to be invoked need not be predetermined. Such paths involving branches based upon logical decisions use the full capability of the ARC system.

Finally, we may mention some additional system-capability of the ARC system planned for future implementation:

Rapid retrieval of stored data sets to verify that they are the particular sets one wants to use for other problems. These data sets will be displayed on remote cathode-ray screens.

In addition to knowing the contents of the data sets, it is useful for the casual user to have at hand a brief description of the data-set contents. This can easily be accomplished via automated report-writing features. It is profitable to generate these reports on a transitory medium such as an image on a remote cathode-ray screen, at the user's option.[42]

We may also foresee in the future the use of a remote console with light pen and extended keyboard to monitor long problems by printing out on request the current status of certain data sets or particular elements of these data sets. Such a facility would provide the capability of data revision during execution in order to expedite long calculations.[42]

The ultimate objective of the ARC system is to provide a flexible computation system of high utility to facilitate solution of any of the problems of interest to the reactor designer. Due to the ever expanding range of interest of the designer, the ARC system will always be in a state of further development.

5 CHANGES OF DIGITAL COMPUTER TECHNOLOGY AND PRICE STRUCTURE

In this section we will analyze in more detail the effects of recent changes in computing technology on reactor design, including the growing capabilities of computers, the changing price structure of computing, and the requirements imposed on the computing industry by reactor calculations. We shall confine our attention to the newest high-speed computers rather than to the whole range of computers which might possibly be used in such calculations.

Computing-capacity requirements

Problems in the analysis of nuclear systems, together with problems in the aerospace industry and weather analysis, belong to that class of scientific calculations for which there has never been a computer with enough speed, storage, or adequate input-output equipment to satisfy the demand. This implies that crude, inexact methods have been used rather than the more accurate, but also more expensive, methods. It has meant that codes for one dimension were used when two and three dimensions were really needed to provide complete and accurate information. It has meant that diffusion theory was used where transport theory was needed. It has meant using a series of time-independent calculations, instead of time-dependent calculations. It has meant running Monte Carlo calculations to low confidence limits. It has meant restricting the number of calculations which have gone into parameter studies for optimization purposes. As the capabilities if computers have increased, some of the restrictions imposed earlier have been removed. But as the computing time for a particular type of calculation is reduced, the calculation is then used as part of a parameter study, thus imposing new quantitative requirements.

The most important restriction, however, has been qualitative in nature. There has always been a class of problems for which the available computers were inadequate—problems which were simply out of question, either because of excessive storage requirements or excessive running time. Calculations which take a full eight-hour shift to complete are annoying to smaller users, since a very large fraction of the total computing capacity is committed to such long problems, and they thereby exclude other calculations.[4]

Storage capacity has also imposed restrictions on the class of possible reactor calculations. An attempt has been made to bypass the storage restrictions by segmenting problems to allow the use of an external storage medium, usually magnetic tape or disk, as an extension of main storage. The effect of this has been to increase the code preparation and debugging time (since the control algorithms are more complex), to increase the probability of error (since no external storage device is as reliable as the main storage), and to increase the running time (since access to any auxiliary storage device is relatively expensive in time).

The increase in computing-capacity over a period of time at the Argonne National Laboratory is demonstrated by the lower part of Fig. 15. The straight line indicates that there has been a long-term increase in computing capacity of about a factor of two every two years.

Each problem type discussed in Section 4 (see Fig. 8) has its own peculiarities and needs. For example, Monte Carlo calculations have only modest

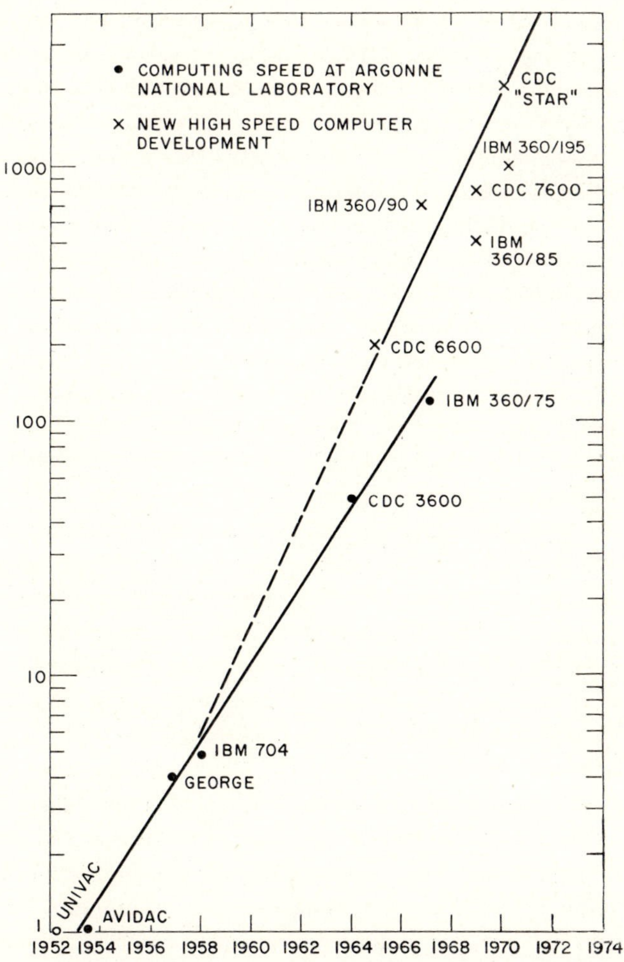

Fig. 15 Increase of Computing Speed from 1952 to 1970 (UNIVAC = 1). Lower line indicates tenfold increase in seven years, upper line tenfold increase in five years (Argonne National Laboratory)

storage requirements, but have long running times; diffusion-theory codes have less stringent requirements for running time, but use large blocks of storage; transport theory calculations use both large amounts of storage and large amounts of running time. Since transport theory calculations present some of the more difficult problems, these requirements will be examined in detail.[45]

Approximate average running times for transport theory calculations of varying dimensionality are shown in Table II, with the CDC 3600 computer used as a reference computer. From this table it is possible to extrapolate the increase in computer speed needed to reduce multidimensional transport calculation time so that these calculations can be used as part of an optimization study. It would require a computer twenty times as fast as the CDC 3600 to run two-dimensional problems at the same rate at which we now run one-dimensional problems. This provides us with a guide for estimating the computing power needed for transport calculations: we gain a dimension when the speed of computers increases by a factor of twenty. It is easily overlooked that the single-calculation time does not always provide a correct estimate of the usefulness of a code, since dozens or sometimes even hundreds of similar calculations are needed in the optimization of a nuclear system.[4]

Table II Approximate running times for transport theory calculations

Computer speed in units of CDC 3600	Execution time		
	1-dimension: minutes	2-dimension: minutes	3-dimension: hours
1	20	400	120
5	4	80	25
10	2	40	12
20	1	20	6

The extremely wide range of storage requirements for transport calculations is indicated by Table III. For a one-space dimensional calculation with 100 mesh points and eight angular directions for the neutron motion (representing a standard problem) 800 mesh points are required, assuming constant neutron energy. For a three-dimensional calculation with 20 mesh points each in two coordinates and 25 points in the third coordinate, and (only!) twelve different angular directions, a total of $10{,}000 \times 12 = 120{,}000$ mesh points are required. If the neutron flux must be calculated for several different neutron speeds, say only ten, the number of mesh points increases to 12,000,000. This storage requirement is usually reduced by telescoping the storage, i.e., each new value of the angular flux is stored in the same storage cell as some previous value, and/or by applying highly sophisticated mathematical methods to reduce the number of space points and angular directions. Mostly, complete information of the fluxes is retained only on the boundaries.[45]

Table III Approximate mesh-point requirements for transport theory calculations with constant neutron speed

Space dimension	Number of space points	Number of angular directions	Total number of mesh points
1	100	3	300
	100	8	800
	100	24	2,400
	100	40	4,000
2	1,000	6	6,000
	1,000	16	16,000
	1,000	48	48,000
	1,000	80	80,000
3	10,000	12	120,000
	10,000	20	200,000
	10,000	60	600,000
	10,000	180	1,800,000

Several conclusions can be drawn from this analysis. First, it is clearly impossible to store the angular flux array in main storage for multidimensional transport problems. This implies that the use of these calculations is tied to efficient auxiliary storage devices with low cost per word, fast access time, and high transmission rate. What the reactor mathematician would like is a storage capacity of fifty million words, an access time of a few milliseconds, a transmission rate matching the main storage, and an equipment cost of a few cents per word in storage capacity. All of these characteristics have been achieved in separate devices, but not in a single device.[45]

A second conclusion from Tables II and III is that the general use of three-dimensional transport calculations awaits (in addition to new and better mathematical techniques) the development of a computer which is several hundred times as fast as the computers which are presently available. The next generation of computers will reduce the calculation time so that a few three-dimensional calculations of high interest can be done. However, they will still consume a large fraction of the total computing capacity.[45]

With regard to computational speed on reactor problems, using the diffusion theory approximation to the neutron transport problem, Westinghouse has carried out a comparison of the ability to solve a particular four-speed, two-dimensional, neutron diffusion problem for the Yankee reactor on a succession of computers with their associated programs. Although these data reflect the combination of machine and program evolution, they provide a reasonably true picture of the capability and economics of the computational systems from the designer's point of view. In Table IV are listed the running times and approximate costs for the calculation as actually experienced on a number of computing systems. It is apparent that such an advance in the capability to carry out detailed analyses has a tremendous impact on the technology of reactor design.[31,33]

Table IV Performance of different computers and codes for the same neutron-diffusion problem (see text)

Computer	Program	Running time (minutes)	Cost	Comments
IBM 704	PDQ 02	360	$1200	Estimated time
PHILCO 211	PDQ 04	85	700	2 Microsecond core
PHILCO 212	PDQ 04	65	650	
IBM 7090	PDQ 03	54	400	Core contained iteration
IBM 7094–I	PDQ 04	53	350	
IBM 7094–I	PDQ 03	34	225	Core contained iteration
IBM 7094–II	PDQ 04	33	225	
PHILCO 212	PDQ 05	19	200	
CDC 3600	PDQ 05	33	200	
IBM 7094–II	PDQ 03	18	150	Core contained iteration
IBM 360/75	PDQ 05	34	200	8 Byte arithmetic
CDC 6600	PDQ 05	18	150	Disk simulates tapes
CDC 6600	PDQ 07	4	35	Disk version

Transport and diffusion theory calculations by no means exhaust the need for larger and faster computers. Reactor dynamics, for example, compounds the difficulties encountered in criticality calculations. A burnup code is a means of running a controlled series of criticality calculations, with changes in composition in reactor operation between refueling. Some kinetics calculations have all the difficulties of criticality calculations, but with the effects of time dependence added.[4]

Trends in computer speed and computing cost

The figures of the previous table are a consequence of the steady improvement of electronic digital computers. Table V summarizes some of the pertinent information about the newest computers. For comparison, we have included Univac and IBM 704.

Table V New high speed computers

Computer system	Approximate monthly rental*	Approximate speed in units of CDC 3600 †	Approximate delivery
Univac	25,000	1/500	1952
IBM 704	30,000	1/10	1957
CDC 3600	50,000	1	6/63
CDC 6600	110,000	4	8/64
GE 635	55,000	2	5/65
Univac 1108	60,000	2	9/65
IBM 360/75	80,000	$2\frac{1}{2}$	1/66
CDC 7600	150,000	16	1/69
IBM 360/85	130,000	$8\frac{1}{2}$	12/69
IBM 360/195	210,000	18	1971

* Rentals are dependent on configuration and can vary considerably. Estimates were obtained from EDP Industry Report 1/68 and AUERBACH Reports 6/68.

† Speed ratios can vary considerably depending on the scientific job mix, and method of evaluating the running time of a stream of problems. Ratios for the CDC 3600, CDC 6600, IBM 360/75, and IBM 360/85 were obtained from execution or simulation of Argonne National Laboratory tests. Other ratios were obtained from various industry reports.

Several facts of interest for reactor calculations are immediately apparent from this table. First, there are now available computers which exceed the IBM 704 speed by a factor of about 160, the CDC 3600 by a factor of about 16. This is rather important since we will gain a dimension in reactor calculations with the delivery of these faster systems. Second, computer rentals and purchase prices have not significantly increased, in particular when inflation is also taken into account for the comparison. Since computer speeds have increased, this means that more problems can be run for the same amount of money than was previously possible, or, in other words, the cost per problem will be considerably lowered by the new computers. Third, significant competition developed in the high-speed computer market. This probably accounts to some extent for the improved price-performance structure being offered to the customers for the new computers. Furthermore, integrated-circuit technology, while over ten years old, is only now making its full impact felt on the industry. Integrated circuits are cheaper to manufacture by about a factor of four, are more reliable, and, because their manufacturing processes can be automated, allow more firms to compete for this market.[4]

In the lower half of Fig. 15 we showed computer speeds plotted against delivery date for computers at Argonne National Laboratory. The trendline is representative for the computing power in general. The upper line indicates the new high-speed development, starting with the CDC 6600. This new trendline indicates a two and a half fold increase in speed every two years.

These trends cannot be projected indefinitely into the future, but it will be interesting to watch the rate of development continue. The length of the transmission lines in computers is already becoming a crucial factor in improving speeds; after all, nature will send information no faster than the speed of light, about one foot per nanosecond. To overcome this limitation smaller computers can be designed so that the signals need not travel so far; or the single transmission time can be split into many shorter, parallel transmission times. The first of these methods lead to the microminiaturization-packaging techniques employed in integrated-circuit technology, the second to the multiprocessing designs. The combination of smaller circuits with multiprocessing systems promises faster computers for more economic reactor design.[45]

The trend to faster computers, then, carries with it the inherent implication of lower cost per problem. It is encouraging to find that a problem which cost $800 to run on the 704 can be run on modern computers for $8![45]

The solid line in Fig. 16 shows the trend in the cost of running a problem. As a figure of comparison for problem cost, we have taken the monthly rental divided by the speed of the computer. Thus, if rentals are equal but one computer is twice as fast as another, twice as many problems can be run for the same amount of money on one computer as compared to the other. This implies that the cost per problem is cut in half. This figure of merit needs to be qualified by a factor which determines the percent of computing capacity which can be profitably used. The figure of merit used here is most meaningful to users who can saturate the capacity of a computer with useful work. Absolute costs become important for those organizations which can use only a fraction of the time they have available on a computer. Superimposed on Fig. 16 is also a line indicating the trend in nuclear power cost, taking as a measure of merit mills per kilowatt-hour of electricity. Although the precise shape of these curves is subject to some argument depending on specific assumptions, it is clear that there has been an impressive rate of reduction in nuclear power costs and that the cost of computation has been decreasing even more dramatically. A quantative evaluation of the dependence of these curves on each other is difficult, but it is obvious that confidence in design techniques acquired from correlating calculations with operational experience has contributed in part to the increase in ratings for reactor plants now being designed.[4,31]

Looking at the enormous rental cost of over $100,000 per month for modern computers, as exemplified by Table V, the reader may wonder what

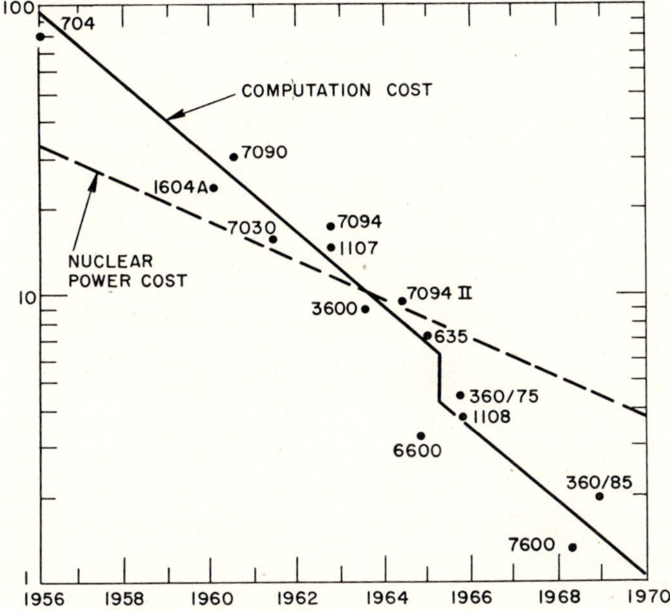

Fig. 16 Comparison of Computing Costs with Nuclear Power Costs. Measure of merit costs are based on rent per computing unit and mills per kilowatt hour (Argonne National Laboratory)

fraction of the total expenditure for nuclear research and development is absorbed by mathematical research and computing. Accurate figures are not available. However, to get a global view on government expenditure, we note that the Atomic Energy Commission had total appropriations for fiscal year 1969 in the amount of $2,570,900,000. Of this amount, $454,600,000 went for reactor development, which represents about one-sixth (17.77 percent) of total AEC activities. In 1968, the allocation for reactor development was $482,100,000 (or 19.21 percent) out of a total AEC appropriation of $2,509,100.000.[47,48]

What fraction of the expenditure for reactor development is allocated to mathematical research and computation? A partial answer can be found for the Argonne National Laboratory in Table VI. The total operating costs at Argonne for the reactor development program for 1968 was $43,181,100, slightly less than 9 percent of the total AEC allocation for reactor development. Of this allocation for Argonne, $771,900 or 1.8 percent was charged

Table VI Operating costs of Argonne reactor development program and its allocation to mathematical research and computer services over a ten-year period

Fiscal year	Operating cost	Mathematics and computer services	%
1968	$ 43,181,100	$ 771,900	1.8
1967	29,624,700	577,100	1.9
1966	29,160,700	569,800	1.9
1965	27,816,800	454,000	1.6
1964	27,268,200	480,200	1.7
1963	23,668,900	530,700	2.2
1962	21,284,300	395,300	1.8
1961	21,840,500	564,500	2.5
1960	22,229,100	805,400	3.6
1959	19,706,700	523,000	2.6

for mathematical research, consultation, and computational services in reactor research and development. Thus, in spite of its enormous impact on reactor design, computation and mathematical research absorb only a very small fraction, namely less than one fiftieth ($\frac{1}{50}$) of the cost for reactor development.

We have chosen Argonne National Laboratory as a representative example because it is one of the largest centers for reactor development and the direct descendant of the Metallurgical Laboratory, where on December 2, 1942, at 3:36 P.M. Chicago time, Fermi brought the first nuclear reactor into operation: with a sliderule in his hand!

References

In preparing the preceding chapter we have drawn material from the following references as well as from direct sources of information available to us. The reader may find the reference list a helpful guide to more detailed information on particular topics of special interest to him. The superior numbers at the end of the paragraphs in the text indicate partial quotation from these references.

1. Allardice, Corbin, and Edward R. Trapnell, *The First Pile*, U.S. Atomic Energy Commission, Division of Technical Information, Oak Ridge, Tenn. (Nov. 1968).
2. Appleby, E.R. (compiled by), *Power Reactor Designs, an Annotated Bibliography*, Battelle Memorial Institute, BNWL–936, Vols. 1 & 2, UC–80 Reactor Tech. (Jan. 1969).
3. Argonne National Laboratory, *Reactor Physics Constants*, published by United States Atomic Energy Commission, Division of Technical Information, Argonne National Laboratory Report ANL–5800, Second Ed. (1963).
4. Argonne National Laboratory, *Proceedings of the Conference on the Application of Computing Methods to Reactor Problems*, May 17–19, 1965, Margaret Butler, General Chairman. Argonne National Laboratory Report ANL–7050 (Aug. 1965).
5. Bareiss, Erwin H., *Flexible Transport Theory Routines for Nuclear Reactor Design*, Research and Development Report, Navy Department, David Taylor Model Basin Report 1030 (Dec. 1956).
6. Bareiss, Erwin H., "A Survey and Classification of Transport Theory Calculation Techniques", *Proceedings of the Second International Conference on the Peaceful Uses of Atomic Energy*, Vol. 16, Geneva, Switzerland (1958), pp. 503–516.
7. Bareiss, Erwin H., "A Spectral Theory for the Stationary Transport Operator in Slab Geometry", *Journ. Math. Analysis and Applications*, Vol. 13, No. 1 (1966), pp. 53–92.
8. Bareiss, Erwin H., and Ibrahim, K. Abu-Shumays, "On the Structure of Isotropic Transport Operators in Space", *SIAM–AMS Proceedings*, Vol. 1, Transport Theory (1969), pp. 37–78.
9. Butler, M.K., Nancy Hollister, Marianne Legan, and L. Ranzini, "Argonne Code Center: Compilation of Program Abstracts", Argonne National Laboratory Report 7411 (Jan. 1968).
10. Butler, Margaret, "Computers and the Nuclear Industry: Yesterday, A Historical Background", *Nuclear News*, Vol. 11, No. 4 (April 1968), pp. 26–30.
11. Clifford, C.E., F.R. Mynatt and E.A. Straker, "Transport Solutions to Shielding Problems: Some Recent Developments", *Nuclear News*, Vol. 12, No. 2 (Feb. 1969), pp. 51–56.

12. Corliss, William R., *Power Reactors in Small Packages*, U.S. Atomic Energy Commission, Division of Technical Information, Oak Ridge, Tenn. (June 1964).
13. Corliss, William R., *SNAP, Nuclear Space Reactors,* U.S. Atomic Energy Commission, Division of Technical Information, Oak Ridge, Tenn. (Sept. 1966).
14. Corliss, William R., *Computers,* U.S. Atomic Energy Commission, Division of Technical Information, Oak Ridge, Tenn. (Nov. 1968).
15. Craven, C. Jackson, *Our Atomic World*, U.S. Atomic Energy Commission, Division of Technical Information, Oak Ridge, Tenn. (May 1968).
16. Crowther, R. L., W. P. Petrick and A. Weitzberg, "Three Dimensional BWR Simulation", *Proceedings of Conference on the Effective Use of Computers in the Nuclear Industry*, Knoxville, Tenn. (Apr. 1969), pp. 344–365.
17. Fermi, Laura, *Atoms in the Family*, The University of Chicago Press, Chicago (1954).
18. Fernbach, S., "The Growing Role of Computers in the Nuclear Energy Field", *Proceedings of the International Conference on the Utilization of Research Reactors and Reactor Mathematics and Computation.* Mexico City (May 1967), pp. 101–125.
19. Froehlich, Reimar, "Computer Independence of Large Reactor Physics Codes with Reference to Well Balanced Computer Configurations", *Proceedings of Conference on the Effective Use of Computers in the Nuclear Industry*, Knoxville, Tenn. (Apr. 1969), pp. 451–468.
20. Glasstone, Samuel, *Sourcebook on Atomic Energy*, Pub. under the auspices of Div. of Tech. Inf., USAEC, D. Van Nostrand Co., Inc., Princeton, N. J. (1950).
21. Glasstone, Samuel, and Milton, C. Edlund, *The Elements of Nuclear Reactor Theory*, D. Van Nostrand Conpany, Inc., New York (1952).
22. Goertzel, G., and M. H. Kalos, "Monte Carlo Methods in Transport Problems", in "Physics and Mathematics", Vol. II of *Progress in Nuclear Energy*, Series 1, pp. 315–369, Pergamon Press, N.Y. (1958).
23. Greenspan, H., C. N. Kelber and D. Okrent, Eds., *Computing Methods in Reactor Physics*, Gordon and Breach, Sci. Pub., New York, London, Paris (1968).
24. Hannum, W. H., and J. W. Lewellen, "Some Thoughts on AEC Funded Coding", *Proceedings of Conference on the Effective Use of Computers in the Nuclear Industry*, Knoxville, Tenn. (Apr. 1969), pp. 119–123.
25. Henry, A. F., "Review of Computational Methods for Space-time Kinetics", *Proceedings of Conference on the Effective Use of Computers in the Nuclear Industry*, Knoxville, Tenn. (Apr. 1969), pp. 3–15.
26. Hogerton, John F., *Atomic Fuel*, U.S. Atomic Energy Commission, Division of Technical Information, Oak Ridge, Tenn. (Nov. 1963; Dec. 1964, Rev.).
27. Hogerton, John F., *Nuclear Reactors*, U.S. Atomic Energy Commission, Division of Technical Information, Oak Ridge, Tenn. (Jan. 1963; Rev. Jan. 1964; Rev. Apr. 1965).
28. H. Kahn, "Use of Different Monte Carlo Sampling Techniques", *Symposium on Monte Carlo Methods*, ed. H. A. Meyer, John Wiley and Sonr, Inc., New York (1956), pp. 146–190.
29. Lyerly, Ray L., and Walter Mitchell, III, *Nuclear Power Plants*, U.S. Atomic Energy Commission, Division of Technical Information, Oak Ridge, Tenn. (Oct. 1968).
30. Martens, Frederick H., and Norman H. Jacobson, *Research Reactors*, U.S. Atomic

Energy Commission, Division of Technical Information, Oak Ridge, Tenn. (Sept. 1968).

31. Minton, George H., "Interrelating Aspects of the Evolutionary Development of Large Scale Digital Computers and Nuclear Power Reactors", *Proceedings of the International Conference on the Utilization of Research Reactors and Reactor Mathematics and Computation.* Mexico City (May 1967), pp. 594–609.
32. Nather, V.A. and W.Sangren, "Reactor Code Abstracts", *Comm. Assoc. Computing Machinery,* Vol. 2(7)6 (1959); Vol. 3(1)6 (1960) and *Nucleonics,* Vol. 19, 154 (1961).
33. Olhoeft, J.E., and G.H.Minton, "Use of Computers in the Design and Engineering of Thermal Power Reactors", *Nuclear News,* Vol. 11, No. 4 (April 1968), pp. 34–36.
34. Peak, J.C., and A.M.Baxter, "Use of Computers in the Design and Engineering of Fast Reactors", *Nuclear News,* Vol. 11. No. 4 (Apr. 1968), pp. 37–39.
35. Radkowsky, A., and R.Brodsky, "A Bibliography of Available Digital Computer Codes for Nuclear Reactor Problems", AECU–3078 (1955).
36. Rashid, Y.R., "On Computational Methods in Solid Mechanics and Stress Analysis", *Proceedings of Conference on the Effective Use of Computers in the Nuclear Industry,* Knoxville, Tenn. (Apr. 1969), pp. 506–509.
37. Roos, Bernard W., and Ward C.Sangren, "Some Aspects of the Use of Digital Computers in Nuclear Reactor Design", *Advances in Nuclear Science and Technology,* Academic Press, New York and London, Vol. 2, Ernest J.Henley and Herbert Kouts, Eds. (1964), pp. 303–317.
38. Schaefer, G.W., and K.Allsopp, "The Calculation of Thermal-Neutron Spectra-Generalized Free-Gas Theory", *Proceedings of the Conference on the Application of Computing Methods to Reactor Problems,* Argonne National Laboratory Report 7050 (May 1965), pp. 33–49.
39. Selengut, D.S., "Digital Computers and Nuclear Design", AEC Research and Development Report, General Electric Hanford Atomic Products Operation, Hanford Laboratories, HW–59679 (Mar. 1959).
40. Spanier, J., and E. M. Gelbard, *Monte Carlo Principles and Neutron Transport Problems,* Addison–Wesley Publishing Co., Reading, Mass. (1969).
41. Stockley, James, *Nuclides and Isotopes,* Rev. 1962 by David T.Goldman, General Electric, Sixth Edition, 1961.
42. Toppel, B.J. (Ed.), "The Argonne Reactor Computation (ARC) System", Argonne National Laboratory Report 7332 (Nov. 1967).
43. Wagner, M.R., "The Application of the Block Inversion Technique to Multi-Dimensional Reactor Calculations", *Proceedings of Conference on the Effective Use of Computers in the Nuclear Industry,* Knoxville, Tenn. (Apr. 1969), p. 219.
44. Weinberg, Alvin M., and Eugene P.Wigner, *The Physical Theory of Neutron Chain Reactors,* The University of Chicago Press, Chicago, Ill. (1958).
45. Worlton, William J., "Computers: Past, Present and Future", *Proceedings of the International Conference on the Utilization of Research Reactors and Reactor Mathematics and Computation.* Mexico City (May 1967), pp. 126–175.
46. American Nuclear Society, Oak Ridge National Laboratory, University of Tennessee (Sponsors), *Proceedings of Conference on the Effective Use of Computers in the Nuclear*

Industry, Knoxville, Tenn. Apr. 21–23, 1969, Betty F. Maskewitz, General Chairman, U.S. Atomic Energy Commission, Division of technical Information, CONF–690401.

47. "AEC's Annual Report for 1968", Nuclear News, Vol. 12, No. 3 (Mar. 1969), pp. 29, 31–32.
48. "AEC Submits Housekeeping Budget for FY–70", *Nuclear News*, Vol. 12, No. 2 (Feb. 1969), pp. 16–18.
49. *Argonne News*, October 1968, Vol. 18, No. 5, p. 7.
50. *Nuclear News*, Vol. 11, No. 3 (Mar. 1968).
51. *Nuclear Reactors: Putting the Atom to Work*, Los Alamos Scientific Laboratory of the University of California, Los Alamos, New Mexico (Mar. 1965).
52. *Nuclear Reactors, Built, Being Built or Planned in the United States as of Dec. 31, 1968*, prepared by the Office of the Assistant General Manager for Reactors; USAEC, Div. of Tech. Inf., TID–8200 (19th Rev.).

CHAPTER 16

Numerical Calculations of Explosive Phenomena

CHARLES L. MADER

1 A GENERAL DESCRIPTION OF AN EXPLOSIVE

FOR ALL PRACTICAL purposes, explosives can be considered to be of two types—burning explosives such as gun powder and detonating or "high explosives" such as dynamite, TNT or RDX. In gun powder the fuel is charcoal and sulfur which are burned by the oxygen contained in sodium or potassium nitrate. In TNT the carbon and hydrogen are the fuel and, in the same molecule, oxygen is combined with nitrogen as nitro radicals to react with the fuel. Gun powder will not produce high pressures unless it is confined, but high explosives always produce high pressures. For blasting, gun powder must be confined in a hole, but high explosives need only to be placed on the top of whatever is to be blasted.

If we start a cylinder of high explosive burning, it will deflagrate with a velocity of 0.01 meter per second. Most rocket fuels are explosives deflagrating unless they accidentally build up to a detonation. If we initiate the cylinder of explosive with a shock wave, instead of a match, it will detonate with a velocity of about 9000 meters per second, or about a million times faster than it burns.

The high velocity of the detonation results in an astonishing power level for the rate of conversion of chemical bond energy to mechanical and heat energy. In a good solid explosive this rate is about 10^{10} watts per square centimeter of front, which may be compared with the total United States electric generating capacity of about 3×10^{11} watts. The energy available is approximately 1500 calories per gram. Pressures of 350,000 atmospheres and temperatures of 3000 degrees Centigrade occur near the front of the detonation wave.

A detonation may be regarded as a strong shock supported by a following chemical reaction, triggered by the high temperatures produced by the nearly instantaneous compression and heating in the shock. The rapidity and violence of the process results in the energy transport being accomplished almost entirely by actual mass motion, as in shock waves, and the usual "transport" effects (heat conduction, viscosity, radiation) can be neglected.

The most easily measured characteristic property of a detonation is the velocity at which the front propagates into the explosive. It is found that as the wave expands from the initiating point in a large piece of explosive so that the front becomes more and more nearly plane, the velocity approaches a limit. From this observation it seems reasonable to make the assumption that a limiting velocity exists, that all the other parameters have limits, and that in this limiting case the chemical reaction takes place in a steadily propagating zone in the explosive. A mathematically tractable problem related to physical reality by the concept of the limiting case is that of plane, steady detonation. It has been generally assumed that its solutions may be expected to describe detonation experiments as the results are extrapolated to infinite size.

It has become customary to treat all detonation problems under the steady-state assumption that the reaction zone differs inappreciably from its plane, steady limit, and to apply the corresponding theory to experiments directly. This assumption seems reasonable when the velocity has come close to the limiting value, which it is observed to do very quickly, and when the radius of curvature is very large compared with the reaction zone thickness. This approach has proven satisfactory for many engineering applications, particularly if some of the parameters, e.g., the equation of state, are chosen to fit measurements made on pieces similar to the ones for which calculations are required. The situation is similar to that in most cases where theory is applied to engineering problems.

Although the detonation front is nearly plane and the velocity is nearly constant as assumed by the theory, close observation reveals a complex, time-dependent structure. It seems likely that this structure is evidence of the processes occurring in the detonation which are responsible for the differences between theory and experiment. It has become clear in recent years that one can not expect exact agreement between experimental observations and the results of a theory that assumes a steady-state time-independent behavior.

The performance of an explosive and the initiation properties of an explosive are amenable to numerical study. These theoretical studies assist in furnishing guidance to and interpretation of the experimental studies of explosive phenomena.

2 THE USE OF COMPUTERS TO PREDICT THE PERFORMANCE OF AN EXPLOSIVE

The performance of an explosive (its ability to push a metal plate or blast a rock) is described by the maximum detonation pressure of the totally decomposed explosive (called the Chapman–Jouguet or C–J pressure), the detonation velocity, and the profile of the pressure of the detonation products as a function of distance and time behind the detonation front. The relationship between pressure, density and temperature (the equation of state) of the detonation products is crucial to the understanding of the performance of an explosive.

It must be pointed out as we start this discussion that using the steady-state assumption it will be impossible to obtain complete agreement with experimental data for all explosives with any equation of state. The observed time-dependent behavior of explosives will not be reproduced by any such theory. Fortunately, the time-dependent behavior is not a dominant feature of the behavior of most explosive systems in the geometries of interest and we include the gross features of this behavior when we calibrate the theory or the equation of state for some particular set of explosives in a particular density and geometry range.

Given these limitations we have found that using the Becker–Kistiakowsky–Wilson equation of state we can reproduce the experimentally observed pressures and temperatures over a wide range of densities and compositions to within 20% and the detonation velocities to within 10%.

The Becker–Kistiakowsky–Wilson (BKW) equation of state is a semi-empirical equation of state based upon a repulsive potential applied to the virial equation of state. It has been the most extensively calibrated and used of the many equations of state proposed to describe the detonation process. The detonation products are assumed to be in chemical equilibrium. Several numerical codes have been developed to perform the complicated calculations necessary to describe the equation of state of detonation products using the Becker–Kistiakowsky–Wilson equation of state. The BKW code was developed at the Los Alamos Scientific Laboratory and is being used at various laboratories in the United States, England, Sweden and Australia. The RUBY code was developed at the Lawrence Radiation Laboratory-Livermore and it has also been used by several other laboratories in the United States. The TIGER code is being developed at the Stanford Research Institute and is designed to be a more general and versatile version of the BKW and RUBY codes. It should be a useful and general purpose code for any laboratory needing to perform equation of state calculations of gaseous and solid products at high pressures or temperatures.

As input data a BKW calculation only requires, for the explosive, its elemental composition, heat of formation, density, and formula weight; and for the detonation products, their elemental compositions, heats of formation, co-volumes (the effective size of the molecules), and fits of their ideal gas entropies as a function of temperature.

If any solid detonation products are present, the BKW calculation also requires the density, molecular weight and the parameters for a suitable solid equation of state.

The BKW calculation computes the equilibrium composition of the detonation products at temperatures and pressures of interest, the single shock detonation state values (Hugoniot curve), the values of the C–J pressure, C–J temperature, C–J density and detonation velocity. The isentrope (constant entropy curve) of the detonation products through the C–J state point is also computed in a BKW calculation.

The BKW code permits us both to estimate the probable performance of a proposed chemical compound and to obtain clues as to why a particular explosive is observed to behave in a particular fashion. It has given us increased understanding of the important features of the detonation chemistry and energetics. For example, after studying various explosive systems with about the same C–J pressure and velocity—but with heats of explosion that were different by factors of two—it was discovered that the C–J performance of an explosive is a very sensitive function of the C–J particle density of the detonation products. The explosive systems with large heats of explosion yielded detonation product molecules that had large molecular weights and therefore the specific C–J particle density was low and the extra energy was present primarily as thermal energy rather than intermolecular potential energy. Since the intermolecular potential energy primarily determines the

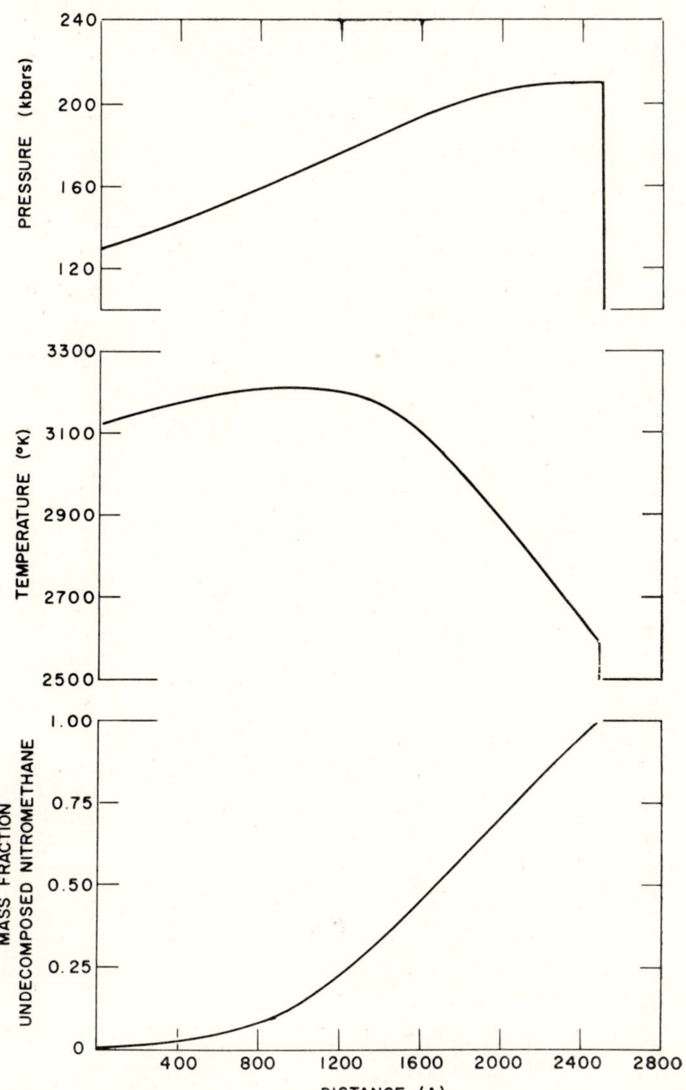

Fig. 1 The steady-state profiles for a nitromethane reaction zone. The pressure, temperature and mass fraction of undecomposed explosive (which is 1.0 when no decomposition has occurred, and zero when the explosive has totally decomposed) are shown as a function of the distance from the C–J state value given in Angstrom units (10^{-8} centimeters)

C–J pressure, the C–J pressures were only slightly affected while the C–J temperatures were considerably higher. The extra energy became available only upon expansion to very low pressures. While the theory is useful there is still considerable disagreement between the experimental observations and the theory. One reason for this disagreement is that real detonations are not steady-state.

To perform time-dependent flow calculations of the detonation process we need to solve the Navier–Stokes equations of fluid dynamics using an equation of state of the detonation products such as the BKW equation of state, an equation of state for the unreacted explosive, an equation of state for mixtures of unreacted and totally decomposed explosive, and an equation that describes the rate at which the undecomposed explosive decomposes to detonation products. This has been performed at Los Alamos using a one-dimensional Lagrangian hydrodynamic code called SIN, a two-dimensional Particle-In-Cell code called EIC, a two-dimensional Lagrangian hydrodynamic code called 2DL, and a two-dimensional Eulerian hydrodynamic code called 2DE.

Using these codes we can calculate the details of the region of the flow where the undecomposed explosive decomposes to its detonation products (the reaction zone). The results of such a calculation of a nitromethane reaction zone are shown in Fig. 1.

Using these codes we can calculate the time-dependent one- and two-dimensional flow of a detonating explosive. The results of such a calculation for nitromethane are shown in Fig. 2. The time-dependent, one- and two-dimensional model of detonation for nitromethane and other explosives exhibits unstable pulsating detonation until the explosives are sufficiently overdriven, and then the detonations become stable.

This complicated time-dependent behavior will have to be included into the calculation of the performance of an explosive if we are to obtain improved calculations and understanding of the behavior of detonations. The assumptions of chemical equilibrium will have to be dropped and detailed chemical kinetics and the time-dependent deposition of solid carbon will have to be more realistically described. Such a study will require three-dimensional geometry and a complete numerical description of these problems must await the development of larger and faster computer hardware.

As increased understanding of the performance of an explosive is achieved, it is reasonable to expect that more powerful and safer explosives can be designed based on such information.

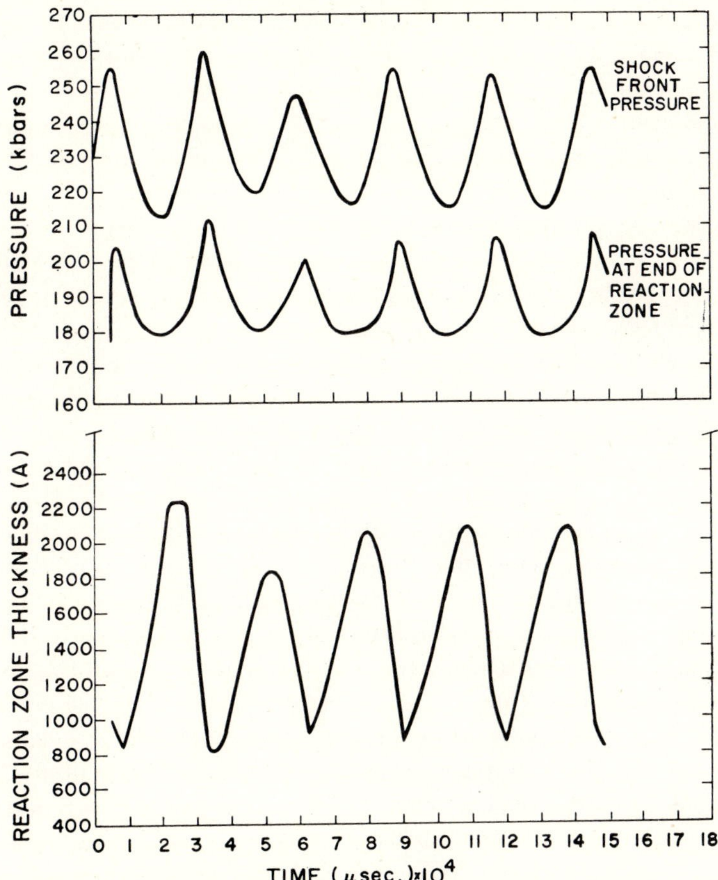

Fig. 2 The periodic, non-steady or "galloping" reaction zone of nitromethane overdriven with a detonation velocity of 6650 meters per second is shown as a function of time

3 THE USE OF COMPUTERS TO PREDICT THE INITIATION PROPERTIES OF AN EXPLOSIVE

An explosive may be initiated by various methods of delivering energy to the explosive. Bulk heating or "thermal initiation" has been treated in a satisfactory manner for engineering purposes by solving the Frank–Kamenetskii equation with Arrhenius kinetics.

Whether an explosive is dropped, burned, scraped, or shocked to furnish the first impulse to start the initiation process, a shock is formed. The initiation of an explosive always goes through a stage when a shock wave is an important feature of the initiation process. Most of the theoretical studies of the initiation properties of an explosive have been primarily a study of initiation by shock waves.

One may initiate propagating detonation in an explosive by sending a shock wave of sufficient strength into the explosive. There is a minimum shock pressure below which propagating detonation does not occur for an explosive of a particular density and geometry.

If one introduces gas bubbles or grit into a homogeneous explosive (such as a liquid or a single crystal) producing a heterogeneous explosive (one containing a density discontinuity), the minimum shock pressure necessary to initiate propagating detonation can be decreased by about one order of magnitude.

The experimental and theoretical studies of the shock initiation of homogeneous explosives suggest that the shock initiation of homogeneous explosives result from simple thermal explosion caused by the shock heating of the bulk of the material. The results of solving the chemically reactive flow using the SIN code are shown in Fig. 3. A shock travels into the explosive and heats and compresses the explosive. The shock heating results in chemical decomposition which accelerates exponentially. Explosion occurs at the rear boundary, since it has been hot the longest, and a propagating detonation develops which travels at the C–J pressure, temperature and velocity characteristic of the shocked (high density) explosive (the "super" detonation). The "super" detonation overtakes the front of the shock wave and decays in time to a normal C–J detonation for the unshocked explosive.

A heterogeneous explosive is one that has voids or any type of density discontinuity which will result in irregularities of the mass flow when shocked. Most explosives of practical interest are heterogeneous explosives. The heterogeneous explosive is initiated by the local hot spots formed in the explosive by shock interactions with the density discontinuities.

To investigate the basic processes in inhomogeneous explosives, the calculations and some of the experiments were simplified to an investigation of the interaction of a shock with a single or a few discontinuities such as bubbles or metal wires in nitromethane.

The formation of low temperature hot spots by shocks in nitromethane interacting with corners of plexiglas, aluminum and gold has been studied using the

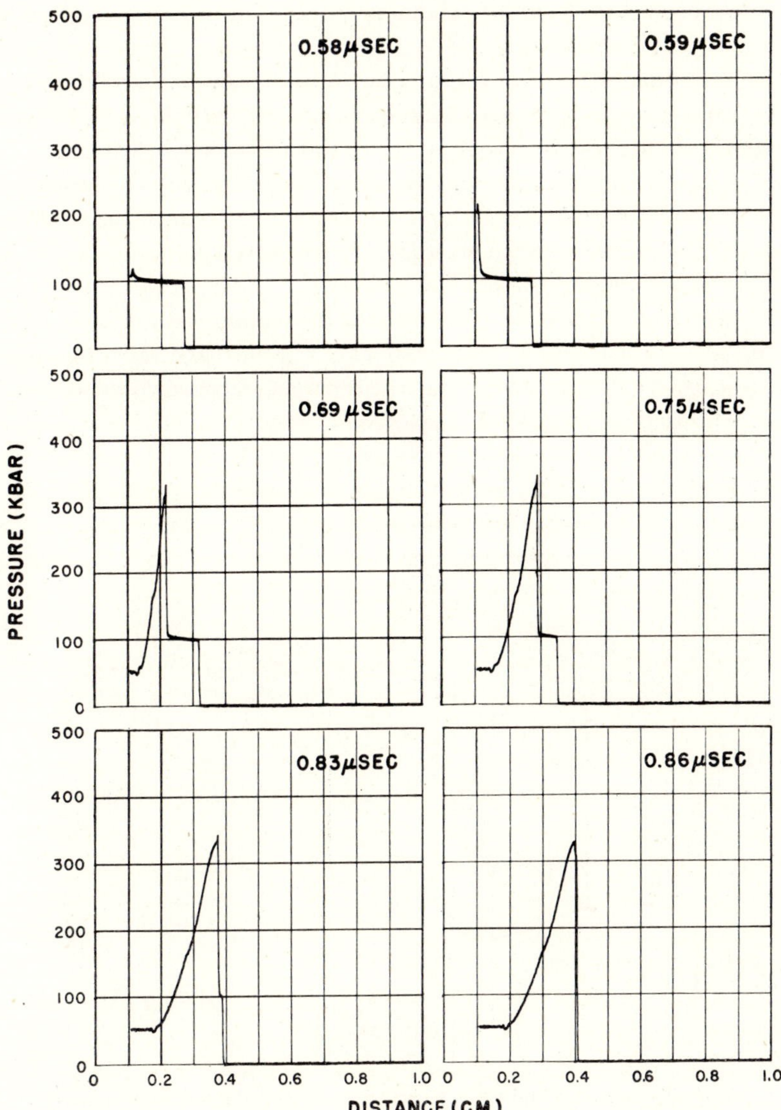

Fig. 3 The pressure-distance profiles at various times for the shock initiation of nitromethane by a 92 kilobar shock

Lagrangian hydrodynamic code for reactive materials, 2DL. The computed formation of a hot spot and the buildup to propagating detonation agree with the experimental observations.

The formation of high temperature hot spots by shocks interacting with air bubbles or voids has been computed using the Particle-In-Cell hydrodynamic code for reactive materials, EIC, and the Eulerian hydrodynamic code for reactive materials, 2DE. One of the basic processes involved in the shock initiation of explosives at high temperature hot spots is the failure of propagating detonation because of side rarefactions cooling the explosive inside the reaction zone.

The failure of a supported nitromethane detonation wave traveling up a copper tube and then into a large container of nitromethane has been studied experimentally with a smear camera. The experimental geometry is shown in Fig. 4. Rarefaction or "failure" waves were observed that ran across the detonation front at about 3700 meters per second. These rarefaction waves extinguished the detonation if the tube was small enough. The smear camera trace of such a failure is shown in Fig. 5 for a 1.02-centimeter-wide, 2.28-centimeter-long, rectangular tube that was 30 centimeters high.

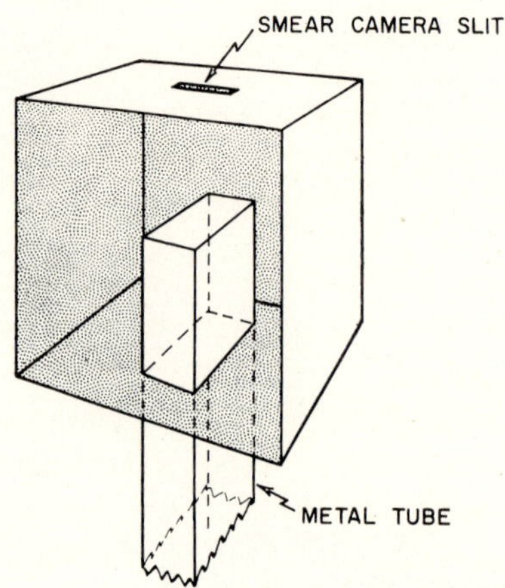

Fig. 4 The experimental arrangement for studying the failure of a supported nitromethane detonation wave. The box and metal tube are filled with the liquid explosive nitromethane. A detonation travels up the metal tube and then into the large container of nitromethane. Failure waves are observed running across the front of the detonation wave and if the tube is small enough the detonation is extinguished

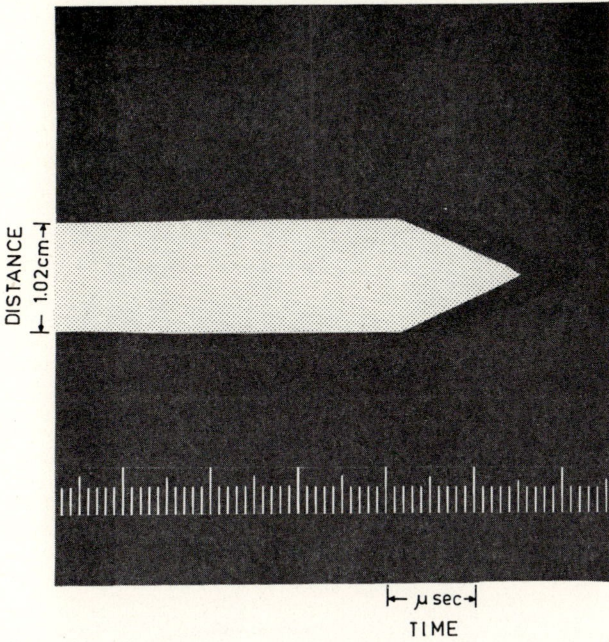

Fig. 5 The smear-camera trace obtained from the experimental arrangement shown in Fig. 4. The detonation is shown proceeding up the metal tube from left to right. The triangular region shows that after the detonation arrives at the end of the metal tube it is extinguished by rarefactions cooling the explosive inside the reaction zone as it proceeds into the unconfined nitromethane

If the tube were larger, the detonation was observed to not extinguish, and narrow failure or dark waves were observed to run into the front getting narrower as the wave progressed. This wave extinguished detonation at its front, but reignition occurred at the rear and caught up with the wave. Under these circumstances the explosive in the box will detonate. These experiments demonstrate the basic processes involved in the failure of detonation, of the failure diameter of explosives, and of the "sputtering" detonation observed for an explosive containing density discontinuities near the critical size.

The experimental arrangement is approximated in the numerical model by a steady-state reaction zone flowing through half of the nitromethane cells near the lower boundary of the calculation to simulate the top of the metal tube. The use of a 300×300 mesh for a total of 90,000 cells is near the maximum resolution possible if one wishes to expend no more than 24 hours of

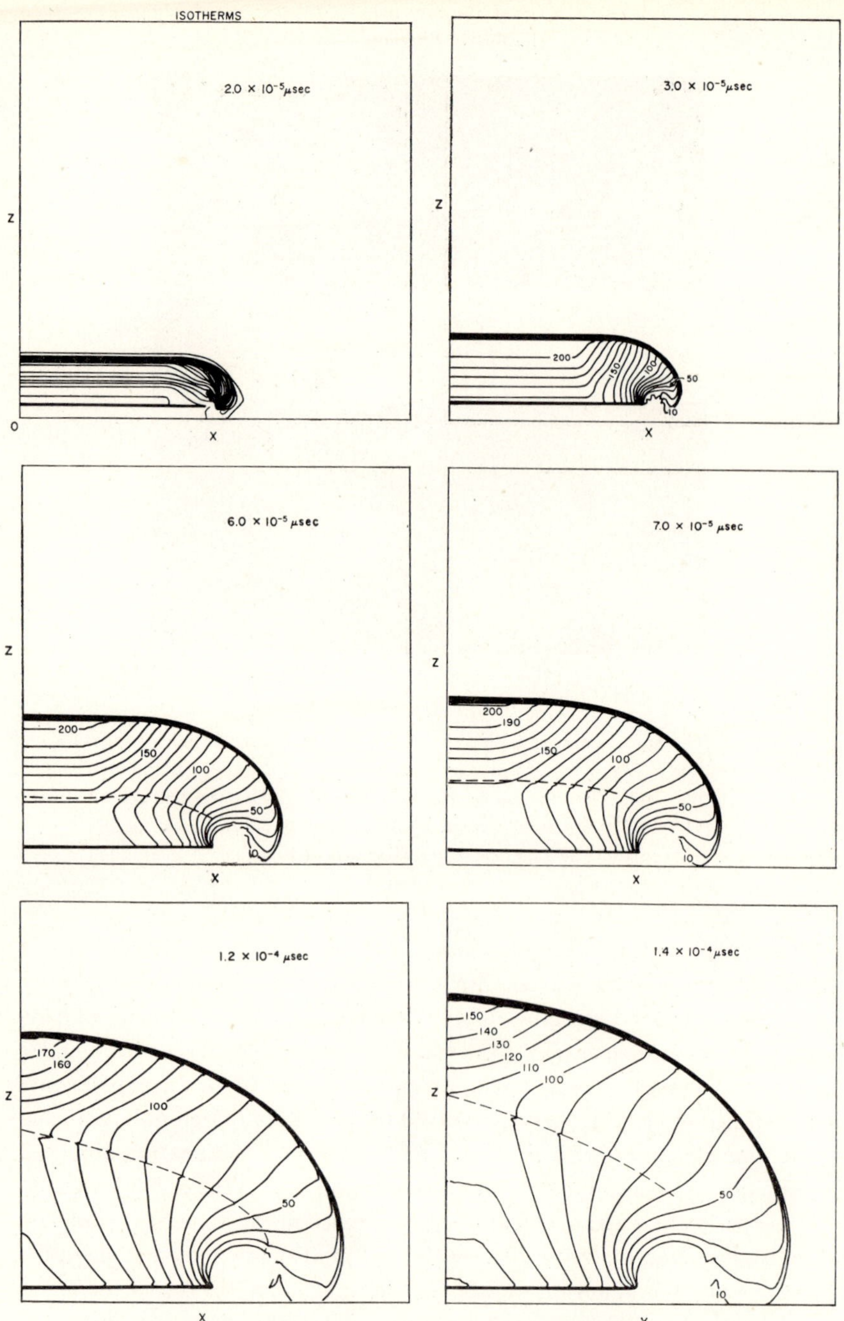

Fig. 6 The results of the numerical model used to approximate the experimental system shown in Figure 4. Shown are the computed isobars (constant pressure lines) for a supported nitromethane C–J detonation passing into a box of nitromethane. The interval between isobars is 10 kilobars. The dashed line shows the location of the end of the reaction zone (the C–J state point). 90,000 cells 40-A-square were used in the numerical calculation

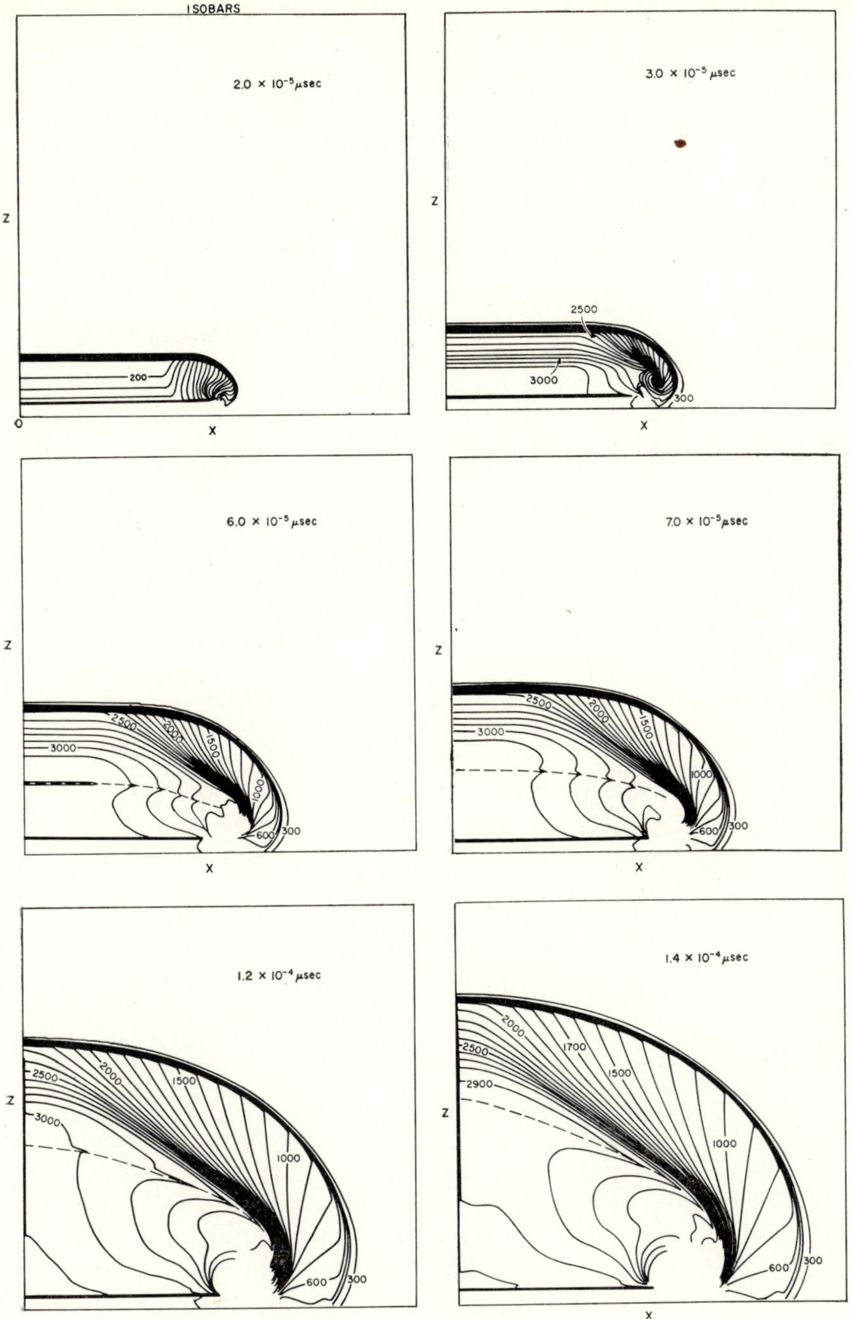

Fig. 7 The computed isotherms (constant temperature lines) are shown for the same system and the same times as shown in Figure 6. The interval between isotherms is 100° Kelvin. The cooling of the reaction zone by the side rarefaction is shown

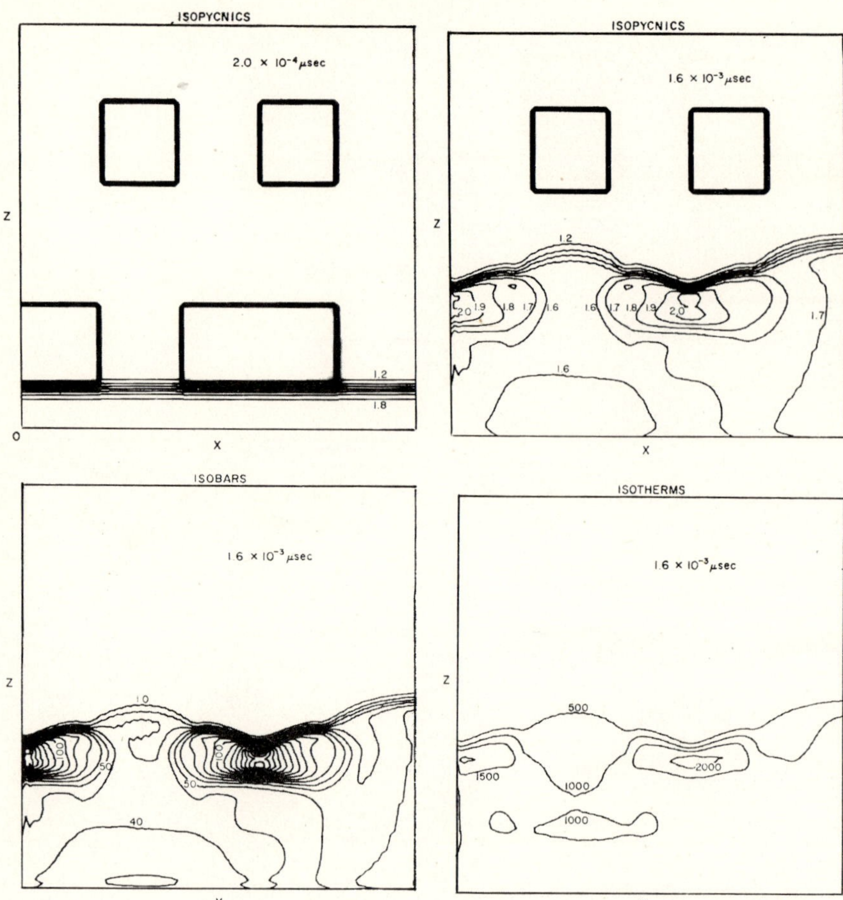

Fig. 8 The interaction of an 85 kilobar shock in a 1.6×10^{-3} centimeter-half-width, 1.6×10^{-3} centimeter-high slab of nitromethane with four rectangular holes. Chemical reaction is not permitted. The interval between isopycnics (constant density) is 0.1 gram per cubic centimeter, between isobars (constant pressure) is 10 kilobars, and between isotherms (constant temperature) is 500° Kelvin. The isopynics show the hole geometry. The hot regions that are formed when shock interact with density discontinuities are shown. These hot regions result in chemical decomposition as shown in the next figure

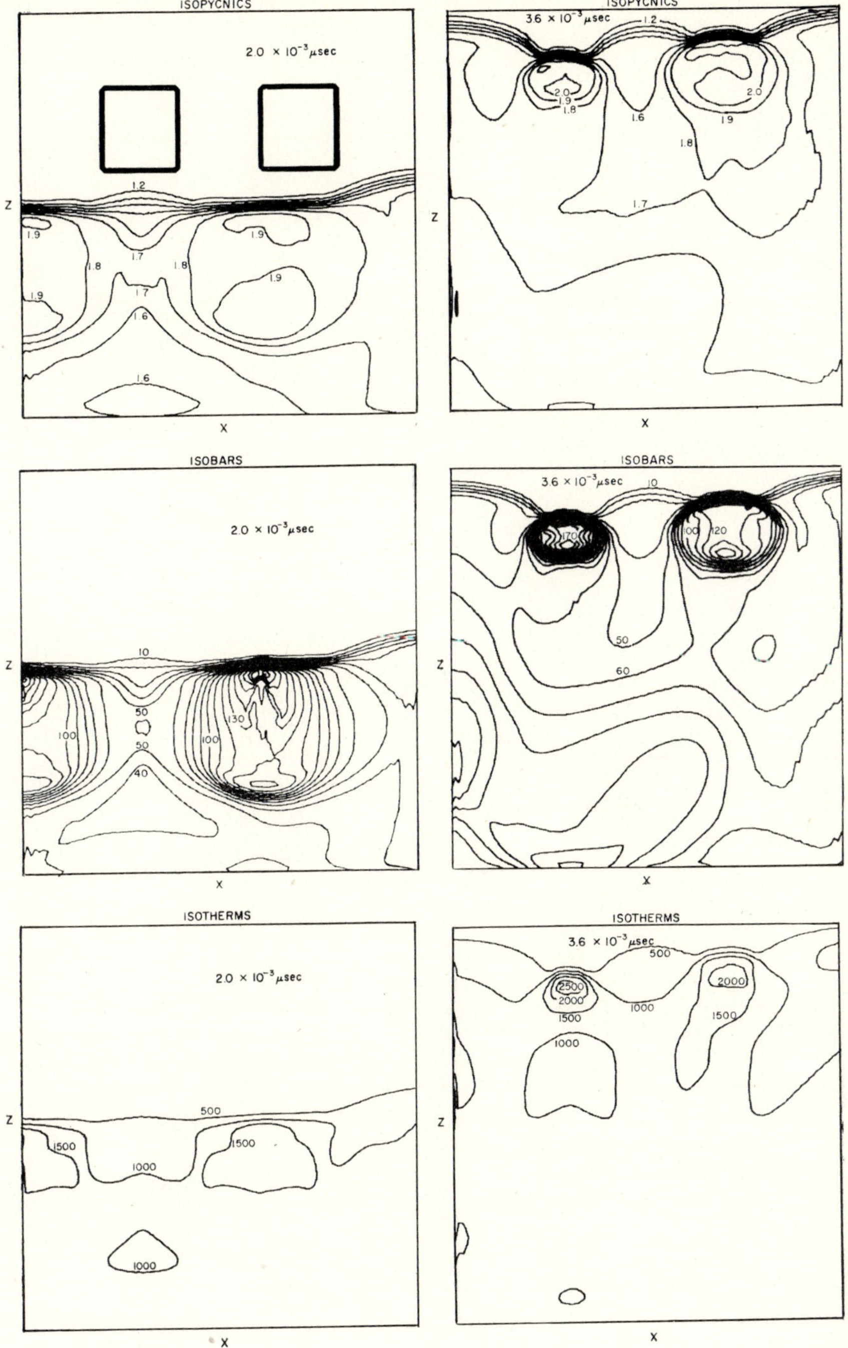
ISOPYCNICS
2.0 × 10⁻³ μsec
1.2
1.9
1.8
1.7
1.6
Z
X
ISOPYCNICS
3.6 × 10⁻³ μsec
1.2
2.0
1.9
1.8
1.6
1.7
Z
X
ISOBARS
2.0 × 10⁻³ μsec
10
50
100
130
40
Z
X
ISOBARS
3.6 × 10⁻³ μsec
10
170
100
120
50
60
Z
X
ISOTHERMS
2.0 × 10⁻³ μsec
500
1500
1000
Z
X
ISOTHERMS
3.6 × 10⁻³ μsec
500
2500
2000
1500
1000
Z
X

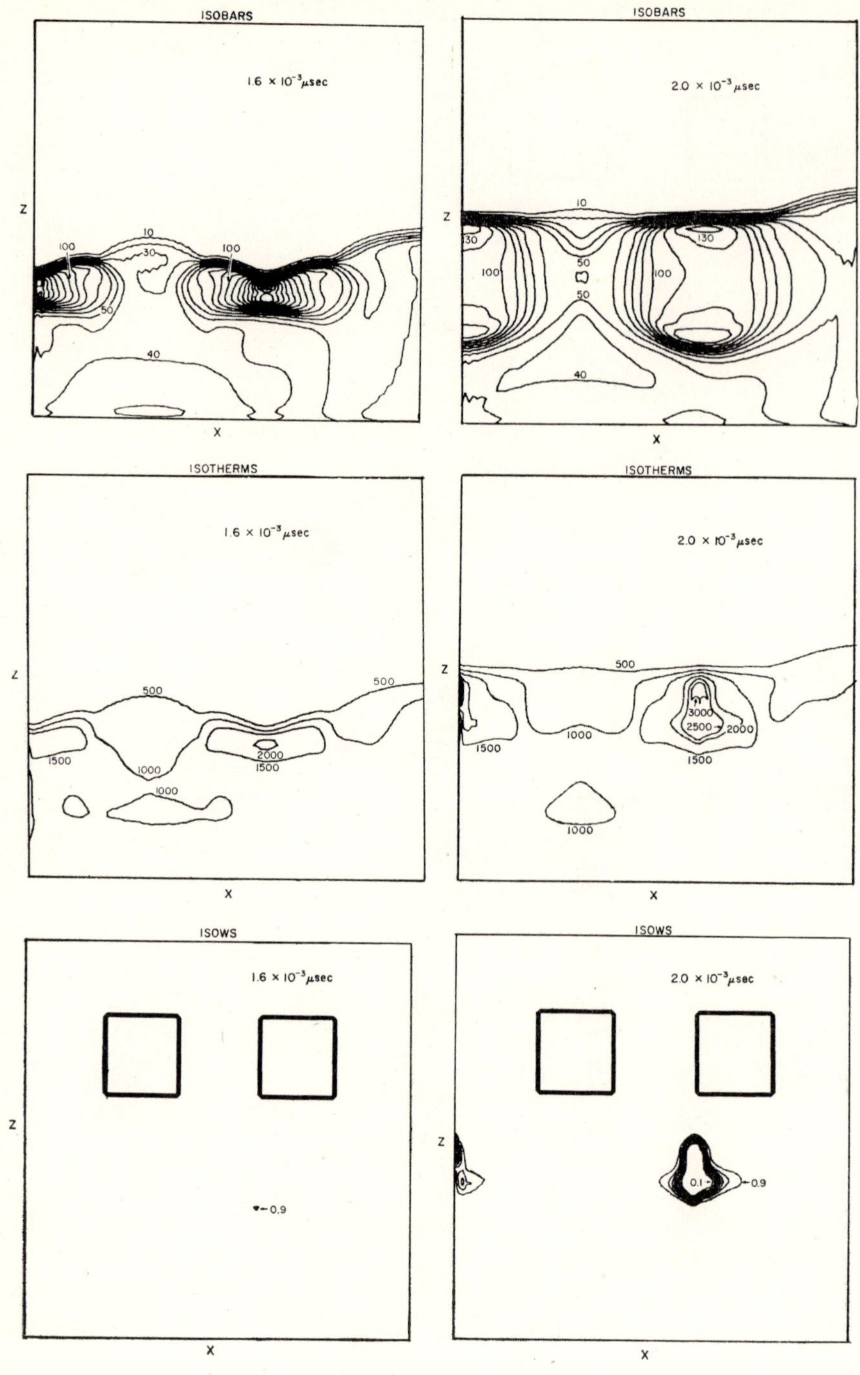
ISOBARS
1.6 × 10⁻³ μsec
10
30
100
100
50
40
Z
X
ISOBARS
2.0 × 10⁻³ μsec
10
130
30
50
100
100
50
40
Z
X
ISOTHERMS
1.6 × 10⁻³ μsec
500
500
1500
1000
2000
1500
1000
Z
X
ISOTHERMS
2.0 × 10⁻³ μsec
500
3000
2500
2000
1000
1500
1500
1000
Z
X
ISOWS
1.6 × 10⁻³ μsec
0.9
Z
X
ISOWS
2.0 × 10⁻³ μsec
0.1
0.9
Z
X

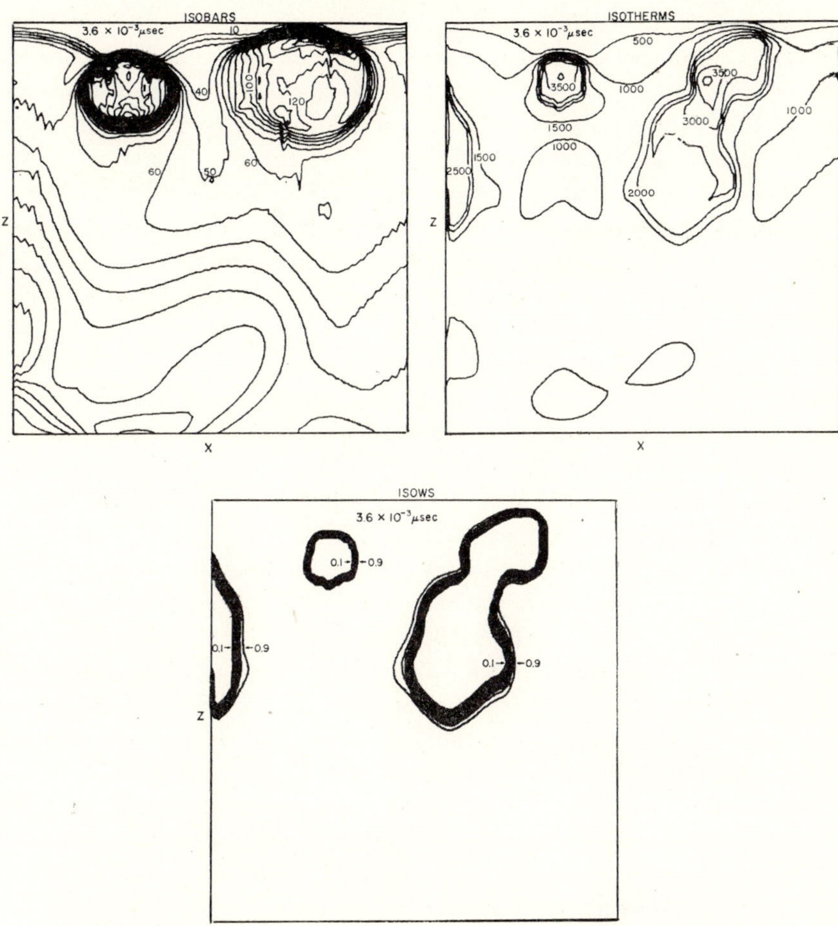

Fig. 9 The interaction of an 85 kilobar shock in nitromethane with four rectangular holes. The calculation is identical to the one shown in Figure 8 with the exception that chemical reaction is permitted. The interval between isows (constant mass fraction of undecomposed explosive) is 0.1. The isows show the area of decomposition of the explosive from the shock heating that results from the interaction of the shock with the holes

computer time using computers of the late 1960's. A reasonably resolved reaction zone requires about 60 of the 300 cells so the calculation can be run only 5 reaction zone lengths. While this is insufficient to study the reignition occurring after the rarefaction wave extinguishes detonation at its front (the "sputtering" detonation), it is sufficient to study the failure of a nitromethane reaction resulting from side rarefactions. The calculated isobars (constant pressure) lines are shown in Fig. 6 and the calculated isotherms (constant temperature) lines are shown in Fig. 7 for a nitromethane detonation failing as a result of side rarefactions.

Having demonstrated that the computational model describes the failure of a detonation wave because of rarefactions from the side cooling the explosive inside the reaction zone, we can now proceed to study the formation of hot spots and the local initiation of fast reaction and subsequent failure which does not result in a propagating detonation in an explosive.

To investigate the basic processes in inhomogeneous explosives further the calculations were simplified to the interaction of a shock with four rectangular voids, the bottom two with a half-width of 3.2×10^{-4} centimeters and the top two with a width of 3.2×10^{-4} centimeters as shown in Fig. 8 for non-reactive nitromethane. The same problem with chemical reaction included is shown in Fig. 9.

While propagating detonation does not occur when the shock interacts with the first two holes, the enhancement of the shock wave by chemical reaction does produce a hotter hot spot upon interaction with the upper two voids. The hot spot is so hot that complete nitromethane decomposition occurs at the shock front. The enhancement of the shock wave is similar to what is observed experimentally in the shock initiation of heterogeneous explosives. These calculations show the basic features of the shock initiation of practical heterogeneous explosives. A shock wave interacts with the density inhomogeneities, producing numerous local hot spots which explode but do not propagate, thereby liberating energy which strengthens the shock so that, when it interacts with additional inhomogeneities, hotter hot spots are formed and more of the explosive is decomposed. The shock wave grows stronger and stronger, releasing more and more energy, until it becomes strong enough to produce propagating detonation.

The basic two-dimensional processes involved in the shock initiation of heterogeneous explosives have been numerically described. The problem that remains is the study of the interaction of a shock with a matrix of holes in three-dimensional geometry. The numerical solution of this problem must await the development of computing hardware that is at least 100 times faster and with memories 100 times larger than those available in the late 1960's. The basic two-dimensional processes involved in the failure of detonation, the failure diameter of explosives, and the "sputtering" detonation observed for density discontinuities near the critical size have been described. The study of the three-dimensional interaction of numerous failures and reignited detonations which is necessary for a complete numerical description of these problems must also await new computing hardware.

As increased understanding of the initiation properties of explosives is achieved, it is reasonable to expect that safer explosives can be developed and that the handling of explosives in use today can be made safer.

Bibliography

Explosive performance

"Calculation of the Detonation Properties of Solid Explosives with the Kistiakowsky–Wilson Equation of State", R.D. Cowan and W. Fickett, *J. Chemical Physics* ***24***, 932 (1956).

"Detonation Properties of Condensed Explosives Computed Using the Becker–Kistiakowsky–Wilson Equation of State", Charles L. Mader, LA–2900 (1963).

"FORTRAN BKW—A Code for Computing the Detonation Properties of Explosives", Charles L. Mader, LA–3704 (1967).

"Detonation Performance Calculations Using the Kistiakowsky–Wilson Equation of State", Charles L. Mader, LA–2613 (1961).

"Detonation Properties of Condensed Explosives Calculated with an Equation of State Based on Intermolecular Potentials" Wildon Fickett, LA–2712 (1962).

Failure of steady-state model

"One- and Two-Dimensional Flow Calculations of the Reaction Zones of Ideal Gas, Nitromethane and Liquid TNT Detonations", Charles L. Mader, Twelfth Symposium (International) on Combustion (1968).

"A Study of the One-Dimensional, Time-Dependent Reaction Zone of Nitromethane and Liquid TNT", Charles L. Mader, LA–3297 (1965).

"The Time-Dependent Reaction Zones of Ideal Gases, Nitromethane and Liquid TNT", Charles L. Mader, LA–3764 (1967).

"Failure of the Chapman–Jouguet Theory for Liquid and Solid Explosives", W.C. Davis, B.J. Craig, and J.B. Ramsay, *Physics of Fluids* ***8***, 2169 (1965).

"Numerical Calculations of One-Dimensional, Unstable, Pulsating Detonations", Wildon Fickett and W.W. Wood, *Physics of Fluids* ***9***, 903 (1966).

Numerical reactive hydrodynamics codes

"FORTRAN SIN—A One-Dimensional Hydrodynamic Code for Problems which Include Chemical Reaction, Elastic-Plastic Flow, Spalling and Phase Transitions", Charles L. Mader and William R. Gage, LA–3720 (1967).

"2DE—A Two-Dimensional Eulerian Hydrodynamic Code for Computing One Component Reactive Hydrodynamic Problems", William R. Gage and Charles L. Mader, LA–3629–MS (1966).

"Three-Dimensional Cartesian Particle-In-Cell Calculations", William R. Gage and Charles L. Mader, LA–3422 (1965).

Initiation properties of explosives

"Shock Initiation of Detonation in Liquid Explosives", A. W. Campbell, W. C. Davis, and J. R. Travis, *Physics of Fluids* ***4***, 498 (1961).

"Shock Initiation of Solid Explosives", A. W. Campbell, W. C. Davis, J. B. Ramsay and J. R. Travis, *Physics of Fluids* ***4***, 511 (1961).

"Initiation of Detonation by the Interaction of Shocks with Density Discontinuities", Charles L. Mader, *Physics of Fluids* ***8***, 1811 (1965).

"Shock and Hot Spot Initiation of Homogeneous Explosives", Charles L. Mader, *Physics of Fluids* ***6***, 375 (1963).

"Thermal Initiation of Explosives", John Zinn and Charles L. Mader, *J. Applied Physics* ***31***, 323 (1960).

"The Two-Dimensional Hydrodynamic Hot Spot—Volume IV", Charles L. Mader, LA–3771 (1967).

"The Two-Dimensional Hydrodynamic Hot Spot", Volume III, Charles L. Mader, LA-3450 (1966). (Also describes the reactive two-dimensional Lagrangian code, 2DL.)

'The Two-Dimensional Hydrodynamic Hot Spot", Volume II, Charles L. Mader, LA–3235 (1964).

"The Two-Dimensional Hydrodynamic Hot Spot", Charles L. Mader, LA–3077 (1964). (Also describes the reactive two-dimensional code, EIC.)

"The Hydrodynamic Hot Spot and Shock Initiation of Homogeneous Explosives", Charles L. Mader, LA–2703 (1962).

CHAPTER 17

Computers and the Optical Model for Nuclear Scattering

T. TAMURA and W. R. COKER

1 INTRODUCTION: ORIGINS OF THE MODEL

WE WOULD LIKE to explain something of the aim and nature of the optical model for nuclear scattering, and the way in which modern, high-speed computers contribute to the model's application. We first need to summarize some basic facts about atomic nuclei themselves, and a little history of nuclear physics.

The nucleus of an atom is a region about 10^{-12} cm (10 Fermis) in radius, lying at the atom's center. The atom's extra-nuclear region is a "cloud" of electrons confined within a volume roughly 10^{-7} cm in radius. The nucleus of an atom with atomic weight A and Z surrounding electrons contains Z protons and $N = A - Z$ neutrons. Since the protons each have a positive charge equal in magnitude to the negative charge e on an electron, the nucleus has total charge Ze.

The usual notation for a specific nucleus is the chemical symbol for the element (which automatically specifies Z) with a left superscript giving A. For example, ^{90}Zr is the $A = 90$ isotope ($N = 50$) of zirconium ($Z = 40$). The isotope ^{96}Zr has 56 neutrons.

The force acting between two protons, a proton and a neutron, or two neutrons is attractive and to a high degree of accuracy (96%) the same in each case, apart from the additional repulsive electrostatic (or Coulomb) force acting between the like-charged protons. Since protons and neutrons also are similar in mass and other properties, they are usually discussed under the single name, *nucleon*. The nuclear force is found to be almost negligible unless the distance between nucleons is about 10^{-12} cm or less.

Because the nucleons are confined in such a small region by the nuclear force, it is clearly extremely powerful over short distances ($\sim 10^{-13}$ cm). At small enough distances, however, the force between two nucleons must eventually "saturate", and become repulsive, else nuclei would collapse into points. As a result of saturation, the density of nucleons in most nuclei is nearly uniform from the center out to the surface, where it rapidly but smoothly drops to zero. The radius of a nucleus of mass A is proportional to the cube root of A: $R_A = r_0 A^{1/3}$, where r_0 is a constant ($\approx 1.2 \times 10^{-13}$ cm). That is, the nuclear density does not depend on A: since the nuclear volume would be $V_A = (\frac{4}{3})\pi R_A^3 = (\frac{4}{3})\pi r_0^3 A$, the density V_A/A is $(\frac{4}{3})\pi r_0^3$.

Essentially all of our knowledge of the physical properties of nuclei has come from scattering experiments or radioactive decay. In the old days of nuclear physics one had available for observation only the products of the natural radioactive decay of a few heavy nuclei ($Z > 82$). Such nuclei decayed, by various routes, into lead (^{208}Pb) by emission of helium (^{4}He) nuclei—"α rays"—or electrons—"β rays". Very high frequency electromagnetic radiation, "γ-radiation", was also observed.

When various types of electrostatic and electromagnetic ion-accelerators became available, it was possible to investigate stable nuclei by bombarding a chemically and isotopically pure sample of material, creating "artifical" radioactivity. A sizable amount of information about the internal structure of nuclei was gained in this way, though the products of the bombardment were still mostly restricted to electrons, and γ rays.

The improvement in particle accelerators following World War II made higher and higher projectile energies available. An electron with a kinetic energy of 100 million electron volts (100 MeV) is, because of the wave nature of matter, localized to within a distance of 10^{-13} cm or less, so that in passing through a thin foil it will be likely to interact with the Coulomb field of at most a *single* nucleus. Since the distance within which a particle is localized is inversely proportional to the square root of its kinetic energy, increasing the energy gives us a greater and greater "magnification" and the accelerator becomes a kind of nuclear microscope.

Suppose nucleons themselves are accelerated in such machines at high enough energies (a few MeV are sufficient, since nucleons have almost 2000 times the mass of electrons) to single out individual nuclei in the target foil. Since the nuclear force is so very powerful, at least over distances comparable to the nuclear radius, it might be expected that usually the nucleon will either be unaffected, passing too far away, or will be sucked into the nucleus, coming too close. In the intermediate case, when the incoming nucleon begins to interact with other nucleons in the near half of the nucleus, those in the back hemisphere are still too far away to influence the projectile's motion. Hence the nuclear collision should be highly complicated, since the many nucleons of the target will tend to interact almost *individually* with the incident nucleon.

Such, at least, was the conjecture made by almost every nuclear physicist prior to a remarkable series of experiments by H.H. Barschall reported in 1952. We cannot go into detail, but the results of Barschall's experiments, in which various nuclei were bombarded with neutrons at energies of an MeV or less, could be explained in a shockingly simple way.

Barschall's experimental data could be readily interpreted assuming the existence of a *single* potential with which the incident projectile interacts. A plausible explanation of how the individual strong, short range interactions of each of the A nucleons in a nucleus can appear as a single interaction, strong over the region of space occupied by the entire nucleus, is far beyond the scope of this discussion. Somewhat inaccurately, we may say that the totality of interactions may be "summed" into an average nucleon–nucleus potential plus a residual nucleon–nucleon interaction, of considerably less strength and range.

An incident nucleon may be deflected from its initial path by this average potential, just as a comet would be deflected by the gravitational potential of the earth, without noticable complication from the residual interaction. An important point is that all the nucleons contribute to the average potential, and they are *individually* unaffected by interaction of an incoming nucleon with this potential, so that the nucleus itself is quite unaffected, and the deflected nucleon continues on its way without loss of kinetic energy. Such a process is called *elastic scattering*.

The role played by the residual interaction is enormously more complicated and varied. Individual nucleons in the target may, through this interaction, take energy from the incident projectile, one or more target nucleons may be emitted into free space, the projectile may be captured, or other nucleons combined with it as it goes away, and so on. Contributions of this kind from the residual interaction amount to the physical removal, from the beam, of nucleons which would otherwise be elastically scattered or not deflected at all.

Thus a certain number of nucleons neither pass through the target foil undeflected, nor are elastically scattered from it. The situation is quite analogous to the classical problem of optics in which a beam of light is incident upon a translucent object. A certain fraction of the beam is transmitted through the object, another fraction is scattered in various directions, and another fraction is absorbed. The absorption can be included by defining the material's index of refraction to be a *complex* number. The larger the imaginary part of the refractive index is, the more effective is the absorption.

The "optical model" for nucleon–nucleus scattering consists, similarly, in defining a single complex nuclear potential, the real part of which is hopefully the overall average potential we have been discussing, and the imaginary part of which is chosen to give the proper absorption. The residual interaction is otherwise ignored.

In Sec. II, we give an account of the application of such a model, including only sufficient detail to make clear how and why the use of computers is important. In Sec. III, a few results of early and modern optical calculations are presented, to show their success in describing experimental data. Finally, in Sec. IV, we discuss ways of taking explicit account of some detailed effects of the residual interaction. It should be kept in mind throughout all of what follows that the aim of all these labors is to reveal the detailed properties of nuclei.

2 BASIC IDEAS OF NUCLEAR SCATTERING, AND COMPUTATION OF CROSS SECTIONS

We have now decided that we can treat the fearsomely complicated interactions between a single nucleon and the many nucleons composing a target nucleus, to some degree of realism, by replacing the many body problem with a one body problem. Treatment of the nuclear scattering process as that undergone by a single particle in some potential, phenomenologically chosen to simulate the actual average nuclear potential, is well within our computing capabilities.

In order to understand what is to be calculated, and how it relates to what can be measured, we must plunge into some detailed considerations. Our problem is quantum mechanical, but we can be guided by classical ideas. In the classical formulation of mechanics, the motion of a particle of mass μ in a region of potential $V(r)$ is determined by the conservation of energy, expressed in the form

$$\frac{\vec{p}^2}{2\mu} + V(r) = E. \tag{1}$$

Here E is the total energy of the system, which must be the sum of potential and kinetic energies. The kinetic energy $\vec{p}^2/2\mu = (\frac{1}{2})\,\mu\vec{v}^2$, since $\vec{p} = \mu\vec{v}$, where $\vec{v}$ is the velocity. Since, if $\vec{r}(t)$ is the time-dependent vector giving the position of the particle in a chosen coordinate system at any time t, we must have $\vec{v} = d\vec{r}/dt$, the equation (1) is seen to be a *differential equation* which can be solved for $\vec{r}(t)$ given V and E.

In quantum mechanics, also, we must solve a differential equation, but not a differential equation in time alone, for we no longer have a mass point at a specific location. We consider $\vec{p}^2/2\mu + V(r) - E$ to be a linear operator acting on a function $\phi(\vec{r}, t)$ of *both* position and time. A differential operator is obtained via the quantum prescriptions $\vec{p} \to -i\hbar\, \partial/\partial\vec{r}$ and $E \to i\hbar\, \partial/\partial t$. The constant $\hbar$ is the fundamental scaling factor of the submicroscopic world, Planck's constant h divided by 2π.

Now we have

$$\left(-\frac{\hbar^2}{2\mu}\left(\frac{\partial}{\partial\vec{r}}\right)^2 + V(\vec{r}) - i\hbar\frac{\partial}{\partial t}\right)\phi(\vec{r}, t) = 0. \tag{2}$$

A traditional method of solving such equations is by separation of variables. Choose $\phi(\vec{r}, t) = \Psi(\vec{r})\, e^{-iEt/\hbar}$. Multiplying through by $e^{iEt/\hbar}$ then leaves

$$\left(-\frac{\hbar^2}{2\mu}\left(\frac{d}{d\vec{r}}\right)^2 + V(\vec{r}) - E\right)\Psi(\vec{r}) = 0, \tag{3}$$

which is a differential equation to be solved for $\Psi(\vec{r})$ given $V(\vec{r})$ and E. (We assume $V(\vec{r})$ does not depend on the time.)

The solution $\Psi(\vec{r})$ must be interpreted as a kind of distribution function. The quantity $|\Psi(\vec{r})|^2$ is proportional to the probability that a particle will be found at the position $\vec{r}$. If we choose spherical polar coordinates, $\vec{r} \to (r, \theta, \varphi)$, Fig. 1, we can separate variables once again. The familiar result from classical analysis is unchanged; one can always write

$$\Psi(r, \theta, \varphi) = \sum_{l=0}^{\infty} \frac{u_l(r)}{r} \sum_{m=-l}^{l} Y_l^m(\theta, \varphi). \tag{4}$$

Here $Y_l^m \propto e^{im\varphi} P_l^m(\cos\theta)$, where $P_l^m(\cos\theta)$ is just the associated Legendre polynomial. The properties of the polynomials $Y_l^m(\theta, \varphi)$, called spherical harmonics, are well known by way of classical physics and mathematics. The sum on l is over all integers: 0, 1, 2 $\cdots$ ∞. The expansion of $\Psi(r, \theta, \varphi)$, as in (4) in terms of functions that depend on l, is called a partial wave expansion.

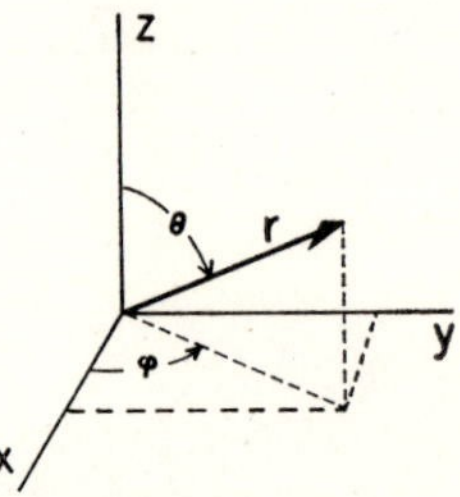

Fig. 1 The spherical polar coordinates (r, θ, ψ), in terms of which the Schrödinger equation is conveniently solved

Substituting (4) into (3), we can separate the equations for $u_l(r)$ and Y_l^m. In the simple case where the potential $V(\vec{r})$ is spherically symmetric, and so depends only on the magnitude of r, we find

$$\frac{\hbar^2}{2\mu}\frac{d^2u_l}{dr^2} + \left[E - V(r) - \frac{\hbar^2}{2\mu}\frac{l(l+1)}{r^2}\right]u_l = 0. \tag{5}$$

Thus separation of variables has reduced the problem to that of solving an ordinary differential equation, (5).

Before discussing equation (5) further, it is important to have in mind the physical significance of Ψ and u_l, and their relation to the scattering process. In quantum mechanics we do not seek to assign an $\vec{r}(t)$ to each particle, because it makes no sense to speak of a trajectory for a submicroscopic object. We seek to assign measurable probabilities, instead, so that we are forced to think in terms of a large number of identical particles incident upon the potential $V(r)$, as would be the case in an actual measurement. A certain fraction of the particles will scatter into a given region of space, and it is this fraction that is ultimately given by $|\Psi(\vec{r})|^2$, or by experiment.

If $V(r)$ vanishes sufficiently rapidly as r becomes infinite ($V(r)$ should vanish at least as rapidly as $1/r^2$ when $r \to \infty$) equation (3) has the form

$$\left(\frac{d}{d\vec{r}}\right)^2 \Psi(\vec{r}) + \frac{2\mu E}{\hbar^2}\Psi(\vec{r}) \approx 0 \tag{6}$$

for large r. For convenience, let $k^2 = 2\mu E/\hbar^2$. If the incident beam is coming along the z axis, we can easily solve (6) to find the corresponding Ψ, which can depend only on z. Equation (6) becomes

$$\frac{d^2\Psi(z)}{dz^2} + k^2\Psi(z) = 0, \tag{7}$$

which is solved by $\Psi_{inc}(z) = A_0 e^{\pm ikz}$. Such a function $\Psi_{inc}(z)$ represents a plane wave of amplitude A_0 moving up (+ sign) or down (− sign) the z axis at the speed $E/k\hbar$, as we can see by recalling $\phi(\vec{r}, t) = \Psi(\vec{r})\, e^{-iEt/\hbar}$, so $\phi_{inc}(\vec{r}, t) = A_0 e^{i(\pm kz - Et/\hbar)}$. The argument of the exponential is in the form $z \pm vt$, with $v = E/\hbar k$ the phase velocity.

To find the solutions at large r which represent the scattered waves, we will try to write down a general function of r and θ satisfying equation (6). The answer cannot depend on φ because the incident beam came down the z axis and if the potential is symmetric in θ and φ no amount of rotation of our experimental apparatus about the z direction would change the answer; we thus require $m = 0$ in eqn. (4). Such a solution is $\Psi(\vec{r}) = f(k, \theta)\, e^{\pm ikr}/r$. It represents outgoing (+ sign) or incoming (− sign) spherical waves. The general solution to equation (6) for very large r should thus be of the form

$$\Psi(\vec{r}) \underset{r\to\infty}{\sim} e^{ikz} + f(k, \theta)\,\frac{e^{ikr}}{r}. \tag{8}$$

Such a function represents a plane wave travelling along the z axis through the scattering potential at $z = 0$, plus a scattered spherical wave expanding

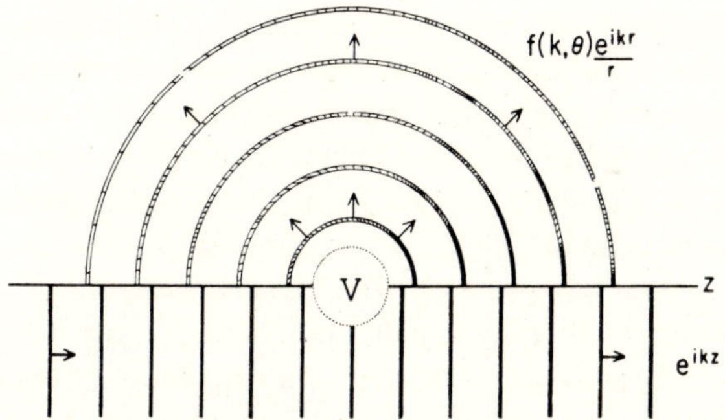

Fig. 2 Very schematic representation of the two terms in the asymptotic scattering wave function [Eqn. (8)]. For simplicity, the incident plane wave fronts are sketched in only below the z-axis, the outgoing spherical wave only above. The shading of the wave fronts suggests the amplitude variation of the waves

outward in all directions, with amplitude $f(k, \theta)$, from the origin, as illustrated in Fig. 2. Note we have chosen the incident amplitude $A_0 = 1$. The function $f(k, \theta)$ is called the *scattering amplitude.*

The only part of the expression (8) which would be changed, if there were no scattering potential at all, is $f(k, \theta)$, which becomes zero when $V(r) = 0$, since then expression (8) must satisfy Eqn. (7). Thus, when $V(r) = 0$ for all r, the complete solution to the problem is just $\Psi_0 = \exp(ikz)$. We can put Ψ_0 in a form comparable to equation (4), by expanding in Legendre polynomials, $P_l(\theta)$. The result of a familiar analysis applied to equation (5) with $V = 0$, is

$$e^{ikz} = \sum_{l=0}^{\infty} i^l (2l + 1) j_l(kr) P_l(\theta). \tag{9}$$

The functions $j_l(kr)$ which appear in (9), and which may be compared with the $u_l(r)/r$ in (4), are called spherical Bessel functions, and are widely tabulated because of their frequent appearance in classical radiation and vibration theory. They satisfy the recurrence relation

$$j_{l+1}(\varrho) = \frac{2l + 1}{\varrho} j_l(\varrho) - j_{l-1}(\varrho) \tag{10}$$

The first few are very simple functions of ϱ. Explicitly, $j_0(\varrho) = \sin\varrho/\varrho$, $j_1(\varrho) = \sin\varrho/\varrho^2 - \cos\varrho/\varrho$, etc. The asymptotic (very small or very large ϱ) values approached by the functions are simple for any l:

$$j_l(\varrho) \xrightarrow[\varrho\to 0]{} \frac{1}{1 \cdot 3 \cdot 5 \cdots (2l + 1)} \varrho^l \tag{11a}$$

$$j_l(\varrho) \xrightarrow[\varrho\to\infty]{} \frac{1}{\varrho} \sin\left(\varrho - \frac{l\pi}{2}\right) \tag{11b}$$

To compare $j_l(\varrho)$ strictly with $u_l(\varrho)$, we will define $F_l(\varrho) = \varrho j_l(\varrho)$. Comparison of equations (4) and (9) shows that $F_l(kr)$ satisfies equation (5) with $V(r) = 0$, since $Y_l^0 \propto P_l$. It is also a solution to equation (5) with $V(r) \neq 0$, for the case of very large r, provided $rV(r) \to 0$ and $r \to \infty$. We see also that $F_l(kr)$ approaches zero as r goes to zero, which is necessary since otherwise $\Psi_0(\vec{r})$ would be infinite at $r = 0$, a meaningless result. For the same reason we shall require always in equation (4) that $u_l(r) \to 0$ as $r \to 0$.

An extremely important point is that for large r, we can express $F_l(\varrho)/\varrho$ as a particular linear combination of incoming and outgoing spherical waves.

We use $\sin x = (e^{ix} - e^{-ix})/(2i)$ and the asymptotic expansion (11b) to write

$$F_l(kr) \xrightarrow[r\to\infty]{} \sin\left((kr - \frac{l\pi}{2}\right) = \frac{1}{2i}\left[\exp\left(ikr - i\frac{l\pi}{2}\right) - \exp\left(-ikr + i\frac{l\pi}{2}\right)\right]$$

$$= \frac{i}{2} e^{i(l\pi/2)}[\exp(-ikr) - \exp(-il\pi)\exp(ikr)] \tag{12}$$

The relative weight of the outgoing-wave part of (12) to the incoming-wave part is $-e^{-il\pi} = (-1)^{l+1}$. Now, the asymptotic form of the solution $u_l(r)$ of the full radial Schrödinger equation (5) can also be put in the form (12). However, in the expansion of $u_l(r)$ we would expect to find a relative weight quite different from $(-1)^{l+1}$, in general, because of the influence of the potential on the portion of $u_l(r)$ passing through the region where $V(r)$ is large. Let us represent this difference between F_l and u_l in the form $e^{2i\delta_l}$, such that we have

$$u_l(r) \xrightarrow[r\to\infty]{} \frac{iA_l}{2} e^{i(l\pi/2)} [\exp(-ikr) - \exp(-il\pi)\exp(2i\delta_l)\exp(ikr)] \tag{13}$$

In equation (13), the constant A_l reflects the additional possibility of amplitude difference in $F_l(r)$ and $u_l(r)$. We will determine A_l later by applying a boundary condition. Now, we can reverse the algebra between equation (12) and (11b), adding and subtracting $(iA_l/2)\, e^{i(kr - l\pi/2)}$ first. We get

$$u_l(r) \xrightarrow[r\to\infty]{} \frac{iA_l}{2} [e^{-i(kr-(l/2)\pi)} - e^{i(kr-(l\pi/2))} + (1 - e^{2i\delta_l})\, e^{i(kr-(l\pi/2))}]$$

$$= A_l\left[\sin\left(kr - \frac{l\pi}{2}\right) + e^{i\delta_l}\sin\delta_l\, e^{i(kr-(l\pi/2))}\right] \tag{14}$$

An alternate expression is also useful. From (13),

$$u_l(r) \xrightarrow[r\to\infty]{} \frac{iA_l}{2} [e^{-i(kr-l\pi/2)} - e^{2i\delta_l} e^{i(kr-l\pi/2)}]$$

$$= \frac{iA_l}{2} e^{i\delta_l} [e^{-i(kr-l\pi/2+\delta_l)} - e^{i(kr-l\pi/2+\delta_l)}] = A_l\, e^{i\delta_l} \sin\left(kr - \frac{l\pi}{2} + \delta_l\right) \tag{15}$$

Comparing (15) and (11b) or (12), we see that δ_l is a *phase shift* of the solution for large r. From (14) we see that $u_l(r)$ possesses an *additional outgoing spherical* wave as compared to $F_l(kr)$, with an amplitude $e^{i\delta_l}\sin\delta_l$. It then is clear that δ_l completely describes the influence of the scattering potential on the outgoing waves at large distances.

We now know two quantities completely specifying the scattered wave at large distances from the region of the potential: $f(k, \theta)$ of equation (8), and δ_l. It seems plain enough, then, that $f(k, \theta)$ can be expressed in terms of a sum over all the δ_l's.

To find the relation between f and δ_l, let us look again at equation (4). Since, as we have mentioned, we can take $m = 0$, it becomes

$$\Psi(r, \theta) = \sum_{l=0}^{\infty} \frac{u_l(r)}{r} Y_l^0(\theta) = \sum_{l=0}^{\infty} \sqrt{\frac{2l+1}{4\pi}} \frac{u_l(r)}{r} P_l(\cos\theta), \tag{16}$$

using the relation between Y_l^0 and the ordinary Legendre polynomial, $P_l(\theta)$. We can now take the limit as $r \to \infty$, and so in (16) replace $u_l(r)$ by its asymptotic expansion (14). Observe that, choosing $A_l = i^l(4\pi(2l+1)/k)$, if we insert (14) into (16), the result falls naturally into the form of equation (8):

$$\begin{aligned}\Psi(r, \theta) \xrightarrow[r\to\infty]{} &\sum_l \left[i(2l+1) \left\{ \frac{\sin(kr - l\pi)}{kr} + e^{i\delta_l} \sin\delta_l \frac{e^{i(kr - l\pi/2)}}{kr} \right\} \right] P_l(\theta) \\ = &\left\{ \sum_l i(2l+1) j_l(kr) P_l(\theta) \right\} + \left[\frac{1}{k} \sum_l (2l+1) e^{i\delta_l} \sin\delta_l P_l(\theta) \right] \frac{e^{ikr}}{r}.\end{aligned} \tag{17}$$

The term in curly brackets is just e^{ikz}. The term in square brackets is a function of k and θ multiplying e^{ikr}/r. Thus, comparing (17) and (8), we have a partial wave expansion of $f(k, \theta)$:

$$f(k, \theta) = \frac{1}{k} \sum_{l=0}^{\infty} (2l+1) e^{i\delta_l} \sin\delta_l P_l(\cos\theta). \tag{18}$$

Equation (18) relates the scattering amplitude $f(k, \theta)$ to the scattering phase shifts, δ_l. Note that it is in the form of an infinite series, which is in practice not a very handy form unless convergence is very rapid, so that we have to know only the first few δ_l's. Nonetheless, our problem is now solved in principle once we have found the phase shifts. The task is to compute the set of δ_l.

Before discussing the problem now at hand, the solution of the radial Schrödinger equation for $u_l(r)$ and ultimately δ_l, let us notice the physical meaning of $f(k, \theta)$.

Consider a typical scattering process with an experimental arrangement as shown schematically in Fig. 3. The scattered particles are detected at an angle θ in the plane defined by the incident beam. The detector is customarily capable of "counting" the number of particles passing into it and measuring their energy (momentum) as well. Since almost all the incident particles will pass through a sufficiently thin target with little or no scattering, one can collect the transmitted beam and take it to be identical to the incident beam in intensity, or in number of particles passing through a unit area in unit time.

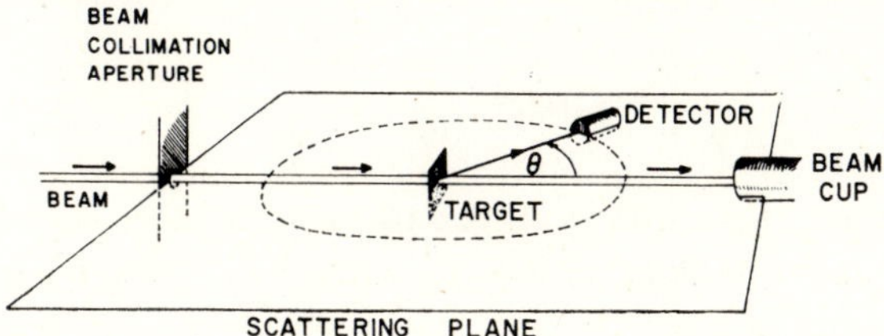

Fig. 3 Typical experimental arrangement for measuring a scattering cross section, $d\sigma(\theta)/d\Omega$. The detector may be rotated to any desired angle

Thus one can measure the ratio $N(k, \theta)/nN_0(E)$, where $N_0(E)$ is the number of particles incident at energy E on a unit area of the target in a unit time, n is the number of possible scatterers (target nuclei) per unit area of the target foil, and $N(k, \theta)$ is the number of scattered particles passing through a unit area of the detector in unit time, with energy $\hbar^2k^2/2\mu$, when the detector is at an angle θ with respect to the incident beam. The ratio is called the *scattering cross section*. Physically it is just the area, surrounding a single target nucleus, through which an incident particle of energy E would have to pass in order to be scattered through an angle θ. But, $|f(k, \theta)|^2$ is precisely this quantity, for if $|\Psi(\vec{r})|^2$ gives in general the probability that a particle is found at position $\vec{r}$, then $|\Psi_{\text{scat}}|^2 = |(f(k, \theta)/r)\, e^{ikr}|^2 = (|f(k, \theta)|^2)/r^2$ is such a probability. Noting that at $\vec{r}$ a detector of surface area A subtends a solid angle $d\Omega = (A/r^2)$, we can readily agree that the differential cross section = (probability of scattering to r, θ)/(unit solid angle subtended by detector at r, θ) $= \{(|f(k, \theta)|^2)/(r^2)\}/(1/r^2) = |f(k, \theta)|^2$. The differential cross section is usually written $d\sigma(\theta)/d\Omega$.

Thus

$$\frac{d\sigma(\theta)}{d\Omega} = |f(k, \theta)|^2. \tag{19}$$

If the theoretical differential cross section $|f(k, \theta)|^2$ actually agrees with the experimental cross section, it means that our theoretical calculation has been successful. It is clear that if a potential $V(r)$ is given, a unique set of δ_l can be calculated. Therefore our theoretical task is to find a *good* potential so that a *good* set of phase shifts is obtained, by which is meant that the theoretical differential cross section, (19), agrees with experiment. We now need to explain how the phase shifts are calculated, for an arbitrary potential $V(r)$.

Clearly, it is hopeless to look for a simple analytic expansion of $u_l(r)$ in terms of know functions, since $V(r)$ must be chosen to describe experimental results, and may have a very complicated functional form. An equation of the form (5) cannot be solved analytically unless the function $V(r)$ is of an *extremely* simple form. Thus we must in almost all cases solve (5) numerically, with a computer. We will now outline the general procedure, with which a computer can be programmed, to solve equation (5). We first rewrite (5) as

$$u_l''(r) = v(r)\, u_l(r). \tag{20}$$

Here $u_l''(r) = d^2u_l(r)/dr^2$ and $v(r) = (-2\mu/\hbar^2)\,[E - V(r) - (\hbar^2/2\mu)\,(l(l+1)/r^2)]$. Note that the solutions $u_l(r)$ will be complex numbers, since the potential $V(r)$ is to be chosen complex. Let us then assume that we know the solution u_l for two values of r, i.e. at r_0 and $r_0 + h$, where h is assumed very small. We then want to obtain $u_l(r)$ at $r_0 + 2h$. A possible way to find $u_l\,(r_0 + 2h)$ is from the equation

$$u_l\,(r + 2h) \approx 2u_l\,(r + h) - u_l(r) + h^2 v\,(r)\, u_l\,(r + h). \tag{21}$$

That (21) is a practical and accurate expression can be shown as follows. Let us expand $u_l\,(r_0 + h)$ and $u_l\,(r_0 + 2h)$ in powers of h, using the Taylor expansion. With the abbreviations $u_l = u_l(r_0)$, $u_l' = u_l'(r_0) = du_l\,(r_0)/dr$, $u_l'' = u_l''(r_0) = d^2u_l\,(r_0)/dr^2$, etc., we have

$$u_l\,(r + 2h) = u_l + 2hu_l' + 2h^2u_l'' + \tfrac{4}{3}\,h^3u_l''' + \tfrac{2}{3}\,h^4u_l'''' + \cdots$$

We can now use Eq. 20, which shows that one can rewrite $u_l'' = vu_l$ $(\equiv v(r)\, u_l(r))$, $u_l''' = (vu_l)' \equiv d\,(v(r)\, u_l(r))/dr$, etc. We thus get, for the left hand side of (21), the expression

$$u_l\,(r + 2h) = u_l + 2hu_l' + 2h^2\,(vu_l) + \tfrac{4}{3}\,h^3\,(vu_l)' + \tfrac{2}{3}\,h^4\,(vu_l)'' + \cdots \tag{22}$$

The right hand side of (21) can also be rewritten in powers of h, and a little algebra shows that it equals

$$u_l + 2hu_l' + 2h^2\,(vu_l) + \tfrac{4}{3}\,h^3\,(vu_l)' + \tfrac{7}{12}\,h^4\,(vu_l)'' + \cdots \tag{23}$$

Comparison of (22) and (23) shows that equation (21) ceases to be exact only at h^4 and beyond. All such terms will be very small if h was chosen sufficiently small. This means that (21) can be a very good approximation for $u_l\,(r_0 + 2h)$, given $u_l\,(r_0 + h)$ and $u_l(r_0)$.

Now, if we know u_l at two points we can use Eqn. (21) to generate the solution at intervals of h for as many points as we need to tabulate $u_l(r)$ completely. We always know u_l at one point: $u_l(0) = 0$. We can then let $u_l(h) = b$, a positive constant. Now we can use Eqn. (21) to find $u_l(2h)$. The procedure can be mechanically repeated as many times as required, until we have u_l at a value of r very large compared to the distances at which the potential $V(r)$ is large. Thus we have performed an outward integration of the differential equation to just beyond some sufficiently large r, R_m, where $V(R_m) \approx 0$.

We have now numerically generated a large number of values of $u_l(r)$, but each is in effect multiplied by an arbitrary constant, through our choice of b. We will have to eliminate this arbitrariness eventually. The values of $u_l(r)$ we have generated will be called the internal solution ($r \leq R_m$).

We also want the derivative (or slope) of the internal solution at $r = R_m$. It can be constructed numerically if we have gone two h-steps beyond R_m in the outward integration, by

$$u_l^{(i)\prime}\,(R_m) \equiv \left.\frac{du_l^{(i)}}{dr}\right]_{r=R_m}$$

$$\approx \frac{1}{12h}\,[8\,(u\,(R_m + h) - u\,(R_m - h)) - (u\,(R_m + 2h) - u\,(R_m - 2h))] \quad (24)$$

where the (i) superscript will henceforth denote the internal solution. This is the widely-used method of numerical differentiation originated by George Rutledge.

The differential equations we have considered so far are all of second order; that is, they contain the second derivative. Such equations are well known to have *two* linearly independent solutions. We have seen that $F_l\,(kr)$ is a solution to (5) with $V = 0$. We thus expect a second solution $G_l\,(kr)$, and indeed $G_l\,(kr) = krn_l\,(kr)$ satisfies (5), with $n_l\,(kr)$ being the spherical Neumann function. It is irregular ($\to \infty$) at the origin, where $j_l\,(kr)$ was regular ($\to 0$), and (to be linearly independent) must approach $\cos\,(kr - (l\pi/2))$ for large r, rather than $\sin\,(kr - (l\pi/2))$. The first few n_l's are

$$n_0(\varrho) = -\frac{\cos\varrho}{\varrho}, \quad n_1(\varrho) = -\frac{\cos\varrho}{\varrho} - \frac{\sin\varrho}{\varrho}, \quad \text{etc.} \quad (25)$$

Looking back at Eqn. (14), we can realize that for large r it should be possible to write

$$u_l^{(e)}(r) = F_l(kr) + C_l(G_l(kr) + iF_l(kr)), \tag{26}$$

for the first term on the right hand side provides the sine, while the second term $\to \cos kr + i \sin kr \to e^{ikr}$. The (e) superscript reminds us that the expansion is useful only external to the potential; i.e., for $r \geq R_m$. The constant C_l must be determined.

Suppose we now match the internal and external solutions by setting equal their magnitude and slope (derivative) at $r = R_m$. Then we will have the function $u_l(r)$ at all points, from the origin in steps of h out as far as we feel we must go.

Thus

$$u_l^{(i)}(R_m) = u_l^{(e)}(R_m), \quad u_l^{(i)'}(R_m) = u_l^{(e)'}(R_m),$$

or

$$\frac{u_l^{(i)'}(R_m)}{u_l^{(i)}(R_m)} = \frac{u_l^{(e)'}(R_m)}{u_l^{(e)}(R_m)}. \tag{27}$$

Now using Eqn. (26), the right hand side is known. Further, the division on the left hand side eliminates the arbitrariness of the internal solution by dividing out the arbitrary constant b. Thus only C_l remains unknown. We have

$$\frac{u_l^{(i)'}(R_m)}{u_l^{(i)}(R_m)} = \frac{F_l'(kr) + C_l(G_l'(kr) + iF_l'(kr))}{F_l(kr) + C_l(G_l(kr) + iF_l(kr))} \tag{28}$$

which can be easily solved for C_l. We find

$$C_l = \frac{-[(u_l^{(i)'}(R_m)/u_l^{(i)}(R_m))]\,F_l(kr) + F_l'(kr)}{[(u_l^{(i)'}(R_m)/u_l^{(i)}(R_m))]\,[G_l(kr) + iF_l(kr)] - [G_l'(kr) + iF_l'(kr)]}. \tag{29}$$

Since we have found C_l, our labor (or that of the computer) is almost at an end, for comparison of Eqn. (26) and (14) shows that $A_l = 1$, and

$$C_l = e^{i\delta_l} \sin \delta_l. \tag{30}$$

So we have found what we need. Eqn. (30) may readily be solved for each δ_l.

But how many do we need? Fortunately, we do *not* need an infinite number. The sum on l in equation (18) can be stopped at some $l_{\max}$, because for $l > l_{\max}$, $\delta_l \approx 0$. The reason for this happy turn of events lies in the effect of the centrifugal barrier term, $l(l+1)/r^2$, in Eqn. (5). As l increases, the barrier becomes larger and the incoming particles of angular momentum l cannot come as close to the target. For sufficiently large l, the particle is always so far out that $V \approx 0$ and so $\delta_l \approx 0$. (This situation can also be understood in terms of classical mechanics, if one notices that in order for a particle with a given linear momentum to have a large angular momentum with respect to the force center, it must follow a trajectory which does not pass near the force center.)

The computational recipe for optical model calculations is now complete for neutrons. If, however, the projectile is charged, $V(r)$ includes not only the short-range nuclear potential, but the repulsive Coulomb potential between projectile and target nucleus. The Coulomb term, even for large r, is

$$V_C(r) = \frac{Z_1 Z_2 \, e^2}{r},$$

where Z_1 is the number of protons in the projectile, Z_2 the number in the target, and e the charge on the proton. Note that $r V_C(r)$ does *not* go to zero as $r \to \infty$. Then in *no* region of space is an expansion of the form (26) valid, and the procedure following it is of no use.

However, there is nothing to prevent us from solving Eqn. (5) with $V = V_C$, and indeed the equation can be solved analytically in such a case, although the solutions are by no means as simple as spherical Bessel and Neumann functions. They can be expressed in terms of confluent hypergeometric functions.

We can identify the regular and irregular solutions of the Schrödinger equation for the Coulomb field, in much the same way we proceeded above for the case $V(r) = 0$ and $Z_1 = 0$. These functions, which we denote as F_l^C and G_l^C are usually called the Coulomb wave functions, and satisfy

$$\begin{aligned} F_l^C(kr) &\xrightarrow[r\to\infty]{} \sin\left(kr - \frac{l\pi}{2} + \sigma_l\right) \\ G_l^C(kr) &\xrightarrow[r\to\infty]{} \cos\left(kr - \frac{l\pi}{2} + \sigma_l\right) \end{aligned} \tag{31}$$

where σ_l is the phase shift due to scattering from the Coulomb potential alone, and is called the Coulomb phase shift. The Coulomb phase shifts are given by the relation

$$\sigma_l = \text{Arctan}\left[\frac{\text{Imaginary Part of } \Gamma(l+1+i\eta)}{\text{Real Part of } \Gamma(l+1+i\eta)}\right]. \tag{32}$$

In Eqn. (32), the parameter $\eta = (\mu Z_1 Z_1 \, e^2)/(\hbar^2 k)$, and the function $\Gamma(z)$ is the gamma, or factorial, function. If z is an arbitrary complex number, with real part >0,

$$\Gamma(z) \equiv \int_0^\infty t^{z-1} \, e^{-t} \, dt \tag{33}$$

It is usually most convenient for the computer to construct σ_l for $l > 0$ by calculating σ_0 explicitly and obtaining the rest from the recurrence relation

$$\sigma_{l+1} = \sigma_l + \text{Arctan}\left(\frac{\eta}{l+1}\right). \tag{34}$$

Having the asymptotic forms of $F_l^C(kr)$ and $G_l^C(kr)$, one can proceed in about the same way as we did for neutrons. The procedure to obtain the internal solution is exactly the same, except that $V(r)$ is replaced by $V(r) + V_C(r)$, so that we wind up with an expression of the same form as (26), with F_l and G_l replaced, respectively, by F_l^C and G_l^C. Making the same replacements in Eqn. (29), we get a new C_l^C which in turn yields a new δ_l^C, through Eqn. (30). Summarizing, we obtain

$$\begin{aligned} u_l^{(e)}(r) &= F_l^C(kr) + C_l^C \left(G_l^C(kr) + iF_l^C(kr)\right) \\ \underset{r\to\infty}{\longrightarrow} & \sin\left(kr - \frac{l\pi}{2} + \sigma_l\right) + C_l^C \left(e^{i(kr - l\pi/2) + i\sigma_l}\right) \\ &= \sin\left(kr - \frac{l\pi}{2} + \sigma_l\right) + e^{i(\sigma_l + \delta_l^C)} \sin \delta_l^C \, e^{i(kr - l\pi/2)} \end{aligned} \tag{35}$$

Comparison of (35) with (14) would seem to indicate that the manipulations connecting (14) and (18) can be repeated exactly, merely replacing $e^{i\delta_l} \sin \delta_l$ by $e^{i(\sigma_l + \delta_l^C)} \sin \delta_l^C$. However, this simple parallelism breaks down, partly because of the additional phase σ_l in $F_l^C(kr)$. If we insert $F_l^C(kr)$ in place of $u_l(r)$ in Eqn. (4) and sum over l, we cannot obtain a plane wave. Fortunately, part of the remedy is straight forward; we surely have to divide out the extra phase, and so should use $e^{i\sigma_l} F_l^C$ in (4). Hence we have to multiply each term of Eqn. (35) by $e^{i\sigma_l}$.

Unfortunately, we still will not have a plane wave as leading term, because $F_l^C(kr)$ is itself a complete scattering solution, and must contain both the incident plane wave and an outgoing spherical wave due to Coulomb-field scattering alone. As a result, substituting $e^{i\sigma_l} F_l^C(kr)$ into (4) gives

$$\sum_{l=0}^{\infty} \frac{e^{i\sigma_l}F_l^C(kr)}{r} Y_l^0(\theta) = e^{i(k_z - \eta \ln(r-z))}\left(1 - \frac{\eta^2}{ik(r-z)}\right) + \frac{f_C(\theta)}{r} e^{i(kr - \eta \ln(2kr))}. \tag{36}$$

The first term on the right hand side of Eqn. (36) indeed resembles a plane wave for large z, although it is distorted by the Coulomb field, through the terms depending on η. However, the second term is the expected Coulomb-scattered outgoing spherical wave, also distorted except at large r (as $r \to \infty$, $kr \gg \eta \ln(2kr)$).

Hence, the analog of Eqn. (18) for the full scattering amplitude will now have the form

$$f(k, \theta) = f_C(\theta) + \frac{1}{k}\sum_{l=0}^{\infty} (2l+1)\, e^{2i\sigma_l} e^{i\delta_l^C} \sin \delta_l^C P_l(\theta). \tag{37}$$

The amplitude $f_C(\theta)$ can be expressed in a fairly simple analytic form,

$$f_C(\theta) = \frac{\eta}{2k \sin^2\left(\dfrac{\theta}{2}\right)} \exp\left(-i\eta \ln\left(\sin^2 \frac{\theta}{2}\right) + 2i\sigma_0\right) \tag{38}$$

and is usually called the Rutherford scattering amplitude, after Ernest Lord Rutherford (1871–1937), whose studies of the scattering of low-energy helium nuclei from thin gold foils in fact led to the hypothesis of the nuclear atom.

Evidently, when $V(r) = 0$, $f(k, \theta) = f_C(\theta)$, since $\sin \delta_l^C = 0$ in Eqn. (37). We can identify $f_C(\theta)$ as the scattering amplitude due to the Coulomb field alone, as represented by V_C. If, on the other hand, $V_C = 0$ and $V(r) \neq 0$, Eqn. (37) is identical to Eqn. (18), since $\sigma_l = 0$, $f_C(\theta) = 0$, and $\delta_l^C = \delta_l$.

Now the description of nucleon elastic scattering from a central potential is almost complete. Only one further complication needs to be discussed. Physically, l is the quantized *orbital* angular momentum of the incident nucleon, $l\hbar$ being comparable to the classical quantity $L = r\mu v \sin \phi$, where ϕ is the angle between $\vec{r}$ and $\vec{v}$. But nucleons also possess an *intrinsic* angular momentum, $\vec{s}\hbar$, of magnitude $\hbar/2$. The intrinsic angular momentum, usually called the spin, can align itself either parallel or antiparallel to a chosen axis. These are the only two possible orientations allowed quantum-mechanically.

Inclusion of the spin in the elastic scattering formalism would be trivial were it not for the fact that experiments which allow determination of the "population" of each spin state in a scattered beam reveal that the true nuclear potential is able to change the number of nucleons having a given spin orientation. However, the total angular momentum $\vec{j} = \vec{l} + \vec{s}$, carried by each partial wave, remains the same, so that if we specify both j and l, we should still be able to expand $f(k, \theta)$ in a form analogous to Eqn. (18) or (37).

Let us be more explicit about the spin effects. Previously, the z-axis has been chosen to be the incident beam direction. We will choose the x-axis as the one along which the spin is to be aligned. In the particular case in which as many particles in the beam have spins along the positive x-axis ("up") as have spins along the negative x-axis ("down"), the beam is said to be *unpolarized.*

As long as the beam is in fact unpolarized, there is nothing special in our choice of the x-axis, and any arbitrary axis in the x–y plane would work as well, so that the independence of the scattering phenomenon from any choice of azimuthal angle φ is retained. But the converse holds also, for we now expect that when the incident beam is *polarized,* so that, for example, more spins point up than down the x-axis, the azimuthal symmetry of the scattering process is lost. The differential scattering cross section measured in the positive y direction ($\varphi = (\pi/2)$, "right") will not be the same as that in the negative y direction ($\varphi = -(\pi/2)$, "left") even for the same θ and k. This effect is known as the right-left asymmetry of the scattering.

To return to our problems with l, notice that to each value of l, there correspond two values of j, the magnitude of the total angular momentum; namely, $j = l + \frac{1}{2}$ (spin parallel to the direction of $\vec{l}$), and $j = l - \frac{1}{2}$ (spin antiparallel). Because the nuclear potential is not quite the same for the two cases, we expect that the nuclear phase shift is not the same. Our expression for $f(k, \theta)$ (Eqns. (18) or (37)) is modified considerably, but we will not restate it save to remark that it will have to exhibit two phase shifts for each l, $\delta_{l,j=l+1/2} \equiv \delta^{(+)}_{l}$, and $\delta_{l,j=l-1/2} \equiv \delta^{(-)}_{l}$, which will not in general be equal.

In order that our phenomenological potential $V(r)$ can produce two such sets of phase shifts, it will have to contain a term explicitly coupling $\vec{l}$ and $\vec{s}$, having a form such as $(\vec{s} \cdot \vec{l})\, V_{SO}$. We do not have space—and the reader is not likely to have patience—for details, but it will be noted that since $\vec{j} = \vec{l} + \vec{s}$, then $\vec{j}^2 = \vec{l}^2 + \vec{s}^2 + 2\vec{l} \cdot \vec{s}$, so that

$$\vec{l} \cdot \vec{s} = \frac{\vec{j}^2 - \vec{l}^2 - \vec{s}^2}{2} .$$

The quantum-mechanical magnitudes of $\vec{j}^2$, $\vec{l}^2$ and $\vec{s}^2$ are just $j(j+1)$, $l(l+1)$, and $s(s+1)$ respectively. Since $s = \frac{1}{2}$ always, $\vec{s}^2$ is just $\frac{3}{4}$. The reader can satisfy himself that for $j = l + \frac{1}{2}$, $\vec{l} \cdot \vec{s}$ becomes l, while for $j = l - \frac{1}{2}$ it is $-(l+1)$. The result is that we obtain *two* equations of the form of Eqn. (5), one for each j, with potentials V differing by the magnitude of the constant V_{SO}, since $[(\vec{s} \cdot \vec{l})\, V_{SO}]_{j=l+1/2} - [(\vec{s} \cdot \vec{l})\, V_{SO}]_{j=l-1/2} = lV_{SO} - (l+1)\, V_{SO} = -V_{SO}$. Each of the two equations can be solved in exactly the same way as we discussed previously, and their solutions give, respectively, the two phase shifts $\delta_l^{(+)}$, $\delta_l^{(-)}$. When $V_{SO} = 0$, we recover Eqn. 18, since the two δ's both become just δ_l.

In the usual scattering experiment, the incident beam is unpolarized, but to be realistic we have to set $V_{SO} \neq 0$. However, the scattering cross section is still independent of φ, as we have explained. The scattered beam will be polarized, to some degree, because of V_{SO}, and we need to say a little about how the polarization is measured.

Assume that we have a *second* target, from which we rescatter the scattered beam. Since the scattered beam is polarized, we expect a left–right asymmetry (φ dependence). If the second scatterer does not introduce an unknown quantity, we can calculate the expected asymmetry from the phase shifts $\delta_l^{(\pm)}$. Consider the situation shown in Fig. 4. The polarization of the nucleons scattered through a certain angle θ relative to the z-axis is

$$P(\theta) = (N_+ - N_-)/(N_+ + N_-) \tag{39}$$

where N_+ is the number of nucleons with spin along the positive x axis, and N_- is the number with spin along the negative x axis. The quantity $P(\theta)$ is obtained by rescattering the nucleons from a second target, usually ^{12}C, and measuring the asymmetry. If I_L and I_R are the number of nucleons scattered through a given angle θ' to the left and to the right of the scattered beam axis, then it can be shown that

$$P = [(I_R - I_L)/(I_R + I_L)]\, P_C^{-1} \tag{40}$$

where P_C is the polarization produced by the second scatterer, and must be known from previous experiments.

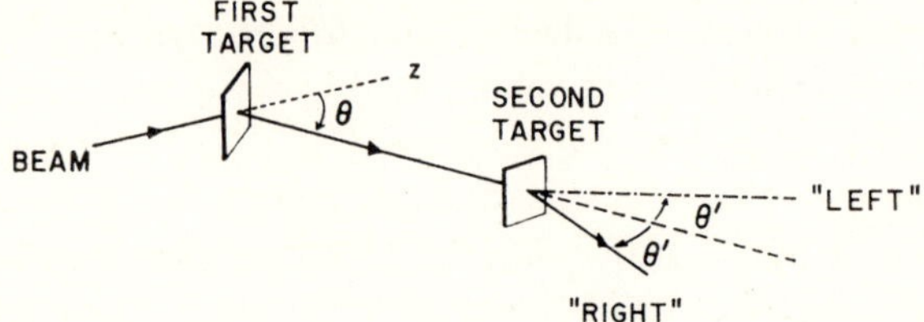

Fig. 4 A simple in-plane double scattering experiment for measuring the polarization, $P(\theta)$

With these details in hand, we may now turn to the application of the model to actual scattering experiments.

3 APPLICATION OF THE OPTICAL MODEL TO NUCLEON-NUCLEUS SCATTERING

A vast number of scattering experiments have been carried out in the past three decades, and a great wealth of differential cross sections is known for scattering of protons and neutrons from nearly all stable nuclei, at incident energies from a few electron volts (for neutrons) up to hundreds of millions of electron volts. What can be made of this data?

The way is clear. Given $V(r)$ we solve for the $\{\delta_l\}$ by the procedure which has been outlined. From the set of $\{\delta_l\}$ we can calculate the cross section. If we get agreement, and the potential chosen is reasonable, the job is over. But what is "reasonable"?

Certainly $V(r)$ must be parameterized, and the form and types of parameters to be used should be suggested by our knowledge of the nuclear force. From electron scattering experiments, it is very well established that a typical nucleus of mass A confines its positive charge within an approximately spherical region of radius $R = r_C A^{1/3}$, where $r_C \approx 1.2 \times 10^{-13}$ cm $= 1.2$ Fermi. The fact that the dependence of R is on A rather than Z, mass rather than charge, suggests that the entire nuclear mass is confined within this region. It is reasonable, then, to assume that the potential $V(r)$ will have a range $R = r_0 A^{1/3}$, where r_0 is of the order of 1.2 Fermi. The fewer additional parameters we are forced to introduce, the less ambiguous will be our results.

In the early days (1950–1954) of the optical model, V was often chosen to be of a very simple form:

$$V(r) = \left\{ \begin{matrix} -V_0 & r \leq R \\ 0 & r > R \end{matrix} \right\} - i \left\{ \begin{matrix} \gamma V_0 & r \leq R \\ 0 & r > R \end{matrix} \right\}. \tag{41}$$

Such a potential is usually called a "square well" though it is not square and bears no resemblance to a well, except as graphed in two dimensions, $V(r)$ versus r—see Fig. 5. Its advantages are: 1. The differential equation can be solved very easily—no computer is needed; indeed, the internal solution is $j_l(K'r)$, with $K' = [(2\mu/\hbar^2)(E + V_0 + i\gamma V_0)]^{1/2}$ and thus is obtained analytically. Then δ_l can be obtained from (29) and (30). 2. There are only three parameters, the constants V_0, γ and R.

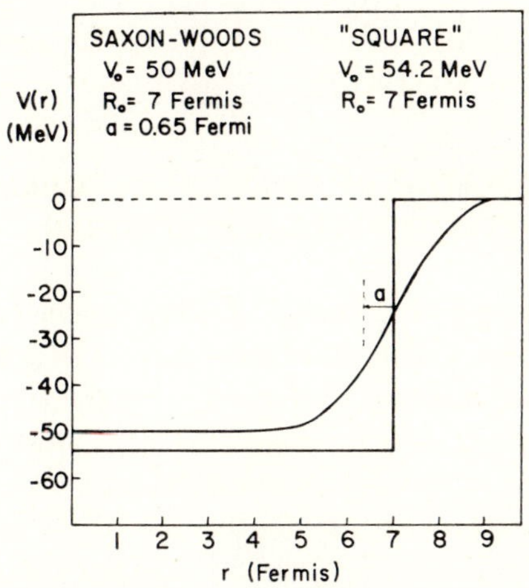

Fig. 5 A comparison of a "square" and a Woods–Saxon potential having identical volume

In the pioneering work of Feshbach, Porter and Weisskopf, it was found that the angular distributions for neutron scattering at incident energies of up to 1 MeV on a very wide range of nuclei ($A \sim 40$ to 250) were reasonably well described. Also well described was the angle-integrated total cross section as a function of incident neutron energy from about 50 keV to 1 MeV, for $A \sim 1$ to 240. The parameters used were

$$V_0 = 42 \text{ MeV}, \gamma \approx 0.05, R = 1.45\, A^{1/3} \text{ Fermi}.$$

An example is shown in Fig. 6.

When interest turned to proton elastic scattering, as well as higher energy neutron scattering, the inadequacy of the simple square well potential became clear. The handling of the additional Coulomb potential for proton scattering is no real difficulty as we showed in Sec. II. The problem is that, because of the

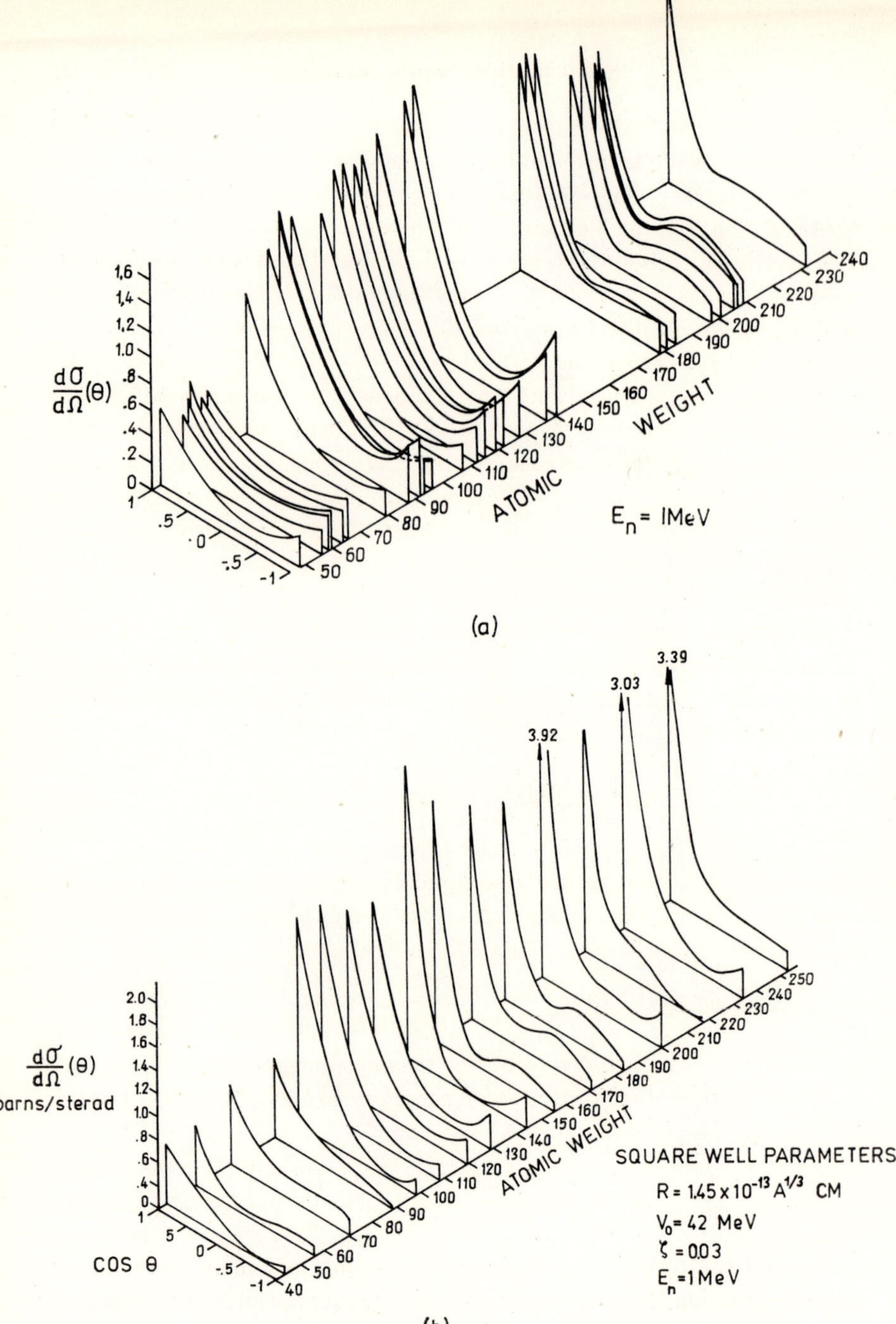

Fig. 6 (a) Experimental differential elastic scattering cross sections for 1 MeV neutrons, incident on various nuclei with $50 < A < 240$, as measured by M. Walt and H. H. Barschall (1954). The cross sections are in units of barns per steradian (1 barn = (10 Fermis)^2 = 10^{-24} cm^2), plotted as a function of $x = \cos\theta$. (b) The calculated cross sections at the same energy, using a "square" well optical potential, as given by Feshbach, Porter and Weisskopf

Coulomb repulsion, informative proton scattering on intermediate to heavy nuclei has to be done at relatively high energies (early data was at ~ 15 to 20 MeV) in order that the protons can overtop the repulsive Coulomb "barrier" and come near enough to be affected by the nuclear potential $V(r)$. The increased energy means also that the incident protons are localized within a smaller region of space, and therefore the scattering is far more sensitive to the detailed shape assumed for the nuclear potential than were the early neutron experiments.

Not surprisingly, Chase and Rohrlich found in 1954 that a square well potential predicts much too large a proton elastic scattering cross section for angles greater than 90°, particularly with heavy nuclei as targets. The reason for the discrepancy was quite clear at the time. The discontinuity in the square well potential at $r = R$ is much more effective at reflecting the incident beam than a realistic potential in which the variation with radius is smooth, and continuous.

A functional form for $V(r)$ decreasing smoothly to zero in a surface region of finite thickness is desirable. The functional form which has stood the test of time, and is most widely used today, is the form suggested by Woods and Saxon in 1954:

$$\text{Real}\,[V(r)] = -V_0 \left\{ \frac{1}{1 + e^{(r-R_0)/a}} \right\}, \tag{42}$$

where V_0 is a constant ($\approx$ the V_0 used by Feshbach, Porter and Weisskopf), and $R_0 = r_0 A^{1/3}$ is the nuclear radius. The length a is the half-width of the surface region, and is usually called the diffuseness.

The imaginary, or absorptive, term of $V(r)$ is not very clearly restricted in radial form by experiment; two fairly common forms taken are

$$\text{Im}\,(V(r)) = -W_0 \left[\frac{1}{(1 + e^{(r-R_I)/b})} \right], \quad \text{or} \quad -\left[\frac{4W_0\, e^{(r-R_I)/b}}{(1 + e^{(r-R_I)/b})^2} \right] \tag{43}$$

where $R_I = r_I A^{1/3}$, with $r_I \neq r_0$ in general, but both usually on the order of 1.2 Fermis. Again, W_0 is a constant and b is the diffuseness of the surface region.

The first possibility is usually called the volume imaginary potential, the second the surface imaginary potential. Both seem to do an equally good job in fitting the data, but there are various reasons for preferring one or the other. Since the imaginary part of the potential describes the absorption of particles from the incident beam, the choice is between absorption throughout the nuclear volume, or at the nuclear surface.

The improvement offered by such potentials, over the square well, in describing proton elastic scattering is illustrated in Fig. 7.

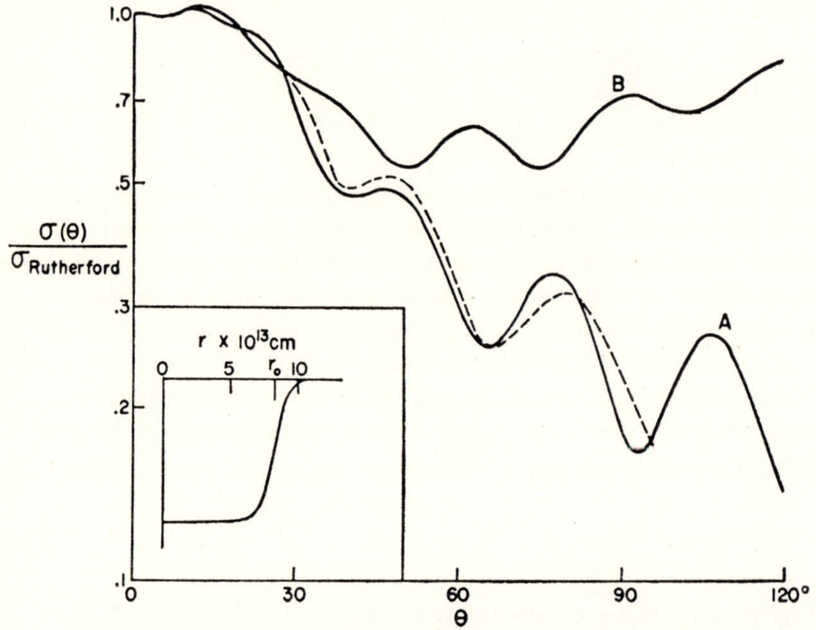

Fig. 7 A comparison of differential elastic proton scattering cross sections calculated with "square" (curve B) and Woods–Saxon (curve A) potentials. Experimental data for *Pt* (*p*, *p*) *Pt* at 22 MeV is shown as the dashed line. For the Woods–Saxon potential, $V_0 = 38$ MeV, $W_0 = 9$ MeV, $r_0 = 1.42$ Fm, and $a = 0.49$ Fm.

The angular distribution of the scattered particles is by no means the sole measurable object of an optical model calculation. The framework we have sketched can also give the total cross section, i.e. the angular distribution integrated over all possible outgoing directions, for neutrons—for protons, the total cross section is infinite, because of the presence of the Coulomb repulsion even at large distances. More important, the optical model potential can be used to predict the *polarization* of elastically scattered particles, as we have described in the end of Sec. II, and the result can be compared with the numerical tables for $P(\theta)$ resulting from double-scattering experiments.

In order to describe such experiments, a term coupling the spin and orbital angular momentum most be included in the optical potential, as discussed at the end of Section II. The simplest such potential is $(\vec{s} \cdot \vec{l})\, V_{\mathrm{SO}}(r)$, usually taken to have the form

$$\frac{(\vec{s} \cdot \vec{l})\, V_{\mathrm{SO}}\, \lambda^2}{\hbar r a} \left[\frac{\exp\left(\dfrac{r - R_0}{a}\right)}{\left(1 + \exp\left(\dfrac{r - R_0}{a}\right)\right)^2} \right] \tag{44}$$

where V_{SO} is a constant and λ is a scaling length customarily put at $\sqrt{2} \times 10^{-13}$ cm, exactly. This potential is in fact chosen to resemble the expression found by L.H.Thomas in 1926 for the energy shift of an atomic electron's binding energy, due to the interaction between the magnetic moment associated with its spin, and the electrostatic potential which binds it. Thus (44) is usually called the "Thomas term".

Adding the various terms (42), (43), and (44), we have the complete nuclear optical potential. While the elastic scattering cross section is particularly sensitive to the terms (42) and (43), the polarization is solely produced by the term (44). Hence both types of experiments are required in general to test a given potential.

The reader will already have noticed a sizable number of parameters in the potential: V_0, W_0, V_{SO}, r_0, a, r_I, and b. Letting the spin-orbit potential have an independent radius and diffuseness adds two more. Hence the physical labor required in repeating the calculation while varying various parameters until a fit to data is achieved can become immense.

As a result, almost all existing optical model programs contained automatic parameter search routines, which vary a given set of parameters in a specific way to minimize the quantity

$$\chi^2 = \frac{1}{N} \sum_{i=1}^{N} [(\sigma(\theta_i) - \sigma^D(\theta_i))/E^D(\theta_i)]$$

where $\sigma^D(\theta_i)$ is the experimental cross section (or data point) at angle θ_i, $E^D(\theta_i)$ is the experimental error associated with the data point, $\sigma(\theta_i)$ is the calculated value of the cross section at the same angle, and N is the total number of data points. A considerable budget of ingenuity has been expended on the construction of such searching subroutines, but they lie far beyond the scope of this chapter.

4 EXTENSIONS AND FURTHER APPLICATIONS OF THE OPTICAL MODEL

One of the principal aims of experimental nuclear physics is the uncovering of detailed information concerning the structure of atomic nuclei. Nuclei as bound systems can exist in a very large number of states, each characterized by a distinct binding (or excitation) energy and a definite total angular momentum, J, and parity, π. The angular momentum is familiar from classical analogy. The parity can be crudely described as an algebraic sign telling us whether the space function, Ψ_J, of the particular nuclear state is an even or odd function of its coordinates.

One of the most useful and simple techniques for obtaining such nuclear structure information is the measurement of angular distributions of products of direct nuclear reactions. As an example, suppose we impinge a beam of deuterons—the nuclei of "heavy" hydrogen, bound states of a single proton and neutron—on a target nucleus.

We will observe, at some given angle, a large number of elastically scattered deuterons. But we can also detect significant numbers of other particles, charged and uncharged: protons, neutrons, α-particles, and so on. These particles are products of *reactions*, and in the majority of cases, if the incident energy is high enough, most reactions can be treated as an abrupt transfer of one or more nucleons from projectile to target, or vice versa. To be specific, a reaction such as ^{90}Zr (d, p) ^{91}Zr, in which a deuteron (^{2}H, or d) is incident upon a ^{90}Zr nucleus, and protons are observed, leaving ^{91}Zr in various states, can be pictured as a direct transfer of a neutron from the deuteron to the target nucleus ^{90}Zr. The point is that the neutron must be captured into one of the specific states of ^{91}Zr. Thus if one observes the protons, one finds many groups of distinctly different energy, their separations matching the energy separations of the states of ^{91}Zr.

But one also gets simultaneously a clue to J^π for each observed state. The differential cross section of a given proton energy group is distinctively different for different values of the orbital angular momentum of the state into which the neutron is captured.

In the calculation of such cross sections, the optical model is invaluable. Since almost all of the deuterons affected at all by the target nucleus are simply elastically scattered, it is a good approximation to use a wave function, calculated via the optical model to fit deuteron elastic scattering, as the total wave function for the system. The outgoing protons can be described, in turn, by a similar wave function calculated using the optical potential for protons elastically scattered at the appropriate energy. With these two wave functions, plus the single-particle wave function of the bound neutron, it is possible to calculate the reaction amplitude $f_p(k_p, \theta)$ for (d, p), as in

$$\Psi_{\text{Tot}} \sim e^{ik_d z_d} + f_d(\theta)\frac{e^{ik_d r_d}}{r_d} + f_p(k_p, \theta)\frac{e^{ik_p r_p}}{r_p} + \cdots, \tag{45}$$

where the subscripts d and p refer to deuteron and proton.

In Eqn. (45) the first two terms are just those discussed in detail in Sec. II, and the third term is the object of our quest. The amplitude $f_p(k_p, \theta)$ can be written, in a somewhat schematic form as

$$f_p(k_p, \theta) = \sum_{l_1=0}^{\infty} \sum_{l_2=|l_1-l|}^{l_1+l} C^l_{l_1 l_2}(\theta) \int_0^{\infty} u^{(p)}_{l_1}(r)\, \phi^{(n)}_l(r)\, u^{(d)}_{l_2}(r)\, dr \tag{46}$$

We will not try to derive or to display in detail the functional form of f_p, but merely to show how the computer and optical model must be used in evaluating it. The functions $u^{(p)}_{l_1}(r)$, and $u^{(d)}_{l_2}(r)$ are the radial scattering wave functions of a proton and deuteron, respectively, with orbital angular momenta l_1 and l_2. They must be calculated as discussed in Sec. II.

The function $\phi^{(n)}_l(r)$ describes the neutron bound to the nucleus which has captured it, in a state with unique orbital angular momentum l. It may be obtained by solving the Schrödinger equation with purely real potential (of the form of Eqns. (42) plus (44) for example), at a negative energy, B_n, the absolute value of which is just the *binding energy* of the neutron, the energy which would have to be applied to release it from the nucleus with zero velocity. Since $\phi^{(n)}_l(r)$ describes a bound system, the neutron cannot be found far outside the potential, and we in fact find that

$$\phi^{(n)}_l(r) \underset{r \to \infty}{\longrightarrow} \exp[-Kr] \tag{47}$$

where $K = (\hbar^2 B_n)/(2\mu)$.

Apart from the different boundary conditions—and the fact that $\phi^{(n)}_l(r)$ can be chosen real and is normalizable in contrast to $u^{(p)}_{l_1}$ or $u^{(d)}_{l_2}$ being complex and not properly normalizable—the computer's task in evaluating $\phi^{(n)}_l$ is much the same as for $u^{(p)}_{l_1}$ or $u^{(d)}_{l_2}$; see Sec. II. Note that, while to describe simple elastic scattering we needed only the phase shifts, we now need the scattering functions themselves, as well as $\phi^{(n)}_l(r)$. Finally, the quantities

$C^{l}_{l_1 l_2}(\theta)$ are functions of angle (related to the associated Legendre polynomials of Eqn. (4)) and the momenta, and are easily tabulated as required.

The integral in (46) must be done numerically, but familiar methods such as Simpson's Rule are adequate. We have to integrate out only to some $r > R_0$ at which $\phi_l^{(n)}(r) \approx 0$. We can stop the sums over l_1 (and l_2) at some $l_{\max}$ because of the centrifugal barrier, as discussed in Sec. II, since for large enough l, $\delta_l \approx 0$ and $u_l(r) \approx 0$.

Hence we compute $f_p(k_p, \theta)$, and obtain the differential cross section for the (d, p) reaction from

$$\left[\frac{d\sigma(\theta)}{d\Omega p}\right]_{dp} = |f_p(k_p, \theta)|^2 \tag{48}$$

For any other reactions in which one or more nucleons are transferred directly from projectile to target nucleus, or target to projectile, the essence of the calculation remains the same. Thus, e.g., for the ^{90}Zr (^{3}He, ^{4}He) ^{89}Zr reaction, we need only have optical potentials for elastic scattering of ^{3}He nuclei from ^{90}Zr and ^{4}He nuclei from ^{89}Zr, at appropriate energies. Then we may proceed to calculate the direct reaction cross section.

Such calculations are usually called distorted wave Born approximation (DWBA) calculations. The Born approximation itself is basically the replacement of the complete, exact state function for the system by the optical model function for elastic scattering of the incident particle. In the past five years such calculations have been performed a truly enormous number of times and an equally enormous amount of information bearing upon the structure of a large number of energy states in a large number of nuclei has accumulated. Publications have hardly kept pace with the explosion.

For a careful and general development of the DWBA transition amplitude, the interested reader is referred to the article by Satchler in the list of references.

Inelastic scattering of nucleons from nuclei can in many cases be described by a DWBA formulation too. In such processes, the incident nucleon excites the target to an energy state different from its initial state and the projectile leaves with an energy loss equal to the energy of excitation of the target.

The DWBA theory, while very useful in a large number of cases, is based on the assumption that the process proceeds through a one-step interaction. This approximation can become poor, however, if the interaction that causes the reaction is very strong. Take the ^{90}Zr (d, p) ^{91}Zr process again as an example. The escaping proton may pick up the neutron from ^{91}Zr and form a deuteron again before it leaves the region of the nuclear volume. The reconstituted deuteron may then again undergo a (d, p) process, and the amplitude of this higher order process must be added to the DWBA amplitude of (46). It is clear that these correction terms are large, if the residual interaction is strong.

Fortunately it seems that, in most of the cases which nuclear physicists have so far encountered, the interaction is weak enough so that the correction due to the higher order processes is small, and thus can be safely neglected, though no one has ever proved it by direct numerical calculation.

The situation is quite different in the case of inelastic scattering. Let us consider, for explicitness, proton inelastic scattering, such as ^{90}Zr (p, p') ^{90}Zr*, where the prime on the outgoing proton symbol implies that it has lost energy relative to an elastically scattered proton, and the asterisk on the residual nucleus symbol indicates it has been excited. As we have said, we may apply DWBA to the process to calculate $f_p'(k_p', \theta)$ in

$$\Psi_{\text{Tot}} = e^{ik_p z_p} + f_p(k_p, \theta) \frac{e^{ik_p r_p}}{r_p} + f_{p'}(k_{p'}, \theta) \frac{e^{ik_{p'} r_{p}'}}{r_p'} + \cdots \tag{49}$$

However, the DWBA will not give a very realistic answer in any case in which the excited nuclear state is one in which a large percentage of the nucleons in the nuclenos share in the excitation energy. Such states are called "collective", and the residual interaction which excites them is (of necessity) very strong.

Thus to evaluate $f_{p'}(k_{p'}, \theta)$ using the DWBA theory in a reasonably exact fashion, we would be *forced* to consider second and even higher order corrections, which could never be neglected. Because these contributions would also be important in $f_p(k_p, \theta)$, we would also have to make corrections to Eqn. (18).

It is, therefore, very fortunate that for inelastic scattering it is possible to obtain $f_{p'}(k_{p'}, \theta)$ and $f_p(k_p, \theta)$ *without* making the DWBA, by solving a set of coupled Schrödinger equations which treat the elastic and inelastic scattering states, or *channels*, on an equal footing. These equations are coupled through the parts of the interaction which induce transitions from elastic to inelastic channels, or vice versa.

Very schematically, we would have

$$\begin{aligned}(T_p + V_p - E_p)\, u_p(r) &= V_{pp'} u_{p'}(r) \\ (T_{p'} + V_{p'} - E_{p'})\, u_{p'}(r) &= V_{p'p} u_p(r)\end{aligned} \tag{50}$$

where T_p, $T_{p'}$ are the kinetic energy operators in the elastic and inelastic channels, and E_p and $E_{p'}$ are the corresponding total energies. The optical potentials V_p and $V_{p'}$ are chosen to describe "pure" elastic and inelastic scattering, while $V_{pp'}$ and $V_{p'p}$ are the parts of the interaction producing transitions from elastic to inelastic channels, and from inelastic back to elastic. We have as many coupled p' equations as there are inelastic channels needing to be considered. Note that only when $V_{pp'} \approx 0 \approx V_{p'p}$ may we drop back to solving the simpler Eqn. (5) of Sec. II.

We obviously cannot go into detail. The reader is referred to the review article by Tamura. The actual numerical solution of coupled equations can be carried through in much the same way as we discussed in Sec. II. Briefly, we can rewrite (50) in the form

$$\begin{aligned} u_1'' &= A_{11}u_1 + A_{12}u_2 \\ u_2'' &= A_{21}u_1 + A_{22}u_2 \end{aligned} \tag{51}$$

when there is only one inelastic channel. Defining the vector $\vec{u}$ and the matrix A by

$$\vec{u} = \begin{pmatrix} u_1 \\ u_2 \end{pmatrix}, \quad A = \begin{pmatrix} A_{11} A_{12} \\ A_{21} A_{22} \end{pmatrix} \tag{52}$$

We can reduce Eqn. (51) to

$$\vec{u}'' = A\vec{u} \tag{53}$$

which is formally the same as Eqn. (20).

Such elastic–inelastic scattering calculations, or "coupled-channels" calculations, have also been performed extensively in the last few years. They have had to await the arrival of large, fast computers, since the solution of a set of coupled equations is time-consuming. Now that computational power is widely available, coupled-channel calculations are being carried out in such numbers as to have clarified the structure of collective states in nuclei to almost the remarkable extent that direct reactions had previously clarified the distribution of single-nucleon states.

Suggested references

General discussion of the model

"A Model of the Nucleus", V. F. Weisskopf and C. P. Rosenbaum, *Sci. Am.* **193**, 84 (December, 1955).

"The Optical Model of the Nucleus in the Light of Present-day Data", I. S. Shapiro, (APS Translation in) *Soviet Physics–Usp.* **4**, 674 (1962).

"The Optical Model of the Nucleon-Nucleus Interaction", P.E.Hodgson, *Ann. Rev. of Nuc. Sci.* **17**, 1 (1967).

Numerical computation

"Nuclear Optical Model Calculations", M.A.Melkanoff, T.Sawada, and J.Raynal, in *Methods in Computational Physics*, Vol. 6, ed. by B.Adler, S.Fernbach, and M.Rotenberg. (Academic Press, 1966.)

Nucleon–nucleus scattering surveys

"Elastic Scattering of 10.5- and 14.5-MeV Polarized Protons from Nuclei and the Optical Model Potential at Intermediate Energies", L.Rosen, J.G.Beery, and A.S.Goldhaber, *Annals of Phys.* **34**, 96 (1965).

"The Calculation of Neutron Cross-Sections from Optical Potentials", D.Wilmore and P.E.Hodgson, *Nucl. Phys.* **55**, 673 (1964).

Optical model for nucleus–nucleus scattering

"Deuteron Optical-Model Analysis in the Range of 11 to 27 MeV", C.M.Perey and F.G.Perey, *Phys. Rev.* **132**, 755 (1963).

"Optical Model Analysis of Heavy Ion Scattering", E.H.Auerbach and C.E.Porter, in *Proceedings of the Third Conference on Reactions Between Complex Nuclei*, ed. by A.Ghiorso, R.M.Diamond, and H.E.Conzett (University of California Press, 1963).

DWBA theory

"The Distorted Waves Theory of Direct Nuclear Interactions with Spin-Orbit Effects", G.R.Satchler, *Nuc. Phys.* **55**, 1 (1964).

Coupled-channel calculations

"Analyses of the Scattering of Nuclear Particles by Collective Nuclei in Terms of the Coupled-Channel Calculation", T.Tamura, *Revs. Mod. Phys.* **37**, 679 (1965).

CHAPTER 18

On-line Use of Computers in Chemistry

DON SECREST

THE ON LINE USER of digital computers becomes involved in many aspects of computer technology which are not encountered by most other users. A program which interacts with apparatus, usually referred to as a *process* program, is an entire software system. It may consist of several independent programs which run on the computer in an asyncronous manner. One or another of the programs may be started before another is finished. This leads to numerous problems in program organization which do not arise in *non-process* programming. Often higher level user-oriented languages are not sufficient to handle these problems and the on-line user, more than any other type of computer user, must work in assembly language. Most computer users are not at all concerned with the electronics involved in computer logic. All of the logic they will use was built into the macine by the manufacturer. The on-line user may have to extend the machine he buys or rents to include his apparatus so he gets involved with AND and OR gates also. Usually he works with commercially available microcircuits which might be called the FORTRAN of computer logic and only rarely with transistors, diodes and resistors or the assembly language of computer design.

A number of on-line uses of computers are obvious and straight forward. Tasks which are routine and tedious or repetitive can be performed by computer. This ability of the computer to reduce the work load of the scientist is what is usually referred to as *automation*. Another function of the computer is the *control* of a process. This could be considered a form of automation also but control functions are often automated without the use of a computer. The computer allows one a more sophisticated control function than can easily be built from operational amplifiers and integration circuits. The most dramatic use of the computer in experimental science and probably the use which will have the greatest influence on the direction which various fields of science will take in the future is *computer-aided experimentation*. The computer can be used as a tool to extend the experimentalist's faculties and put him closer to the phenomenon he is studying. The computer gives the researcher extra hands and eyes and faster reflexes. The philosophy of computer-aided experimentation is just the opposite of automation. When the computer is used in automation it is a device to remove those aspects of the experiment which are uninteresting from the scientist's attention and leave him free to concentrate on the relevant points. In computer-aided experimentation the computer's job is to help the scientist follow in more detail the progress of the experiment.

The effect of the digital computer on chemistry has been phenomenal in recent years and promises to have an overwhelming influence in the direction taken by many branches of chemistry in the near future. The effects have been most strongly felt in the area of automation of industrial processes and in the institution of automatic analytical procedures. The computer can respond faster than a human operator so it often is possible to carry out, for example, an analytical process faster under computer control. This often leads to great savings in time and allows increase through-put on a given facility. In a continuous flow process increased speed can lead to vast saving by allowing corrective action to be taken earlier when, for example, a complex chemical reactor drifts out of adjustment.

Many industrial and analytical procedures are not limited by human response time and cannot be speeded up by digital computers. Even in operations of this sort, however, great savings are realized by assigning routine operations to the computer which must otherwise be laboriously performed by hand. Many operations which are encountered in a chemistry laboratory require collecting large amounts of data. Often in such an event a small computer will collect data from one or more apparatus and feed it to a larger computer for processing.

The control of a large process may consist of watching a number of indicators and making adjustments from time to time when some variable in the process gets out of range. In a well-adjusted process no response from the operator may be required for hours. Nevertheless, it must be watched closely so that adjustments can be made as soon as necessary. This can be a very boring job and human beings tend to get careless under such conditions. A computer can tirelessly watch such a process indefinitely and be employed in other tasks at the same time. Thus, the computer can serve as a tireless and reasonably reliable watchdog. Of course, computers do fail occasionally and cannot be relied on exclusively in critical processes. Often such a computer is programmed to reset an external timer at an interval in its scanning process.

Computer interface for a shock tube. Professor R. L. Belford is shown checking the setting on the computer interface for his shock tube. The shock tube is the large horizontal tube in the background. Data is taken, converted to digital, and placed in a fast core memory at the shock tube. A central processor, located in the same building two stories away, processes the data collected in the core memory. There is no processor located at the shock tube (Photo by Professor John Lombardi, University of Illinois, Chemistry Department)

If the timer is not reset in a specified length of time an alarm is sounded to indicate that the computer is not functioning properly. Even with such precautions the process must be manually checked from time to time. Computers have been known to give up all functions except reseting the alarm timer.

These are all tasks which could be performed without the use of a computer though they might be more laborious or slower or less accurate when carried out manually. There are a number of automatic procedures which are routinely done under computer control, however, which would be completely impractical without it. These are operations which require accumulating immense quantities of data, or the control of large numbers of variables at a rate which is incompatible with human response times. Examples of such operations are time averaging of signals to enhance the signal-to-noise ratio and Fourier inversion techniques which are becoming common in various types of spectrometry. I say a computer is necessary for these applications, but I must be careful to spell out what I mean by a computer. For any process which is routine and will be repeated often we can build a special piece of apparatus which will handle the operation. If the apparatus works with digital numbers and performs digital arithmatic and makes decisions as to what to do next based on the results of these operations, I refer to it as a computer. Many of the computers used in automatic data control are special purpose computers with wired-in programs which do only the operations they were designed to do. Many very small computers designed for use with chemical apparatus have a machine language and a repertoire of instructions which are to be wired in. These are programmed by the user and the program is entered into the computer with a soldering gun. Such computers can be faster than stored program computers and for some applications their added speed is necesary. In many applications in which a wired, or hard programmed, computer is used an ordinary general purpose computer would serve as well. But if the computer is to be dedicated to the application and reprogramming is seldom required the advantages of a general purpose machine are lost. A hard programmed computer is far cheaper than a general purpose computer and is adequate in many cases.

The general purpose computer comes into its own in the field of computer assisted experimentation. This field is just developing and I feel that it is in this area that computers will make the greatest impact on chemistry and science in general. In the course of studying new phenomenon the experimenter will need to change his mode of interaction with the phenomenon often and it is here that the versatility of the programmable general purpose computer is an invaluable asset to the chemist. In work of this sort the chemist must be familiar not only with programming computers but also with the design and construction of *interface* logic. By interface we mean all of the hardware needed for communication between the apparatus and computer. Before discussing a few of the problems which arise in such applications it would be well to look at some of the more conventional on-line applications in some detail.

AUTOMATION

It is in the area of automation that the on-line use of computers is probably most in evidence at the present time. For a number of applications complete computer-apparatus systems are commercially available. Automated gas chromatography columns are very common in industrial laboratories, often with a large number of columns tied to a single computer which transmits a report of the analysis to the site of the column and the billing and bookkeeping involved with the run is automatically entered into the company records. Mass spectrometers are another group of instruments which are commonly interfaced to computers. A mass spectrometer identifies a molecule in a mixture by measuring its mass. The molecules are ionized and the ions of different masses are sorted by various techniques depending on the type of spectrometer. Usually when molecules are ionized they break up into fragments and the output of the mass spectrometer is a reading which depends on the fragment masses. The computer's job is to collect the data and then identify fragment mass from the reading. This may require a considerable computing effort if the mass spectrometer is very high resolution and can sort masses which lie close together. The computer output is then a list of fragment masses with, possibly, a list of possible fragment compositions. The chemist is still faced with the job of putting the fragments together in hope of determining the original compound or mixture of compounds. This step may also be aided by the computer. The computer can be given the job of producing a list of compounds which could produce such fragments. The jigsaw puzzle produced by a mass spectrometer frequently has no unique solution and other information is required for complete identification.

The number of automated instruments one might find in use is too large to allow even a cursory description here so I shall describe two different types in some detail, one involving the collection of great quantities of data and the other repetitive data sampling.

One of the important techniques for determining the structure of large molecules is the use of X-ray crystallography. This is a particularly interesting problem for computer application as it has not yet been completely solved. The structure of molecules cannot yet be determined completely automatically.

The problem of data collection has been aided by the on-line use of computers, however. The amount of data required to determine the structure of a large molecule is immense. A molecule capable of having its structure determined by this technique must have one or more crystalline forms. In the crystal the molecules form an orderly array which is determined by the crystal type. When a beam of X-rays is passed through the crystal the electrons of each atom absorb and re-emit the X-rays in all directions. In certain directions the X-rays from each particular atom in a single molecule

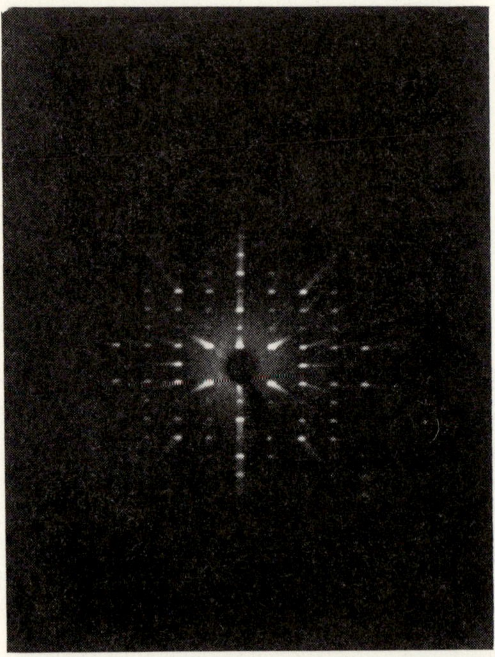

Fig. 1 X-ray photograph. This is a photograph such as one might use for starting a computer scan of a diffraction pattern. If measurements are to be made from the photograph a higher quality film is needed and many times the number of spots visible here could be measured (Photo courtesy of Professor Iain C. Paul, University of Illinois, Chemistry Department)

reinforce each other and produce a particularly dense stream of X-rays in that direction. If the amplitude of X-rays traveling in these preferred directions were known the molecular structure could be immediately computed by performing a Fourier inversion. Unfortunately, the amplitude, which is a complex quantity, cannot be measured. Only the intensity of the stream, which is proportional to the square of the absolute value of the amplitude, can be measured. The phase is lost. The problem of determining the phase is the major difficulty encountered in this technique. In practice, the direction and intensity of the X-rays traveling in the preferred directions are measured. One possible technique for accomplishing this feat is to place the crystal between the incident X-ray beam and a photographic plate. Thus, the x-rays passing through the crystal are diffracted in preferred directions making a pattern of discrete spots on the plate (see Fig. 1). These diffracted beams are usually referred to as reflexions (spelled with an *x*) in the jargon of X-ray crystallography. The angles of diffraction can be determined simply from the position of the spots on the plate. Since the intensity must be measured a special plate is used which consists of a pack of 6 films. The intensity of the beam forming the spots can be determined approximately from the density of each spot on all 6 films of the plate. One photograph contains data for only one orientation of the crystal with respect to the incident X-ray beam. A number of different orientations of the crystal are needed to determine the structure and this means that data must be obtained from a number of photographic plates.

Much more accurate intensity data can be obtained by replacing the photographic plate by an X-ray detector at the position at which spots would appear on the photographic plates. Positioning the detector without the guidance of a photographic plate is a long and tedious job. Commercial computer interfaces for a number of computers exist for accomplishing this task. A photograph of one of the machines in use in the Chemistry Department at the University of Illinois appears in Fig. 2. The interface must be able to move the detector through two angles. The computer searches for a spot and receives the output of the detector and then commences to search for a new spot. In this way the computer can accumulate the data in a much more accurate form than can be obtained from a photograph. In practice a crude photograph is usually made first and the approximate locations of some of the reflexions are given to the computer. With this start the computer can compute approximate locations of more reflexions and a great amount of time is saved in the collection process. After completing the scan for one orientation of the crystal the computer can rotate the crystal, and again it must be able to rotate the crystal through two angles to achieve all possible orientations. For each orientation of the crystal the new reflexions must be found and measurements taken. The intensity of the reflexions differ by many orders of magnitude, so the detector must be sensitive. At an intense reflexion a sensitive detector will become saturated. The computer must detect this condition and insert the proper filter into the refracted

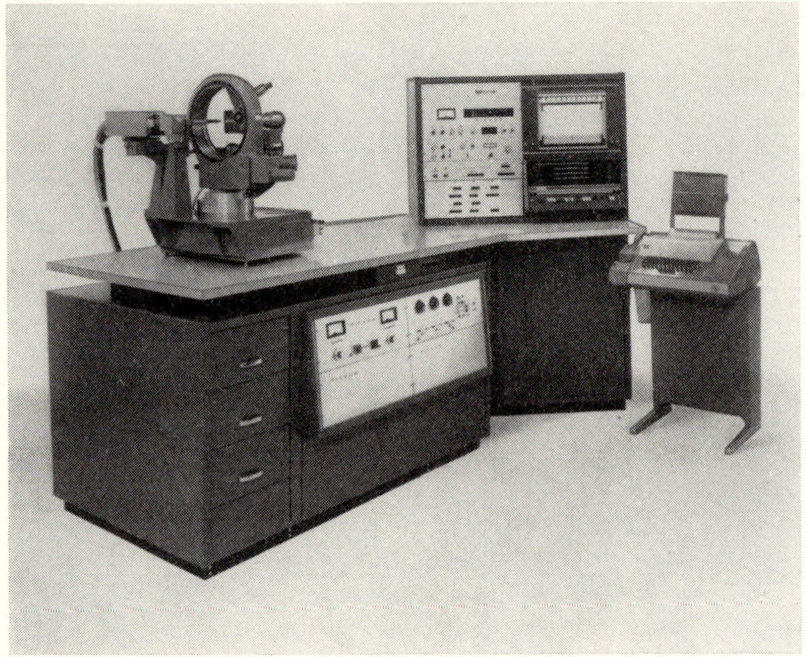

Fig. 2 A computer automated x-ray diffractometer system. This is a commercially available system complete with computer interface and software. The computer in the lower right of the control panel is a minimum system. Input and output are through the attached teletype which is supplied with a paper tape reader and punch. Since data is produced slowly by the X-ray machine this is sufficient *I/O* for this application. Input of programs is slow, however, and paper tape, while fast enough to keep up with the x-ray machine, is not ideal for use on the large central computer which must eventually process the data. At the University of Illinois, this system will be interfaced to a larger time-shared computer which will make available to it a high speed card reader and punch and afford a better output medium for the large computer (Photo courtesy of Picker Nuclear Corporation)

beam and then correct the intensity reading. All of this would be extremely laborious if it were carried out manually. For a complete structure determination of a large molecule with the aid of a computer and approximate positions of the reflexions from a photograph, the data collections may take as long as two weeks of continuous operation. The data is collected usually by a very small computer or an intermediate sized time-shared computer and output is produced in a form suitable for processing on a large computer. This may be on cards, paper tape, magnetic tape or by a direct connection to the large computer. The processing of the data for a large molecule will require computation times of the order of hours on the most powerful of computers presently available.

The data collection would not be impossible without the use of a computer. Such work is still being done in many laboratories manually. But the use of a computer frees the chemist of a long tedious and unenlightening task and his time may be spent in more creative pursuits. There are many examples of this type of computer usage in chemistry. Many procedures requiring simple manipulations which can be performed mechanically and producing quantities of data are being automated.

Another kind of computer usage which might come under the heading of automation are problems which require repetitive measurements at precisely determined points of some independent variable. Such tasks would often produce unmanageable quantities of data without the aid of a computer, but with a computer's services the data can be processed as it is collected and the chemist need concern himself only with the processed data.

An example of such an application is the technique of signal-to-noise enhancement by time averaging. The top plot in Fig. 3 is a typical result of a scan of a nuclear magnetic resonance machine. In the horizontal direction the external magnetic field on the nucleus is plotted against the absorption of electromagnetic radiation by the nucleus. The high peaks occur at field strengths at which the nucleus is in resonance with the radiation. The position of the peak gives one information about the magnetic environment of the nucleus in the compound and from this one can learn something of the structure of the compound. The small peaks are mostly due to random noise. Very weak peaks may be buried in the noise. Since the noise is random, if another scan is made the noise pattern will be entirely different but the resonant peaks will always be present. Thus, if many scans are made and the intensity at each field strength is added for all scans the random noise, which is sometimes in a positive direction and sometimes negative, will tend to average to zero. This is easily accomplished by computer. It is only necessary for the computer to be able to start and stop the scan and collect data. During each scan the data collected at each field strength is added to that collected at the same field strength on previous scans. The result after a number of scans can then be rescaled and plotted. The second plot in the illustration shows the result after 4 scans have been averaged. One can see peaks appearing at positions where only a hint of a peak occurred before. The last plot shows

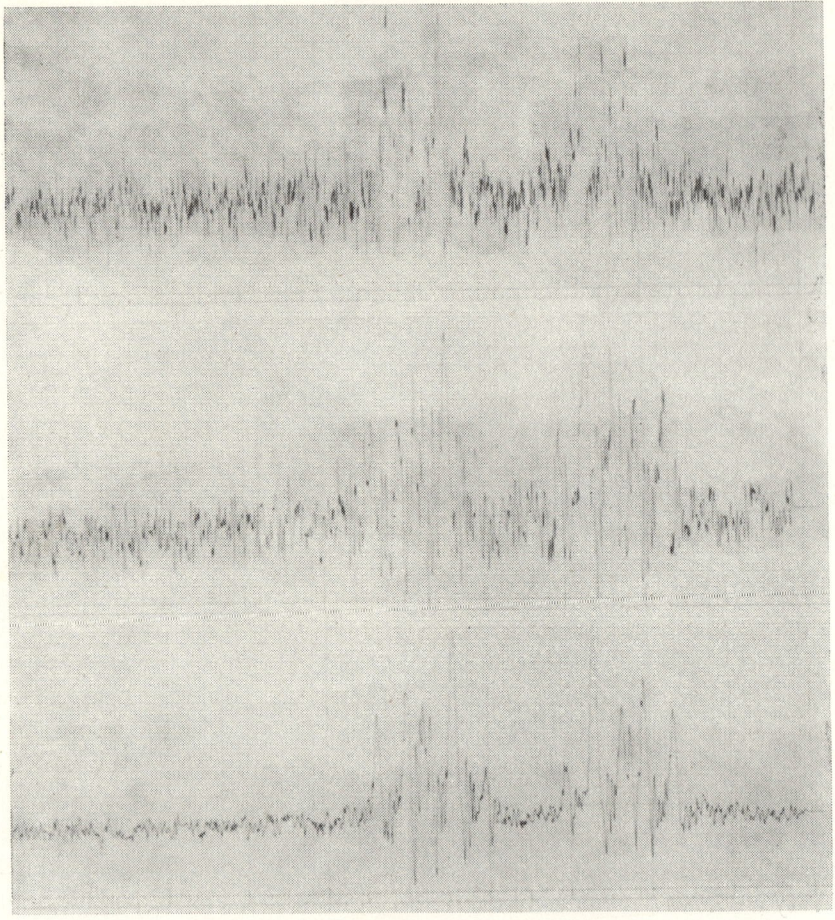

Fig. 3 Signal-to-noise enhancement by time averaging. The top plot is produced by a single scan of a nuclear magnetic resonance spectrum. The tall peaks contain information about the molecule being studied and the small peaks are mostly noise. The noise may hide some weak peaks. The noise is random and on each run the noise spectrum will be different. The second plot is the result of 4 scans whose amplitudes have been added together by computer and then renormalized and plotted. Under this treatment the noise will tend to average to zero. One can see peaks beginning to appear which were not evident with only 1 scan. The last plot is the result after 32 scans. The new peaks now show distinctly above the noise (Photo by Professor John Lombardi, University of Illinois, Chemistry Department)

the same spectrum after averaging 32 scans. This appears to be enough in this case. In some work thousands of scans are required to produce usable spectra. The mass of data never accumulates even in the computer, however. If 500 points define a single scan, on each succeeding scan, the new points are added to the old as they are collected and there are never more than 500 points in the computer.

Since time averaging is a common technique, not only with NMR but with other types of apparatus used in science, a number of special purpose computers are available to do just this task. Such a computer cost a fifth as much as a general purpose (programmable) computer capable of doing the same thing.

THE COMPUTER IN SCIENTIFIC RESEARCH

These two examples, X-ray data collection and time averaging, are typical of many uses of the on-line computer to carry out tedious and routine tasks which the chemist does not want to be involved with himself. The scientist making use of these techniques does not need to get involved to any great extent with computer technology unless he is developing a new system. The scientist using a computer as a tool for studying a new phenomenon, however, is in a different situation altogether. He will not find a system already built which will handle his problems and so he must plan the entire system himself. This includes developing and writing or at least directing the writing of the software and planning and building the interface to the apparatus. Software for on-line application has many aspects to it which are unique. The problem of building an interface is not unlike that of building a special purpose computer. A discussion of some of the problems which arise in the software and hardware is given in the next two sections.

The computer allows the research chemist to tackle larger and more complicated problems than he is able to handle manually. It is difficult to keep more than a few variables under observation at a time during an experiment. If control is to be exercised over several variables during the process it rapidly becomes unmanageable for one person to handle and more than one have trouble coordinating. Frequently the measured quantity is not simply related to the phenomenon being studied and much work is required after the experiment to determine what has happened. If it is necessary for the experimentalist to know what is happening during the experiment to exert the proper control over it, he must use a computer to process the data on-line and feed it back in an easily comprehensible form continuously such as by means of an oscilloscope or a strip chart recorder.

Chemists and other scientists are just awakening to this computer–experiment interaction chiefly due to the recent availability of small computers cheap enough to be accessible to the experimentalist. Chemists are rapidly learning the new techniques required to implement such systems and a number of schools around the country are preparing to introduce these techniques into the chemistry curriculum. The use of on-line computer techniques in basic research is now in a period of rapid growth and we may expect to see a computer induced shift in the type of research being done in the near future due to new areas becoming accessible through computer-aided experimentation.

SOFTWARE CONSIDERATIONS

When a computer is used to collect data and possibly control an apparatus in real time, the organization of the program is quite different than it would be for a non-process program. The two most obvious jobs of the computer are to collect data and control the apparatus. The computer must collect the data when it is ready. The precise time when the data is ready to be collected cannot be determined by the computer. It must be determined either by the apparatus itself or by a clock. In some situations the computer decides when the next useful data point can be collected using information obtained from previous data points. Even so the computer must use some sort of timing device to determine when that time has arrived. Thus, the time to collect a data point is indicated by a mechanism which is not part of the program itself. Most modern computers are equipped with one or more *interrupts*. I would say all computers suitable for on-line use are so equipped. An interrupt is an asyncronous jump which causes the computer to stop executing instructions in one part of the memory and begin at another point. The location at which the computer was operating when the interrupt occurred is saved automatically. In the simplest computers this is all that is saved by the hardware and registers such as the accumulator and index registers must be saved by an interrupt handling program. Computers exist which save more of the registers by hardware, which is of course faster but not necessary for most applications.

Most data acquisition computers, even the smallest have a number of distinguishable interrupts. That is to say that when an interrupt occurs the computer starts executing in a new location. At that location will be the beginning of a program which we call an interrupt program. It is possible for that program to determine what caused the interrupt. In some computers the determination of the cause of the interrupt is automatic since each unique interrupt condition causes a jump to a different place in memory, allowing a different interrupt program for each interrupt condition. Most computers use a combination of these techniques so that there must be several interrupt programs, each handling a small number of interrupts. A common variation on this theme is priority interrupts. Various interrupts may be assigned different priorities. Then if a condition occurs which causes an interrupt the normal sequence is stopped and the interrupt program which services that interrupt is begun. If while this program is in execution a condition occurs which causes an interrupt of higher priority that program will stop and the higher priority program will be executed (see Fig. 4). When it is finished the registers will all be restored and the lower priority program will be resumed. If an interrupt of still lower priority occurs it will go into a wait state and be

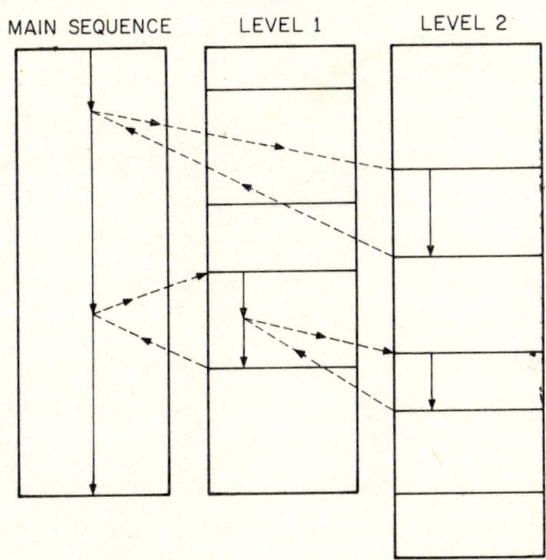

Fig. 4 Priority level interrupt programming. The flow of a process program may look something like this. The main sequence is interrupted by a program on level 2. When this finishes the main sequence is continued. Later it is interrupted by a level 1 program which in turn is interrupted by a level 2 program. It is convenient sometimes to consider the main sequence an interrupt level with priority zero

recognized when all higher priority conditions are serviced. When all interrupt conditions have been serviced the normal sequence will be resumed.

Data gathering is usually accomplished by an interrupt program. The interrupt condition can be caused either by the apparatus or by a timer or some other device. This influences the way in which the program to handle data is written. While a data point is being collected no interrupt at the same or lower priority can be recognized, thus, it is imperative that not much time be spent in the interrupt program. The higher the priority of the program the shorter it must be. Thus, such a program usually reads the point, which may consist of several values such as a signal intensity, a field strength and a temperature, and stores it somewhere and returns. The points are then processed within the normal sequence or on a lower priority interrupt level. Usually it is possible to cause an interrupt through programming. This is convenient if a point must be read at a precise instant and, thus, must be read on a very high priority level but the point needs considerable processing before the next point is read. The processing must be done reasonably soon and there is no guarantee that the normal sequence will resume in time due to other interrupts. When the data point is collected on the high level an interrupt may be programmed to an appropriate lower level for processing the point. Then when the high level interrupt program is left the lower level interrupt will be recognized as soon as all other high level problems have been serviced but possibly before some of the conditions on still lower levels are serviced.

Usually one adopts conventions prescribing the length of time programs on each level may consume. For example, one might limit all programs on the highest priority level to 50 microseconds, the next lower level to 400 microseconds, the next level to 2 milliseconds, etc. This is particularly important on a machine which is shared among a number of users, but it also helps with organizing the programming effort on a dedicated machine if the process being implemented is at all complex.

Usually there are standard routines which are used on a number of different interrupt levels. It is convenient to be able to use the same subroutine on many different levels. A subroutine which can be used in this way is called *re-entrant* and must be specially coded for this purpose . Imagine that a program on an interrupt level calls a program to compute a square root. While the square root is being computed a higher level interrupt occurs and the program on that level calls the same program to compute a square root of a different number. If the square root program were coded as it usually is for a computer used only for computation, all of the information given to it by the first interrupt program would be lost when the second interrupt program entered it. One way of avoiding this is to have a different square root program on each interrupt level on which it is needed. This is wasteful of space and usually process computers are either small or are used by a number of different experiments and in both cases space is at a premium. The technique of re-entrant programming at least for commonly used routines avoids this difficulty. A re-entrant program must make provision for saving the data and return address of the caller.

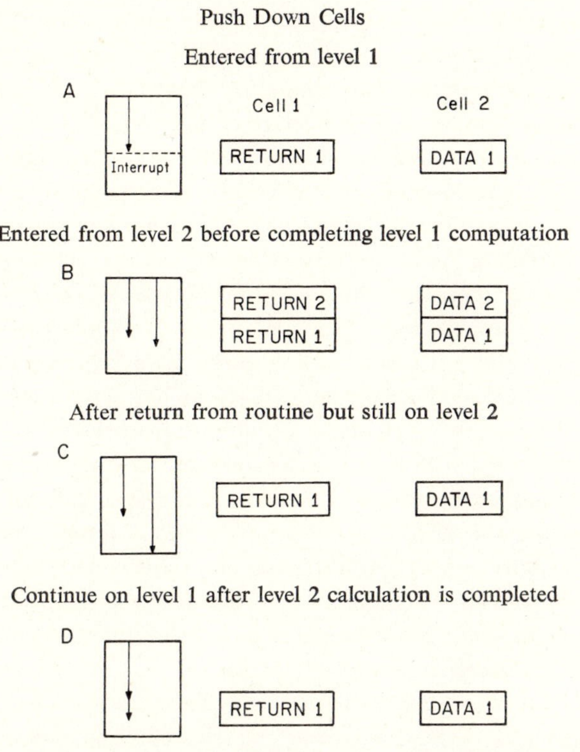

Fig. 5 The use of push down cells in re-entrant programming. In A the routine is interrupted in performing a computation on level 1. In B it is doing a computation on level 2. The level 1 information has been pushed down. In C the program has just finished its computation on level 2 and popped up the level 1 results. It could be entered again on level 2 before level 1 is reactivated and again the level 1 results would be pushed down. Finally in D, the level 1 computation has been resumed

There are a number of techniques for accomplishing this. One possible method is to keep the return address, data and intermediate results in *push down cells* (see Fig. 5). Push down cells are very like stacks of trays in a cafeteria. A number is placed on the top of a cell. It can then be retrieved when needed and when the routine is finished with it the top number is removed and the next one pops up. Thus, if the routine is called to compute the square root of 5 it is placed in on a push down cell. If the program is interrupted and it is asked to compute the square root of 7, that is placed on top of the 5 and the routine works on 7. The return and all intermediate results are placed on other push down cells. When the square root of 7 is completed all cells are popped up and the routine is left in the state it was in when interrupted. When it is continued after the higher level interrupt is complete it will have the proper data for continuing.

Another technique for achieving re-entrant coding is to use work areas for each interrupt level. When this technique is used a pointer is set by the interrupt handler which points to the work area. This pointer may be kept in a base register or an index register. The re-entrant program then stores returns, data and intermediate results in the work area by means of the pointer and when interrupted it will use a different area automatically.

There are advantages and disadvantages to both of these techniques. Most computers do not have build-in push-down cells, so they must be simulated by software. This is time consuming and since there are usually time limits on interrupt levels this is an important consideration. The work area technique is much faster since the occurrence of an interrupt requires only the changing of the pointer to save the old state of the routine and then the program operates as fast as if it were not re-entrantly coded. This technique is wasteful of space, however, since work areas for each level must be defined large enough to accommodate the needs of the largest program used by that level. This space is not aveilable even when it is not being used while any space not presently occupied by a push-down cell is available to any other program. Combinations of these two techniques are sometimes useful.

A third approach to re-entrant coding which is convenient in some environments is to keep temporary values for all intermediates in the computation including the return address in an area supplied by the user. A pointer to this area is kept in a nonvolatile register which is saved in the interrupt process.

Since the main sequence may also use the square root routine it is also advantageous to consider it an interrupt level of lowest priority.

Another area in which process programming differs considerably from non-process is in the use of non-shareable facilities. There are a number of facilities in any computer system which can be used by only one program at a time. In the process environment in which many programs are activated asyncronously by interrupts, provision must be made to avoid conflicts in the use of these facilities. An example of such a facility is the printer. If one program, for example, has the job of watching the temperature of some part of the process and taking action to keep the temperature in limits, it may be programmed to make a report every hour or so on the stability of this variable and of corrective action taken. It is not desirable for the report to be printed in the middle of some other programs output or for output from another program to come in the middle of the temperature report. This problem is particularly important on a machine which is shared among a number of other experiments but it is also a consideration on a dedicated machine. For example, suppose that the temperature monitoring program is associated with the only apparatus controlled by the computer. There may be another program like the temperature program which monitors a magnetic field for this apparatus. These two programs both print reports from time to time and provision must be made to avoid the possibility of both of them using the printer at once. An obvious solution is for the program to test if the printer

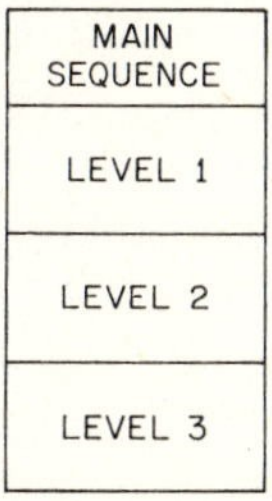

MAIN SEQUENCE
LEVEL 1
LEVEL 2
LEVEL 3

Fig. 6 Printer usage list. To use the printer on level 2 the program on level 2 puts a busy marker in the list at the level 2 position. It then tests if any other level is busy. If another level is busy it removes its busy marker and makes a note to try again later. If no other level is busy it uses the printer and removes its marker when it is finished

is busy before printing. This must be done carefully, however. If one writes the program to test for the printer being busy and then starts printing if it is not, he may find that another printing program interrupted between the time he performed the test and the time he started printing. One way to accomplish such a test is to use a busy table. This table could be a special type of push-down cell or a simple list such as pictured in Fig. 6. If one wishes to use the printer he sets a busy indicator in the printer busy list at a position corresponding to his program. He then checks to see if there are any other busy indicators on. If there are, he removes his indicator and makes a note to try later. If there are not other indicators he uses the printer and removes his indicator when finished. Placing his own busy indicator in the table before testing will insure that if he is interrupted before he starts printing, the interrupting program will find the printer busy. It is important that if he finds the printer busy he remove his own busy indicator. It would not be a good plan to attempt to leave his busy indicator on until all others go off. If two programs were to attempt this, what would happen?

Occasionally a more satisfactory technique for handling this problem is the use of *queues*. Program queues are sets of programs which are to be run one after another. Each program in the queue is run to completion before the next queued program is started. The order in which the programs are run may be the order in which they were entered into the queue or in some cases a priority is assigned to each program on the queue, and that of the highest priority will be run next. The priority queue is handled differently than priority interrupt, however, in that if a high priority program appears on the queue the presently running lower priority program is completed before the high priority program is started. If queues are used in our example of the temperature program, when the temperature program or any program which writes reports on the printer want to use the printer it puts its report writing program in the queue and it will be executed when its turn comes up. If all programs which use the printer are queued there will be no problem of interference.

If the computer is not equipped with an instruction for inhibiting interrupts, one might have trouble adding a program to the queue table. If a program is interrupted in the process of modifying the queue table and the interrupting program modifies the table confusion might result. Thus, a "busy list" similar to that described above might be used to avoid confusion in modifying the queue table.

THE INTERFACING PROBLEM

For a number of years it has been necessary for chemicts to be able to program computers for theoretical prediction of chemical properties, processing of data and evaluation of data. Usually these have been more or less complex numerical computations and in most cases FORTRAN, PL1 or some version of ALGOL such as MAD have been sufficient. For on-line processing it has become necessary for chemists to become involved with assembly language. Some problems are so complex and intimately involved with chemistry that it is impossible to have them programmed by non-chemists. It is not worthwhile for the chemist to do all of his own programming but it is necessary for the chemist to be familiar with programming to be able to direct his programmers.

A similar situation has developed with the design of the computer–apparatus interface. A number of processes has been automated and this can be done entirely by an electrical engineer with the chemist knowing little or nothing about the electronics involved.

When a new experiment is being designed, however, and there are a number of variables which must be controlled on a time scale which is too fast for the chemist to handle manually a computer interface must be designed to handle it. If automation is the reason for the apparatus–computer interface one would like the computer to take care of the problem completely. With a new phenomenon the experimentalist wants to known what is going on with every aspect of the experiment and either the computer or the interface must feed back information in a form which is easy to comprehend quickly. It is also important that the information come back to the experimentalist in real time so that he can mentally correlate what is happening at all times. In a new experiment one is never sure what will happen and it would be self defeating if an interesting feature of the phenomenon were missed because it was buried in the computer and never came to the scientist's attention. It is highly likely that the experimentalist will want to change the interface often during the development of the experiment. In such a situation it is well worth the chemist's while to learn enough electronics to be able to design his own interface logic. The chemist could hire an electrical engineer to build the interface, but unless the chemist knows something about electronics or the electrical engineer knows a lot about chemistry, there will be a problem in communication.

The design of interface logic is no more difficult than computer programming. In fact, there is a great deal of similarity between logic design and computer programming. One first draws an overall flow chart in logic symbols. Then one must collect the integrated circuits necessary to construct the interface. This consists of looking through the catalogs of the manufacturers of such circuits. These commercially available circuits may be thought of as the instruction set for the interface "program". There is one important difference between interface design and computer programming, however. There are only certain specified instructions in the repertoire of any computer but the number of integrated circuits available is constantly increasing. One manufacturer of integrated logic circuits produces a new circuit every week so that it is never possible to produce the optimal circuit design. As soon as the solder is cold on the interface new circuits are available by means of which a more optimal interface could be built. A commercially available integrated circuit is shown in Fig. 7. These circuits can be soldered to a standard board such as that shown in Fig. 8 and connected with wires according to the design. If a particular circuit is repeated a number of times it is worth

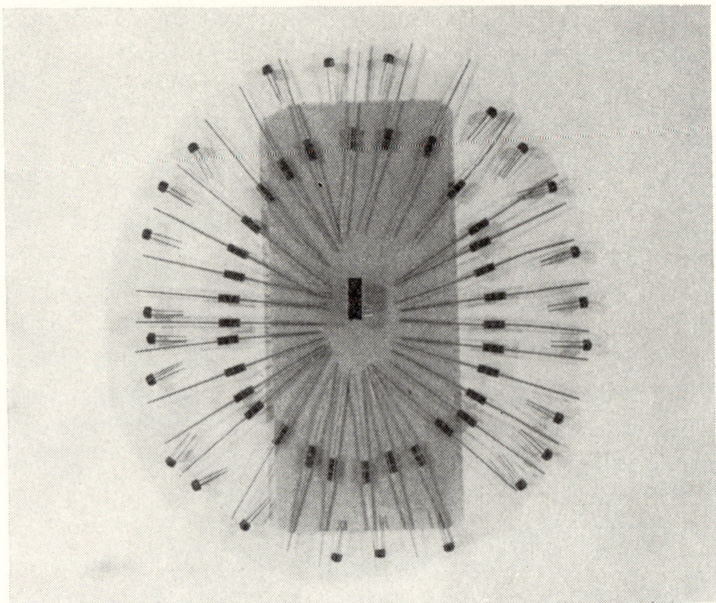

Fig. 7 Integrated circuit module. In the center of the IBM card is an integrated circuit, or microcircuit, typical of those commercially available for interface logic construction. The integrated circuit is the FORTRAN of logic design. The small package may contain a number of logic elements. These elements are built up in this case from transistors and resistors. Shown around the integrated circuit module are the resistors and transistors which would be required to make circuits equivalent to those in the module (Photo by Professor John Lombardi, University of Illinois, Chemistry Department)

Fig. 8 Circuit boards. The top left board is a blank standard board in use at the University of Illinois Chemistry Department. The lower right is a printed circuit board. This board was printed in the Chemistry Department electronic shop for use in a computer interface. The other two boards are standard boards with microcircuit logic elements wired in. The printed board requires a minimum of wiring. No wiring is shown here in the picture of a printed board. If many identical circuits are needed it is more economical to print them than to wire standard boards. For a complicated circuit, three identical circuits would justify printing a board (Photo by Professor John Lombardi, University of Illinois, Chemistry Department)

having the connections printed on the board. This is a simple procedure and at the University of Illinois it may be done in the Chemistry Department electronic shop. The process is no more difficult than making and developing a photograph (see Fig. 9).

In order to indicate what is involved in interface construction let us design a simple interface. For the computer to control the apparatus it must be able to send information to the apparatus. The computer operates in binary so the output from the computer is a number of binary digits which are either 0 or 1. They are represented at the output connections of the computer as voltage levels; one level for 0 and another for 1. These levels can be used to control lights, motors, relays, valves or anything which can be

Fig. 9 Art work for printed circuits. The first step in making a printed circuit board is shown here. The connections are laid out in black tape on a white surface four times actual size. The layout is photographed with a special camera and then etched into printed circuit board material (Photo courtesy of J. A. Saldeen, Assistant Electronics Engineer University of Illinois, Chemistry Department)

electrically controlled. Let us assume that a positive voltage corresponds to 1 and zero voltage corresponds to 0. On a twelve bit computer there are 12 different 0 and 1 levels available for any output instruction corresponding to the twelve bits in the computer word. The problem of the interface is to decode these 12 bits and energize the particular light or motor which that bit configuration indicates. If there were only a total of 12 things one wanted to do the problem would be trivial and one bit could be assigned to each job. Usually there is far more to be done than this. With 12 bits it is possible to select 2^{12} different operations with one output. By using a sequence of outputs an even greater selection is possible. The selection of position of a crystal for an X-ray diffractometer may easily require far more than 12 bits of information and, thus, several outputs are necessary. Thus, a means of manipulating the digital outputs of the computer are required. This is provided by the logic elements found in integrated circuits. In Fig. 10 a few logic symbols are shown. The flow chart for the circuit is made up of symbols such as these. The inverter is the simplest of these. If a 1 appears at *A* then 0 will appear at *Q* and vice versa. The Nor gate is another commonly used element. If 0 appears at *A*, *B*, and *C*, a 1 will appear at *Q*. If a 1 shows up at any of the inputs the output *Q* will be 0. The name Nor comes from symbolic logic in which a 1 represents a true statement and a 0 a false. The output of the Nor gate states that neither *A* nor *B* nor *C* is true. It this statement is true the gate is true if it is false the gate is false. For simplicity let us assume that we have a three bit computer, and that we wish to control two motors and two valves from the computer. The motors have an electronic switch and will run while voltage is applied to the switch and stop when it is no longer applied. Each valve has two inputs. If a voltage is applied to one input the valve will close if it is not already closed. When a voltage is applied to the other input, it will open if it is not already open. It is obvious that all of this could be trivially handled with 6 bits. Usually, however, we would want to do far more than this and we would quickly run out of bits on any reasonable sized computer. To keep the problem simple and still illustrate some of the concepts involved we have assumed that we have only 3 bits available. Let us further complicate the problem by assuming that occasionally we wish to keep a motor running while we open or close a valve. The diagram in Fig. 11 shows such an interface using only inverters and Nor gates. The three bits from the computer are labeled 0, 1, and 2. When an output from the computer occurs we will assume that they come on according to the computer word that is output and stay on until the next output instruction is given. That is, if the computer outputs a 3, or 011 in binary, bits 0 and 1 are on and bit 2 is off. In the interface shown when bit 2 is off bits 0 and 1 control the motor. When bit 2 goes on, gates 1, 2, 5 and 6 go off and gates 9 to 12 are controlled by bits 0 and 1. Thus, bit 2 of the computer word indicates whether motors or valves are to be controlled by this output and bit 0 and 1 indicate what motor (or valve) 1 and 2 are to be doing. The circuit formed by gates 3 and 4 remembers what motor 1 is to be doing while the valves are being set. To indicate this let us assume the computer outputs a binary 001. This will turn on motor 1. This happens as follows; since bit 0 is on gate 1 goes off. Gate 2 goes on since bit 2 is off and inverter 0 is off since bit 0 is on. Gate 2 will turn off gate 3, and since gate 1 is off, gate 4 will come on turning on motor 1. If a valve is to be set now bit 2 will come on and gates 1 and 2 will both be off.

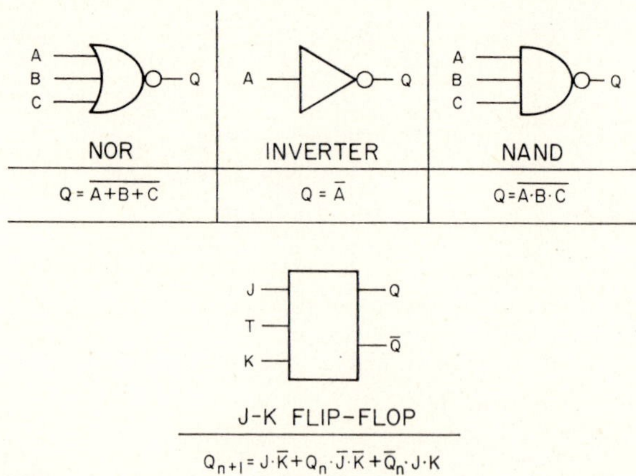

Fig. 10 Logic elements or gates. The symbols and logical values for a few of the logic operations commercially available in microcircuits. The flip–flop is a triggered gate which changes its state when the trigger T changes from 1 to 0. The subscript n indicates the state prior to triggering and $n + 1$ the state after triggering

Gate 4 is on keeping gate 3 off which in turn keeps gate 4 on and the motor running. Gate 4 will not go off until gate 1 goes on which only happens when both bits 2 and 0 go off. When this happens gate 3 will come on and keep gate 4 off. The dashed line encircling gates 1, 2, 3 and 4 is to indicate that they are all four available in one integrated circuit package. This entire interface consists of four integrated circuits costing about 60 cents apiece.

This is an over simplified example which avoids the question of matching the computer output to the logic, which is a trivial problem with most process computers which are built for that purpose, and the more difficult problem of matching the logic output to the apparatus. It does give some idea of the control which can be exercised by micrologic, however. Digital information can be fed back to the computer by similar techniques.

Aside from operating on-off devices such as motors and valves it is possible to convert digital output signals to analog voltages. That is to say, a digital word output by the computer can be converted to a voltage by means of a digital-to-analog-converter (DAC) and the voltage can be used to drive a recorder as is the case with the NMR machine mentioned earlier. The recorder is part of the NMR apparatus. During a normal scan of the spectrum when the apparatus is not connected to the computer the output of the machine is on the recorder. The first plot in the NMR sequence on page 446 was produced by the apparatus without the aid of the computer. When the computer is driving the apparatus the output which would normally go to the recorder is converted to digital by an analog-to-digital converter (ADC). The operator can ask to see the result at any time during the process of time averaging and the computer will output to the interface which will convert its digital information to analog and plot it on the same recorder. Thus, if the computer is remote from the apparatus the experimentalist may get his results without ever leaving the apparatus.

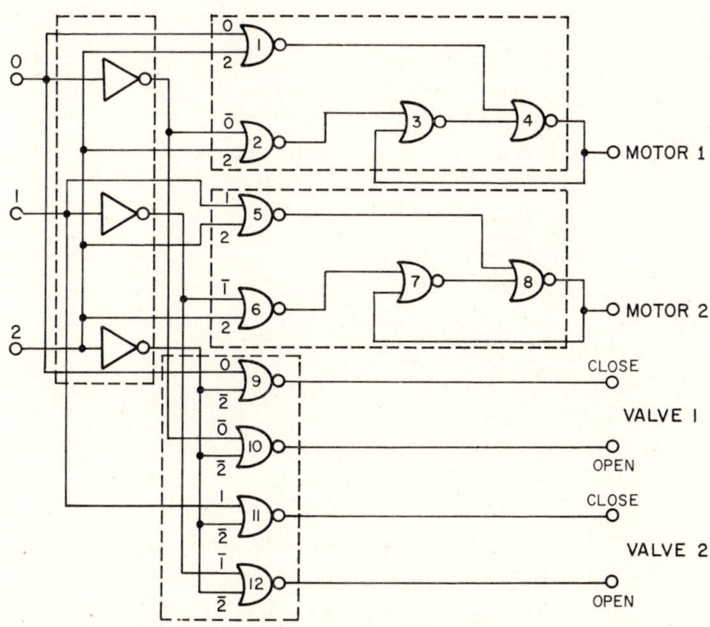

Fig. 11 An example of a simple logic interface. The dashed lines encircling various groups of logic elements indicate that the entire group may be obtained in a single microcircuit module

This sort of feed back is of particular importance when a new phenomenon is being investigated. It often happens that the quantity one measures is not simply related to the phenomenon he is studying. When a computer is not used on line it may still be necessary to process the data by computer to determine exactly what happened during the experiment. If the measured quantity is computed and fed back to the experiment in a few milliseconds on a plotter, allowing the experimentalist to observe the response of the measured quantity to the controls on the apparatus, much more can be learned than could possibly be gleaned from data studied 20 minutes after the experiment is finished.

TRENDS

A knowledge of the use of the computer has been an asset to chemists for some time now and most universities are training chemists in computer programming, if not as undergraduates, in their graduate programs. With the growing importance of on-line usage and the special techniques involved it is now manditory that universities train their chemists not only in programming but in system design and interface hardware. A number of universities are developing undergraduate laboratory experiments which will interact with the computer. The students will design and build a simple interface from microcircuits. At Illinois this project is being developed as part of the physical chemistry laboratory. At other schools it will possibly be a part of the chemist's training in some other branch of chemistry. At least one school plans to institute such a program in the analytical chemistry division. The computer has achieved a position of importance to a chemist comparable to that of the analytical balance. Modern chemistry has entered the computer age and has been expanded and enriched by it.

Acknowledgment

This work is supported in part by a grant from the National Science Foundation.

CHAPTER 19

The Role of Computers in the Analysis of Bubble Chamber Film

JAMES N. SNYDER

BUBBLE CHAMBER DATA ANALYSIS is a young field; only a decade has passed since 1958 when the Franckenstein projecting microscope system along with the computer analysis programs PANG and KICK were developed in the Alvarez Group at Lawrence Radiation Laboratory, Berkeley. The growth of this field during these ten years has been phenomenal; much money, the efforts of many people, and the time of many computers have been absorbed in the development. A broad consideration of this growth, its initial hopes and aims, its current realization, its portent for the future constitute the main theme of this article. One conclusion which will emerge must be stated and stressed at the outset. Just as spark and bubble chambers are crucial as detectors allowing the proper exploitation of accelerators for experimental research in high-energy physics, so computers have become central in solving the data analysis problems associated with the torrential flow of data from these detectors. Their expense and the frustrations of their programming must be borne and their even greater fractional involvement must be faced if the experimental output in high-energy physics is to be maintained or increased.

Approximately one-half dozen different types of film data analysis systems have been developed and are in productive use or soon to be. Excellent descriptions of all hardware systems have been given at several International Conferences on Instrumentation for High-Energy Physics. Proceedings of these conferences as well as several other review articles have been published[1-8]. These form an adequate basis for gaining a technical insight into the details of the hardware aspects of the various systems.

Hence, rather than concentrate on the different hardware measuring devices, it will be more novel and instructive to examine the role computers have played in the various systems and the increasing role they are playing and will play. Although the evolution of the various systems occurred under the scrutiny of the high-energy community, most debate and attention has been directed toward the measuring (or scanning) hardware front-ends of the total systems; on the other hand, the financial and personnel investments in the computer portions of the systems, while not so evident or so often discussed, are as great or greater. To illustrate this, each of the data analysis systems (or a generic example of each) will be examined in turn, looking first at the original hopes and goals for a system and then at the degree to which they have been fulfilled and with what difficulty. A word of caution must be inserted here. The quantitative estimates which will be attempted are subject to error. There are many reasons for this: in the face of the stress involved in getting a system operating, inadequate records of computer time spent and personnel time spent on programming and debugging are kept; in addition to well identified programmers and data analysts, many other people (for example, the experimentalists themselves) get deflected into programming tasks and are not credited toward this function; it even seems characteristic, once the job is done, to underestimate the effort which went into the computer and programming areas. It is very difficult to determine exactly when a system makes the transition from the development stage to the production stage. Months can elapse between the time at which the first event is dragged through laboriously "by hand" and the time at which the system is processing events on a mass production basis with an acceptable and understandable rejection rate. It is surprising how sketchy was the information that could be elicited on these points either from the literature or from personal inquiry. In spite of this, estimates will be undertaken, although their accuracy cannot be guaranteed. An attempt to defend them can be made only on the basis of long experience in the field and personal involvement in several of the systems.

The first analysis system, conceived and developed at Berkeley became operational in 1958–59. It involved scanning on conventional tables, measuring on projection microscopes, with subsequent computer analysis of events *off-line*. This system or minor variants thereof has been adopted by many groups and is familiar to all. With several exceptions (and until only several years ago), it is the system responsible for the majority of the bubble chamber events analyzed. Some statistics on this system can be obtained by looking at the national output during 1964 since by this time it was in effective operation while other systems had not yet entered the production phase. It is estimated[9] that the year 1964 saw 1,000,000 events processed. Involved in this process were 50 scanners, 125 measures, 125 programmers and analysts, 125 Ph.D. physicists and an unstated number of graduate students and degree candidates. Allowing again for the traditional underestimate of programming involvement and the fact that the physicists and graduate students probably spend a substantial portion if their efforts in this area, it should be noted that probably over one-third of the personnel was devoted to the computer and programming aspects of the system. Note further that 65 projecting microscopes at $50,000 were involved. This represented a capital investment of about $3,000,000. Although hard to estimate, this number events probably saturated a computing capacity represented by 3–5 IBM 7094's (a convenient unit). This represented a capital investment of some $10,000,000 or an annual operating expense (at $100 per hour) of close to $2,000,000.

It would be difficult to estimate precisely the original programming effort necessary to create the off-line analysis system used in processing this output. Most of the development was carried out in Berkeley during 1958–62 when the basic programs comprising the total system finally became fully usable. Note the considerable period over which this effort was extended. Even since that period, a considerable investment is involved in modifying, improving, and extending programs for new experiments. Effort is involved at other installations which acquire and use these basic programs since the system must be modified or partially translated to be compatible with the procedures in effect at these installations. For example, Argonne reports about five man-years necessary to convert the HGEOM–GRIND–SUMX system to their installation. In addition each installation must commit up to some three man-years per year to the maintenance of such a system. Parenthetically, it is interesting to note that one result of these efforts on modification and maintenance is the continual increase in program size. The original 8000 word Berkeley KICK (a kinematic analysis program) has grown to 50,000–100,000 words in its Brookhaven version.

Several groups have undertaken major revisions of the PANAL–PANG–KICK–EXAMIN–SUMX program sequence originally used in the Alvarez group at Berkeley. The FOG–CLOUDY–FAIR sequence of the Powell group at Berkeley and the THRESH–GRIND–BAKE sequence developed at CERN are examples. Each in turn has its adherents and its variants. As a further example, Brookhaven is developing a comprehensive and unified analysis sequence called QLOD which will incorporate various improved analysis programs such as the three-view geometry program (TVGP). 20–25 man-years of programming have gone into QLOD. In total as much as 400 man-years of programming effort must have been invested to date in these "conventional" analysis systems.

The computing costs involved per million events per year as discussed above should be carefully noted. This type of *post measurement* stereoreconstruction, kinematic analysis, and summary statistical analysis will be common to all other systems discussed. Faster computers and more efficient programs have allowed the computing costs devoted to this phase of the analysis to be cut by a factor two every two to three years. This has allowed the output of events to be increased by over a factor four in the past five years without a similar increase in costs.

The spiral reader was conceived in 1958 by McCormick[10] who was then at Berkeley. Originally, it was intended as an analogue device particularly suited for the automatic measurement of single vertex events and was, of course, to be operational within a year or so. It suffered several radical changes in design and even operating concept (McCormick himself left Berkeley in 1960) so that its development time and cost understandably grew. It became a production system in late 1964-early 1965 and has to date grown in output capability to nearly 200 events per hour (principally single vertex events). This is impressive and allows the instrument to act as a basis for a system which comes close to fulfilling an often quoted goal of one million events per year. The device is fast enough and the one vertex events for which it is used occur frequently enough that the system is operated without scanning.

In this system we meet for the first time the use of computers in the measuring process itself, over and above their subsequent off-line use for conventional stereoreconstruction and kinematic analysis. A small PDP–9 computer is attached on-line to the spiral reader for controlling film advance, film positioning, automatic fiducial measurement, and formatting of output data onto magnetic tape for subsequent further analysis. Here are the rudiments of *automatic measurement*—the computer control of the measurement process.* In addition to this, the data from the spiral reader must be pre-

* It should be noted in passing that this degree of automatic measuring via a control computer has been applied to conventional projection microscope systems at several installations including Argonne National Laboratory.

processed by a pass through a CDC 6600 filter program prior to conventional analysis. This program removes the unwanted signals from the non-event background tracks seen by the rotating slit.

Realistic cost estimates for the system are virtually impossible to arrive at since the drawn out and often changed course pursued during development must have doubled or trebled them. The computer programming necessary for the on-line control function, filtering, and calibration must have involved of the order of 10–15 man years, which is smaller than that involved in the automatic measurement portion of other such systems.

The flying spot digitizer (FSD) was conceived and prototyped during 1959–60 by Hough and Powell[11], then at CERN. A collaboration between CERN, Brookhaven (to which Hough had gone) and Berkeley was formed in 1960 in order to produce three similar systems each capable of processing 1,000,000 events per year. Involved would be the FSD device itself, estimated to cost $250,000, along with 4–8 digitized scanning tables at $40,000 on which "roads" containing the tracks of interest would be roughly digitized. This information would be available to the IBM 7094 (which operated the FSD in the automatic measuring mode) so that signals from the tracks of interest could be gated and so that filtering could be done. By 1964 it had become abundantly clear that although the measuring hardware was capable of the goals set, the computer and programming problems involved in control and filtering were of greater difficulty than originally anticipated. It is to Hough's credit to have realized this clearly and to have applied ingenuity and massive effort to the solution of this problem. Two years were spent before an experiment of 10,000 events (which had been run through analysis several times before) finally was run in the Spring of 1966 and passed Hough's stringent criteria. In this automatic measuring mode, the IBM 7094 spends approximately 15 seconds per view operating the FSD and receiving its data; the filtering of each view is done in series with this and occupies another 15 seconds for a total of $1\frac{1}{2}$ minutes of computer time per triad. While operating, the FSD saturates the 7094—this is in addition to the subsequent conventional analysis now carried out on a CDC 6600. A unique "fix-up" procedure can be applied to events which fail to pass the analysis programs. The recorded data can be re-displayed via the CDC 6600 so that a human operator can identify the salient features of the event for the analysis programs. Some 25 man years of programming effort have gone into the automatic measuring mode of FSD used at Brookhaven; the experience must be similar at Berkeley and CERN.

In 1961 Pasta, Marr and Rabinowitz[12] at Brookhaven suggested a further extension of the FSD system into as full *automatic scanning and measuring* system. They demonstrated that a 7094 would be able to accept the FSD's full data output of about 10,000 digitizations per second, initiate tracks, predict new hits, and make the appropriate associations. Efforts were immediately undertaken in the Applied Mathematics Department at Brookhaven to develop this idea into a production system. After an investment of about 25 man years of programming the first production experiment was undertaken in 1966. This constituted a set of approximately 10,000 small angle proton scatters, an experiment very appropriate for the system. It should again be noted that the FSD used in this automatic scanning and measuring mode will saturate a 7094 computer while running. As much time is spent in scanning (and measuring if an event is found) in this mode as in measuring alone in the automatic measurement mode. Hence, unless events of interest occur in almost every exposure, the useful output of events will be less than in the automatic measuring mode with no concommitant reduction in computer facility needed. It is not clear that a completely automated analysis system complete with pattern recognition and computer scanning can, in the face of the amount of computational facility required, compete with partially automated systems involving human scanning and human monitoring and guidance of partially automated systems.

As a hardware device the scanning and measuring projector (SMP) was conceived by Alvarez[13] in 1960. In 1961, the idea[14] of extensive on-line computer control was added thereby achieving an integrated analysis system. The usual prediction of operation within a year or so was given. Due to the fact that the device was modest and a basic simplification of previous devices, and due to the initial realization of the central nature of the on-line computer and the programming problems which would be entailed, rapid progress was made. The system entered its production phase both at Illinois and Berkeley in late 1963.

SMP's are sufficiently cheap (about $35,000 each) to permit complete integration of scanning and measuring into one continuous operation controlled by an executive program within the computer which in an interactive conversational mode leads the operator, checks and filters results, requests remeasurement, and formats the data onto an output tape for later conventional analysis. This executive function probably adds about 10% to the total computer needs of the system. SMP's have achieved measuring rates of 10–15 events per hour. Between 10 and 15 SMP's would saturate a 7094 and yield a system approaching an output of 1,000,000 events per year. However, the executive function can be handled in a smaller computer (conventional analysis still being handled on the larger 7094); the Berkeley system, prior to its displacement by the faster spiral reader system, was operated on an IBM 7040, the Illinois system on a home-built CSX–1 (of capacity similar to that of a PDP–1), and the Purdue system on an IBM 1401. The original programming investment in the SMP executive, filter and calibration routines, was approximately 12–15 man years; it required about 2 man-years to convert the programs for the Illinois system; approximately 2 man-years per year are used at Illinois for program maintenance and development.

The precision encoder and pattern recognizer (PEPR) was conceived by Pless[15] early in 1961. Integration of an on-line computer into this system for automatic scanning and measuring at a hoped-for rate of 1,000,000 events per year was contemplated from the very outset. In fact two on-line computers were to be used. The first, the controller, was a special purpose, wired-program digital device which positions and sweeps a CRT line element in both position and angle, accepts and buffers data, and performs some of the corrections for optical distortions thereby freeing the second computer from these repetitive tasks. The second, a PDP–1, was to be programmed to solve the pattern recognition problem: construction of track segments from hits, construction of complete tracks by track following, identification of true vertices, rejection of non-relevant tracks and background, etc. Programming began about 1963 in collaboration with Rosenfeld of Berkeley and Taft of Yale. By early 1965, it was clear that the full pattern recognition problem was sufficiently difficult that to continue to pursue it as a first goal would only serve to unduly delay the use of PEPR as a useful production instrument. Hence, a three phase program was embarked upon. The first phase, point guidance mode, involves the prescanning of film on image plane digitizers (IPD), seven of which were provided, in order to identify events of interest and provide a rough coordinate on each track at which the PEPR itself can later initiate its search and reconstruct the track by standard track following methods. The second phase, grid guidance, involves prescanning film on an IPD and providing PEPR with the rough coordinates of the vertices of interest by means of a superimposed grid on the table. True automatic scanning (and measuring) with programmed pattern recognition would comprise the third phase. Note that phases two and three are parametrically related; phase three, true pattern recognition, can be approached as the limit of grid guidance as the grid size approaches the film frame size. Point guidance began to work in 1966 and a production system quickly evolved. Over 40 man-years of programming effort have been invested in PEPR so far; several more calender years and at least 20 man-years of programming would be necessary before phase three could be reached. Before attempting such an automatic system, the three view PEPR which can measure and track by examining all three pictures of a stereo triad simultaneously and which will become available in 1969 will be exploited in the point guidance and vertex guidance modes.

Due in part to the controller concept, the PEPR measuring hardware is extremely fast and efficient; in the point guidance mode it outmatched the originally used PDP–1 by a factor of approximately 20. In this mode 30 seconds per frame ($1\frac{1}{2}$ minutes per triad) was required, yielding an event rate of 40 per hour. As a result, a PDP–6 (in the same class as an IBM 7040) was acquired which along with program improvements has improved this mismatch to a factor 5–6 and has raised the event rate (point guidance mode) to 120 per hour. By the end of 1969, it is hoped that the PDP–6 will be replaced by a PDP–10 and that the event rate will increase by a factor 1.3–2. Thus the oft-desired 1,000,000 event per year system is being gradually approached. It is interesting to note that the PEPR hardware (CRT plus controller), costing barely $300,000 is probably capable of very nearly saturating a computer in the CDC 6600 class, costing $3,000,000–$4,000,000. It should further be noted, that given a limited amount of computer capacity, the event output rate of PEPR will drop in progressing from phase one through phase two to phase three since the latter phases impose a greater computational burden. In this progression scanning becomes first simpler and then non-existent. The event output rate in phase three further depends upon the frequency of events per frame since each frame (even those containing no events) will have to automatically scanned. Again serious questions as to the economics of such completely automatic systems are posed.

The POLLY, developed[22] at Argonne National Laboratory over the past several years, represents a unique synthesis of some of the better features of earlier systems. It measures by means of a spot on a CRT but like PEPR allows a programmable positioning and sweeping. An on-line computer (a PDP–7 for POLLY–I, a Sigma 7 for POLLY–II) carries out the measuring sequence on pre-scanned film and carries out basic consistency checks. An operator guides and monitors this process; the operator can guide the computer through troublesome events. Measurement rates of 80–100 events per hour have now been achieved.

A novel approach to the bubble chamber data analysis problem as a specific example of the much larger problem of visual pattern recognition[16] was formulated by McCormick during the period 1959–60. The growing realization that computers and digital techniques were central to these problems led to the conclusion that the problem could be best solved by a general purpose pattern recognition computer[17,18,19] oriented specifically to such problems in a manner analogous to the orientation of all past and currently serially operating computers toward numerical arithmetic problems and serially organized logical and data manipulations. Such a computer would contain many parallel operating elements organized into an array which could be loaded with a portion of a digitized photograph and which would process it for recognition of track elements, vertices, and points, etc. The computer complex would also contain other units operating in a list processing fashion which could form associations and set up and analyze the tree-like graphs to which visual recognition problems can be reduced. An arithmetic organ is, of course, to be provided to handle the arithmetic tasks arising in conventional analysis. Of crucial importance would be the extensive and flexible input-output system consisting of channels, on-line visual monitors, consoles, massive back-up stores, etc. This would allow concurrent input-output operations, on-line monitoring, interactive use of the system, information retrieval, and efficient data management which will become of increasing central importance to any bubble chamber data analysis system as experiment size and complexity grows with the consequent growth in the amount of data which must be retained, organized, and manipulated at any one time. Actual logical design of this computer started in 1962; construction began in 1963. A disastrous fire early in 1967 has slowed progress but the first components of the system should become operational toward the end of 1969. The system will be provided with four fast programmable CRT scanners for input of photographic information for the automatic scanning mode of operation. These will also have a calibrated precision sweep mode similar to either or both the FSD and PEPR for automatic measuring of the events found in the scanning mode. Of the approximately $5,000,000 invested in this Illiac III so far, less than 10% can be attributed to these scanning and measuring front-end input devices.

The programming of this computer is underway but presents special problems. Special techniques[20] must be developed for this new field of visual pattern recognition and ways to communicate conveniently with such novel parallel-operating instruments must be devised. Since this is a "home-built" computer, these assembly, compiler, monitor and data management programs must be written in-house. Some 50 man-years of programming have gone into this area to date; at least 25 more will be needed. Only then can the specific application system of programs for the analysis of bubble chamber photographs be undertaken. A complete system would probably require about 50 man-years of programming. The design target of the device, if exploited for film data analysis would be the processing of 1,000,000, stereo triads per year requiring a rate of approximately 300 events per hour.

The recent realization that fully automatic analysis systems offer no economic advantage over semi-automatic man-machine interactive systems for film data analysis has served to remove this goal as one of the main motivating factors for the design and construction of Illiac III. Thus, the thrust of the design and programming effort is being directed toward the more general problems of pattern recognition and the processing of visual and photographic information, irrespective of the specific field of origin.

This completes the generalized and very brief histories of the existing or contemplated analysis systems and brings us up to the present. Although only one development in each family was described, it should be noted that some of the various systems have been implemented at other installations and that there have been variants of several of the systems developed, e.g. the impressive CHLOE system at Argonne is a CRT implementation of the FSD concept.

Some interesting observations on these developments up to the present time can be made. All of the advanced systems which have become or are now becoming operational (SMP, FSD, FSD with automatic scanning, PEPR, Illiac III) were conceived in the 1959–1961 period, a scant several years after the birth of the entire field. It is equally surprising that in the eight years since that period no new basic concept for bubble chamber data analysis systems has appeared. It is also generally true that the development times of these systems and the problems to be overcome were all seriously underestimated. The most glaring underestimate seems to be with respect to the computer and programming facilities needed to implement the systems. It is interesting to note that there is surprisingly little variation in the cost of the front-end *measuring hardware* necessary for a 1,000,000 event per year system; $400,000 appears to be a good, rounded estimate. Of further interest is the fact that the cost of the measuring hardware never exceeds 25% of the cost of the computing equipment necessary to both operate it and analyze its output data. In retailing parlance, measuring instruments are the loss leaders for the computers which they are capable of saturating.

No mention has been made so far of spark chambers and the data analysis systems pertinent to them. Only a brief mention will be made since the problems in this area seem less pressing and difficult and since they do not lend themselves to the same type of general approach as that which has been applied to bubble chambers. The origin of this difference lies in the basic differences between the detecting instruments themselves. The bubble chamber is capable of much greater resolution, hence the analysis system must be capable of much greater precision which in turn requires more involved measuring devices and more involved and time consuming computer programs. Bubble chamber configurations are fixed while spark chambers are modular and lend themselves to very flexible spacial configurations which can vary from experiment to experiment. Hence, a massive fixed analysis system pertinent to a fixed format and a fixed sequence of computations is not feasible; new analysis programs almost need to be written for each new experiment. The shorter sensitive time of spark chambers eliminates much of the background, thereby simplifying both the measuring process and the pattern recognition and analysis programs within the computer. A spark chamber can be monitored by counter arrays and triggered after the event. This allows good discrimination toward obtaining only these events pre-selected to be of interest. Thus, analysis programs can be simpler; they need not try so many hypotheses to understand the event. Of course, this selectivity also discriminates against the one-of-a-kind chance discovery which has allowed the bubble chamber to discover so many new particles.

So, in summary, analysis programs for spark chambers have to be written more often and with more particularization, but they need not be so general or so complex nor need they maintain such high precision. Since they tend to be straight-forward, the use of procedure oriented languages such as FORTRAN and the accumulation and exploitation of a good library of useful subroutines and computing aids can go far in easing the burdens of programming. Several successful spark chamber analysis systems are in operation, for example at Brookhaven and Argonne. The Argonne system, CHLOE, required about seven man-years of programming for its creation and requires about three man-years per year of programming for maintenance and development.

In addition, spark chamber analysis has been tending toward the use of smaller dedicated on-line computers for data acquisition, partial analysis of events on-line, and continual monitoring of results so that these can be fed back to affect the course of the experiment and so that equipment malfunction can be detected early. Even these small dedicated installations require substantial investments in programming. Based on experience at Brookhaven and the Stanford Linear Accelerator, the system and monitor programs (not the applications programs for a specific experiment) can consume from 5 to 10 man-years of programming effort.

Returning to bubble chamber analysis systems and having arrived at the present, what portends for the future? It might seem that some of the anecdotes and some of the comments have been critical of the past developments in this field. This is a mere superficial appearance. It is clear that data analysis systems are absolutely crucial to progress in high energy physics; they will become even more crucial in the future as larger accelerators become available, as more spark chamber and bubble chamber detectors come into use, and as larger and more precise experiments are undertaken. Data analysis has been and is expensive; it will become even more so. However, it is still cheap compared to the costs of either building or operating accelerators and cheaper yet compared to the valuable and scarce human resources which analysis systems serve to amplify.

Thus any remarks made here, which may be interpreted as critical, are not made to criticize the investment of time, money, or manpower in analysis systems. These investments have paid off handsomely. The plea is for realism, and the application of foresight rather than hindsight. An analysis system is sufficiently important and sufficiently complex to implement that there is no necessity for the a priori estimates of time, money, and personnel needed being set at inadequate levels. More should be asked for rather than less. If accelerators are running at one-third billion dollars, why not face the fact that effective analysis systems may run into several millions?

All of these caveats apply even more strongly to the computer and program portions of analysis systems. Here, a priori estimates have also been at fault; history shows that these aspects of the systems are larger and more important than either estimated or expected. Let it be remembered that the measuring and scanning front-end input device is only a fraction of the total system. It does not seem profitable to continue the debate concerning the merits of this or that input device and worry whether the one for \$270,000 is to be preferred over that for \$280,000; each will have to be supplied with up to \$2,000,000 worth of computing equipment and of the order of 50 man-years of programming.

Particularly, should the costs of programming, the scarcity of programmers, and the time programming consumes not be underestimated. Two interesting examples serve to illustrate these problems. First, IBM is currently still constructing and developing operating systems for its new line of 360 computers. Granted that this is a task far larger than the construction of an analysis system, it is staggering to note that the degree of visible activity points to a programming budget in the range of 40–60 million dollars per year—the precise figure not having been released. Second, the effective output of a programmer is hard to measure; it can fluctuate widely depending on such variables as program size, complexity, etc. However, one study[21] by the System Development Corporation finds an average output of 322 machine instructions per man-month programming directly in machine language and 555 machine instructions (after compilation) per man-month for programming in a problem oriented language such as FORTRAN. These results are averages over programs of various complexity; the type of programming involved in massive analysis programs certainly must lie toward the upper end of the spectrum of complexity. Thus a program of 60,000 words, not an unusual size for one of the several segments of an analysis system, can easily require at least 10 man-years of programming effort, even if done in FORTRAN.

More specifically, some of the considerations which will have to be made and the problems which are sure to arise in the very important computer portion of analysis systems can be discussed. The number of events[9] analyzed should be and probably will be increased by a decimal order of magnitude within the next several years. How will this affect computing costs? As illustrated in the histories of the various analysis systems, all of these systems are really barely past their infancies; many use their computational facilities inefficiently, which is certainly forgivable at this stage of their development. Their efficiencies have and are bound to continue to improve thereby gaining some additional capacity even within the computing facility currently allotted to them. Beyond this, some have argued that computer speeds, sizes, and efficiencies have been increasing at a much greater rate than computer costs and hence the necessary increased computer capacity can be obtained for approximately the same cost as the currently used capacity. This has been true for the computer developments of the past 10 years; in fact the square root law has held quite well—the cost of the newer faster computers increased only as fast as the square root of their increased speed. However, such a law can remain valid only so long as the increased computer capacity is obtained by increases in raw circuit speed. Limits imposed by the speed of light and size are now being approached so that future computer capacity increments will have to be achieved by more complicated logical organization, more hardware and parallelism. It is hence expected that the square root law will begin to shift more toward a linear law; it is too early in the third generation of computers to determine the exact functional dependence. As computers become faster and their memories become more capacious, the demands of the users similarly increase. More ambitious and more complex demands are made on the analysis programs and their sizes grow to again strain the larger memory sizes. Furthermore, the computer time required for a complete experiment is not expected to be a linear function of the number of events in the experiment. The repetitive scanning of large data bases and the bookkeeping operations involved imply a dependence stronger than linear. All these effects argue that in spite of the faster and bigger computers becoming available, the computer costs and problems involved will increase significantly (hopefully not linearly) if the number of events scanned, measured, and analyzed per year is increased significantly.

Another problem which will have to be faced is that of efficient data manipulation. As experiments grow in size, so do the data bases on which they rest. These are problems also faced in the information retrieval area. No adequate solutions are yet in hand. For this reason, a significant fraction of the research going into current computer projects (Illiac III and Illiac IV can serve as examples) is being directed toward this aspect of the total analysis problem.

One of the currently most popular concepts in the computer field is the interactive use of computers via time-shared consoles. This man-machine interaction, as exemplified by Project MAC at MIT, needs no description here; it has been the subject of numerous speeches, symposia, newspaper and magazine articles and even some articles in learned journals. The advantages offered by such a system need no extensive supporting arguments since they are so obvious: turn-around-time can be reduced; a large computer offering tremendous peak capacity can be shared by many small tasks, no one of which by its *average* use could justify such a large computer but each of which could from time to time employ the *peak capacity* of such a computer; rapid interactive diagnosis can be applied to the programming task, thus reducing its now too-long duration; human feedback can be introduced into difficult problems, thus introducing the new dimension of human insight into the traditional computer loop, etc. Will this concept be applicable to bubble chamber data analysis? The answer is an obvious yes. This concept is no stranger to the field of bubble chamber data analysis having been introduced in the SMP system in 1961 (Project MAC saw its inception circa 1963).

The concept has been extended and employed in all of the modern analysis system, for example it is used in the OFF MIF procedure which is part of the FSD system. In this process, FDS data is recorded on a drum, is displayed on a CRT, and is modified by an operator via a light pen in order to aid the tracking and filtering procedures. The HELP mode built into the PEPR system similarly allows an operator to use a light pen and a CRT in order to guide the system through any one of the subroutines making up the system. The concept finds its culmination in the Illiac III in which visual time-shared monitor displays and consoles will be an integral part of the hardware system and HELP modes of operation an integral part of the programming system. One aspect of this interactive time-shared console use of computers should be clearly realized. It will be more expensive than conventional batch processing use since it imposes an extra (often-large) bookkeeping overhead. The extra expense will be warranted due to the conveniences to be expected and the conserving of valuable human resources to be achieved, but the extra cost must be realized and faced honestly, not buried in emotional irrelevancies.

Thus, the field of bubble chamber data analysis, scarcely ten years old, has arrived at the present; much has been accomplished, often more had to be accomplished than had been anticipated. The field will grow, especially the computer aspects of it; the needs must not be underestimated, they must be faced realistically and without timidity or undue diffidence.

References

1. A. H. Rosenfeld, "Digital Computer Analysis of Data from Hydrogen Bubble Cham bers at Berkeley", *Proc. of the International Conf. on High-Energy Accelerators and Instrumentation*, CERN, 1959.
2. *Proc. of the International Conf. on Instrumentation for High-Energy Physics*, Berkeley 1960, Interscience Pub., New York, 1961.
3. *Proc. of the International Conf. on Instrumentation for High-Energy Physics*, CERN 1962, North-Holland Pub. Co., Amsterdam, 1963.
4. "Instrumentation for High-Energy Physics", *Nucl. Instr. Methods*, 20, 367–463 (1963).
5. *Proc. of an Informal Meeting on Track Data Processing*, CERN, Geneva, 1962.
6. A. H. Rosenfeld and W. E. Humphrey, "Analysis of Bubble Chamber Data", Annual Review of Nuclear Science, Vol. 13, 1963.

7. Y. Goldschmidt-Clermont, "Progresses in Data Handling for High-Energy Physics", *International Conf. on High-Energy Physics*, Dubna, 1964.
8. *Methods in Computational Physics*, Volume 5, Nuclear Particle Kinematics, Academic Press, New York, 1966.
9. A. H. Rosenfeld in Hearings before the Subcommittee on Research, Development, and Radiation of the Joint Committee on Atomic Energy of the Congress of the United States on High Energy Physics Research, U.S. Government Printing Office, Washington, 1965.
10. B. H. McCormick and D. Innes, in reference 2, page 246.
11. P. V. C. Hough and B. W. Powell, in reference 2, page 242.
12. R. B. Marr, J. R. Pasta, and G. Rabinowitz, "A Preliminary Report on Digital Pattern Recognition at BNL", BNL Report No. 6866, 1963.
13. L. W. Alvarez, "A Proposed Device for the Rapid Measurement of Bubble Chamber Film", Lawrence Radiation Laboratory, Physics Note 225, 1960.
14. J. N. Snyder, "Some Remarks on a Data Analysis System Based upon the Scanning-Measuring Projector (SMP)", Lawrence Radiation Laboratory, Physics Note 326, 1961.
15. No adequate published reference embracing the original PEPR concept exists. The initial ideas were first expressed by I. Pless and L. Rosenson in a "Proposal to the U.S. Atomic Energy Commission for the Development of a Precision Encoder and Pattern Recognition Device (PEPR) in the Massachusetts Institute of Technology Laboratory for Nuclear Science" (February 10, 1962).
16. A. Graselli, B. H. McCormick, S. J. Penney and J. N. Snyder, in reference 2, page 249.
17. B. H. McCormick and R. Narasimhan, in reference 3, page 401.
18. B. H. McCormick, "The Illinois Pattern Recognition Computer—Illiac III", *IEEE Transactions on Electronic Computers*, Vol. EC-12, No. 5, page 791.
19. B. H. McCormick, "Illiac III: A Processor of Visual Information", *Proc. of the IFIP Congress 65*, Vol. II, Spartan Books, Washington, D.C., 1966.
20. R. Narasimhan, "Syntax-Directed Interpretation of Classes of Pictures", *Communications of the ACM*, Vol. 9, No. 3, page 166, March 1966.
21. G. F. Weinwurm and H. J. Zagorski, "Research into the Management of Computer Programming: A Transitional Analysis of Cost Estimation Techniques", TM-2712/000/00, System Development Corp., Santa Monica, California, November 1965.
22. R. Barr, R. Clark, D. Hodges, J. Lohen, W. Manner, B. Musgrave, P. Pennock, R. Royston, R. Wehman, "POLLY–I: An Operator-Assisted Bubble Chamber Film Measuring System", Argonne National Laboratory, ANL/HEP 6806, May 1968.

CHAPTER 20

Recognition of Visual Data in High-Energy Physics

L. KOWARSKI

1 EXPERIMENTS AND DATA IN HIGH-ENERGY PHYSICS

The subject of high-energy physics is the study of the properties and behavior of physical entities ("elementary particles") which have been observed as present in the inner structure of atomic nuclei, or in their interaction and break-up. An observable physical "event" involving such particles can be initiated by one of them moving at great speed. Primary projectiles of this kind can be found in cosmic radiation, occurring naturally, or—usually and more conveniently—they are produced by accelerating machines in laboratory conditions.

Collisions between these primary carriers of kinetic energy and other material atoms (which may be put in the way of a beam of accelerated particles as a target for their impact) liberate other particles which may pick up some of the energy brought into the collision, or released by it. These fast-moving secondaries can be detected by suitably sensitive instruments. Many of these are capable of showing the path taken by the particle in its motion across the sensitive volume. In the oldest, and now outdated, instrument of this kind, the cloud chamber, the passage is marked by drops of condensing vapor; in bubble chambers, by a string of bubbles formed in a superheated liquid. In spark chambers, gas-filled gaps between a series of parallel charged plates show spark-like discharges along the trajectory of the particle across the plates. These strings of drops, bubbles or sparks can be seen and photographed; their quantitative properties such as lengths, angles of intersection and curvatures—if the particle is moving on a curved path owing to the presence of a magnetic field—constitute the meaningful physical information produced in the experiment.

Numerical values defining the geometry of the visible tracks can be easily measured on the photographs of a detecting chamber, taken during its sensitive periods of operation; the physical meaning of these values usually becomes apparent only after intricate computations which, in to-day's practice, are performed on digital computers.

2] TRACK POINTS AND THEIR COORDINATES

For the purposes of computation, each bubble or spark track is represented in the memory of the computer by the coordinates of at least as many points as are necessary to define it unambiguously in space. In their raw input form, the coordinates refer not to the track itself, but to its projections in the 2 or 3 stereoscopic plane images which are recorded by the film cameras. Each of the plane-coordinate pairs fed into the computer expresses the measured position of a black point on the film (if the tracks are visible in black-on-white; it is convenient to talk of "black" points, even if the image is actually white-on-black).

In this article we do not have to go into the details of the measuring process itself. In all measuring systems used at present, there is an optical mark moving across the image; when it comes into coincidence with a black point, a signal is sent either by a human observer or by a photoelectric device. The motion of the mark is synchronized with some kind of a spatial or temporal digitized sequence (clock pulses, countable fringes etc.) enabling each signal to carry with it a coordinate value.

A black point, in addition to its position, possesses a far more subtle characteristic which could be described as its physical "meaning". There is no useful meaning in a point which is due to some accidental impression or blemish on the film, or which belongs to a track left by some stray particle. Many of the tracks visible on bubble pictures record simply the passage of the particle beam which had been sent into the chamber, and once recognized as such, they are not worth any further study. Even if the track is of the kind we do wish to study, it may have resulted from some trivial or uninteresting kind of physical event. These four categories of rejects have to be filtered out; only the points "recognized" as meaningful will be retained for the final numerical processing. It is this "recognition of visual data" which is our problem here; we are going to describe various approaches to its solution at various stages of development and automation.

Let us note at once that many of the items of black-point information which, in a given picture, have to be filtered out, are recognizable as such at a glance, before any measurement. Some of the tracks or full events can, however, be recognized as irrelevant only after measurement and computation. The term "candidate" tracks or events has been conveniently used[1] to establish a distinction between the task of recognizing candidates (before measurement) and that of recognizing the truly *relevant events* among the candidates (as a result of the measurement). We shall see that these two kinds of recognition, which we may call pre-metric and metric, have evolved along very distinct paths and only rather recently began to be considered as joint aspects of a single problem.

3 BEFORE THE AUTOMATION

Why automate at all? A trained human observer can spot promising tracks or events just by looking at an image, preferably projected on a screen with a suitable magnification. In the early days of particle physics the cloud chamber was the only visual track detector and the computer was a human person. Later on, mechanical and electronic calculators came to be used; their input was the written-down result of human measurement of the point coordinates or even directly of lengths, distances, angles and curvatures, measured with the help of divided rules, protractors and templates. A few events per day could thus be treated by a human processor. Looking at an image and deciding what, if any, tracks in it are worth measuring would take only a fraction of a minute, often a very small fraction; the measurement could take an hour or more. As long as the recorded output of a cloud-chamber experiment was slow enough compared to these possibilities of human processing, there was no problem. Faster-producing chambers could still be used in those experiments where the chance of finding an interesting-looking event on a given picture was rather low, since the necessity of inspecting ("scanning") even a hundred or more pictures before finding and measuring a candidate would not affect the pace of the whole processing. With fast-producing chambers and a high candidates-to-pictures ratio, the limitations of fully human picture processing began to be felt; since the cloud chamber (unlike the bubble chamber) *can* be triggered to operate only (or nearly only), when an interesting event is there to be recorded, the trend in cloud-chamber physics developed towards the study of rare events, selected by elaborate triggering, rather than towards "high-statistics" phenomena (i.e. those which occur often and whose physical meaning is statistical rather than derived from the observation of a single event).

4 ENTER THE BUBBLE CHAMBER

Bubble chambers suitable for high-energy physics began to be developed and to supersede the cloud chambers around 1954. Spark chambers came about six years later, at a time when the basic problems and methods of handling the information contained in bubble pictures were already well understood.

The pioneering initiative in this direction was due to L. W. Alvarez, whose early interest in bubble chambers led him immediately to notice[2] that "certainly one day of bubble chamber operation could keep a group of cloud-chamber analysists busy for a year". Automation offered the hope of a speedier processing and its possibilities appeared to be worth looking into, whatever the difficulties. Fortunately, Alvarez and his group realized at once that the most time-consuming aspect of the processing, that is the measurement, was also the easiest to mechanize, at least partially, whereas the pre-metric recognition was far less in need of speeding-up and could therefore be left to the human faculties, thus postponing the immensely hard task of developing a pattern-recognizing automaton.

This realistic approach prevailed until 1960 or so and, during that time, the still current habit of viewing picture processing as logically divided into three distinct and successive stages (scanning, measuring, computation) became firmly established[3]. Since scanning comprises all pre-metric recognition, and is entirely human, we may consider this whole period as prehistoric, insofar as we are concerned mainly with automatic *recognition*.

5 THE FRANCKENSTEIN-IEP SYSTEM

The human scanner would recognize the tracks to measure and produce a sketch, of which Fig. 1 offers a somewhat idealized specimen; apart from being given a conveniently engineered projector, he would work in substantially the same way as his predecessor from the cloud-chamber era. The effective speeding-up was due entirely to the automation of *measurement*; in this process the role of the human operator is reduced to the recognition of

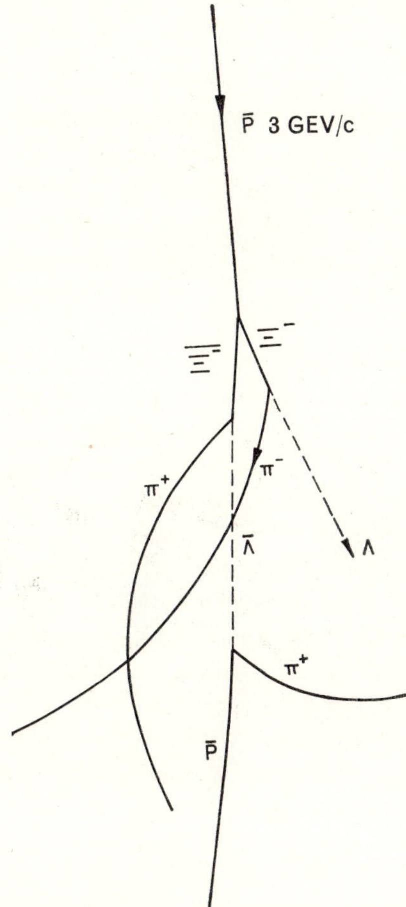

Fig. 1 Sketch of a bubble-chamber event prepared by a human pre-scanner

successive points to be measured and the precise positioning of the optical mark in coincidence with each of these points. The operator then presses a pedal and all tasks of recording, assignment of a pair of digitized coordinates and transmission to the computer are entirely assumed by a machine (Fig. 2).

Machines working on these principles were put in operation at Berkeley in 1957 or 1958 under the name of Franckenstein (after J. Franck, their first constructor in the Alvarez Group). Due to an initiative, first expressed by O. R. Frisch in England[4] and then taken up by a group at CERN (Geneva)[5], roughly similar machines were built in Europe under the name of Iep ("Instruments for evaluation of photographs"). They enabled the processing of

Fig. 2 A measuring machine in operation

bubble pictures to proceed at a rate of less than 10 events per hour (in practice often only 3 or 4 owing to operating mistakes, difficult configurations etc.). With a few machines, worked every day by a succession of short-shift operators, a single laboratory could hope to process some 100,000 events per year—an enormous progress compared to previous unmechanized methods, and yet a clear invitation to further development, since a single bubble chamber working on a 2–3 second cycle is capable of producing several million pictures per year.

Attempts to speed up the Franckenstein system were made in two different directions: either the manual positioning of the measuring mark, which is the slowest link of the whole chain of operations, was helped by fairly primitive devices enabling the mark to move by itself, approximately along the track; or the first computed results, obtained by the computer in the early stages of the processing of an event, would be quickly fed back to the operator, thus enabling him to correct false moves and diminishing the need for subsequent re-measurements. In the "track-following" Franckenstein, a crude recognition by machine helps the operator to position the mark on points of a humanly pre-recognized track; in the "Franckenstein on-line" (to a computer), a measure of automatic metric recognition corrects the mistakes of a prior human recognition and judgment. Thus a system originally based on wholly human recognition began to rely in part on recognition by machine. (It is interesting to note that the idea of "on-line dialogue" was first proposed by J. Snyder in connection not with the Franckenstein, but with a more sophisticated machine; see below.)

By 1968 the manually operated measuring machines were no longer dominating the bubble-processing market; yet one of the most accomplished systems of this type[6] claims for each of its machines an average production rate of 20 events per hour. It comprises all of the above-mentioned improvements and corresponds well to the term "semi-automation" which is currently used in U.S.S.R. to describe this kind of processing. Several (4 to 6) measuring machines are served on-line by a single medium-sized computer (at Berkeley, an IBM 7044).

6 MECHANIZED MEASURING SYSTEMS

Attempts to remove the need for manual positioning began to appear around 1960. In a system first proposed by Alvarez himself[7], the scanning and the recognition of bubble tracks are still fully human, but the positioning aims only at enclosing a portion of the track somewhere in a fairly wide luminous spot (several millimeters on the projected image). Several points lying on this portion will find themselves in positions where their precise distance to one of the bench marks, forming a fixed grid superimposed on the projected image, can be easily and automatically determined by the operation of an ingenious optical device. The operator only has to move the spot on a broad trajectory enclosing the track. In this "Scanning and Measuring Projector" system (S.M.P.), the two first phases of the three-stage approach were supposed to be practically merged; and the constant recourse to the information obtained on-line from metric recognition was meant to involve the third stage as well. In practice it was found that this automation of measuring permitted to speed up the overall processing only by a factor 2 or 3, partly due to the fact that a quick preliminary scanning (distinct from measuring) was found to be still useful. The most decisive innovation—the on-line dialogue—was quickly adopted by the Franckenstein system; the speed advantage of S.M.P. was thereby further whittled down, making the continuation of development on this line hardly worthwhile.

In an even earlier (1958) initiative taken by B. McCormick[8] the claim on human recognition and positioning is reduced even more significantly than in the S.M.P. system. The "Spiral Reader" requires, for preliminary guidance, only the human knowledge of whether and where the image contains a promising-looking configuration of tracks and the position of the point (the "vertex") from which the tracks, constituting the event, appear to be radiating. The picture is swept by an oblong aperture (about 250 $\mu \times 10$ μ on the film scale). The sweeping motion has a rotating component around a center which has to be exactly positioned at the vertex (by a human operator) and a linear component in a radial direction away from the center. It was found constructionally convenient to produce this motion by means of movable mirrors. Thus the aperture describes a spiral (Fig. 3), crossing each track several times. Since its length is oriented radially, each crossing covers a portion of the track as long as the track itself remains approximately radial (curved tracks, at some distance from the vertex, tend to lose their radiality and their coincidence with the aperture becomes increasingly imperfect). Every approximate "hit" produces a digitized signal; in this way several thousand crossings are collected from each operation around a vertex. A small computer is used to operate the system; filtering-out of spurious hits and assembling the true hits into meaningful tracks is done off-line on a much larger computer.

Using our terminology, we may state that the Spiral Reader requires rather little human (pre-metric) recognition and adds a sizeable amount of automatic recognition which is also pre-metric, i.e. non-dependent on numerical results of the measuring process. Reliance on post-metric feedback is, in principle, less essential. The successful reduction of the human contribution

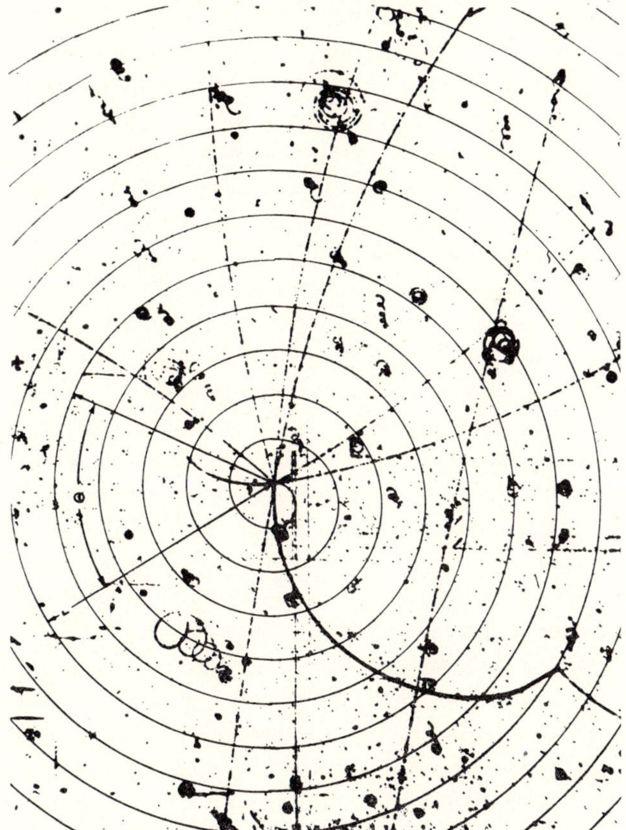

Fig. 3 Exploration of an event by the Spiral Reader

enabled the Berkeley group to achieve, by the mid-1960's, steady production rates of the order of 100 bubble events per hour; it was found however that, for optimum efficiency, the role of the human recognizer could be somewhat expanded, for example by adding one human-measured point to the information collected about each track.

7 TOWARDS FULLER AUTOMATION

In all systems described previously, human recognition remains necessary for a certain minimum amount of prescanning and positioning of measuring devices. Production rates of, at best, a few events per minute could be hoped to be reached in this way, but hardly as high as an event every one or few seconds: for *that* kind of speed the pictures should be left alone with a fully automatic recognizing machine.

Two very different ways towards this goal were initiated in 1960–1961. The Flying Spot Digitizer, proposed by Hough and Powell[9]—now usually known as the HPD, sometimes as the FSD, system—comprised a fully automatic measuring machine which required no human positioning at all and, for the interpretation of its digitized measurements, relied on a certain amount of human pre-metric recognition. It was hoped that the system would develop so as to give a gradually increasing role to the metric recognition by computer, and so to eliminate the human intervention altogether. This approach—a machine measures first, then another machine recognizes the meaning of the measurements—was in a way the exact reversal of the original Franckenstein concept in which the (fully human) recognition was to *precede* the (human-operated) measurement.

The PEPR System ("Precision Encoding and Pattern Recognition"), originated by I. Pless[10], aimed from the very beginning at a fully-automated recognition. For this purpose the picture has to be explored by a moving spot produced by a cathode-ray tube. Each hit (i.e. the spot encountering a black) is reported to an on-line computer; acting on information received, the computer instructs the spot as to where it should move next. This kind of recognition can be considered as metric (since the computer can deal only with digitized, i.e. metric information); yet the amount of meaning reported from each reported hit is strictly limited to what is necessary to direct the spot towards the next hit with a minimum of straying-off. The programmed logic of successive instructions is such that a succession of hits amounts to the recognition and following of a track. To eliminate spurious hits, the spot has the shape of a small, narrow segment (2 mm long on the film); it can rotate so as to choose the angle in which it is oriented most closely along the track (the hit is then strongest; cf. the somewhat less crucial reliance on the same principle in the Spiral Reader). Figure 4 shows how visual information from a bubble-chamber picture, in terms of such hits, can be assembled and redisplayed by a computer.

To achieve the PEPR aim, the spot-moving hardware has to be far more versatile than in the HPD. The latter puts the accent mainly on precision and stability; the spot is produced by a slotted wheel (Fig. 5) and is capable only of covering the image in successive lines ("TV scan"). The spot is pointt-like

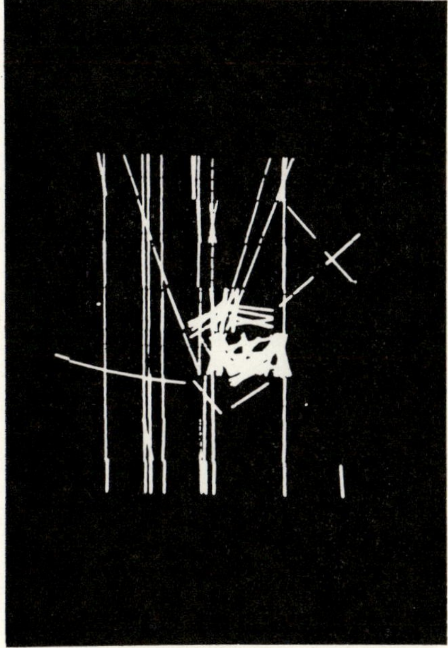

Fig. 4 Exploration of an event by PEPR

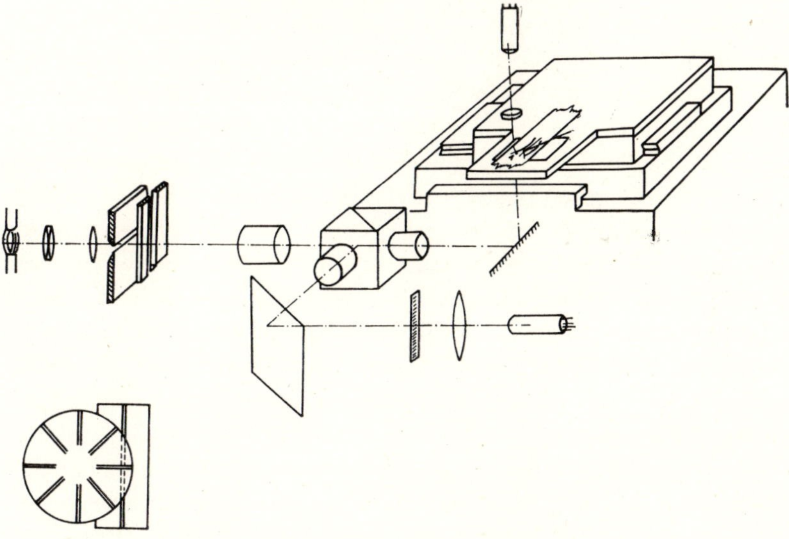

Fig. 5 Schematic view of the HPD hardware

(about 15 μ in diameter) and reports to the computer the position of *every* black point it encounters. To make sense out of some 50,000 digitizings received from a picture, the computer should preferably (but, in principle, not necessarily) be given some information as to where the interesting tracks lie.

In the original version of the HPD system each picture had to be pre-scanned by a human observer, who would measure with a low precision the position of three points of each interesting track. A relatively wide "road" (several hundred μ of width on film) would be defined by these three rough digitizings and the computer would be told to retain only those HPD-measured coordinates which lie inside the roads. After the application of this "gating" program, the "filtering" would start. In most laboratories the computer is programmed to assemble the retained digitizings in successive slices along the road (Fig. 6); some work has also been done on other approaches.

HPD systems operated in this "road guidance" mode are at present in production, or about to start production, in some 20 laboratories in various parts of the world and, taken together, represent a greater amount of experience in partially automated recognition (millions of processed events) than that offered by any other system of bubble picture processing. Their production rate appears to be of the same order as that of the Spiral Reader; road-making constitutes an obligation payable either in terms of a loss of overall throughput or—if enough road-makers are put to work so as to keep pace with the HPD—in terms of extra manpower.

It was found in practice that sophisticated programming and the obvious clustering of digitized points along the tracks enables the computer to find its way, even if the only advance information it receives from a human pre-recognizer is an affirmation that there *is* an event worth measuring and an indication of the approximate position of the vertex. This "minimum guidance" mode has been developed mainly at CERN[11]. It is not yet in routine production, but appears promising and is an important step in the direction of a "zero-guidance" system which would do away with human pre-scanning altogether. Attempts to develop such a fully automatic system, without waiting for the experience of minimum guidance, have been in progress since 1961 at Brookhaven, Columbia and notably at Berkeley[12]; if, some day, a faster-producing FSD is developed, leaving no time for any kind of human intervention, some sort of a zero-guidance system will offer a valuable practical advantage. For the time being, the HPD practice seems to evolve in a different direction (see below).

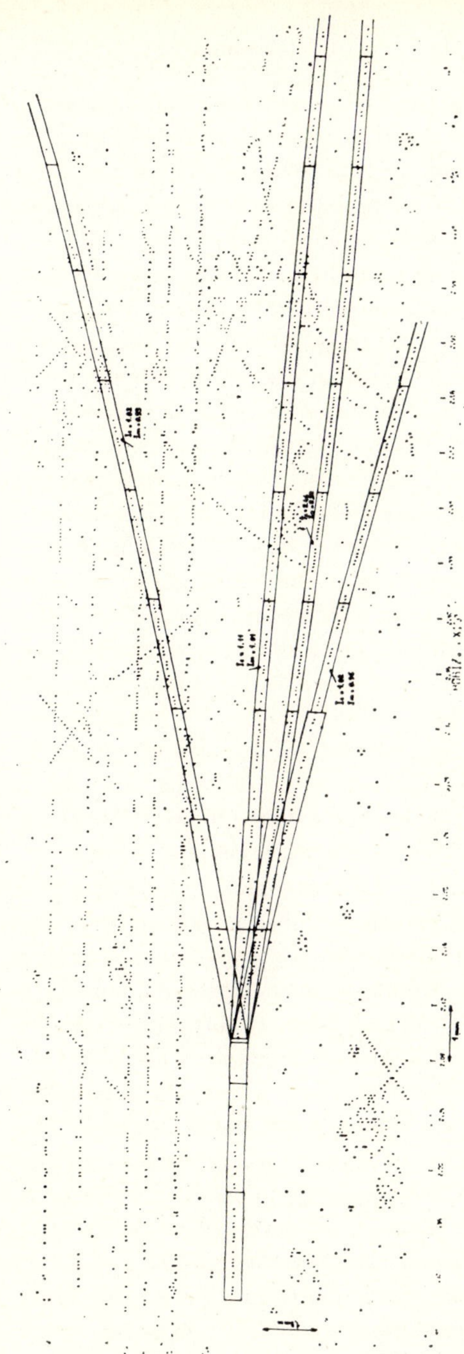

Fig. 6 Slice reconstruction by computer (an explanatory sketch from a computer display; CERN HPD system)

In both the ideal PEPR and the ideal zero-guided HPD, the machine receives no warning as to the physical value of a track of which it may get hold. Special routines are provided for an early identification and elimination of beam tracks (see Section 2 above).

Taken as a whole, the recognition process in these systems is entirely metric, i.e. derived by the computer from numerical data supplied by the measuring machine, but in HPD all measurement is performed in a rigid and undiscriminating way, whereas in PEPR the measuring hardware is flexible and works in close conjunction with the computer, thus taking an essential part in the recognition process. The HPD hardware aims at simplicity and stability (hence the mechanical generation of the spot) and leaves all recognition to software and a rather big computer (IBM 7094 or so). PEPR needs a CRT-produced spot, capable of moving according to running instructions, but achieves the recognition of tracks with a smaller computer. Experience gained with PEPR as a production machine is, at the time of writing, not quite sufficient to warrant a valid comparison between these two approaches; a more sophisticated form of PEPR proposed by M. E. Sekely at Princeton[13] is, at present, even further removed from the possibility of a judgment based on performance.

8 STRENGTH AND WEAKNESS OF AUTOMATED RECOGNITION

The main reason for developing an automated system of recognition was the promise of greater speed; another advantage became noticeable as the development began to bear fruit. Recognition by machine is free from many of the human biases and momentary lapses of attention; machine-assembled statistics present sharper and less distorted peaks. Ample evidence to this effect is presented on every conference on advanced methods of data processing in high-energy physics. Even for the delicate measurement of bubble density, essential for some particle studies, some of the automatic recognition systems promise to be more accurate than visual extimates by human observers[14].

All automatic systems are, however, prone to a practical weakness, whose extent could be assessed only on the basis of accumulated experience. Natural events sometimes produce unexpected and abnormal-looking configurations of tracks; they may comprise abortive tracks too short to be conveniently analyzed, and yet physically significant. After an intersection, especially at a small angle, the identity of the continuing track may become confused. In all such complications human recognition is usually more efficient than the necessarily more limited set of rules governing recognition by machine. Even a human observer is incapable of disentangling the configuration in a few percent of all cases. The proportion of such "rejects" is noticeably higher in automated systems and it is highest when the automation is fullest, to the point where the rejects become about as numerous as the acceptances.

Just as the Franckenstein had to evolve so as to add metric-recognition help to its essentially human pre-metric recognition, the experience with the automated systems showed the need for re-introducing human assistance, specifically to deal with cases in which the machine, after a first attempt, gives the signs of hesitation or incompetence. In this new kind of man-machine dialogue, the operator has to watch a graphic display of what the computer has found in an event. If the computer has been successful, the display looks very much like the event itself and is quickly erased, as the system passes to the examination of the next event. If the display looks confused, it also tends to linger and the operator perceives it as "a yell for help"[15]. A trained operator can form his conclusions, pass them as additional information to the computer and watch for a new attempt. If such cases are relatively rare, the slowing-down for a long series of pictures is not significant; if the proportion of first-attempt rejects is high, the slowing-down is worthwhile, since a very incomplete series of processed pictures would be of doubtful scientific value.

9 RETREAT FROM FULL AUTOMATION

In systems where automatic recognition was supposed to proceed with only a *prior* help from human recognition (as in Spiral Reader or in road-guided HPD) or with no such help at all (as in the original PEPR, or in zero-guided HPD), either the hardware or the software or both had to be made sufficiently elaborate to deal with all foreseeable emergencies and to reduce the rejection rate to the barest possible minimum. As the necessity of a human a posteriori guidance came to be recognized, some of this elaboration began to appear unnecessary. Both HPD and PEPR were originally supposed to function without any intervention on line; now that this ban is being relaxed here and there in some phases of the operation, the PEPR hardware and the

HPD software could stand some simplification with, possibly, resulting gain in reliability.

As an alternative to such post-factum redesigning, new lines of approach were imagined in mid-1960's. Once the relationship between man and machine was recognized as a symbiosis rather than a rivalry, it became obvious that the unique cognitive faculties of a human operator should be relied upon both at the beginning and the end of the recognition process, that the operator should assume only those functions which he is apt to perform easily and quickly, and that the machine should do all the plodding rest.

The Polly system[16] began to operate at the Argonne National Laboratory in 1967, after a long line of development which was started in 1961 by J. Butler and D. Hodges[17]. It uses a CRT-generated point spot which—like the linear spot of PEPR – can be positioned by a computer. From this first position, the spot can explore a small rectangular portion of the image, typically an area of 2 mm by ½ mm, within limits set by the computer (Fig. 7). From the hits achieved within this small area, the computer decides where to position and how to orient the next rectangle so as to follow the track. Both the positioning and the orienting are far more loose than those which the ideal PEPR was supposed to achieve. Within the small rectangle, however, the TV scan proceeds with its full accuracy. It can be said that Polly hardware has to be able to work both as a PEPR and as an HPD, but each of these two abilities is given a lighter task.

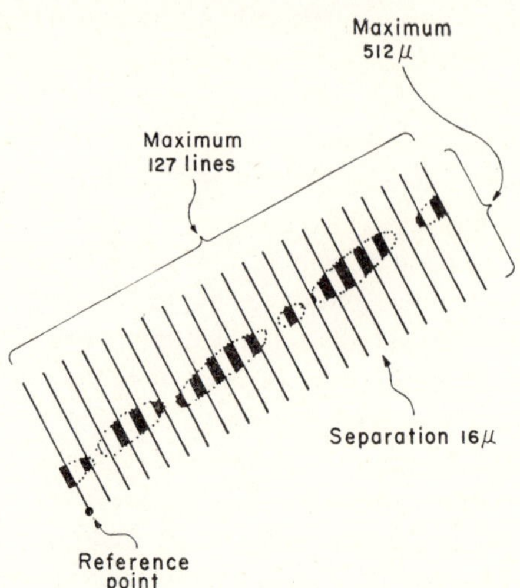

Fig. 7 Principle of the area scan by Polly

A human pre-scanner has to recognize whether and where (approximately) a given picture contains an interesting vertex. This information, recorded in advance for a whole roll of film, is given to the machine which proceeds to recognize and measure the tracks on its own, until it finds itself compelled to "yell for help" (see above). After the first few weeks of Polly operation it was found that help was needed in about 50% of all cases. With accumulated experience and improved programming, this proportion is expected to fall, and the overall rate of production (currently about 1 event per minute) to approach that offered at present by more extensively proven systems such as the Spiral Reader or the HPD.

A somewhat less ambitious system (the "Sweepnik") has been recently developed by a group led by O.R.Frisch[18], whom we have already mentioned as one of the earliest pioneers of mechanized picture processing. The spot is laser-generated and, as in PEPR, has the shape of a short (1 mm) line segment; its digitized positioning is performed by steering mirrors controlled by a small on-line computer and its sweeping ability is exercised over a small annular area (Fig. 8). Pre-metric human recognition plays an appreciably more important role in this system than in those described above; in this respect Sweepnik resembles more closely the S.M.P. or the most sophisticated versions of the Franckenstein. The operator has to recognize each interesting track and to position the spot closely enough to prepare a possible hit, as the spot sweeps the annulus. The track following, directed by the computer after the first hit, is easy and entirely automatic; it is hoped that, with improved programming, the machine will eventually be able to recognize all the tracks radiating from a vertex after an initial human positioning on only one of them. Post-metric human intervention is still needed in difficult cases, but owing to the strong pre-metric human help, their frequency may well be lower than in other post-guided systems. On the other hand, the reliance on pre-metric help reduces the production rate by a factor of 2 or 3 compared to the other, more fully automated systems.

In their present state, both Polly and Sweepnik work better with certain types of events than with others (the same is true of the Spiral Reader when it has to deal with curved tracks curling away from a radial direction). This may be a natural consequence of the tendency to design a system tailored to a certain set of a priori requirements, and requiring only a relatively small computer (both the Sweepnik and the prototype Polly were able to produce track information with only a PDP–7 on line). A more universal system such as HPD is, by the same criterion, inevitably over-capable and over-designed.

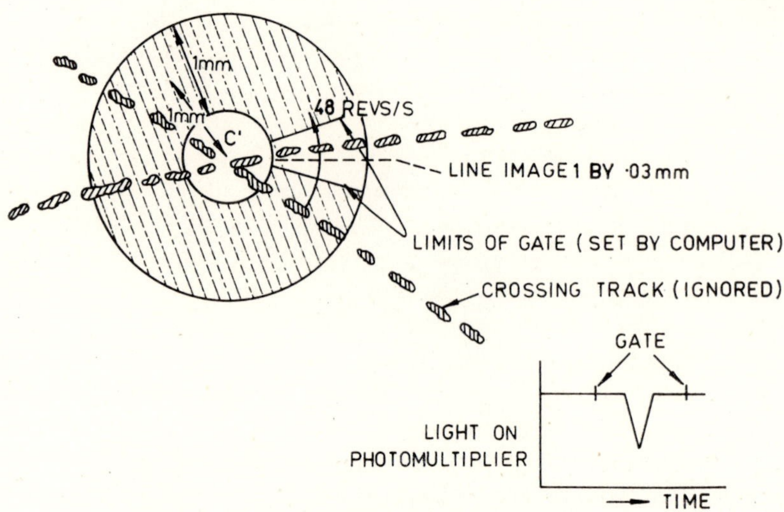

Fig. 8 Principle of the area exploration by Sweepnik

In its symbiotic mode (aimed at a combination of minimum guidance and post-guidance), it has been probably more difficult to master than some of its newer and more closely designed rivals, but its redundant capabilities (at present furthest developed by P. Hough at Brookhaven) and its bigger computer may yet become essential in some future context.

10 SPARK CHAMBERS: A SIMPLER PROBLEM

Spark chambers are made in a great variety of shapes and dimensions, and by using suitable photographing optics, views of several chambers taking part in the same experiment may be recorded side by side in the same film frame. The total number of sparks in a picture may vary by several factors of ten from one experiment to another, but at its highest, it remains far below the number of bubbles in a typical bubble picture. It is *a priori* obvious that a bubble image (containing millions of bits of recorded information) must be far more difficult to process than a spark image, whose useful content may be as low as a few hundred bits (in the simplest chambers with few spark gaps and no magnetic field) and hardly ever higher than 100,000 or so. Sparks occur only along a few parallel "gaps" whose precise location is known in advance; track-following from gap to gap is an easy task compared to the variety of bubble-track locations and orientations.

Because of these *a priori* regularities, automatic recognition of tracks out of an unsorted heap of spark coordinates is a relatively easy task. The sorting-out procedure can be either built into the way of sensing the film by a moving spot, or into the programs which begin to apply *after* all spark positions visible on the picture have been digitized and transmitted to the computer. The first approach was historically tempting to those physicists who were more familiar with electro-optical devices than with programming; the second came as a logical sequel to the adoption of the TV scan in bubble processing, as it operates, for example, in the HPD system.

Possibilities of full automation began to be explored in the early 1960's not only because they looked temptingly easy, but also because the need for a very speedy way of spark picture processing was felt with a great urgency. Spark chambers have a far shorter recovery time than bubble chambers, and are potentially capable of producing 100,000 pictures per day or more (depending on the mechanics of filming). An ambition of processing one picture per second was, therefore, not at all unnecessarily high, especially as—there being no time for any kind of human pre-scanning—every picture, whether containing a candidate event or not, had to be processed. Moreover, the proportion of empty pictures promised to be low, since the triggering of spark chambers can be made very selective and therefore the frequency of picture-taking can be kept close to that of the occurrence of interesting events.

The first successful automatic system[19] was in principle capable of approaching the event-per-second rate. A human supervisor had to intervene only when the operation would jam, or start producing obviously inept results. At this early stage of development, the system relied mainly on track-following by a computer-guided CRT spot and used a small computer (PDP–1), guided by rather simple programs. Accordingly, it could be used only in very simple spark-chamber configurations[20]. A somewhat more sophisticated approach, adopted in many laboratories, favored a spot-guiding hardware embodying a pre-set information on the position of spark gaps (Fig. 9). The spot then does not need any track-following capability and the computer can work with an even simpler set of programs. Human recognition—or rather precognition of the narrow lines on which sparks are likely to appear—intervenes in such a system only at the beginning of a long series of pictures and, after that, the processing can proceed at its full automated rate.

For greater variety and complexity of events, the principle of giving to the hardware only the simplest (TV-scanning) task, and of concentrating *all* recognition on its metric stage appears tempting even more than in the bubble processing, since genuine spark digitizings can be easily identified by simple

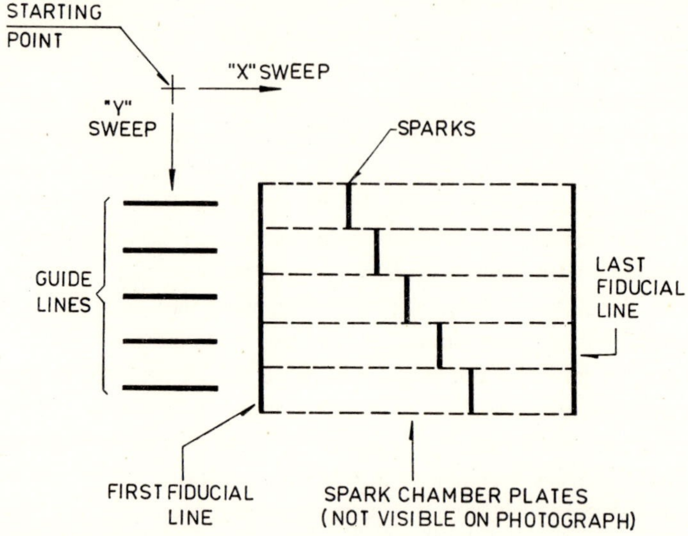

Fig. 9 Principle of the spot-guiding system for a simple spark chamber (CERN, 1963)

programmed criteria and filtering out of irrelevancies is relatively easy. CERN's early experience with HPD provided an encouragement to start in this direction[21]; soon spark pictures were processed at CERN by hundreds of thousands[22], first using the HPD hardware, and then a somewhat less precise, but easier to operate, CRT spot generator, named Luciole[23]. Other devices of the same kind have been constructed elsewhere; some of them offer a compromise between the stability of the pure TV scan and the flexibility offered by even a modest measure of spot guidance by computer. Pre-metric recognition (by human pre-scanning) and post-metric correction by a human supervisor are resorted to wherever the spark pictures tend to present unusually complex or diversified features, but on the whole the human part remains relatively unimportant and processing is fast even in fairly intricate cases (typical production rates for Luciole: 600–1000 pictures per hour, actually obtained at CERN with a time-shared computer).

11 FUTURE TRENDS AND DEVELOPMENTS

Fast-acting bubble chambers will soon create a demand for systems combining the present-day bubble-processing precision and complexity with the present-day spark-processing speed (and independence from human recognition). Fast spot generation will be offered by electronic devices (such as cathode-ray tubes) and possibly also by mechanical ones, but the former will probably be preferred because of their inherent flexibility and programmability, of which some use will be made (as it is now in Polly), in order to ease the formidable tasks of the unguided metric recognition by software. The qualities of precision and stability of non-mechanical spot generators have been already considerably improved since the late 1950's, when both the Spiral Reader and the HPD were first conceived; further development will no doubt solve this problem completely.

Photographic cameras and precision CRT's will be replaced by precision TV cameras (vidicons) as soon as possible, since the essential information contained in an optical view of a chamber will anyhow have to be transferred onto a digital storage medium, and the optical storage of film will become unnecessary. Vidicon viewing of spark chambers, requiring less precision, became possible as early as 1961[24]. Its development, both in America and in Europe, has not gone very far because of the growing competition of wire chambers. In this mode of detection the location of sparks is directly evaluated and recorded by a system of electromagnetic sensors; the optical and photographing stage is omitted altogether.

If, in the processing of data from optical spark chambers, it may be thought desirable to eliminate all human recognition in the interests of speed, the same elimination becomes necessary in wire-chamber practice, simply because in their whole chain of processing the data remain inaccessible to direct human observation. Human supervision, occasional checking of the physical meaning and occasional help in coping with difficult cases may be less urgently required in spark techniques than in dealing with bubble-chamber records, but this requirement is by no means negligible. Once they have passed through the computer, wire-chamber data can however be displayed on a CRT screen as easily as if they *had* previously passed through a film-recorded stage[25]. The recognition of visual data thus becomes useful even in cases when the original data are *not* visual at all!

New types of computer hardware may come into use in order to increase both the speed and the reliability of the automatic recognition. In particular, for the systems using a complete TV scan and relying on recognition by software, devices have been proposed with a purpose of predigestion and easing the difficulties of filtering; see, for example, Ref. 26 and 27. In two more ambitious projects additional hardware is intended to perform specific aspects of filtering, such as track construction in three dimensions (SATR,[28]) or reduction of points to linked track elements (BRUSH,[29]). A more radical innovation would consist in using a computer specially designed for recognition of visual patterns. The Illiac III processor[30], at present being developed by B. McCormick (the original inventor of the Spiral Reader), is intended to handle an array of 32×32 point-like informations in parallel, which would give it a topological competence far in excess of the present-day sequential methods.

New basic principles for developing recognition software may become necessary for the same reason which led the pioneers of the automatic translation of languages to take a new look into general linguistics. The "Picture Calculus" recently proposed by W. F. Miller and A. C. Shaw[31] intends to offer a new logical insight into the aims and methods of automatic recognition and a chance to put future developments on a new and more systematic foundation.

12 AN "ECONOMY OF ABUNDANCE"

Economists tell us that many of our mental habits have been inherited from the times of economic scarcity and may no longer apply to the present-day affluence of the industrially advanced countries. In the same way, Alvarez's and Frisch's original insistence on the data processing bottleneck as a limitation on the full use of track chambers, has shaped the physicists' approach to these methods of detection in a way which is no longer consistent with the present and future efficiency of data handling. Some changes in the way of designing a track-chamber experiment, and even some soul searching as to the very meaning of physical experimentation as a concept, are already under way. We have already mentioned the trend away from the search for rare events towards high statistics: the experimenter is no longer afraid of finding many candidate events in a given number of bubble pictures, nor of having to measure many candidates in order to retain one valid event.

In spark chamber techniques, the need for elaborate and rigid triggering is no longer as pressing as it used to be; with a looser system of triggering, the experiment becomes easier to set up, has fewer built-in biases and offers a greater chance of occasional unexpected finds. At the same time, the possibility of almost instant checking of the physical meaning of the collected data, as they emerge from the detector, enables the physicist to modify and adapt the course of his experiment before it has been completed or even consumed a substantial fraction of his allocated time at the accelerator.

New possibilities begin to be available also in the aftermath of the data collection in real time. The physicist is left in possession of a filmed or taped record as detailed and as abundant as he may wish. He may examine it at his leisure and in his usual academic surroundings rather than in the enforced vicinity of the accelerator[32]. If a new idea strikes him as to what really has been happening in his detecting chambers, he may take another look at his recorded data. In addition to these new kinds of "portable" and "retrospective" high-energy physics, the experimentation may yet acquire a new kind of space-independence through the use of digital data links in real time and at long distances.

As a further development of the trend towards looser triggering and richer variety of observed events, the setting of a detecting chamber in an accelerator-produced beam may become an exploring excursion into the world of high-energy particles similar to the sending of a dredge down to the floor of an ocean, or of an instrumental balloon up in the atmosphere. In its early days, particle physics had to rely on cosmic rays as the only available source of high-energy events; the new opportunities for observing cosmic radiation in extra-terrestrial conditions will no doubt make good use of the increasingly efficient techniques of data processing.

References

1. A.H. Rosenfeld, Testimony for Joint Committee on Atomic Energy (U.S. Congress), Hearings on High Energy Physics, March 1965.
2. "The Bubble Chamber Program at UCRL"—text of the original proposal submitted by L.W. Alvarez in April 1955; available at the Lawrence Radiation Laboratory, Berkeley.
3. See Ref. 1 and J.R. Pasta's testimony at the same Hearings.
4. Private communication received by the present writer from O.R. Frisch in December 1955; also: "Evaluation of Bubble Chamber Photographs", an unpublished report by O.R. Frisch, dated January 1956.
5. Y. Goldschmidt-Clermont, G. von Dardel, L. Kowarski, C. Peyrou, *Nuclear Instrum.* **2**, 146 (1958).

6. "The Cobweb Data Reduction System"—report presented by R. W. Birge at the International Conference on Advanced Data Processing for Bubble and Spark Chambers, Argonne National Laboratory, October 1968.
7. L. W. Alvarez *et al.*, UCRL Report 10109; for a more complete and up-to-date description, see R. I. Hulsizer, J. H. Munson and J. N. Snyder, *Methods in Computational Physics*, vol. 5 (1966).
8. Reported by L. W. Alvarez at the Informal Meeting on Track Data Processing, CERN 1962.
9 P. V. C. Hough and B. W. Powell, *Nuovo Cimento* **18**, 1184 (1960).
10. Proposal submitted by I. Pless and L. Rosenson in February 1961; first detailed description by I. Pless *et al.*: *Proceedings of the International Conference on High Energy Physics*, Dubna (1964).
11. See the reports presented by G. Moorhead *et al.* at many recent international conferences on data processing in high-energy physics.
12. See the reports presented by H. White *et al.* at the same conferences.
13. Princeton–Pennsylvania Accelerator Report PPAD 628 E (July 1967).
14. Reports from several laboratories presented at the Argonne Conference (cf. Ref. 6).
15. Private communication received by the present writer from D. Hodges in 1967.
16. "Polly I", by D. Hodges *et al.*, *Review of Scientific Instruments* **39**, 1556 (1968).
17. Reported by J. Butler at the CERN meeting (cf. Ref. 8).
18. O. R. Frisch, G. Street and D. Davies, report completed in May 1968 and available at the Cavendish Laboratory, Cambridge (England).
19. M. Deutsch, Proceedings of the Conference on Multiparameter Analyzers (Grossinger 1962), Report CU (PNPL)–227; see also Ref. 20.
20. As reported by M. Deutsch at the Conference on Programming for Flying Spot Devices (Bologna 1964), CERN Report 65–11.
21. L. Kowarski, "Guiding Lines for a Spark Picture Processing Project", internal CERN report, April 1962.
22. P. Blackall, B. W. Powell, P. Zanella—Bologna Conference (cf. Ref. 20); see also P. Blackall, G. Macleod, P. Zanella, *IEEE Trans. Nuc. Sci.*, vol NS–13 (June 1966).
23. H. Anders, T. Lingjaerde, D. Wiskott—Proceedings of the International Symposium on Nuclear Electronics, Paris 1963, published by IAEA, Vienna 1964.
24. H. Gelernter, *Nuovo Cimento* (1961).
25. See the report by P. Zanella *et al.* presented at the Argonne Conference (cf. Ref. 6).
26. O. P. Fedotov, internal CERN report, May 1967.
27. B. Equer *et al.*, presented by G. Reboul at the Argonne Conference (cf. Ref. 6).
28. See the reports presented by M. A. Thompson *et al.* at various recent conferences, in particular Argonne (cf. Ref. 6).
29. H. Billing, A. Rüdiger, R. Schilling, presented at the Argonne Conference (cf. Ref. 6).
30. B. McCormick and R. Narasimhan, *Nuc. Instrum. and Methods* **20**, 401 (1963).
31. Report SLAC–PUB–358 (October 1967).
32. See L. Kowarski, "An Observer's Account of User Relations in the U.S. Accelerator Laboratories", CERN Report 67–4 (1967).

For recent developments up to the Spring 1970, see the Proceedings of the Conference on Data Handling in High-Energy Physics, Cambridge 1970, CERN Report 1970.

CHAPTER 21

Molecular Structures

E. CLEMENTI

Abstract

In the introduction we briefly analyze why there is an intimate connection between computers and chemistry. In the second section we put forward the theoretical foundation of the electronic structure of molecules and we subdivide the progress towards the rigorous solution in steps; the first step is a crude one-electron approximation (LCAO), the second step is the accurate one-electron approximation in the field of all the electrons for the system (SCF–LCAO or Hartree–Fock technique), the third step is a solution to the many-electron problem (MC–SCF–LCAO). The third step is one of the most debated areas since no formulation at present has proven to be "operationally" acceptable. For this reason the problem is exposed and analyzed in a different way in the fifth section of the paper. In the remaining sections a simpler and intuitive interpretation of molecular wave functions and electronic energies is given together with our attempt to forecast some of the future progress in chemical computation for small, medium and large molecules.

1 INTRODUCTION

ONE WAY TO BRIEFLY comment on the relations between chemistry and computers is to attempt a succint definition of both.

Therefore, let us start by asking what is chemistry. Many and different answers can be given, with emphasis on the historical development of the fields or on their present subdivision following the academic curricula, or on their impact on the economy of the society as known today. I prefer to answer the question by using an implicit definition and stating some conditions to be satisfied.

Intellectually, chemistry is primarily concerned with understanding of matter, in its solid or liquid or gaseous phase, either as isolated system or as interacting systems, either on a macroscale or on a microscale, either predominantly classical in behavior or quantized in behavior, either so-called dead (not self-reproducing) or so-called alive (self-reproducing).

Technically, chemistry devises methods for shaping matter so as to have it respond in a controlled way to special conditions (like responses to temperature, pressure, electric or magnetic fields, or to other chemicals) and with desired qualities (plasticity, viscosity, purity, low cost, etc.).

Understanding and ability to first qualitatively predict and later quantitatively predict, as well as ability to produce in a planned mode, these are three connected phases in the spiral which can visually represent the evolution of the study and control of matter in chemistry.

Another overall division of techniques in achieving the above goals has emerged, namely theoretical methods and experimental methods. In the latter, I shall combine the entire spectrum of work done both by research scientists, experimentally oriented, and by production or plant engineers.

Let us now briefly define, if possible, what is an electronic computer. Very crudely, it was (relatively speaking) a very fast adding machine. However, to the ability of doing the four basic mathematical operations, from the very beginning there was the added possibility of performing a few logical operations, by branching a given path of sequential operations in response to "true" or "false" queries. Finally, the computer was capable of storing both "list of operations to be executed" and "data" on which operations are performed. The storage can be in fast or slow devices like core, memory, drums, disk files, tapes, etc.

Within a short time, from the appearance of the computers, libraries of utility routines were made available, whereby the computer user was given the option of avoiding being concerned with some of the inner detail of the computers logical structure, or with the detailed performances of simple and specific mathematical tasks. Within a short time the user was in a position to choose between a variety of languages in order to communicate with the computer.

This way, we reached today's situation where the hardware is only one part of the "computer", where we talk of "computer systems", namely a system which provides us the possibility of solving a variety of problems for a variety of tasks, accomplished either directly by the hardware or by softwares, or a combination of both.

The main evolution in computers has been in two directions, one is increase hardware performance (and concommitantly decrease of the cost per operation); the second is progressively cementing hardware and softwares in a single "system". These two directions have been developed in a complementary mode and today's computer systems are characterized by their extreme versatility in performing a large number of generic tasks.

Meanwhile, revolutions upon revolutions in hardware technology have occured, making circuits much faster and cheaper, smaller and more reliable. The reliability is critical, for now more circuits can be put in the same box to increase performance.

Ten years ago the 704 was certainly a powerful machine and the 7090 had not come out yet. Now the 360/91 is available with operation speed about 500 times that of the 704, and 100 times the 7090. In the next 10 years we can expect another 100 fold increase in CPU performance.

It follows as no surprise that computers, being such a general purpose instrument, have become more and more familiar tools for a chemical laboratory.

One type of use takes primarily advantage of the high speed in performing arithmetic operations and branching. This is the field of simulation of the basic quantum mechanical laws in many electron systems, be it an atomic or a molecular system. The binding of atoms in forming a molecule, the bond eenergy, the vibrational and rotational frequencies of a molecule, the energy required for transitions from a given electronic state to another, the value and sign of a dipole moment, of quadrupole moment, of polarizability, etc., are exactly determinable quantum mechanical quantities. Therefore, at considerable expenses in developing appropriate theory, algorithm and computers program, such quantites are now within reach of accurate computations without need of any experimental apparatus but the computers. We are reaching the time when taste, cost and availability of proper equipment will dictate whether, for example, a given spectroscopical property should be either obtained by standard experimental techniques or on the computer via simulation of the quantum mechanical laws. We shall refer to this area as "Large Scale Computation in Chemistry". Since the goal is to simulate chemistry from first principle, this area has been often referred to as "ab initio" computations.

Another mode of computer usage in chemistry and chemical engineering looks at the computer as an interface between the scientist and an experimental apparatus or a production plant. This has been referred to as laboratory automation, or process control. In the large Scale Computation mode, the computer is the experimental apparatus or the production plant in itself, insofar as it simulates the entire phenomena. In laboratory automation the computer is used for data analysis, for data reduction and for controlling a given environment via the computer ability of giving appropriate response to a given situation (interrupt facilities) with high reliability and speed. Computer usage in analytical chemistry for atomic emission or absorptions, for data analysis of ultra centrifuges, for data analysis of x-ray diffraction for gas chromotography, for mass spectrometry, for distillation plants, for control in a cement factory or in a steel mill, are examples of laboratory automation.

This list of examples here is given only as an indication, since clearly it can be expanded locically to all fields where quantitative data are recorded and analyzed as intermediate part of a process whether in research or in industrial applications. Whenever and wherever one needs a chart recorder, or a manometer or a pyrometer, or any measuring equipment in a chemical laboratory or a chemical plant, there you collect quantitative data, which are either end-results of a given task or intermediate-results, needed for reaching a decision for a subsequent task. In both instances one asks himself whether the task could be done either more efficiently or more economically or more reliably by using computers (and computer programs).

Large scale computations and laboratory automation are clearly the two extremes with a full set of intermediate possibilities. Likely, in time, the greatest impact of computers in helping chemistry and chemical engineering will be exactly in merging the two extremes. Full simulation of a process by computers will be accomplished by computer control in the real process, with feedback between the two phases.

Perhaps one of the most dramatic effects of computers on the future of chemistry (and on the future of our society, in general) is that the computer systems could bridge the gap between ever-increasing need of generalization in science and technology demanded by our society and the always present drive to specialization and depth which characterize us more and more as individual scientists. Specifically, the specialized scientist and engineer of tomorrow should be in a position to readily and quickly move into fields and techniques outside his own specialization by simply calling the appropriate program or programming system. By so doing he could be free from a large amount of tedious and narrow scope tasks and he could be more avai lable in translating intellectually and technically the goals of chemistry and chemical engineering into reality.

It is not science fiction but near reality, the situation whereby a chemist will use large scale chemical computer-programs in chemical synthesis, to determine thermodynamic properties of a compound; then he will use the computed data to expedite and optimize the experimental synthetic process and extract with computarized techniques these data which will allow him to set up a production line. Simulation, prediction, production: the cycle of the modern chemist.

One more step after this is to let the computer help in finding out which compound will have pre-required physical or chemical properties, then let the computer synthetize the compound and follow it by the previously described cycle.

There are obstacles to this goal, however, and these are mainly associated with human inertia, with learning limitation, both in amount, depth, and speed. On one side, the computer industry has provided us with enormously capable instruments; but it could have done more. On the other side, the scientist and engineer is not sufficiently trained by our institutions of learning on what he could produce if he would have a better training in computer use, and a broader knowledge of the progress in simulating physical phenomena.

2 THEORETICAL FOUNDATIONS

Whereas it is pointless to remind ourselves that molecules are built up of atoms, still there are divergent opinions among us chemists concerning the extent of the usefulness in formulating theory whereby a molecule is assumed from the start to be built from a set of atoms or ions (of course somewhat deformed in their electronic structure).

Pauling,[1] in the early thirties (and later, his school) did very much with the basic assumption that it is worthwhile to retain the electronic structure of the atoms as the starting point in the electronic structure of molecules. Pauling's theory is referred to as the valence bond approximation. An x-ray analysis of a chemical compound seems to be the clearest proof that indeed this is the best conceptual approach.

On the opposide side, Mulliken[2] and his school, immediately reject the necessity of the above starting point by introducing from the onset the concept of "Molecular Orbitals", i.e. of electrons which share the space around not a single nucleus, but as many as energetically convenient.

Attempts to put together the two points of view have been made frequently. For example, one can go back to Moffitt's "Atoms in Molecules" technique.[3]

Others attempted to define the range in which one approach is better than the other. Here we refer, for example, to Kotani's analysis.[4]

Clearly, an appropriate extension of Pauling's viewpoint (not existing today) and presently available extensions of the "Molecular Orbital" approximation should give identical results. We refer to Coulson's book[5] on this subject (but we must point out that Coulson's work was written prior to much work in the field of theoretical chemistry in the last decade and, therefore, does not represent an up-to-date analysis).

In order not to give an incorrect representation of the "Molecular Orbital" approximation we hurry to state that this model does not claim that one should disregard completely the parent atoms in a given molecule, but rather that you do not have to do so, that the molecular orbitals will go naturally to atomic orbitals as the molecule dissociates. The main point is that most of the atomic structure of the separated atoms is so drastically changed by the time we are at the equilibrium configuration geometry, that it is more of a hindrance than a help to consider the separated atoms and their concomittant deformation when we study a molecule.

The concept of heavy distortion in the electronic cloud when one proceeds from the separated atoms to the molecule is very well within chemical tradition. Indeed, much before quantum theory chemists assumed (and correctly) that, for example, the carbon atom has several ways to bind corresponding to different arrangements of the surrounding atoms. Thus, the concept of trigonal and tetragonal hybridization. Today we would rather speak of "valence states", but this is no more than remembering that in a given external field the valence electrons (in particular), and the inner shell electrons (to a smaller degree) of a particular atom experience, not only a central field of force alone, but the external field as well.

It is important to note that an electronic theory of molecules should in principle and in practice be applicable to the limiting case of a single atom. Electrons do not change nature from atoms to molecules and the same should hold for any model which describes the electronic structure either of atoms or molecules.

We shall not elaborate further on the above differences between the two starting points. We simply wish to remind the reader of some historical evolution in theoretical chemistry. Nor shall we claim that one technique is correct and the other is incorrect, since, in principle, one can have the same results from both. Pragmatically, we shall devote our attention to the Molecular Orbital approach and its improved assumptions, since via such technique we can today perform computations of increasing sophistication to the point that we can quantitatively predict some of the chemistry and understand some more of it.

"Esthetical" considerations, namely economy in the number of basic assumptions very much influence my choice and, in addition, it is frankly admitted to the reader that I have a biased preference toward the Molecular Orbital point of view.

Rather dogmatically, and in part for sake of levity, I start by stating that the Schroedinger equation, namely today's foundation of quantum mechanics, is good enough for chemistry. I also state, but in a somewhat more guarded way, that today's understanding of relativistic corrections is good enough for chemistry. From this it follows that it is worthwhile to attempt a solution of the Schroedinger equation. As known, this is the problem; a problem not too much of principle, but very much of practice, of substantial numerical difficulty for the parts and of extreme numerical difficulties for the whole. Therefore, let us make simplifying assumptions in a step-like fashion and let us first solve the larger part of the problem and then let us go to smaller and smaller details. The good sense and logic of the approach seems to be commendable. However, as we shall see later, this approach did require a large amount of work, indeed several decades of hard work, to reach a few years ago a stage with which the chemist was most dissatisfied since, in general, no quantitative agreement with experimental data was obtained neglecting very few exceptions. Therefore, since the chemist is an eminently empirical scientist, continuous effort took place toward finding short-cuts, regardless of the basic theoretical validity of such attempts. Lately, another factor appeared, namely the electronic computer (and the cost of computations!). Here again the short-cuts were and are intriduced on the ground that theory "must be cheap" to be good.[6] In the last three to four years, with the help of computers, the stage was finally reached whereby quantitative agreement with experiments is obtained using exact theoretical formalisms.

Other techniques have been proposed in addition to the valence bond and the molecular orbital approximation. Of special interest are those computational techniques, which can be applied to molecular systems with few electrons. For example, Kolos and coworkers[7] have obtained very accurate wavefunctions for a number of electronic states in the H_2 molecule at many internuclear distances. Kolos' computer program can definitely not only supplement, but compete with, most sophisticated spectrographs. The resolution of his program is within fractions of wave number and the use of the program presents so little difficulty as to be correctly handled by any "technician". The results of Kolos and coworkers provide a concrete, although partial example of what one wishes to obtain from theoretical chemistry. However, most molecules have more than 2 electrons and exact wavefunctions are presently not easily obtainable. Nevertheless, a number of important steps have been made and there is good reason for optimism in the future.

A pleasant characteristic of the molecular orbital theory is that each progressive improvement, or step, has a natural physical explanation. Rather arbitrarily we shall present the molecular orbital theory as a five step evolution.

I shall outline some of the steps which have been taken, their accomplishments and limitations.

The first step is the LCAO–MO approximation. There are actually two approximations in the above step: the first is the MO approximation, the second is the LCAO approximation of an MO. As known, the short notation LCAO–MO stands for "linear combination of atomic orbitals—molecular orbitals".

The MO is a one-electron function which is factored into a spatial component and a spin component. The expression "one-electron function" means that only the coordinates of one electron are explicitly used in a given MO. This factorization into spatial and spin components is permissible, since generally one uses a hamiltonian which does not explicitly contain spin dependent terms. The MO's are the exact analog of the atomic orbitals, which describe the electrons in an atom to a first approximation. Indeed, one can read several chapters of the classical works of Condon and Shortley,[8] replace the word "AO" with the word "MO" and one will read a book on molecular physics instead of atomic physics.

This situation has some important consequences: namely, a large amount of testing and development for molecular wavefunction techniques can be done with atoms. For this reason atomic and molecular examples are freely mixed in the following sections of this review.

If the molecule contains $2n$ electrons (let us consider a closed shell case for simplicity), the MO approximation will distribute the electrons in $2n$ molecular orbitals $\varphi_1, \varphi_2, \ldots, \varphi_{2n}$. Since there are two possible spin orientations (α and β spins) a space distribution function has either spin α or β and, therefore, the $2n$ electron system is described by n spacefunctions and $2n$ spin-orbitals. Thus φ_1 and φ_2 will have the same space distribution (will depend on the coordinates of one electron alone), but, in accordance with the Pauli exclusion principle, will have different spin functions. It is stressed that the one-electron model is justified only because it simplifies the treatment. Indeed, in the very beginning of quantum theory, Hylleraas introduced a wavefunction for the He atom in which one orbital is described in terms of the coordinates of *both* electrons.

The total wavefunction Ψ of the $2n$ electron system is then

$$\Psi = \frac{1}{\sqrt{(2n)!}} \begin{vmatrix} \varphi_1(1) & \cdots \varphi_1(2n) \\ \varphi_2(1) & \cdots \varphi_2(2n) \\ \vdots & \vdots \\ \varphi_{2n}(1) & \cdots \varphi_{2n}(2n) \end{vmatrix} \tag{1}$$

or $\Psi = ((2n)!)^{-1/2}\,\{\phi_1^{(1)}, \phi_2^{(1)}, \ldots, \phi_{2n}^{(2n)}\}$ where the $\{\ \}$ notation indicates the determinant, explicitly given in (1). This determinant wavefunction guarantees that any interchange of two electrons (i and j) brings about a sign change in the wavefunction. This is the Pauli constraint for fermions. The energy for such a system is given by the relation

$$E = \langle \Psi^* \,|H|\, \Psi \rangle$$

where the hamiltonian H is

$$H = -\sum_{ia} \frac{1}{2} \Delta_i^2 - \sum_{ia} \frac{Z_a}{r_i} + \sum_{ij} \frac{1}{r_{ij}} - \sum_{ab} \frac{Z_a Z_b}{R_{ab}}.$$

The first term is the kinetic operator for the ith electron, the second term is the potential operator between the ith electron and the ath nucleus (with charge Z_a), the third term is the electron–electron potential between the ith and the jth electrons at a distance $r_{ij} = r_i - r_j$, and finally the last term is the nucleus–nucleus potential with R_{ab} the distance between the ath and bth nucleus of respective charges Z_a and Z_b.

The first and second terms are subsequently referred to as the one-electron hamiltonian and will be indicated as h_0. The total energy for such a determinant was given by J. C. Slater, and it is

$$E = 2\sum_i h_i + \sum_{ij} (2J_{ij} - K_{ij}) + E_{NN}$$

where

$$h_i = \langle \varphi_i^* | h_0 | \varphi_i \rangle$$

$$J_{ij} = \langle \varphi_i(1)^* \varphi_j(2)^* | r_{12}^{-1} | \varphi_i(1) \varphi_j(2) \rangle$$

$$K_{ij} = \langle \varphi_i(1)^* \varphi_j(2)^* | r_{12}^{-1} | \varphi_i(2) \varphi_j(1) \rangle$$

$$E_{NN} = \sum_{ab} (Z_a Z_b / R_{ab}).$$

As known, J and K are usually referred to as coulomb and exchange terms, respectively.

What form should the MO have? Clearly, the molecular orbitals are subjected to symmetry constraints (as in the case of atomic orbitals) and any molecular orbital will transform as an irreducible representation of the molecular symmetry group. This statement, however, is not a sufficient one; indeed it tells us mainly how the molecular orbital should *not* be. In principle we could insist on the anology between atomic one-electron functions and molecular one-electron functions and "tabulate" the MO in a way analogous to the method of Hartree and Fock in the 1930's. This would ensure that we have the best possible molecular orbitals. It is noted that numerical Hartree–Fock functions for diatomic molecules are a somewhat tempting possibility; this, however, has not seriously been explored at the present.

Nevertheless, chemistry is concerned with more than only diatomic molecules. An answer is provided by the LCAO approximation, in which the MO's are built up as linear combinations of atomic functions. We refer to R. S. Mulliken's classical series of papers for the early development and application of the LCAO MO approximation.[2]

The *second step* in the evolution of quantum theory is the introduction of self-consistency. Again, the physical model is provided by atomic physics, namely by the Hartree–Fock model. The LCAO approximation to the MO requires the best possible linear combination: this is what one intends for self-consistency. A good review paper on this subject is the one by C. C. J. Roothaan.[9] There the self-consistent field technique in the LCAO MO approximation (SCF LCAO MO) is systematically exposed for the closed shell case.

Up to now we are strictly in the one-electron approximation. The electrons interact among themselves only via the average field and the MO has no explicit electron–electron parameter. Fortunately, the Pauli principle keeps electrons with parallel spin (in different MO's) away from each other, but it has nothing to offer to electrons with anti-parallel spin in the same MO. The full catastrophe might be appreciated by recalling that in the SCF LCAO MO approximation, two fluorine atoms are incapable of giving molecular bonding when brought together; i.e., the SCF LCAO MO does not recognize the existence of the F_2 molecule.[10] Of course, it does not require a computation of F_2 to realize this point. For example, when Roothaan's work appeared (1950), another less familiar paper was written by Fock[11] to a large degree solving the problem and introducing the concept of two-electron molecular functions or "geminals", as they are called today. At the same time Lennard–Jones and collaborators[12] put forward a classical series of papers in which part of the correlation problems was tentatively solved, but at the expense of drastic orthogonality restrictions. For a variety of reasons, neither of the two avenues was numerically explored and in the meantime a third possibility slowly emerged (1) *the third step*r.

Hylleraas,[13] and later Boys,[14] proposed the possibility of using not only one determinant, but as many as needed. This technique is known as the configuration interaction or superposition of configuration technique; since the first designation is more common, I shall adopt it hereafter (C.I. for short).

If we denote a finite set of determinants by $\Psi_0, \Psi_1, \ldots$ etc., the resulting C.I. wave function is

$$\Psi = a_0\Psi_0 + a_1\Psi_1 + a_2\Psi_2 + \cdots$$

By optimizing the orbitals in each function and by variationally selecting the C.I. coefficients $a_0, a_1, a_2, \ldots$ we shall have a solution necessarily as good as, or likely better than, Ψ_0, and if the series of the above equations is sufficiently long, then we shall reach an exact solution. The only trouble is that the necessary series is *too* long. The slow convergence of the series is due to the fact that in most cases one insists on using a $1s$ orthogonal to the $2s$ and to the $3s$, a $2p$ orthogonal to the $3p$, etc. with $2p$, $3p$, $3s$ functions overlapping very little the $1s$ and $2s$ functions. If the added functions overlap very little, they will interact very little and correlate equivalently.

However, let us assume that when we construct Ψ_0 we construct Ψ_1, Ψ_2, etc. *at the same time*, and we do not insist on the best possible Ψ_0, but on the best possible Ψ; then the variational principle, used simultaneously on both the a's (the C.I. coefficients) and the ϕ's (the atomic orbitals), will ensure that the Ψ_i overlap as much as possible. This is accomplished in *the multiconfiguration SCF LCAO MO* technique (MC SCF LCAO MO).

Let us now stress somewhat more the details of the Multiconfiguration SCF approximation. The MC SCF theory seems to have been first proposed by Frenkel[15] (1934) and by D.R.Hartree, W.Hartree and Swirles[16] (1939), and Yutsis[17] (1952). Recently, it has been re-analyzed and applied by Yutsis, Vizbaraite, Strotskite and Baudzaitis[18] (1962), Veillard[19] (1966), Veillard and Clementi[20] (1966), and Clementi[21], and Das and Wahl[22] (1966).

We shall first consider the simpler case of two configurations and[19] expand it later to many configurations.[20]

Let us consider a configuration of the type $1s^2\ 2s^2\ 2p^n$ (called configuration A) and a configuration of the type $1s^2\ 2s^0\ 2p^{n+2}$ (called configuration B). States of like symmetry from A and B will interact and the resultant function

$$\Psi + A\Psi_A + B\Psi_B \tag{3}$$

will be The SCF theory can be used in solving first Ψ_A and then Ψ_B, and a secular equation can be solved for $\Psi = A\Psi_A + B\Psi_B$. This is standard configuration interaction. However, the problem can be solved in one step, i.e., an optimal Ψ_A and Ψ_B can be found so that when the two interact, an optimal Ψ is given. In other words, for a given basis set in Ψ_A and Ψ_B an optimal two-determinant combination can be obtained. The standard SCF guarantees an optimal Ψ_A or an optimal Ψ_B, the MC SCF guarantees an optimal Ψ, but not an optimal Ψ_A or Ψ_B.

The correlation energy is defined as the difference between the experimental energy and the sum of the Hartree–Fock and relativistic energies. In the following table we shall give the correlation energy ($E_{\text{Corr.}}$) for the three atoms and the experimental ($E_{\text{Exp.}}$), the Hartree–Fock (E_{HF}) and the relativistic ($E_{\text{Rel.}}$) energies:

	$E_{\text{Exp.}}$		E_{HF}		$E_{\text{Rel.}}$		$E_{\text{Corr.}}$		
Be (^{1}S)	−14.6685	−	(−14.5721	−	0.0022)	=	−0.0942 a.u.	=	−2.563 eV
B (^{2}P)	−24.6580	−	(−24.5278	−	0.0061)	=	−0.1241 a.u.	=	−3.378 eV
C (^{3}P)	−37.8557	−	(−37.6869	−	0.0138)	=	−0.1550 a.u.	=	−4.217 eV

Introduction of the second configuration lowers the correlation energy error by 0.0424, 0.0311, 0.0173 a.u., respectively. The reamining error is partly due to the $1s^2$ electrons and this is about 0.0443, 0.0447, 0.0451 a.u., respectively (these values are taken from the two-electron isoelectronic series). In the following table we shall give the percent error of the Hartree–Fock energy, the Hartree–Fock plus relativistic correction, of the two-configuration SCF calculation plus relativistic correction, and of the two-configuration calculation plus relativistic correction and the $1s^2$ correlation energy contribution.

Table I Energy contributions to the total energy

	HF	HF + R	MC SCF + R	MC SCF + R + $E_{\text{Corr.}}(1s)$	Remainder
Be (^{1}S)	99.3428%	99.3578%	99.6474%	99.9495%	0.0505%
B (^{2}P)	99.4720%	99.4967%	99.6228%	99.8041%	0.1959%
C (^{3}P)	99.5541%	99.5906%	99.6362%	99.7554%	0.2446%

We shall now extend the MC SCF theory to the case of n configurations for a closed shell ground state.

We assume that the $2n$ electrons of a given closed shell system are distributed in n double occupied orbitals $\phi_1 \cdots \phi_n$ and we shall refer to this set as the "(n)" set. A second set of orbitals $\phi_{(n+1)} \cdots \phi$ is used and this will be referred to as the "($\omega - n$)" set. We consider *all* the possible excitations from the (n) set to the ($\omega - n$) set, i.e., we consider n ($\omega - n$) configurations. A given excitation from the (n) set to the ($\omega - n$) set will be indicated as $t \to u$ where t is a number from 1 to n and u is a number from $n + 1$ to ω.

We shall designate as the Complete Multiconfiguration-Self-Consistent Field (CMC–SCF) technique the one where a given orbital of the (n) set is excited to all orbitals of the ($\omega - n$) set; if an orbital of the (n) set is excited to one or more, *but not all* orbitals of the ($\omega - n$) set, then we shall describe the technique as *Incomplete MC SCF* (IMC SCF).

The wavefunction of the system in the CMC SCF approximation is

$$\Psi = a_{00}\psi_{00} + \sum_{t=1}^{n} \sum_{u=1}^{\omega - n} a_{tu}\psi_{tu}. \tag{4}$$

We shall use t or t' as indices for the (n) set and u or u' as indices for the ($\omega - n$) set.

Let us introduce a coefficient A_t representing the "fraction of an electron" which is excited from the ϕ_t orbital of the (n) set to the ϕ_u orbitals of the entire ($\omega - n$) set. The coefficient B_u represents the "fraction of an electron" in the ϕ_u orbital of the ($\omega - n$) set as a result of the excitation from the entire (n) set. It is, therefore, tempting to re-examine the configuration structure of a $2n$ electron system. The standard electronic configuration for the

$2n$ electrons is a product of n orbitals. For example ψ_{00} has configuration

$$\phi_1^2\phi_2^2 \cdots \phi_n^2. \tag{5}$$

Let us call such a configuration a "zero-order electronic configuration". The MC SCF LCAO MO function will be a set of $(\omega n - n^2)$ zero-order configurations with appropriate coefficients, a_{tu}. It is rather difficult to visualize in a simple way the effect of such a rather long expansion. However, we can make use of the Al and B_u coefficients and write the following configuration

$$\underbrace{\varphi_1^{2(1-A_1)}\,\varphi_2^{2(1-A_2)} \cdots \varphi_n^{2(1-A_n)}}_{(n)\ \text{set}}\;\underbrace{\varphi_{n+1}^{2B_1}\,\varphi_{n+2}^{2B_2} \cdots \varphi_\omega^{2B(\omega-n)}}_{(\omega\ -\ n)\ \text{set}} \tag{6}$$

which we shall refer to as the "complete electronic configuration". The set of (n) orbitals has a fractional occupation equal to $(1 - A_t)$ for the orbital ϕ_t, whereas the remaining orbitals [the ϕ_u's of the $(\omega - n)$ set] will have, in general, relatively small fractional occupation values, B_u. Clearly, the sum of the fractions of electrons annihilated from the (n) set is equal to the sum of the fractions created in the $(-n)$ set, since $\sum_t Al = \sum_u B_u$.

Whether the MC–SCF–LCAO technique is the final solution to *step three*, is it not established at present. There are many alternatives to it. We shall deal with one of the most useful alternatives in the next section (Section IV) of this paper.

A *fourth step* in the molecular orbital theory is the inclusion of relativistic effects. There is little work done in this area at present (and this is not only true for molecular functions, but for atomic functions as well). Recent advances in metallo-organic chemistry, with heavy metals as constituents, demand a relativistic interpretation of the electronic structure. Even in molecules containing low Z atoms, the importance of spin-orbit effects in transition intensities is demanding more studies and computations in this area. It is gratifying to note that a simple perturbation treatment on atoms (Clementi and Hartman,[23] Clementi[24]) gives energies as good as a full relativistic Hartree–Fock treatment.[25]

Finally, the "electronic structure" of molecules should always be considered a limiting case of the vibronic structure of molecules. Real molecules vibrate (and rotate and translate, too) and, therefore, the question of how much we can rely on the Born–Oppenheimer approximation should not be ignored. This is a *fifth step* and I shall refer to the work of Kolos and coworkers for more details.[26]

3 INTERPRETATION OF THE CHEMICAL BINDING

The computed wave function can be analyzed indirectly via a study of the physical properties (examples are dipole, quadrupole moment, field gradient, polarizability, vibrational or rotational frequencies, etc.).

Mulliken[27] proposed a simple formal way to analyze the wave functions and to subdivide the molecular charges into atomic components. Mulliken's technique, known as "electron's population analysis" is arbitrary in two ways: first, his specific algorithm for breakdown is one among many possible alternatives (however very intuitive and simple) and secondly, *any* subdivision of a molecular field into atomic components is arbitrary by definition. To put it simply, the atomic densities are not quantum mechanical observables in a molecule. However, the advantages for modeling and for thinking derived by the population analysis far exceed the somewhat "impure" foundations of the analysis.

But the computation not only provides us with a wave function, but also with an energy. This is a single number, which, however, contains much information. In order to extract this information we have proposed an energy breakdown by considering "atomic–atomic" interactions. Thus from Mulliken's "population analysis" we can have a simple analysis of the wave functions and from our work an equally simple (and "impure") analysis of the total energy.

Let us analyze, for example, the molecule guanine (this molecule is one of the four bases of the DNA molecule, which as is well known, contain the genetic code for reproduction). The electron's population analysis, condenses the rather complex expression of the molecular wave function in a few and easily understandable data. In Fig. 1 such results are given. At each atomic site a charge is given which is the computed difference of the electronic charge from the atom at infinite separation from guanine, to the same atom in the proper site in the guanine molecule. For example, an oxygen atom (8 electrons), has charge -0.36836 in guanine (position 0 of Fig. 1). This means that the oxygen atom has gained 0.36836 of an electron. This gain of electrons is made at the expense of other atoms. For example, the two hydrogen atoms designated as $H(1)$ and $H(2)$ in Fig. 1 have charge $+0.21306$ and $+0.38013$ or have lost 0.21306 and 0.38013 of an electron respectively. The electrons population analysis tell us in addition that the distribution of the electrons in the 0 atom is 1.998, 1.790; 3.269, 1.284 electrons for the 1*s*, 2*s*, 2*p* (planar), 2*p* (perpendicular) orbitals (see Table 2). Therefore, the oxygen atom does not have in the molecule 2 electrons in the 2*s* orbitals, but only 1.790 and that these are hybridized with 3.296 electrons of the 2*p*-planar type. It tells that the original 4 electrons of 2*p* type in the free oxygen atom are increased to 3.296 + 1.284 = 4.580. This type of information tells in a simple way what variation in electronic density occurs when a molecule is formed from atoms. It tells how electrons transfer from one atom to a different one (charge transfer) and how they are mixed on a given atom (hybridization). Finally, it should be noted that this information is expressed in a very simple language that any chemist can understand. The same information, of course, is available in the wave function (or in its square, i.e., in the electronic density distribution), but in such a complex way as to be of little use to the chemist not specialized in the theoretical chemistry field.

Let us briefly discuss the "bond energy analysis" as we have formulated it.[21]

We shall make use of a number of bond energy classifications which are

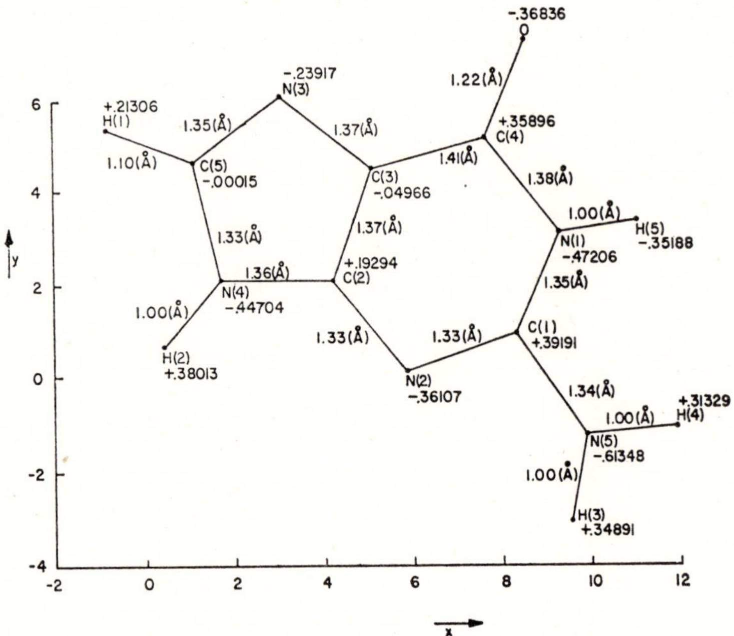

Fig. 1 Geometry and Gross Charges for the Guanine Molecule

Table 2 Hybridization: guanine

Atom	1s	2s	$2p_\sigma$	$2p_\pi$
0	1.998	1.790	3.296	1.284
N (1)	1.997	1.336	2.413	1.726
N (2)	1.997	1.440	2.599	1.324
N (3)	1.997	1.554	2.596	1.091
N (4)	1.997	1.352	2.474	1.624
N (5)	1.997	1.367	2.424	1.824
C (1)	1.999	0.917	1.811	0.881
C (2)	1.999	0.931	1.865	1.012
C (3)	1.999	0.934	1.905	1.211
C (4)	1.999	0.907	1.801	0.933
C (5)	1.999	1.011	1.901	1.089
H (1)	0.787		*	*
H (2)	0.620		*	*
H (3)	0.651		*	*
H (4)	0.687		*	*
H (5)	0.648		*	*

* Data presently not available due to choice of basis set.

derived at first from the usual SCF LCAO MO energy expression for the total energy (see equation (1).

$$E = 2\sum_i h_i + \sum_{ij} P_{ij} + E_{NN}. \tag{5}$$

The above energy expression can be written as

$$\begin{aligned} E = {} & 2\sum_i\sum_A h_{iA} + \sum_{AB} h_{iAB} + \sum_{ABC} h_{iABC} \\ & + \sum_i\sum_j\sum_A P_{ijA} + \sum_{AB} P_{ijAB} + \sum_{ABC} P_{ijABC} \\ & + \sum_{ABCD} P_{ijABCD} + \sum_{AB} E_{NA,NB} \end{aligned} \tag{6}$$

where A, B, C, D are indices running over the atoms and where h_{iA}, h_{iAB}, h_{iABC} are the one-, two-, and three-center components of h_i, $E_{NA,NB}$ is the two-center component of E_{NN}, and P_{ijA}, P_{ijAB}, P_{ijABC}, P_{ijABCD} are the one-, two-, three-, and four-center components of the electron–electron interaction energy.

We shall then denote as the zero-order energy diagram

$$E_0 = \sum_A E_A = 2\sum_A\sum_i h_{iA} + 2\sum_{ij} P_{ijA}, \tag{7}$$

as the first-order energy diagram ($A \neq B$)

$$E_1 = E_{AB} = \sum_{AB} \sum_{i} h_{iAB} - E_{NA,NB} - P_{ijAB}, \tag{8}$$

as the second-order diagram ($A \neq B \neq C$)

$$E_2 = E_{ABC} = \sum_{ABC} \sum_{i} h_{iABC} - \sum_{ij} P_{ijABC}, \tag{9}$$

and finally as the third-order diagram ($A \neq B \neq C \neq D$)

$$E_3 = \sum_{ABCD} E_{ABCD} = \sum_{ABCD} \sum_{ij} P_{ijABCD}. \tag{10}$$

The E_0 should be compared with the sum of the energy of the separated atoms. The correlation correction within E_0 can be estimated directly from atomic energy computation.

E_1, the first-order diagram energy, is the first "bond" energy, and links any two atoms in a molecule, two at a time. Clearly, the *classical chemical formulas are a representation of the first-order diagram*—an incomplete one, however, since in them only some of the nearest neighbor atoms are connected. So we usually do not write a bond between neighbor hydrogen atoms in benzene, although there is clearly an interaction between them. The first-order energy differs from the other bond energies in that it includes the nuclear–nuclear repulsion, clearly present only in E_1. This quantity is numerically quite significant; it is part of the potential energy and, therefore, in view of the virial theorem, one can expect that E_1 will be the dominant part of the binding energy. A large number of other evidences, both theoretical and experimental, are known to support this point. As is the case for E_0, E_1 is composed of a number of terms which will satisfy molecular symmetry consideration. In other words, the equivalent atoms in E_0 will have equal energy, so the equivalent bonds in E_1 will be equal in energy. The correlation energy correction associated with E_1 for the atoms A and B will clearly depend on the electronic density between atoms A and B. But here we should be careful in the use of atomic correlation energy computations, since no bond analogy can be uniquely drawn between the electrons in an atom and those in a molecule (see however Wigner's work for a possible analogy).

E_2, the second-order diagram energy, is the three-atom interaction less the direct atom–atom pair interaction. This term does not contain a nuclear–nuclear repulsion term, and the one-electron energy (kinetic and potential) is relatively small.

E_3, the third-order diagram energy, is the only term which does not include one-electron terms; therefore, it is totally a potential energy (coulomb and exchange) term. This term includes, by definition, mainly long range interaction. E_3 as well as E_2 are in effect neglected in many semiquantitative computations (like the Pariser–Parr approximation).

The population analysis, as previously described, has an energy analog in E_0 and E_1, not in E_2 and E_3. Classical chemistry formulas have a partial analog in E_0 and E_1.

The present breakdown of the total energy should represent a natural frame for transferability of bond energy for which there is a large body of thermodynamic evidence within families of compounds. It provides a framework which will not change by introducing more and more configurations in the MC SCF LCAO MO formalism. It seems to offer advantages to ab-initio vibrational analysis and to offer easy interpretation of the overall constancy of group frequency in different molecules. Finally, it maintains the basic ideas of the molecular orbital theory but it associates it intimately with our basic intuitive approach in chemistry, i.e., that there are atoms in molecules.

This breakdown does not follow traditional ideas on the number of bonds one would like to see associated with an atom. So we could have in a molecule of N atoms a hydrogen with $(N-1)$ first-order bonds. However, this number will be reduced (a) by consideration of the nearest neighbors and (b) by quantitative consideration of the energy associated with each bond.

It is noted that there are at most four atomic bonds in our analysis: This is an effect of having chosen a basis set centered at the atoms. Alternatively, one could use localized orbitals and this would alter the number of atoms involved in bonds.

In the following we give as an example the bond energy analysis for CH_2F_2. The zero order energies are -0.2202 a.u. for the hydrogen, -99.3880 a.u. for the fluorine, -36.06633 for the carbon atom. The two center energies are -0.76538 a.u., -1.11522 a.u., 0.10199 a.u., 0.075929 a.u., 0.01029 a.u. for C—F, H—F, H—H, F—F, and H—F respectively. This allows us to say that there is a bond (in the traditional sense) between C and H and between C—H. The other bonds (H—H, F—H and H—F) are repulsive in nature. The three center energy are 0.18716 a.u., 0.15937 a.u., 0.18382 a.u., 0.00709 a.u. and 0.01343 a.u. for CFF, HCF, HCH, HFF, HFF respectively. Since these are all positive quantites, they *oppose* molecular formation. These can be interpreted as two dimensional bonds (whereas the two center are one dimensional). Finally, the four center contribution are -0.03496 a.u., -0.03949 a.u. and $+0.00507$ a.u. for HFFC, HHFC and HHFF respectively. The first two are contributing to molecular binding, but not the latter. It is noted that the four centers (three dimensional bonding) are re-enforcing the two center effects. How much this is a general result or not is not known, since the above values are the first bond energy in literature. It is noted that this energy breakdown tells us also about bond energy additivity. For example, to assume that the FCF group has the energy of the individual one centers plus twice the energy of C—F and once the energy of F—F would be in error by the three center diagram FCF (0.18716 a.u.).

The above discussion brings about the following conclusions:

(1) a chemical bond is an arbitrary concept and can be defined in as many ways as one wishes; (2) of the many possible representations, some are more useful than others, and (3) the "chemical bond" concept is to a certain extent simply a bookkeeping device, however, of great practical importance.

One might prefer to have an exact correspondence between electron population analysis and bond energy analysis. However, the starting point of the electron population is $(\varphi_{\lambda i})^2$ and this can lead only to a subdivision involving one or two centers, whereas the bond energy analysis, in view of our definitions, involves one, two, three and four centers.

One of the expectations in proposing the bond energy diagrams partitioning of E is that, by a systematic comparison of a number of molecules, a simple correlation will emerge which will put "transferability of bond energies" on a sound basis. It is noted that in the early attempts at studying molecular kinetics, the approach of using Morse-type potentials between any pair of atoms was often adopted. The present analysis in some respects does exactly that, if one considers only zero and first-order diagrams. In addition it does much more. Therefore, it is expected that this type of analysis will be of help in the formulation of a theory for reaction mechanism.

Other problems, like vibrational analysis, the study of the barrier to internal rotation, and charge transfer can be explained quite naturally in this framework of analysis. We shall return later on to these points.

4 PRESENT STATUS OF COMPUTATIONAL CHEMISTRY AND EXTRAPOLATION TO ITS NEAR FUTURE

This section attempts to give a tentative analysis of the present status in molecular computations. This work does not intend in any ay to be either a complete review nor a limited review, but rather a study and interpolation on past-present work so as to be in position to extrapolate the (near) future of some direction for quantum mechanical computations of molecular wave functions.

Somewhat arbitrarily I shall divide molecules by their size as follows: (1) small molecules, (2) medium size molecules, (3) large molecules without translational symmetry and (4) large molecules with translational symmetry. For "size" I mean the number of electrons and indirectly the number of atoms. In addition, the "size" will be proportional to mathematical complexity of the problem. For example, an atom with n electron is considered to be of smaller size than a molecule with n electrons. The concept will be more clearly defined afterwards.

The "size" is, of course, a not too "traditional" chemical division. Traditionally, the division goes according to functional lines, like aromatic and aliphatic molecules or like organic and inorganic molecules, etc. To take the "size" as division has, however, some advantages. First of all, it is somewhat more quantitative a division than those above and it parallels more closely the depth (or at least the possible amount of it) for a *detailed* knowledge available either theoretically or experimentally. Certainly it parallels what we can do in computations. In addition, it parallels the extent of feasibility in assignment of vibrational frequencies, electronic transitions (experimentally) x-ray structural determination, to give a few examples.

From the point of view of computational quantum chemistry it forces us to ask ourselves, what might be usefully done so as to avoid relatively useless work done simply because of technical and financial power. We shall ask, for example, "Given infinite facilities, do we know today what to do with an exact wave function for a very large molecule, say, an enzyme?" Neglecting considerations on how to print or store or retrieve such data, what would we do with it? Moreover, the interest today in large molecules is often an interest in biological functions: this means essentially reaction mechanisms. Therefore merely a wave function, however accurate, on a specific reactant could be too inadequate.

This is the type of question I shall tentatively address in the fourth and fifth sections of this paper.

Finally, some comments will be made on the problem of correlation energy estimates and/or computations. In the last ten years this problem has been sufficiently publicized among chemists and now techniques are becoming available, so that I would venture to state that this work is finally much less glamorous than in the past, and slowly theoretical chemists are facing again the old traditional problem: "how to handle large molecules", since, here, theoretical generalizations of a purely formal nature seem to find their first encounter with the real world of chemistry. The formal solution of a problem and the numerical solution of the same problem are two quite distinct processes. In the past the latter has not been sufficiently recognized since by tradition any complicated numerical solution would have been abandoned. Today, with computers, the situation is different.

A Small molecules

Small molecules are defined as those molecules with up to twenty electrons and four or more atoms or up to forty or fifty electrons and two atoms. For example, the Mn_2 molecules does not present computational difficulties exceeding those one would encounter in Mg_2. A computation at the *Hartree–Fock level* for Mg_2 requires inclusion of polarization effects, therefore, *d* and *f* type functions. The orbital exponents of the basis set require quadrature formulas (if one uses Slater type functions) capable of handling with sufficient numerical accuracy both the case of high orbital exponents, high quantum numbers and relatively large internuclear distances. This list of requirements essentially indicates a full control of two electron two center integrals. However, the complexity for a twenty electron molecule and four to five atoms (without the restrictive characteristic of linear symmetry) requires as much labor as Mn_2 because of the many (versus two) center integrals for the two electrons matrix elements. In addition, for both types of computations, one is well outside the possibility of containing the data within the fast core memory and has to face a "data handling" problem. This turns out to be a not n egligible task and in some respects one could say that whereas during the last ten years there was considerable effort in developing solutions for many centers —two-electron integrals now the problem is one of information-retrieval, since many millions of integrals are standard requirement.

The above mentioned difficulties are within full control by now for small molecules. For example, the Tables of Linear Molecules Wave Functions by A.D. McLean and M. Yoshimine, clearly stand as a proof that this level is at hand.[28] It might be worthwhile to note that the McLean and Yoshimine work was done with an IBM 7094. This type of computer is well superceded by more powerful ones available today. For a rather complete bibliography of ab initio computation of small molecules we refer to M. Krauss' recent compendium.[29] For a review paper of the limitation and successes of the Hartree–Fock approximation in small molecules, we refer to a review paper by R.K. Nesbet.[30]

For small molecules even a large fraction of the correlation problem seems to be more and more under control. For example, the work of Davidson[31] on first row hydrates is already a step towards mass production of substantially correlated wave functions. The work by Wahl and collaborators,[32] by F. Harris and collaborators,[33] and by F. Boys and collaborators[34] are all indicative of the fact that the correlation problem for small molecules is nearly under control. It is rather easy to extrapolate that in the next few years we shall have more small systems accurately computed (within one or two kilocalories of the exact dissociation energy, for example). The expectation values such as dipole and quadrupole moments, field gradient of the nucleus, polarizability, a vibrational frequency and transition probabilities are reaching computationally the limited accuracy presently available via experimental techniques.

The field of small molecules is, therefore, the place where we can expect to obtain the first series of "payoff" in the field of computational quantum chemistry.

On this assumption, I shall add a few comments. For example, I would like to note that the justification for future computations in small molecules should not be, as often in the past, restricted to attempts to reproduce well known experimental data, but rather in predicting new facts, new data.

As a second comment, the testing ground for correlation energy computational techniques will slowly shift from atoms to small molecules, since problems due to localization can be studied properly only in molecules.

It seems that the field of small molecules computations has finally entered maturity and in a relatively short time. For example, it can be noted that the work by Ransil,[35] McLean,[36] Clement;[37] and others at the Laboratory of Molecular Structure and Spectra in Chicago, a first broad beginning, was started in the late fifties or early sixties. This annotation could also provide some incentive for speculation on the rate at which progress can be made.

Finally, between 1960 and 1970 we have assisted to essentially three evoluntionary steps in the digital computer's capacity growth; another fact that should be carefully remembered.

As this field reaches maturity, the advisability to introduce more and more the experimentalist to computational techniques becomes clear, since he will soon use the theoretical tool as one more apparatus at his disposal. The work by B. Bak[38] is a welcome step in this direction.

B Medium size molecules

Here we would like to collect polyatomic molecules with up to one hundred electrons; either diatomic with inclusion of relativistic correction or polyatomic molecules with a most third row atoms and limited to the non-relativistic Hartree–Fock level. Accurate wave functions (to the Hartree–Fock limit) are not available at present; however, computer programs capable to do the job have been available for some time. The lack of such computations is mainly due to the unavailability of adequately large computers. Since computers of the class of the IBM 360/91 or CDC 7600 are now available, we expect that this situation will improve radically within the next few years. It is noted that what is now defined as medium size molecules, was previously defined as "large molecules" by Clementi and Davis.[39] That was at the "IBM 7094 type" computers time.

Some of the papers in this series deal with medium size molecules. A few years ago the difficulty was simply in obtaining a wave function within the LCAO–MO–SCF approximation. (With present day computers the entire work could be recomputed in one day of computer time.)

In addition, we are now slowly learning that the molecular orbital approximation is certainly powerful enough to cope with the problem of reactive molecules so as to predict useful reactions surfaces. We shall comment somewhat more on this point in the final section of this paper (section V).

This extension of the use of the molecular orbital approximation brings us to some reflection on the main concepts, on which the entire theory of the electronic structure of molecules resides. Two main concepts stand out in describing the electronic structure of molecules: *the concept of ionic binding* [in its various manifestations like the indirectly related ideas on covalent binding and the directly related ideas on electronegativity, charge transfer (inter or intra-molecular), hydrogen bonding, etc.] and the concept of *hybridizations* as directional binding ability (including the precursory stage of valency states). With these two concepts, we feel in the position to make far reaching statements about chemical bonds. Both concepts can be reconciled within the molecular orbital framework. Both concepts recognize the atomic parentage of a molecule, but this is an after effect in the molecular orbital approximation, rather than any assumption. In addition, this "after effect" is not needed, unless one wishes to build on the present mainstream of chemical concepts.

Less immediate is the connection of the molecular orbital theory to the concept of "resonance", which was for thirty years so important to chemical thinking.

The fact that the two previous concepts remain as basic today as yesterday and the fact that we speak less and less of "resonance" at today's theoretical chemistry meeting, indicate (to the author at least) that chemistry at large has slowly rediscovered the concept of "field" in the Slater, Hartree and Fock sense. Of course, this is in large part due to the work of Mulliken in his insistence to understand chemistry so as a physicist attempts to understand atoms. The "reference field" for ionic binding is the separated atomic or ionic field; the reference field for hybridization is the neutral atom in a point charge field (valency state).

In this context the many body problem (or correlation energy problem) can be looked upon simply as a "technicality", of course full of very important quantitative implications. But if the implications are mainly quantitative rather than qualitative, then clearly we better concentrate on computational rather than formal solutions for this problem. In this context, we note that the concepts of molecular orbitals can be most readily modified so as to include correlation correction. Therefore, one can state that the concept of ionicity and hybridization—in this broad sense—are capable of explainint qualitatively and quantitatively "what electrons are doing in molecules".

Clearly, if we feel that the main stream of quantum chemistry is based on extension of the concept of "electronic field" in the Hartree–Fock sense, then we should be most careful in avoiding to devote our attention to the field of some electrons rather than others; therefore, my insistence on all-electron computation. Not only π electrons, not only π and valence σ electrons, but all-electrons.

As a third comment we would stress a clear need of a re-examination of our use of symmetry concepts. Most molecules are systems devoided of symmetry, with exception of *relatively* few examples; however, some part of a molecule has often higher symmetry than the entire molecule. The use of "local symmetry" as distinct from overall molecular symmetry could bring substantial computational savings. It clearly brings new possibilities in the understanding of functional groups (like —OH, —NH, —NH_2, —COOH, etc.), new possibility in future chemical libraries, a natural division for more accurate work in a given region of a molecule.

The interest in localized orbitals is certainly a useful step in this direction. Thus, we are looking forward to all-electron computations *with the possibility of computing part of the molecule at one degree of sophistication and other at higher degree of sophistication.* In the author's opinion, local symmetry and localization are two important techniques to such a goal.

Concluding this section, we feel that forthcoming study on the medium size molecules will yield a selection among the variety of present chemical concepts with stress on "usefulness" for predictive purpose and "reference" in the sense of moving towards a unified theory in chemistry.

C Large molecules without translational symmetry

Introduction

In this group we include any molecule larger than those previously defined as medium size. For "molecule" we do not necessarily mean the traditional definition of molecule, but rather system of atoms either weakly or strongly bound. The emphasis is on the existence of a field: As long as the field is not zero, we shall speak of "molecules".

It seems worthwhile at this stage to briefly point out that the quantity "total energy of the system" is not very interesting in large systems (incidentally, this is true even for smaller systems, but there we can afford the luxury of stressing the total energy). Clearly, for a system like CH_3—$(CH_2)_n$—CH_3 the sum $(2 \times n + 6)\,0.5 + (n + 2)\,37.8$ gives the total energy of the system in atomic units, accurate to about four digits if $n = 100$. To improve the accuracy, say to six places, any table of C—H and C—C bond strength would suffice. Yet, it is clear that such number would have little practical value. For such a system *a minimum* of eight figures accuracy is required in order to yield useful data. However, this would represent a rather complicated task, even with large computers. Rather, we are interested mainly in energy differences when we consider the energy of a large system. In small molecules, the difference with respect to the separated atoms is an obvious reference. In large molecules this is not necessarily the case; likely energy difference between different conformations would be a more appropriate reference.

At present, we feel that the main avenue available for studies in the electronic structure of large systems *(without translational symmetry)* is one which makes use of empirical correlations.

The empirical scheme I am referring to has not as yet been fully developed, but there are now sufficient evidences to make it at least plausible. This scheme claims that knowledge of the "*geometry*" of a molecule should be sufficient in order to obtain (1) charge distribution, (2) orbital energies, (3) relative energy among different configurations for a system.

In a very broad sense, this is in accordance with the main stream of chemical thinking. In a more restricted sense it can be noted that whatever success might be attributed to the Huckel method, this was due to careful parameterization of minimum amount of data for given "families" of molecules. But this is tantamount to talk of molecular geometry.

Table 3 supplies evidence that knowledge of molecular geometry should be sufficient for the prediction of atomic charges. In analyzing the nitrogen atom in adenine, cytosine, guanine and thyamine,[40] it is clear that the three arrangements =N:, =N—H, and —N=H_2 have quite different charge distribution. In addition, the maximum charge deviation is very small for =N—H and —N=H_2 group; however, this is not equally true for a nitrogen with a lone pair, =N:. However, by considering the neighboring charges in the lone pair cases, we can empirically find trends which allow us to rationalize the deviations.

Therefore, we would like to put forward the hypothesis that (1) knowledge of structure (geometry) of a molecule should be sufficient to allow us to make accurate and *immediate* prediction of gross atomic charges; (2) know-

Table 3 Gross charges and geometry (for nitrogen)

Case	=N:	=N−H	−N=H_2
Adenine	−0.268	−0.455	−0.602
Adenine	−0.307		
Adenine	−0.309		
Cytosine	−0.335	−0.463	−0.598
Guanine	−0.361	−0.447	−0.613
Guanine	−0.239	−0.472	
Thyamine		−0.463	
Thyamine		−0.459	

ledge of geometry and atomic charges will make possible prediction of orbital energies (and all the implicit consequences such knowledge can bring about) (3) knowledge of geometry, gross charge and orbital energies should give us the required data for an empirical determination of energy differences between various configurations.

D Large molecules with translational symmetry

Clearly, we are speaking of polymers. There are several open questions regarding the electronic strcuture of polymers. For example, we wish to know the variation in the electronic structure from the isolated monomer to one of the monomers in the polymer. Likely in a polymer one can recognize the monomers as much (or as little) as one can recognize an atom in a molecule. If this is the case, and in view of the present amount of reliable published material on the electronic structure of polymers, one might conclude that indeed our present knowledge on the electronic structure in polymers could welcome some improvement.

In addition, there is a rather pressing and practical reason for developing this field, namely, the present trend in polymer chemistry whereby one attempts to synthesize new types of polymers to be used, not so much for their mechanical properties but for their optical and magnetic properties. Since a large amount of knowledge has been gained in solid state from the band structure models specifically in understanding optical and magnetic properties, it follows that we should attempt to make use of such knowledge in understanding organic polymers. It seems that one way to ensure a dialogue between this new field in theoretical chemistry and the more seasoned field of solid state is to have a "common language".

In addition, it seems that the study of polymers should benefit as much as possible from our knowledge in molecular computations. Therefore, if the tight binding approximation is the best one for molecules (and the LCAO–MO approximation is the tight binding approximation with neglect of translational symmetry), so likely it should be for organic polymers.

Indeed the concept of hybridization, charge transfer, and ionic character certainly will not lose their importance in going from small to larger molecular systems; however, new *collective* characteristics will appear to be of importance.

The work of J. M. André[41] is certainly an important step in opening the field of electronic structure of polymers. In his work it was realized that one can use the Hartree–Fock technique in polymers and carry over most of our knowledge of molecular electronic structure by simply redefining the molecular orbitals and the Fock supermatrices.

The standard molecular orbital of the monomer are transformed into "polymer orbitals" by imposing on them a periodicity factor in the Block sense. The standard molecular orbital is computed by solving the Fock equation for the monomer. However, since the monomer in the polymer experiences the field of the neighboring monomers, then each super matrix element contains not only the standard contribution arising from the orbitals of the monomer among themselves, but also all the contribution of the neighbors as well.

In concluding this section, one would like to expect in the near future, a real beginning in computation of the electronic structure of polymers. In general, one could expect many more and new examples of cross fertilization between the fields of theoretical chemistry and solid state.

5 RELATIONS BETWEEN "STEP THREE" AND STATISTICAL METHODS

The physical reasons for the need of improving on the Hartree–Fock model (step two) was given for example by E. Wigner in 1934.[42] If we ask the question: What is the statistical relation between the position of two electrons, say electron 1 and 2, from the wavefunction (1) we have to solve for the integral

$$\int \Psi_0 * \delta(x - x_1)\, \delta(x - x_2)\, \Psi_0 \prod_i^{(1,2)} d\tau_i \tag{11}$$

where x stands for the cartesian and spin coordinates, and the integral is carried over all electrons coordinates except for those of electron 1 and 2 in consideration. The above integrals yield (by indicating with σ the spin variables)

normalization factor x ×

$$\times \sum_{i=1}^{2n} \sum_{j=1}^{2n} [1 - \int \varphi_i(1)\, \varphi_i^*(2)\, \varphi_j^*(1)\, \varphi_j(2)\, d_{\sigma i}(1)\, d_{\sigma i}(2)\, d_{o j}(1)\, d\sigma_j(2)] \tag{12}$$

For the case of two electrons with both parallele spin, the second term above is, in general, different from zero. For the case where the two electrons have opposite spin the second term of the equation above is zero. This means that the single determinant wave function allows a pair of electrons with parallel spin to "feel" each other's relative position, to be correlated, whereas for the case of electrons with antiparallel spin the wave function allows any position with equal probability. Thus, the Hartree–Fock determinant introduces correlation in pairs of electrons with parallel spin, but does not correlate those of anti-parallel spin. Since the second term of equation (12) comes about because of the antisymmetry of the wave functions required for fermions, the second terms effect is referred to as "Fermi hole". Thus, in the Hartree–Fock wave function we have Fermi potential holes for the system of parallel spin, but no hole between two electrons with antiparallel spin. Since clearly two electrons with opposite spin should never occupy the same position simultaneously, the Hartree–Fock function lacks a mechanism to provide for a potential hole experienced by an electron with spin α, when in the neighborhood of an electron with spin β. The hole not present in the Hartree–Fock function, is referred to as "Coulomb hole".

The energy gained by the system in introducing the Coulomb hole in the wave function which represents such a system is the *correlation energy*. This name was introduced by Wigner in the work previously quoted.[42] This energy gain, can be re-defined therefore as the energy difference between the exact nonrelativistic energy and the Hartree–Fock energy.[43]

The energy gained by the system in introducing the Fermi hole in the wave function which represents such a system, is called *pre-correlation energy*. This energy can be re-defined, therefore, as the energy difference between the Hartree and the Hartree–Fock energy.[44] To introduce the Fermi hole, the form of the wave function was changed from a simple product (Hartree) to a determinant (Slater). To introduce the Coulomb hole, we again have to change the form of the wave function. All of this is, of course, very well known. A large number of papers are devoted to this subject since the early 1930's and we have no intention to review the extremely abundant literature. We note that one possibility put forward as early as 1934, namely to add corrections to the Hartree–Fock energy using wave functions of the form

$$\Psi = \Psi_0 + \lambda\phi \tag{13}$$

where Ψ_0 gives the Hartree–Fock energy and $\lambda\phi$, the correlation correction[45] represents one of the starting points of today's computations. In this section we shall present some new data on the use of Wigner's type relation for a semiempirical estimate of the correlation energy.

We note that the total density distribution given by Hartree's functions is not far different from the density distribution given by Hartree–Fock functions. It is, therefore, reasonable to assume that the density distribution variation caused by the introduction of a Coulomb hole (relative to the Hartree–Fock density) is also small. Therefore, we can use the Hartree–Fock density as the correct density and attempt to extract relations which should give the correlation energy from knowledge of the Hartree–Fock density. As known, this task was solved by Wigner[42] who did propose a relation between density and correlation energy.

The statistical model of Wigner has been revised a number of times in attempts to extend its validity to the case of high density.[46] Gombas[47] attempted to give an additional expression which covers both the high and low density region. His expression is for the correlation energy E_c

$$E_c = \int \varrho^2 \varepsilon_c(\varrho)\, d\tau \tag{14}$$

$$= \int \varrho^2 a_1 \varrho^{1/3} (a_2 + \varrho^{1/3})^{-1}\, d\tau + \int \varrho^2 b_1 \ln (1 + b_2 \varrho^{1/3})\, d\tau \tag{15}$$

where $a_1 = 0.0357$, $a_2 = 0.0562$, $b_1 = -0.0311$, $b_2 = 2.39$ are Gombas' constants.

It is important to note that since $\varrho = \sum \varrho_i$ (the Hartree–Fock densities) the previous equation can be written as

$$E_c \cong \int (\textstyle\sum \varrho_i)^2\, \varepsilon_c(\varrho)\, d\tau = \sum_i \int \varrho_i^2\, \varepsilon_c(\varrho) + 2 \sum_{i>j} \int \varrho_i \varrho_j\, \varepsilon_c(\varrho)\, d\tau. \tag{16}$$

This equation is clearly one of the first steps toward pair separability. We note that the separability is not fully obtained because $E_c(\varrho)$ can not be easily separated into pairs (analytically). However, one can most easily see that ε_c numerically will contribute only in the region where ϱ_i is non zero. It is clear that since ϱ_i in a molecule can be expressed in the LCAO form, *then the above relation furnishes a clear link between transferability of correlation data information from atom to molecules.* This has been done often in the past but with insufficient knowledge on its early literature references.

If in the first term of eq. (15) we put $a_2 = 0$, then this term contributes to the correlation energy E_c by a constant amount, a_2, per electron. In Table 4 the value of the correlation correction using Hartree–Fock function is given for the first and second row atoms.

The overall agreement with experimental correlation energies is not too bad as seen by comparing column 1 (the experimental correlation energy)[48] with the sum of the first and second term (columns 4, 2 and 3 respectively).

Table 4 Correlation energy computed by adapting statistical models

Atom		Correct values	Gambas			Modified relation	
			First term[a]	Second term[b]	Total	Best a_1^c	Corr. energy[d]
He	^1S	−0.0421	−0.0316	−0.0263	−0.0579	0.0237	−0.0421
Li	^2S	−0.0453	−0.0448	−0.0393	−0.0841	0.0180	−0.0476
Be	^1S	−0.0944	−0.0610	−0.0551	−0.1161	0.0276	−0.0947
B	^2P	−0.1240	−0.0782	−0.0738	−0.1520	0.0285	−0.1263
C	^3P	−0.1551	−0.0958	−0.0958	−0.1917	0.0289	−0.1580
C	^1D	−0.1659	−0.0956	−0.0954	−0.1910	0.0309	−0.175
C	^1S	−0.1956	−0.0953	−0.0948	−0.1901	0.0366	−0.175
N	^4S	−0.1861	−0.1137	−0.1208	−0.2345	0.0292	−0.1882
N	^2D	−0.2032	−0.1135	−0.1202	−0.2337	0.032	−0.209
N	^2P	−0.2274	−0.1134	−0.1199	−0.2333	0.035	−0.209
O	^3P	−0.2339	−0.1314	−0.1475	−0.2789	0.0318	−0.2408
O	^1D	−0.2617	−0.1313	−0.1472	−0.2785	0.0355	−0.265
O	^1S	−0.2985	−0.1312	−0.1467	−0.2779	0.0406	−0.2649
F	^2P	−0.3160	−0.1493	−0.1768	−0.3261	0.0378	−0.2980
Ne	^1S	−0.381	−0.1672	−0.2084	−0.3756	0.0406	−0.3594
Na	^2S	−0.386	−0.1807	−0.2302	−0.4109	0.0381	−0.347
Mg	^1S	−0.428	−0.1967	−0.2537	−0.4504	0.0388	−0.4417
Al	^2P	−0.459	−0.2130	−0.2774	−0.4904	0.0384	−0.4667
Si	^3P	−0.494	−0.2303	−0.3030	−0.5333	0.0383	−0.4910
Si	^1D	−0.505	−0.23	−0.3026	−0.5326	0.0392	−0.533
Si	^1S	−0.520	−0.2298	−0.3022	−0.5320	0.0404	−0.532
P	^4S	−0.521	−0.2479	−0.3300	−0.5779	0.0375	−0.5132
P	^2D	−0.539	−0.2477	−0.3297	−0.5774	0.0388	−0.559
P	^2P	−0.555	−0.2475	−0.3295	−0.5770	0.04	−0.558
S	^3P	−0.595	−0.2655	−0.3582	−0.6237	0.0400	−0.5818
S	^1D	−0.606	−0.2654	−0.3580	−0.6234	0.0407	−0.631
S	^1S	−0.624	−0.2653	−0.3576	−0.6229	0.042	−0.630
Cl	^2P	−0.667	−0.2833	−0.3879	−0.6712	0.0420	−0.6545
A	^1S	−0.732	−0.3013	−0.419	−0.7203	0.0433	−0.7310
Li^+	^1S	−0.0435	−0.033	−0.0353	−0.0683	0.0237	−0.0438
Be^{+2}	^1S	−0.0443	−0.0337	−0.0421	−0.0758	0.0237	−0.0447
B^{+3}	^1S	−0.0448	−0.0341	−0.0478	−0.0719	0.0237	−0.0453
C^{+4}	^1S	−0.0451	−0.0344	−0.0525	−0.0869	0.0237	−0.0456
N^{+5}	^1S	−0.0453	−0.0346	−0.0566	−0.0912	0.0237	−0.0459
O^{+6}	^1S	−0.0455	−0.0347	−0.0602	−0.0949	0.0237	−0.0460
F^{+7}	^1S	−0.0456	−0.0348	−0.0635	−0.0983	0.0237	−0.0462
Ne^{+8}	^1S	−0.0457	−0.0349	−0.0664	−0.1013	0.0237	−0.0463
Kr^{+36}	^1S	−0.0470	−0.0355	−0.1038	−0.1393	0.0237	−0.0471

a) First term is the contrubution to the correlation energy from $-0.0357 \int \varrho^{4/3} (0.0562 + \varrho^{1/3})^{-1} d\varrho$.

b) Second term is the contribution to the correlation energy from $-0.0311 \int \varrho \ln (1 + 2.39\varrho^{1/3}) d\varrho$.

c) Best value of the constant a_1 (see equation) using only the first term and fitting the correct correlation energy (first column).

d) Best empirical value of the correlation energy using equation.

If we consider the statistical model as a useful fitting formula then we can improve the situation. First, we put $b_1 = 0$, since the b_1 term was introduced only for high density region.

Then we obtained the best value of a_1 for $a_2 = 0.0562$ (second last column of Table 4 for the atoms of the first and second row. This done we expressed the function a_1, analytically and we end up with the following modified relation

$$E_c = \alpha \int \varrho^{4/3} (0.0562) + \varrho^{1/3})^{-1} \, d\tau \tag{17}$$

where α is a numerical constant obtained from the relation (for an n electron system):

$$\alpha = 0.0237 + (0.0279n - 0.08176)(n + 3.45)^{-1} \tag{18}$$

which fits (but not too well) the best values of a_1. It is noted that by neglecting the b_1 term, we re-introduce Wigner's original equation for low density case, but wish somewhat different numerical constants.

The computed correlation energies using equation (17) are given in the last column of Table 4. In Fig. 2 we report the experimental correlation energy and the "statistical" correlation energy computed either with equation (15) or with the empirical set of constant (equation (17)).

Since equation (17) is very near to the original formulation of Wigner, a part of the structure included by the author and Salez[49] in the constant a_1. Therefore, in principle good estimates of the correlation energy *would have been easily available for the last three decades if one would have had Hartree–Fock functions.*

Rather than solving for equation (17) we can obtain the correlation energy by a different technique introduced by the author.[50] This technique uses the Hartree–Fock formalism, but adds to the Hartree–Fock operator, F (i.e., the average field effect of all the electrons) an operator of the form $\sum J'_{ij}$ where i and j refers to particle. The operator $\sum J'_{ij}$ does not allow the electron i to go nearer to the electron j than a prescribed amount. This prescription is used for electrons with antiparallel spin. The radius of the sphere which limits the access of electron i to electron j is proportional to an inverse power of the density, following Wigner.[42] It is noted that in order to determine the radius of the excluded volume, we have not made use of the Wigner or Gombas relations but we have obtained empirically the relation by fitting for the He and the Ne atoms cases. The results of such technique are presented in Fig. 3. It is clear that the overall agreement with experimental data is not too bad. This technique again uses the *Hartree–Fock density* as a

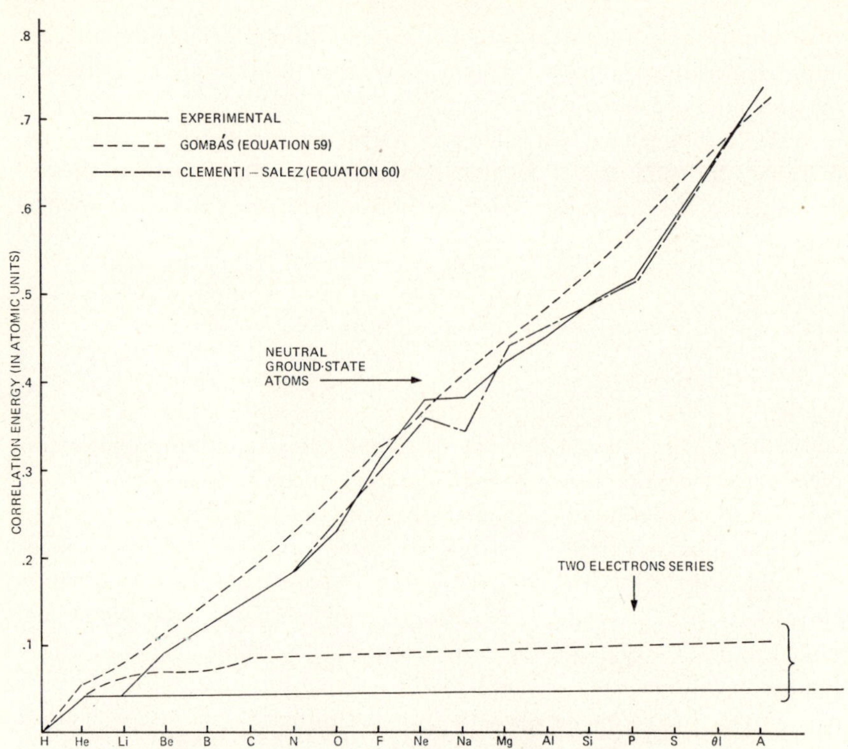

Fig. 2 Correlation Energy for statistical models. The solid line gives the experimental correlation energy for neutral atoms from helium to argon and for the isoelectronic series He, Li^+, Be^{+2}, ..., A^{+16}. The dash–dash and the dash-line reports the correlation energy using Gombas or Clementi and Salez relations. For the two electron isoelectronic series, the Clementi–Salez data are in exact agreement with the experimental one

starting point, and does not provide correlated functions but only estimates of the correlation energy. The method we have introduced gives the correct correlation energy for the two electron series He, Li^+, etc. up to KR^{+34}. The statistical methods previously discussed (with exception of the semi-empirical modification by Clementi and Salez) tend to give a much larger correlation for the highest ions in the two electron series.

It can, therefore, be concluded that there are sufficient indications to prove that knowledge of the Hartree–Fock density is sufficient for obtaining a good estimate of the correlation energy following the work available (but in part forgotten) for the last 35 years.

From our previous discussion on the atomic correlation energy it is clear that the correlation energy in a molecule should be related to the correlation energy in the component atoms.

The energy difference obtained by subtracting from the correlation energy of a molecule the correlation energy of the component atoms (in the correct dissociation state) has been called MECE, Molecular Extra Correlation Energy (Clementi[11] 1962). The name implicity contains the connotation of an increase in correlation energy from the sum of those belonging to the corresponding atoms. The main reason for such an increase is that molecule formation always increases the number of electron pairs. This increase is two-fold: the most important one is due to the fact that electrons previously unpaired in the separated atoms, pair during molecule formation. For example, the hydrogen atom has one unpaired spin, the fluorine atom has 9 electrons, 8 paired and one unpaired, but in the hydrogen fluoride molecule the two unpaired spin (the one of H and the one of F) pair up

$$[H(^2S) + F(^2P) \rightarrow HF\ (^1\Sigma^+)].$$

Extreme cases of pairing in diatomic molecules clearly correspond to extreme unpairing (higher spin multiplicity) of separated atoms like N_2 from $N(^4S)$, P_2 deom $P(^4S)$, Mn_2 from $M_n(^6S)$.

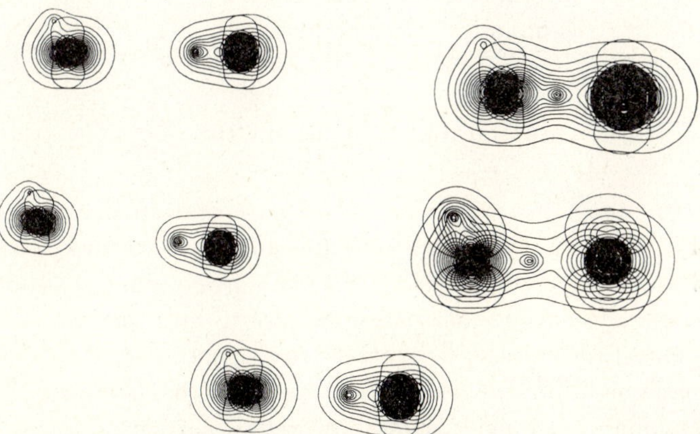

Fig. 3 Density contours for cross section of the NH_3 + HCl reaction and formation of NH_4Cl. The Nitrogen–Chlorine distances are 11.4 a.u., 9.4 a.u., 7.4 a.u., 5.4 a.u. and 4.4 a.u. respectively (see text for additional explanations)

The second factor in increasing the number of pair more inter-pair correlation energy contributions will occur, because of the increase in the number of electrons.

The molecular extra correlation energy can be a substantial fraction of the total dissociation energy.

Let us take a few examples of dissociation energy. For the N_2 molecule in its ground state ($^1\Sigma^+$) the experimental dissociation energy is 9.90 e.v. The computed dissociation energy in the Hartree–Fock approximation is 5.18 e.v., therefore, the Molecular Extra Correlation Energy is 4.72 e.v.

Let us now consider HCN in its ground state ($^{1+}$). The computed dissociation energy is 8.85 e.v., the experimental dissociation energy $D = 13.55 \pm 0.05$ e.v. and, therefore, the MECE $= 4.7 \pm 0.05$ e.v. It is noted that the two molecules, N_2 and HCN, are isoelectronic and that also for HCN we form three new pairs in the reaction $H(^2S) + C(^3P) + N(^4S) = HCN\,(^1\Sigma)$. The agreement of the extramolecular correlation of 4.72 e.v. for N_2 and 4.70 ± 0.05 e.v. indicates how transferable are data within molecular systems. It is noted that MECE for N_2 is somewhat larger than MECE for N_2: the reason is that in HCN the electronic charges are more localized than in HCN.

In the past few years an increasingly large number of data are available so that it is expected that we shall be very shortly in position to estimate from atomic data and Hartree–Fock dissociation energy, the molecular extra correlation energy within a few percent of error.[51]

6 CONCLUDING REMARKS

In the previous sections of this paper, we have indicated that there is a satisfactory theoretical framework which is sufficiently accurate for quantitative chemical prediction (MCSCF). Alternatively one can stop at the Hartree–Fock approximation and use empirical rules derived from the statistical model for corrections to the Hartree–Fock energy.

We have, in addition, stressed the fact that the rather complex output of the computation can be made more intuitively available by the use of the population and bond energy analysis techniques.

Finally, we have noticed that computers of the CDC 7600 or IBM 360/195 class are of sufficient size for doing rather complex computations. For such computers computer programs for chemistry are available.

It seems, therefore, that "computational chemistry" can finally be more and more of a reality. Before addressing the remaining part of this paper to the expectations of the near future (next five years) for the computational aspect of molecular chemistry, I would like to point out some possible obtstacle, which would, if not controlled, negate the successful progress in this field.

Clearly, the large computer of today (and even more, those of tomorrow) require large capital, both for the hardware as well as for the maintenance and supervision. Where as in high energy physics, the scientific community was willing to pull together their financial resources, today the scientific community in chemistry is very far from such mentality. The periodic attempts made in this direction (in occasion of scientific meetings) has never been more than timid and wishful whispers with no real organization and follow-up. This is true despite the availability of compatible machines and languages, which should make much easier, regional or national or perhaps international collaboration. Unless some positive steps are taken for this new mentality for the chemist, the financial problem will be sufficiently acute so as to frustrate progress in the field.

A second rather serious obstacle is the degree of preparation in computer sciences as well as in numerical analysis provided by the chemistry curricula. The last twenty years has witnessed a widespread and often well-thought increase of quantum mechanics in the chemistry curricula. However, no equivalent effort was initiated in order to provide future chemists with a sufficient knowledge in the use of one of the most important tools in molecular quantum mechanics.

Neglecting the above two possible stumbling blocks, the future of theoretical chemistry computation is rather astonishing in its potentiality.

In the field of material science the obvious areas are high temperature chemistry. Much information on these species can be already obtained today at equivalent cost via computation rather than via laboratory techniques.

In meteorological science and space sciences and in some aspects of astrophysics data related to stability, cross section, lifetime again can be competitively done by computations today.

The expectation in polymer science, along the lines mentioned in the fourth section of the paper are particularly intriguing since problems of superconductivity in organic materials are hard to be obtained experimentally.

Finally, the field of biological sciences is a natural candidate for the good use of computational chemistry. For example, we have mentioned in this work some computation on cytosine, adenine, guanine and thyamine. These are the four constituents of the well known macromolecule DNA. We are now studying the hydrogen bonding between guanine and cytosine. This will help to solve the problem of mutual interaction of hydrogen bonds. It will, in addition, give numerical data for the tunneling rate from one molecule to the next. Lowdin has proposed that this tunneling rate could be responsible for the occurrence of spontaneous mutations and other basic phenomena.

This is only a beginning. Let us note that in the next 10 to 20 years more and more scientists will turn their effort to the *molecular aspect of information retrieval and storage in human brains*, since the electro-chemical aspect is now sufficiently understood. Computational chemistry can certainly help very much in this extraordinary important new field.

A final field where computation will bring about major advances is in chemistry of reaction mechanism. For example, a few years ago I studied the mechanism for reaction for the $NH_3 + HCl \leftrightarrow NH_4Cl$ process. Via computations, the molecular geometry as well as its thermodynamic properties were predicted. Some of the predictions ware later confirmed experimentally, primary among them, that the NH_4Cl molecule is a stable species in the gaseous phase.

The practical importance of predictive computations in reaction chemistry computations is so evident that it does not require comments. We would like, however, to stress that our knowledge and intuition of a chemical reaction will expand even more. For example, the computer can not only give us the energetic balance of the reaction, but can show us the detail of the electronic process which accompany the molecular formation. In Fig. 3, we give five pictures taken from a visual display tube, totally under computer control, which depicts the computed electronic density of the NH_3 and HCl molecule at 11.4 a.u., 9.4 a.u., 7.4 a.u., 5.4 a.u. and 4.4 a.u. separation (measured from the nitrogen nucleus to the chlorine nucleus). These five cross sections of the $NH_3 \leftarrow HCl$ process, have been taken from a continuous motion picture. The detail and information contained are very revealing, and can not be obtained by any other instrument but computers. And this is perhaps the last advantage of chemical computation I wish to stress: even if computations increase in complexity, the results are becoming not only of increasing reliable predictive value but also more easily interpretable.

References

1. L.C.Pauling, *Nature of Chemical Bond.* Cornell University Press, Ithaca, New York, 1960.
2. R.S.Mulliken, For example, "Electronic Structure of Molecules", *J.C.P.* **3**, 375 (1935).
3. W.Moffit, *Proc. Roy. Soc. (London)* **A210**, 245 (1951).
4. M.Kotany, *Rev. Mod. Phys.* **32**, 266 (1960).
5. C.A.Coulson, *Valency*, Oxford University Press, 1959.
6. Whereas there is no doubt that technology should be related to economy, nevertheless, one should be careful in accepting the rule which attempts to appraise the value of knowledge only or predominantly in terms of cost. The justification for such comments in this theoretical exposition of the electronic structure in molecules, is prompted by the existence of a vocal current among chemists, which is very insensitive to the last two generations efforts in theoretical chemistry and would not mind to eliminate its support, paradoxically, now, when preliminary and solid evidences are finally available that computational chemistry is developing into a practical tool of enormous consequence for future chemistry.
7. W.Kolos and L.Wolniewicz, *J. Chem. Phys.* **41**, 3663 (1964), *ibid.* **43**, 2429 (1965).
8. E.V.Condon and G.H.Shortley, *The Theory of Atomic Spectra*, University Press, Cambridge (1957).
9. C.C.J.Roothaan, *Rev. Mod. Phys.* **23**, 69 (1951).
10. A.C.Wahl, *J. Chem. Phys.* **41**, 2600 (1964).
11. V.Fock, *Izvest. Akad. Nauk SSSR, Ser. Fiz.* **18**, 161 (1954).
12. A.C.Hurley, J.E.Lennard-Jones, and J.A.Pople, *Proc. Roy. Soc. (London)* **A220**, 446 (1953).
13. E.Hylleraas, *Z. Physik* **54**, 347 (1929); **65**, 759 (1930).
14. F.S.Boys and G.B.Cook, *Revs. Mod. Phys.* **32**, 285 (1960).
15. J.Frenkel, *Wave Mechanics, Advanced General Theory*, Clarendon Press, Oxford, England (1934).
16. D.R.Hartree, WHartree and B.Swirles, *Phil. Trans. Roy. Soc. (London)* **A238**, 223 (1939).
17. A.P.Yutsis, *Zh. Eksperim. i. Teoret. Fiz.* **23**, 129 (1952); *ibid.* **24**, 425 (1954); A.P.Yutsis, *Soviet Phys.–JETP* **2**, 481 (1956). See in addition T.L.Gilbert, *J. Chem. Phys.* **43**, S248 (1956).
18. A.P.Yutsis, Ya.IVizbaraite, T.D.'Strotskite and A.A.Bandzaitis, *Optics and Spectroscopy* **12**, 83 (1962).
19. A.Veillard, *Theoret. Chim. Acta* **4**, 22 (1966).
20. A.Veillard and E.Clementi, *Theoret. Chim. Acta*, **7**, 133 (1967).
21. E.Clementi, *J. Chem. Phys.* **46**, (3842), 1967.
22. G.Das and A.C.Wahl, *J. Chem. Phys.* **44**, 87 (1966).
23. H.Hartman and E.Clementi, *Phys. Rev.* **133**, A1295 (1964).
24. E.Clementi, *J. Mol. Spec.* **12**, 18 (1964).
25. Young-Ki Kim, *Phys. Rev.* **154**, 17 (1967).
26. W.Kolos and L.Wolniewicz, *J. Chem. Phys.* **41**, 3674 (1964).
27. R.S.Mulliken, *J. Chem. Phys.* **23**, 1833, 1841, 2338, 2343 (1955).

28. A.D. McLean and M. Yoshimine, Tables of Linear Molecule Wave Functions. Supplemental volume, *IBM Jrnl. Res. & Devel.* **12**, 206 (1968).
29. M. Krauss, National Bureau of Standards, Technical Note No. 438, 1967.
30. R.K. Nesbet, *Advances in Chemical Physics*, Interscience Publishers, New York, 1966, Volume 9.
31. Paper presented by Prof. Davidson on first row hydrates at the Sanibel Island, Quantum Chemistry Meeting, 1968.
32. G. Das and A.C. Wahl, *J. Chem. Phys.* **44**, 87 (1966).
33. F. Harris, computation on selected first row atoms, to be published (private communications).
34. For updated references see B. Handy, *Proceedings of the Snibel Island, Quantum Chemistry Meeting* (1968).
35. B. Ransil, *Rev. Mod. Phys.* **32**, 245 (1960); **34**, 1468 (1961).
36. A.D. McLean, *J. Chem. Phys.* **32**, 1595 (1960).
37. E. Clementi, *J. Chem. Phys.* **36**, 33 (1962).
38. Work on the LiCN and CNLi molecule presented at the Sanibel Island Quantum Chemistry Meeting (1968).
39. E. Clementi and D.R. Davies, *J. Comp. Phys.* **1**, 223 (1966).
40. E. Clementi, J.M. André, M.Cl. André, D. Klint, D. Hahn, Study of the Electronic Structure of Molecules. X. (To be published.) This paper reports all-electrons computations in adenine, cytosine, guanine and thyamine. *Hungarica Physica Acta*, **27**, 493 (1969).
41. J.M. André. (To be published.) See IBM Technical Report RJ 527.
42. E. Wigner, *Phys. Rev.* **46**, 1002 (1934).
43. P.O. Lowdin, *Advances in Chemical Pyhsics*, Vol. 2. Interscience Publishers New York (1959).
44. E. Clementi, *J. Chem. Phys.* **38**, 2248 (1963).
45. E. Wigner did not write equation (13) as such, but his formulation (see his equations (5) and (9) is not too different in form, and, not at all in the physics.
46. Note in particular the work by K.A. Brueckner and collaborators, e.g. K.A. Brueckner, *Phys. Rev.* **96**, 508 (1954); **97**, 1353 (1955); **100**, 36 (1955).
47. P. Gombas, Pseudo potentiale, Springer–Verlag, New York (1967).
48. E. Clementi and A. Veillard, *J. Chem. Phys.* **49**, 2415 (1968).
49. The numerical work related to equation (17) was done by Mr. C. Salez of the French Atomic Energy Commission. It is my pleasure to acknowledge his help and that of his manager, Mr. Giulloud, in securing this collaboration.
50. E. Clementi, *IBM Journal of Res. and Dev.* **9**, 1 (1965).
51. E. Clementi, *J. Chem. Phys.* **38**, 2780 (1963); *ibid.* **32**, 487 (1963). K.D. Carlson and P.N. Skancke, *J. Chem. Phys.* **40**, 613 (1964). Aa an early example. See in addition, the review paper by R.K. Nesbet: *Advances in Quantum Chemistry*, Edited by P.O. Löwdin, Volume 3, Academic Press, New York, 1967 for complete set of references.

CHAPTER 22

The Use of Computers in Studies of the Earth

BRUCE A. BOLT

Abstract

The history of geophysics is marked by a struggle with two severe problems: complexity of the appropriate mathematical analysis and the vast number of observations. These problems are the inevitable consequence of the complex geological structure of the Earth and of the variety of physical processes which are forever changing its properties.

Both these problems have been greatly relieved by the accuracy, speed and storage capacity of the modern digital computer. On the theoretical side, more realistic models of Earth structure have been explored. For example viscous and non-homogeneous and non-isotropic properties of rocks and soils are now included in the analysis where they are significant (as is often the case, particularly in seismology). The reduction and interpretation of large amounts of geophysical measurements are now performed with a speed and reliability far beyond the cleverest manual methods. This capacity has revolutionized the experimental and field work in significant parts of solid-earth geophysics, satellite-orbit reduction, oceanography and mineral prospecting.

The use of computers in the Earth sciences has also had more questionable consequences: machine methods have reduced significantly personal numerical analysis which has been one historically major ingredient in the process of discovery in geophysical problems; and even vaster quantities of raw data have been amassed—sometimes beyond the handling capacity of the most recent digital computers.

1 ACTIVITIES OF GEOPHYSICISTS

GEOPHYSICS IS THE STUDY of the physical properties of our planet and the Moon and the other terrestrial planets. It has been found already that the Earth has a shape something like that of an eggplant, mainly spherical but with more flattening at the South than the North Pole. The mean radius a is 6371 km and, based on the best laboratory determination of the Newtonian constant of gravitation* G, the mean density ϱ is 5.52 g/cm^3. Recent fine estimates of the shape of the Earth (which depend mainly on computer reductions of satellite orbits) have given 0.3308 Ma^2, where M is the Earth's mass, as the revised moment of inertia I of the Earth about a polar axis.

The broad structure of the Earth's interior, shown in Fig. 1, is known from the study of earthquake waves (see Bullen, 1967). There is a thin crust which has an order of thickness under continents of 40 km and under oceans of 5 km. Below the crust there is a (generally) solid rocky mantle down to depth 2895 km, give or take perhaps 10 km. At this depth there is a liquid shell some 2250 km thick whose presence has far-reaching consequences. Its hydrodynamical properties are believed to affect the Earth's rotation and motions within it are thought to be the cause of the terrestrial magnetic field. Near the center, a separate inner core exists, which is about the size of the Moon and may be solid.

Most geophysical results, like those above, depend on inferential argument. Because direct sampling of rocks can only be made to depths of a few kilometers, the physical properties of most of the interior comes from indirect evidence. For this reason, special attention to scientific procedures must be used in geophysics. Of the infinity of conceivable interiors of the Earth, only those are worthy of study which both satisfy relevant physical observations and allow predictions which can be checked by later work. Such methods demand that many physical models be worked out in detail; then, by successive checks against available observations, the number of allowable models is whittled down.

Geological and geophysical observations make plain that, in detail, the Earth is much more structurally complicated than is shown in the approximation given by Fig. 1; this complexity makes the computation of realistic Earth models time-consuming and numerically tedious. In such circumstances it is not surprising that the availability of high-speed digital computers has revolutionized the study of the Earth. Mathematical models with many varying parameters can now be considered to an extent not previously dreamed

* The first close experimental determination was made by Henry Cavendish in 1798; modern values are based on work of C. V. Boys (in 1895), P. R. Heyl (in 1930), and Heyl and P. Chrzanowski (in 1942). The precision of G relative to many other physical constants remains poor.

of. This revolution in geophysics is turning the subject about; unlike a few years ago, lack of precision in drawing inferences comes often, nowadays, from the lack of reliable observations from suitably designed experiments, rather than shortage of adequate theoretical models.

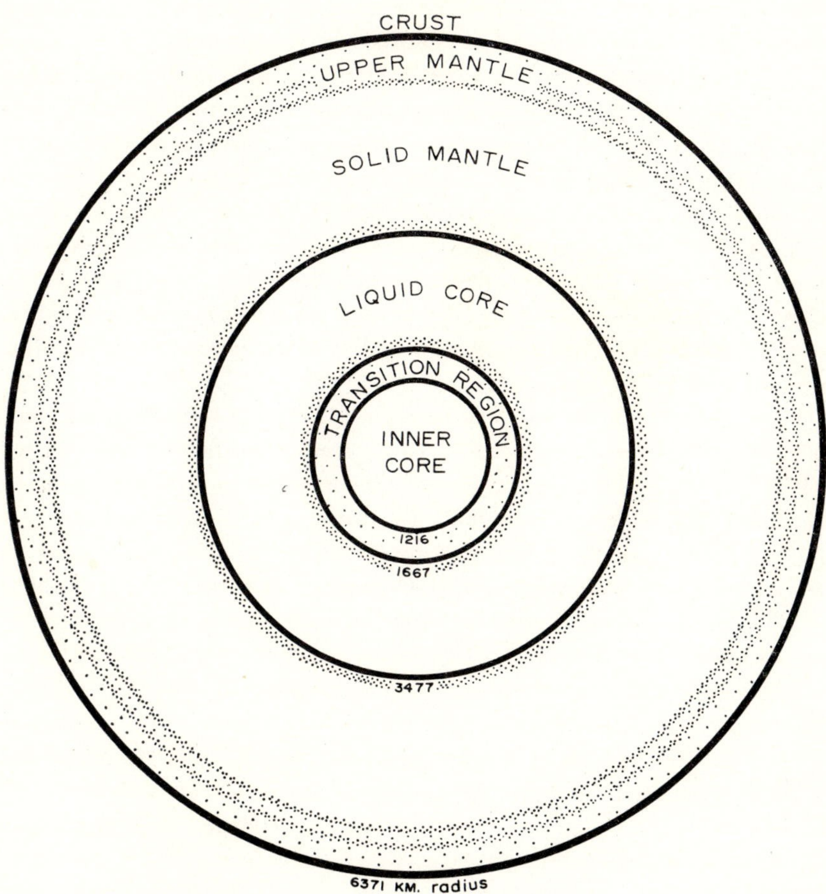

Fig. 1 Main regions of the Earth's interior

Theoretical models of a planetary interior endeavor to reproduce in mathematical forms, such as differential equations, the cardinal features of the observed mechanical and electromagnetic behaviour. In this light, theoretical geophysics is a special branch of applied mathematics; it depends mostly on classical theories of the mechanics of deformable media and electromagnetic theory, including, especially, modern developments in nonlinear visco-elasticity, magneto-hydrodynamics and solid-state physics.

In addition, special methods in statistical numerical analysis and high-speed computation are required for the adequate treatment of geophysical observations. To give an example, in order to record earthquakes there are more than 1000 seismographic stations distributed around the world. Many of these record ground motion continuously on six or more sensitive pendulums. There are thus over 5000 seismograms each day which must be scanned for earthquake waves; a certain number of seismic phases are read from each detectable earthquake and transmitted to seismological data centers. One such center is the International Seismological Centre in Edinburgh, Scotland (the successor of the International Seismological Summary*, Kew). For the year 1964, the Edinburgh center located, using electronic computers, some 10,000 earthquakes around the world and catalogued 10^5–10^6 separate readings of seismic phases.

Great as the scope of this global seismological data processing would seem, it hardly compares with the number of pieces of geophysical information which are now recorded each year, for later analysis, in certain branches of geophysics, notably oceanography, oil prospecting, and meteorology. A fully-equipped oceanographic vessel may untertake on any one day, more-or-less continuous recordings of ocean depth, salinity, temperature, variation in gravitational attraction and in the Earth's magnetic field, as well as echo soundings of the sediment layers on the ocean bottom and meteorological data on air temperature, barometric pressure and wind speed. The storage of such enormous quantities of data on paper charts for real-time visual scanning is not feasible and magnetic tape or, sometimes, punched-paper tape storage is needed. The processing of such material could hardly be attempted without high-speed computers. The best equipped oceanographic vessels nowadays carry computers with them which allow the use, while underway, of geophysical and meteorological analysis routines and on-station oceanographic processing routines. In 1967 and 1968, for example, the U.S. Coast and Geodetic Survey (ESSA) had at sea two new oceanographic vessels,

* The International Seismological Summary from the year 1954 onwards was prepared using high-speed computers to make least-squares estimations of earthquake locations (Bolt, 1960). By 1963–64, all seismological centers were using such methods.

SS Oceanographer, SS Discoverer, each exceeding 3000 tons displacement, and equipped with Westinghouse PRODAC computing systems utilizing a UNIVAC 1218 as central processor (16 thousand word, 18-bit core memory). This allowed digital acquisition and on-line processing of geophysical and oceanographic data as well as real-time engine-room monitoring.*

The history of theoretical geophysics shows that it is one of the most fruitful parts of applied mathematics. From the beginning, it has attracted such leading mathematicians as Isaac Newton, P.S. Laplace, C.F. Gauss, G.C. Stokes, Rayleigh (J.W. Strutt), G.H. Darwin and Harold Jeffreys†. In the study of the properties of the Earth, these and other theoreticians found an array of problems which were fully capable of testing their ability and ingenuity. Some of the classical geophysical work which involves great computations will be touched on in the following sections and an indication given of the way that the older developments have been broadened since the coming of high-speed computers.

Because geophysics has many aspects it is not possible to give an exhaustive account. Instead, with the reader who is not a geophysicist in mind, I have selected a small number of examples of the role of digital computers in geophysics, which are particularly familiar to me. (Analogue and hybrid (i.e. mixed analogue-digital) devices have proved valuable for some special geophysical work but it is high-speed digital computers which have had most impact on the subject.) Within these bounds it is possible to explain not only the computer application but also a little about the geophysical problem itself. Because of my own interests and research, the examples deal mainly with broad properties of the Earth's interior rather than, for example, marine studies, mineral exploration, or studies dealing with the atmosphere, iono-

---

* According to Dr. H.B. Stewart, Jr., in 1968 each computer on the SS Oceanographer and SS Discoverer logged approximately 6×10^6 bits of geophysical data and about the same amount of sea water data. On a geophysical survey measuring gravity and magnetic data only, the computer system can acquire information at a rate of 156 bits per minute.

† Newton (1642–1727) really started theoretical geophysics. He estimated the density of the Earth relative to the Sun and thought it would be "five or six times greater" than water; he found a reasonable value for the Earth's flattening. Laplace (1749–1827), in his studies of the gravitational field, considered various forms of density distribution in the Earth. Stokes (1819–1903) showed how the shape of the globe could be found from the gravity variation over it; in his 1849 paper, on the Dynamical Theory of Diffraction, he unknowingly gave the first mathematical model of an earthquake. Contributions of Gauss, Rayleigh, Darwin and Jeffreys are mentioned later.

sphere and satellite geodesy. It must be stressed, however, that oceanography, geophysical prospecting*, meteorology and geodesy and related fields are among the first-ranked users of computers.

2 NUMERICAL WORK IN GEOPHYSICS

Geophysicists were not slow to perceive the advantages of high-speed computers with large memories. In the first place, computers were used to greatly extend data storage and retrieval processes. (This was particularly the case in the oil industry where the initial high costs could be met more readily.) Storage on punched cards and on paper tape had been replaced generally by about 1965 by magnetic tape storage. Most geophysical centers have now achieved the next step: the ability to convert continuous recording on magnetic tape to digital samples in a format which allows direct use of the digital tapes on such computers as the IBM 7094, IBM 360, and the CDC 6000 series. Many types of automatic analogue- to-digital electronic converters are now marketed. Hybrid computers have also been developed at a few places to handle special types of both analogue and digital geophysical data. A discussion of such a computer at the Seismological Laboratory at the California Institute of Technology has been given by Smith (1965). Recent manufacture of extremely high-speed digital computers with more adequate output display makes it unlikely that hybrid machines will become very widespread.

There is little need to dwell on this obvious computer usage. One example from seismology may be of interest. In an attempt to discriminate between underground nuclear explosions and natural earthquakes, from about 1961, relatively large amounts of money became available for seismological instrumentation and research in a number of contries. It was recognized that signal-to-noise ratios of recordings of seismic waves could be greatly enhanced if large arrays of seismometers were used as spatial filters. Seismic array design has taken various forms; the most elaborate is the Large Aperture Seismic Array (LASA), arranged in a circular geometry, 200 km across, in Montana. The basic configuration consists of 21 sub-arrays, each of 25 short-period vertical seismometers. The digital signals from each channel

* Dr. Sidney Kaufman, Shell Development Company, Texas, informs me that the major oil companies in the United States, and their major contractors, invest about $40 million per year in computers.

may be pre-filtered to reduce microseisms and the signals from each of the sub-arrays simply added. This procedure increases the seismic signal relative to the random noise. Each sub-array signal may then be given an appropriate time-delay and the delayed signals added in order across the array. This procedure enhances a particular seismic wave pulse of specified direction and speed. Such mathematical treatment, carried out by an on-line computer, is equivalent to steering an antenna to receive electro-magnetic waves.

A fast-growing application of computers in geophysics is rapid display of empirical and other data. There are still only a small number of geophysical research centers (excluding mineral exploration geophysics) which have really adequate facilities for automatic graphical or photographic output such as is provided by peripheral XY plotters (see Fig. 2). There is also unevenness in the present distribution of such facilities between geophysical disciplines. For many years the display facility of modern computers has been much exploited, for example, by meteorologists to follow graphically many atmospheric processes in detail. In oil prospecting, most elaborate digital computer systems have been designed to process the seismic records from spreads (or arrays) of geophones. The computer edits each trace before making an optimum output-display for analysis. Machine programs allow, among other options, polarity reversal of the signal, various types of filtering and trace-by-trace equalization of the wave amplitude. Similarly, reduction of gravimetric and magnetic survey data is carried out mainly by a computer system.

Not only can maps showing the contours of various geophysical parameters be produced automatically by modern computer systems, but, also, numerical solutions of mathematical models can receive a variety of graphical treatments with little manual effort; mathematical tables can be published by direct offset of the computer output, almost eliminating errors in tables. The main drawback is the lack of variety and accuracy of pen motion on most XY plotting machines. Up till the present, most available automatic plotting devices have provided figures in published papers (cf. Figs. 2, 3, 4) with little contrast or drafting elegance. It is often most suggestive in the study of complicated geophysical processes to make movie-films from a succession of XY plots or photographic (cathode ray tube) output.

Geodesy provides an example of a section of geophysics which involves the reduction of great quantities of field measurements without the need for very complicated algebra. Under normal circumstances, computations, based on levelling and triangulation measurements, can be performed adequately by

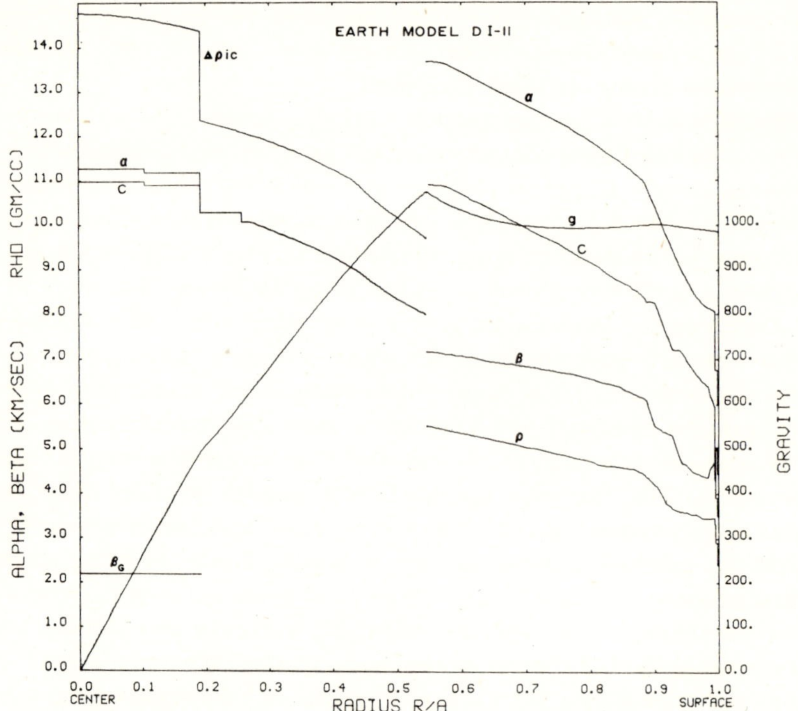

Fig. 2 Mechanical properties of a model Earth recently constructed using extensive machine computations by J. Derr. The properties, graphed using an XY mechanical plotter, are density ϱ, gravity g, P and S seismic velocities α and β, and the parameter C which equals $\sqrt{k/\varrho}$ where k is the bulk modulus

rather low-cost electronic desk computers with small stores and card programmers such as the Wang 370/380 system. Even so, geodesists are turning more and more to large computers not only for data storage on magnetic tapes but also for more complete time-dependent estimates of regional strain, usually involving the inversion of large matrices. Similar comments could be made about the combinatorial procedures in the estimation of gravity at different places; differences in gravity are usually determined by pendulums or gravimeters and then referred by a least-squares adjustment to a standard value at a place where an absolute determination has been made. The *interpretation* of gravity surveys has been greatly facilitated by rapid computations of topographic and Earth curvature corrections to the gravity field on computers. In particular, the effect of three-dimensional topographic contours can be found almost routinely (see Tobin and Talwani, 1966). Magnetic surveys now receive analogous treatment.

The new field of satellite geodesy, including satellite photogrammetry, grew up with electronic computers and has from the beginning depended on them exclusively. Computer programs are used to predict satellite orbits by numerical integration, including the effect of solar radiation pressure on balloon-type satellites. Tracking of satellites such as PAGEOS has made possible a world-wide geometric satellite triangulation network.

A great deal of statistical analysis of geophysical data is often necessary. Many instances have already arisen where computers have allowed quite extensive treatment of uncertainties in the observations—particularly in geophysical time series such as fluctuations in the magnetic field recorded by a magnetometer at a magnetic observatory. Two recent developments need special emphasis. The first is the growth of what might be called *global analyses* of geophysical data. In other words, world-wide measurements of a particular geophysical parameter such as gravity, flow of heat through the Earth's surface, or direction of the magnetic field, are analyzed as a distribution over the surface of a sphere. An example is given in some detail in Section 3; the method of analysis by expansion in spherical harmonics has been employed since Gauss but electronic computers have removed the great labor previously involved. The main stumbling block nowadays is usually the poor distribution of observations—often caused by the lack of land areas in the southern hemisphere. The second development, which is yet to be fully exploited, is the use of statistical *analysis of variance* to treat measurements of geophysical parameters which permit grouping in different ways. This allows variances within one group and within another to be estimated as well as the variances from interaction between groups. It has been shown already, for instance, that earthquakes can be located more accurately if they (and the stations recording them) are treated in regional groups, rather than one by one; this makes available sufficient equations of condition to estimate not only latitude, longitude, focal depth and origin time for each earthquake but also the regional deviations in travel-times from the mean due to geological inhomogeneities. In such analysis of variance problems, solution by least squares normally requires the inversion of very large-sized matrices and machine capacity remains a problem.

On the theoretical side, it is clear that much more realistic models of the Earth can be made amenable to analysis by the use of the latest generation of computers. Until recent years, most attempts at explaining geophysical phenomena have relied on the results of quite simplified physical situations. As an illustration, the variation of such parameters as electrical conductivity and velocity in the mantle and core of the Earth have generally been restricted to functions of the radius only. Computers have allowed a start to be made on problems involving lateral structural variations as well as anisotropic properties. Two other consequences are important: mathematical models once thought to be too complicated for further analysis are now being re-examined. An illustration comes from the working out of the dynamo theory for the generation of the Earth's magnetic field. When the theoretical models were first considered, at a certain stage of development, it was thought by Bullard* (to whom with J. Lamor and W. M. Elsasser much of the scheme is due) no longer practical to continue the analysis because of the great algebraic and computational complexities (Bullard and Gellman, 1954). Lately, integration of the equations on a computer has been used to develop the theory further and the interaction between the convection in the liquid outer core and the magnetic field has been simulated numerically. Time-dependent studies of three dimensional dynamos require speeds and capacities beyond the present generation of computers.

The second consequence is that certain mathematical developments have been abandoned in favor of others which are more suitable for numerical work; analytical treatments, perhaps more elegant, have given way in favor of such approximate methods as finite differences and finite elements. The old seismological problem of the prediction of strong motion of the ground during earthquakes exemplifies this change. In this problem, local geological structures and soil conditions are of cardinal importance. Recently, machine programs based on finite element and lumped-mass methods have been shown to be more effective (see Idriss and Seed, 1968) than analytic formulae based on plane-parallel layering of continuous elastic media. The finite element method permits assumptions on plasticity to be introduced through special elasto-plastic, stress–strain relations. The elastic continuum is replaced by a system of discrete mathematical elements which are interconnected at

* E. C. Bullard (1907–) has missed few opportunities to exploit high-speed computers in geophysics. As an example, in 1957, he computed sets of density distributions in the Earth using the DEUCE computer at the National Research Laboratory, U.K.; in 1965, with J. E. Everett and A. G. Smith, he made a detailed numerical investigation of the degree of fit of North and South America with Europe and Africa—the fit, at the edge of the shelf, was close.

nodal points. The models approximate an infinite degree of freedom system by one which has a finite number of degrees of freedom. The partial differential equations of a continuum are then replaced by ordinary differential equations. With proper choice of elements, there is little restriction on the structural geometry of the region.

Now about the form of geophysical problems. Many geophysical problems can be stated as follows: given a set of readings on the Earth's surface, derive the interior properties of the Earth which would give rise to them. Such boundary value problems have been called "inverse" problems (see Keilis–Borok and Yanovskaya, 1967). Such questions are commonplace in seismology. If the velocity structure in the Earth's interior were known as a function of depth (assuming no lateral variations) then it would be straightforward to calculate, using a form of Abel's integral equation, the travel-times of seismic waves between any two points on the Earth's surface. The inverse problem is to derive an internal velocity distribution which is consistent with the observed times.

Such problems are among the most fascinating in science; they involve questions of uniqueness and of precision and are linked, through tests of significance, with probability theory (Jeffreys*, 1961). Even if the surface data were known exactly, there is usually no proof available that iteration (by successive approximations, steepest descent or other methods) leads to a *unique* solution. Actually, the observations are not exact but are samples from some population of random variables. Because memory capacity is lacking, computers have only begun to be used to locate optimum ways of solving inverse problems. The usual procedure is to attempt to fit Earth models by successive adjustments of their internal structures, using trial-and-error or generalised matrix inversion, often after linearization of constraints by Taylor's expansion. Work on inverse probability suggests that it may be better to compute sets of rather different *a priori* models and then, along the lines of Bayes' theorem, to use the observations as likelihood functions to derive *posteriori* models. If the observations are adequate the final models should closely resemble each other. Such attempts appear feasible only if the labor is done on a very large computer.

It would go beyond the bounds of this article to discuss the training of

---

* The fruitful interaction of theoretical, statistical and numerical methods in geophysics is, perhaps, best seen in the work of H. Jeffreys (1891–). The publication of his book "The Earth" in 1924 (now in its fifth edition) was a milestone in the subject.

students in geophysics at universities. Numerical work of the kind outlined above is so essential to the subject, however, that no geophysics student should graduate without some training in numerical analysis and high-speed computing. In my view, this should go beyond merely writing programs in, say, Fortran or Algol for one or two problems. The spectrum of numerical methods in geophysics should be introduced, together with a critical analysis of inferential procedures used throughout the history of Earth sciences. A course for geophysics majors which attempts to provide such a syllabus was introduced in the upper division in the Department of Geology and Geophysics, University of California, Berkeley in 1964.

3 FOUR EXAMPLES OF COMPUTER-AIDED STUDIES OF THE EARTH

Geophysicists have become great users of computers indeed. Four cases are described below to exemplify how this situation came about. The examples are: Fourier analysis of geophysical time series, free oscillations of the whole Earth, frequency response inversion for geophysical structures, and Gaussian spherical harmonic analysis of global geophysical parameters. The first and the last examples involve classical methods which have been used with the expenditure of great desk-computer labor for over a century. The other two examples have only been attempted on a wide scale since the availability of electronic computation.

Let us consider first the problem of decomposing data into discrete sinusoidal waves or harmonic components. As known by Fourier, under certain conditions, functions can be expressed as an infinite series of sine and cosine terms*. The idea is a fundamental one in science both for theoretical solutions of differential equations and for the treatment of observational material.

In the study of ocean tides, for instance, the water heights in a certain harbor may be measured by a tide gauge at, say, intervals of one hour. These hourly measurements represent a function which can be thought of as including terms of various periodicities. There will be waves caused by the attractions of the Moon and Sun leading to periods near twelve hours (semi-

* J. Fourier (1768–1830) treated the matter in 1807; the first assertion that any function can be represented by the addition of trigonometric terms was probably made by D. Bernoulli about 50 years before.

diurnal harmonics) and near 24 hours (diurnal harmonics). There may also be some significant harmonics corresponding to free periods of oscillation of the water in the harbor perhaps derived from local winds or storms at sea. The observed values can be fitted to trigonometric series and the amplitudes of various harmonics determined.

The next step is to determine from a statistical treatment which harmonics are significant so that inferences on tidal behaviour in the harbor can be made over a long time-span. (Such methods provided successful prediction of tides before the advent of high-speed computing machines. Certain tidal problems defied solution, however, particularly those associated with shallow water basins and estuaries. Now digital models defined by the hydrodynamical equations are used to estimate tidal elevations where little, previously, was known.)

The computations involved in harmonic analyses of this kind are lengthy and, until electronic computers were available, various short cuts, often of an analog nature*, were developed. Because of the finiteness of the sample and for other reasons the basic Fourier theory has been extended in recent years. In this regard, publications of Blackman and Tukey (see 1958), growing out of work in communications engineering, had considerable impact on periodogram analyses in geophysics from about 1960–61. These developments, of what was called "power-spectral analysis", were particularly valuable because the authors faced up to the computational problems and limitations of periodogram analysis. Subsequently, the growth of the demand for frequency spectral estimates in geophysics has stimulated further theoretical developments, independently of electrical engineering problems.

Even with electronic computers with extensive memories, however, both the power spectral method and classical harmonic analysis take uneconomical amounts of computer time in many geophysical contexts. To this end, computing algorithms, called "fast Fourier transforms", have been derived recently which greatly decrease the machine time. In a method of Cooley and Tukey (1965), for n discrete samples, the computational time is reduced by a factor $n/\log_2 n$ apart from programming differences. Alsop and Nowroozi

* Ingenious mechanical devices, called harmonic analyzers, were substituted for hand calculations at an early stage. From 1885, tidal amplitudes around the United States were calculated by a Tide Predicting Machine at the U.S. Coast and Geodetic Survey, Washington, D.C. It did not *analyze* tides but *synthesized* them. The last model was shut down, in favor of electronic calculation, only in 1966. A more refined model, planned in 1956, was not carried through as a result of digital computer availability.

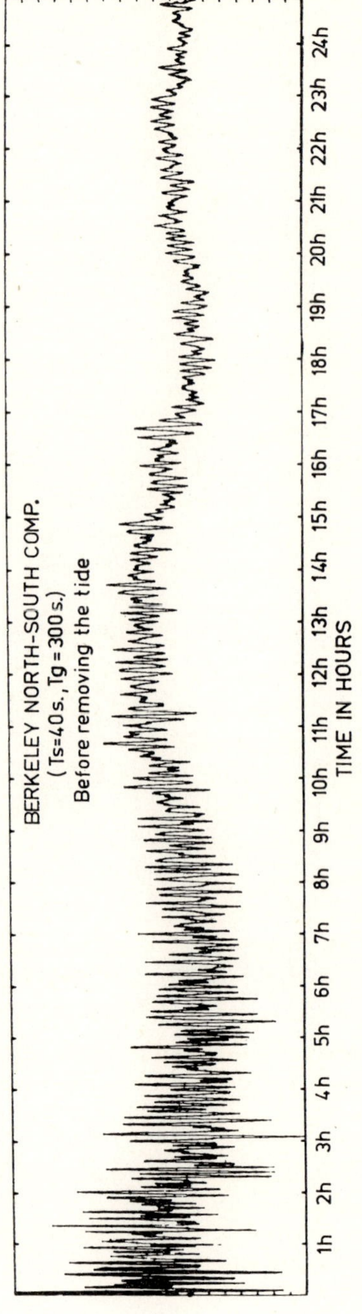

Fig. 3 XY-plot of the digital data from a photographic seismogram of the ground motion at Berkeley (north-south horizontal component) after the 1964 Alaskan earthquake. The seismograph used had a pendulum with free period 40 s and a galvanometer with free period 300 s

(1966) compared a conventional Fourier analysis program with the factor algorithm; 2048 data points from a seismogram of a large earthquake were spectral analyzed on an IBM 7094 in 1567.8 sec and 2.4 sec, respectively!

Recent practical experience in seeking periodicities hidden in complicated signals contaminated with noise has been gained in geophysics through the analysis of the vibrations of the Earth's surface after great earthquakes.

Consider the ground motion recorded by seismographs at Berkeley following the 1964 Alaskan shock. The energy released in this earthquake was so large that the whole Earth was set into its natural periods of oscillation. Fig. 3 shows the Berkeley record extending for 20 hours after the earthquake. This record was analyzed by A. A. Nowroozi by first taking digital samples of the amplitude on the record each half minute for 25 hours. The resulting 3000 samples were then subjected to both Fourier and power spectral analysis using the IBM 7094 computer. The resulting frequency spectrum from the harmonic analysis is shown in Fig. 4. One important outcome of this work (involving analysis of a number of seismograms) was the measurement of the mean amplitudes of the vertical ground displacement at Berkeley for the so-called spheroidal modes of oscillation ${}_0S_7$ to ${}_0S_{15}$ excited by the Alaskan earthquake; these ranged from 0.03 cm for ${}_0S_7$ to 0.009 cm (${}_0S_{10}$). These values were used by Kovach and Anderson (1967) to estimate a total energy of order 10^{23} ergs in this range of spheroidal modes. (Other methods have led to a total seismic energy release of about 5×10^{24} ergs.)

In 1961, I made a similar analysis, using EDSAC 2 at the Mathematical Laboratory, Cambridge University, of waves recorded after the great 1960 Chilean earthquake. The recording was achieved by a very long-period pendulum built by Professor A. Marussi at Trieste (Bolt and Marussi, 1962).

The second example follows naturally from the discussion in the last paragraph. Any finite elastic body such as the Earth has a certain set of free periods of oscillation (normal modes of vibration). The vibrations of an elastic sphere were studied by Poisson, Rayleigh* and others; in 1882 Lamb showed that there were two classes of oscillation possible. One class involves no vertical displacement of the Earth's surface, but only twisting or shearing about

---

* Much of the work of Rayleigh (J. W. Strutt, 1842–1919) found application in geophysics. An important class of observed seismic waves along the surface of the Earth was discovered theoretically by him in 1887; the waves are now called Rayleigh waves. For layered media, the mathematics of surface waves is hard; a break-through came in the nineteen fifties with numerical solutions on digital computers.

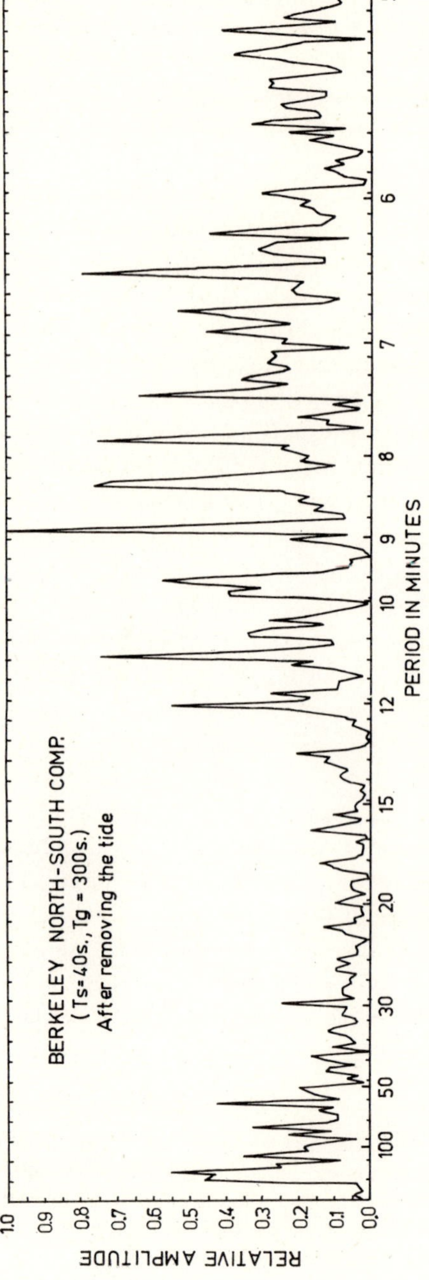

Fig. 4 XY-plot of a Fourier analysis of the data plotted in Figure 3. Significant peaks may be seen between periods of 6 and 12 minutes. These correspond to certain free oscillations of the Earth

an axis through the center of the Earth. These are called *torsional* oscillations; the sense of geometrical displacements in the longest (or gravest) period of torsional oscillations is shown at the top of Fig. 5. The second type of oscillation, named *spheroidal*, involves vertical displacements and fluctuations in local gravity. The two gravest modes are illustrated at the bottom of Fig. 5.

For a solid sphere, in which the elastic properties are simplified to be functions of the radius only, the differential equations which govern the motions of the spheroidal modes reduce to three second-order ordinary differential equations. In his standard textbook "Mathematical Theory of Elasticity", A. E. H. Love gives some of the theory and quotes his 1911 calculation that for a homogeneous sphere having the radius of the Earth, constant density, and the rigidity of steel the gravest mode would have a period of about one hour. In 1920, L. M. Hoskins treated somewhat more realistic Earth models by assuming simple algebraic variations in properties with depth. However, the numerical work was still so heavy he obtained few numerical results.

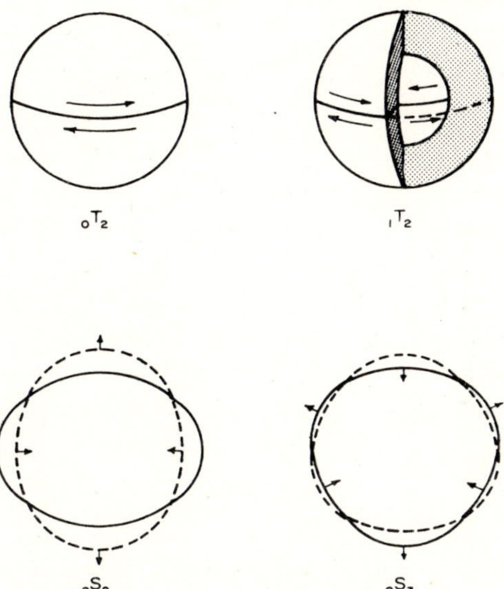

Fig. 5 Shapes and sense of motions in the longest periods of free oscillations of the Earth. The motions can be thought of as vibrations of an elastic rubber ball. The upper diagrams illustrate the fundamental torsional oscillation $_0T_2$ and its first overtone $_1T_2$. The lower diagrams show the deformation in the two gravest fundamental spheroidal vibrations $_0S_2$ (the "football" mode) and $_0S_3$

Much credit for evaluating complete sets of frequency spectra from the differential equations for much more realistic Earth models goes to C. Pekeris and his colleagues (see Pekeris, 1966) at the Weizmann Institute in Israel. It is clear that they felt able to make a new attack on this old problem because they had access to the high-speed computer WEIZAC*. The method of location of the normal frequencies of the Earth is similar to that used in 1928 by D. Hartree in studying the wave mechanics of an atom. Since the work on WEIZAC, many solutions have been calculated for various models of the Earth, Moon and terrestrial planets in an attempt either to narrow the fit between observed periods of free oscillations and those computed for a particular model or to design crucial experiments (Bolt and Derr, 1969). Indeed, in a recent tour de force F. Press (1968) at MIT examined some millions of Earth models and calculated the free oscillation spectra of those which passed certain plausibility tests. The whole analysis was handled by Monte Carlo (i.e. random) variation of various physical parameters such as velocity and ϱ at certain depths (see Fig. 2); rejection criteria were applied automatically in the computer before proceeding to the next computation.

The above is an example of an inverse problem in geophysics defined in the last section. For this particular problem a more controlled method of attack has been used by J. Derr. He computed tables of first derivatives of the frequency functions for each mode of oscillation with respect to the various free physical parameters such as β (shear velocity) and ϱ (density). He then made linear adjustments to the provisional physical parameters in order to minimize the sum of the squares of the differences between the observed periods and the computed ones.

In Table 1, the differences (residuals) are listed between the graver observed periods and the periods computed for the model DI–11 which was reached at the last iteration (see Fig. 2). The overall root-mean-squared deviation for 135 modes is only 0.61 sec, a remarkably small value. This work involved the computation of dozens of complete solutions on both IBM 7094 and

---

* WEIZAC, much used for theoretical geophysics, was a machine developed at the Weizmann Institute. It was one of the class of computers whose design was initiated at the Institute for Advanced Study, Princeton, New Jersey by the late Professor J. von Neumann. ILLIAC built at the University of Illinois was another machine of this class. SILLIAC, at Sydney University, was copied from ILLIAC plans. In 1959, J. Butcher and I (1960) computed, on the Sydney machine, one of the first suites of dispersion curves for Rayleigh waves in a medium with two layers. It was used about the same time by E. A. Flinn and L. Howard to provide automatic locations of earthquake epicenters. I have no record of ILLIAC being used for geophysical work.

Table 1 Free periods and residuals for earth model DI–11

Mode	Computed period sec	Residual sec	Mode	Computed period sec	Residual sec
${}_0S_0$	1227.19	0.5			
${}_0S_2$	3232.39	0.9	${}_0T_2$	2638.75	1.9
${}_0S_3$	2134.08	−0.5	${}_0T_3$	1706.55	1.3
${}_0S_4$	1545.71	1.5	${}_0T_4$	1305.88	−1.5
${}_0S_5$	1190.57	−1.3	${}_0T_5$	1076.66	−0.4
${}_0S_{10}$	579.81	0.2	${}_0T_{10}$	618.94	0.4
${}_0S_{20}$	346.84	0.6	${}_0T_{20}$	360.11	0.4
${}_0S_{30}$	261.57	0.4	${}_0T_{30}$	258.11	−1.3
${}_0S_{40}$	212.16	−0.1	${}_0T_{40}$	201.54	0.3
${}_0S_{50}$	178.25	−0.4	${}_0T_{50}$	165.34	0.2
${}_0S_{75}$	126.26	−0.3	${}_0T_{75}$	114.06	0.1

CDC 6400 computers; for some modes, double precision arithmetic (28-decimal place accuracy) was needed for the root-location method used.

A recent fruitful method in geophysics could hardly be attempted without the use of high-speed computers. It is applicable to the study of any linear physical system, that is, one in which superposition holds. The name of the method, which can be traced back to the Operational Calculus of Oliver Heaviside (1850–1925), varies somewhat but in seismology it can generally be recognized under the name "frequency response function" technique. It provides the third specific example of the contemporary role of computers in geophysics.

Consider the source of a wave which might be seismic, electromagnetic or acoustic, for example. The wave travels through the terrestrial medium and is recorded as a function of time by a receiving instrument. Then, for a linear system, the recorded signal $f(t)$ can be thought of as a function of *frequency* $F(\omega)$ which is the simple algebraic product of three separate functions

$$F(\omega) = S(\omega)\, M(\omega)\, R(\omega). \tag{1}$$

The first function, $S(\omega)$, represents the source output in the frequency domain, the second, $M(\omega)$, is the transfer or response function of the medium and $R(\omega)$ denotes the frequency response of the receiver.

In practice, all functions in (1) are known only in numerical (digital) form. The procedure is applicable to geophysical prospecting, where the frequency spectrum of the explosive source as well as the frequency response of the geophone which records the waves are known. The recorded signal can be treated, using the above relation, to obtain numerically on a computer the response function of the geological strata.

By Fourier theory, a product of functions of frequency can be transformed to an equivalent function of time so that variation in the time domain can be predicted if required. For example, suppose for simplicity that $R(\omega)$ is unity in (1), then $F(\omega)$ is equivalent to the superposition (or convolution) integral

$$f(t) = \overline{F}(t) = \int_{-\infty}^{\infty} \overline{S}(\tau)\, \overline{M}\,(t - \tau)\, d\tau \tag{2}$$

where overbars denote Fourier transforms. Integrals like (2) are tedious to evaluate without automatic computers.

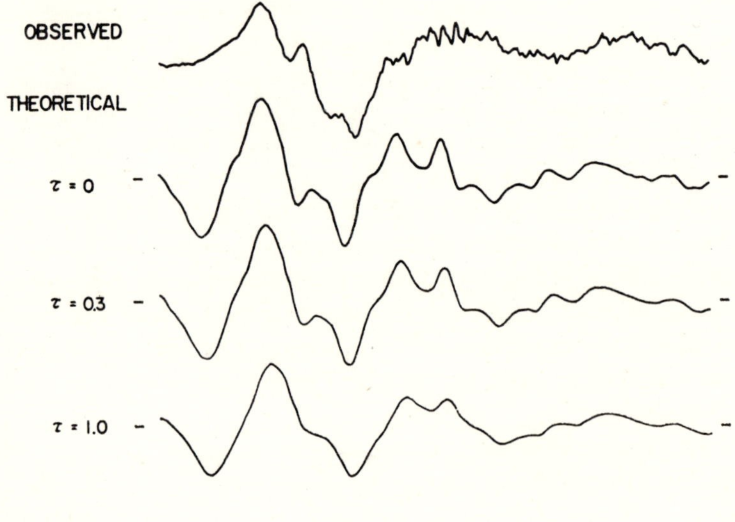

Fig. 6 A comparison of barograms (continuous measurement of air pressure). The top trace is the recorded variation in pressure at Berkeley during the passage of an atmospheric acoustic wave generated by the 1964 Alaskan earthquake. The lower three traces are computed barograms for theoretical models; τ is the time constant (in minutes) of ground deformation

Another application has been worked out in a novel problem which arose following the great 1964 Alaskan earthquake previously mentioned. The vertical motion of the ground near the earthquake source was evidently so rapid and extensive that a pressure acoustic wave in the atmosphere was generated near the Prince William Sound region of Alaska. The resulting dispersed acoustic wave was recorded very clearly by a microbarograph at the Berkeley observatory (Bolt, 1964) and elsewhere. The form of this wave, $f(\tau)$ say, is shown at the top of Fig. 6.

The observed frequency spectrum of this barogram, $F(\omega)$, can be decomposed, as in equation (1), into a product of $S(\omega)$, the source pressure function, $D(\omega)$, the source dimension factor, $A(\omega)$, the atmospheric transfer function, and $B(\omega)$, the barograph response. The factor $D(\omega)$ which is a measure of the areal extent of the regional vertical displacements can be estimated from the post-earthquake geodetic field measurements in Alaska.

Now, a good deal is known about the physical properties and layering of the atmosphere. It is possible, therefore, in this case to take a realistic model of the atmosphere and compute its response function $A(\omega)$. Because the response of the barograph, $B(\omega)$, was known, the function $S(\omega)$, (and hence the equivalent source time-function), could be estimated. This research was carried through by T. Mikumo (1968) with great success; numerical work was done partly by a HITAC 5020 computer at the University of Tokyo. He was able to show that rapid surface uplift and depression over a region of dimensions about 700 by 400 sq km could explain the observed barogram. In Fig. 6, three computed barograms with different time constants τ are drawn. The condition $\tau \geqslant 1$ min introduced inconsistencies between the theoretical and observed acoustic waves thus setting a bound on the speed of vertical ground motion.

The final example illustrates further the way in which computers have allowed classical methods of analysis to be used more effectively. The example concerns the variation of the Earth's magnetic field over a considerable interval of geological time.

In his mature years, Gauss* extended Fourier analysis methods to the three-dimensional problem of the sphere by carrying out a spherical har-

---

* The ubiquitous mathematical genius Gauss (1777–1855) published memoirs on geomagnetism in 1832 and 1838. He invented the bifilar magnetometer for use in terrestrial magnetism and the heliotrope (and "least squares") for geodesy. He saw great importance in improved computational methods but did most of his figuring mentally.

monic analysis of the observed magnetic field of the Earth. In this pioneer work he showed that the origin of both the regular and the irregular parts of the field was almost entirely internal to the Earth. In this type of analysis, the sine and cosine functions of Fourier analysis must be replaced by a series of spherical harmonics which satisfy Laplace's equation.

The expansion for the magnetic potential V at radius r is

$$V = a \sum_{n=1}^{\infty} \sum_{m=0}^{n} P_n^m(\theta) \left[\{c_n^m z^n + (1 - c_n^m) z^{-(n+1)}\} A_n^m \cos m\varphi + \{s_n^m z^n + (1 - s_n^m) z^{-(n+1)}\} B_n^m \sin m\varphi \right], \tag{2}$$

where θ and φ are the co-latitude and longitude, $z = r/a$, $a = 6371$, $0 < c_n^m < 1, 0 < s_n^m < 1$, and P_n^m is the associated Legendre function. The constants c_n^m, s_n^m, A_n^m and B_n^m are constants to be estimated from the observations of magnetic field intensity and field direction. The first pair partition the field between internal and external sources.

Spherical harmonic analyses of the present field indicate that the major contribution to V (more than 90 per cent) comes from the dipole terms, $n = 1$. Some non-dipole terms ($n > 1$) can be detected but fluctuate in intensity. It is a crucial question in geomagnetism to determine if the geomagnetic field has been mainly dipolar throughout geological time. The answer lies in the fossil magnetism which is found in many rocks. Lava, for example, will acquire a magnetization which is representative of the strength and direction of the local geomagnetic field at the time of solidification. Thus careful geological sampling of lavas in time and space (with due allowance for deformation and physical and chemical changes since the rocks formed) should permit a reconstruction of the past geomagnetic field. In such paleomagnetic work it turns out that directions of permanent magnetization can be measured rather precisely but intensity cannot in general.

Measurements of the direction of magnetization are now available for rocks from a number of sites on the various continents. In a recent study, J. Wells set out to determine the configuration of the Earth's field for the geological periods Quaternary (recent to 3*my* ago), Upper Tertiary (3 to 25*my* ago), Lower Tertiary (25 to 63*my* ago). He assumed that the source of the ancient field is internal to the surface $r = a$, so that $c_n^m = s_n^m = 0$ in (2). Difficulties arise in the statistical evaluation of coefficients in (2) because of the sparseness and nonuniformity of sites around the globe; difficulties in interpretation of the analysis of paleomagnetic data combined from different continents arise from possible continental drift and wandering of the magnetic pole.

The series (2) was terminated with the octopole ($n = 3$) terms so that a maximum of 15 coefficients were estimated numerically. The best estimates were taken to be those which minimized a certain sum S of the residuals between observed and predicted field directions. The evaluation of S on a CDC 6400 required lengthy calculations. S is non-linear in the spherical harmonic coefficients A_n^m, B_n^m and its minimization raises the usual difficulties of distinguishing between local and global minima of the hypersurface. In order to test the sensitivity of the method and the degree of improvement in fit when non-dipole terms were included many synthetic sets of paleomagnetic data were also generated by Monte Carlo methods and harmonic analyses made of these.

The main conclusions adduced by Wells were that (a) for the Quaternary, all 15 terms are needed to satisfy the data (from 47 sites) but the dipole field dominates, (b) for the Late Tertiary, the field was predominately an axial dipole, and (c) for the Early Tertiary, (with only 17 sites) there was a suggestion of scatter in measurement in excess of that expected from quadrupolar terms alone.

4 A CRITIQUE OF THE USE OF COMPUTERS IN GEOPHYSICS

The high-speed electronic calculating machine has, as we have demonstrated, opened the way to a new level of geophysical activity by restoring to the geophysical worker the ability to create without being overwhelmed by calculations; it has freed his interest and his intelligence for acts of insight, criticism and pattern-perception.

At an earlier time in geophysics, creative acts often went together with hand computational work. In this respect, there was a close kinship with the methods and procedures in astronomy and it is no accident that scientists such as Laplace and Gauss contributed to both astronomy and geophysics. Capital examples of geophysical research involving massive hand calculations can be found in the study of the yielding of the whole Earth to period disturbing forces such as the attraction of the Sun and Moon. In the first rank are G.H. Darwin's* monumental studies of tidal friction. The difficult gene-

* Jeffreys dedicated "The Earth" to G.H. Darwin (1845–1912). Darwin's scientific papers receive little study these days mainly because high-speed computers have allowed the treatment of more realistic Earth and ocean models by simpler mathematical methods.

ral problem has been advanced by the work of H. Takeuchi (1950), M. S. Molodensky (1953) and Jeffreys and R. O. Vincente (see 1957)*. In each case much intricate hand computation was involved such as numerical integrations; the agreement between observed and predicted motions of the axis of rotation (variation of latitude and nutations) is still not wholly satisfactory.

Some would argue that the necessity for personal hand computation leads to a sounder treatment of the geophysical problem, that important mental interactions occur during the processes of numerical analyses and the drawing of geophysical inferences. In short, there is a likelihood that properties will be overlooked if analysis is not carried through personally. Whatever the truth on this point, it is likely to be the case that the use of high-speed computers will remove the possibility of geophysicists making much contribution in numerical analysis. High-speed computation does not encourage the seeking of clever, sophisticated short-cuts in arithmetic; often quite elementary arithmetic methods may be programmed and the speed and large number of decimal places used provide adequately accurate solutions.

It is generally agreed that the place of computations (and mathematics) in geophysics is to throw fuller light on the Earth. There is, however, a danger that the adequacy of more elementary numerical methods in much (but not all) high-speed computing will lead students in geophysics to become less knowledgeable of numerical analysis in general. For example, nowadays some students see no need to know any more about numerical integration than Simpson's rule! Perhaps more important, the trend to simplicity of arithmetic work may make a thorough understanding of the classical geophysical treatises more difficult. The extent to which this will be a loss to the subject is not yet clear.

The use of electronic computers may also react on scientific method in the Earth and planetary sciences. For many geophysical problems, order of magnitude calculations are very useful and may be, in fact, all that can be justified, given the vagueness of the observations available. Such order of magnitude arguments have been developed and used with great effectiveness in the subject. They often lead to crucial tests which rule out competing geophysical theories. In Jeffreys' words, "the method of exhaustion of alternatives is specially useful in geophysics, because incorrect geophysical hypotheses usually fail by extremely large margins".

---

* An account of Earth tides and the use of electronic data processing in this research can be found in the book by P. Melchior (1966).

In geomagnetism, for example, the theory that the Earth's magnetic field arose from uniform magnetization of the interior could be ruled out when the effect of high temperatures on ferromagnetic substances became known. At a particular temperature, the Curie point, ferromagnetic properties are lost. For iron, the Curie temperature is 750 °C which must occur above a depth of 30 km in the Earth. Uniform magnetization (unless the Curie point increases with increasing pressure in the deep interior) may thus be rejected even though the temperature distribution in the Earth is not known very precisely. As another simple case, the empirical result $I = 0.3308\ Ma^2$ for the Earth's moment of inertia makes quite implausible any interior structure which does not allow a fairly strong density increase with depth. (For a uniform sphere, $I = 0.4\ Ma^2$.)

A very recent illustration of the effectiveness of rather general arguments comes from the global tectonic theory concerning ocean-floor spreading which has recently received much attention. The strength of the theory stems from new oceanographic observations which show very symmetric patterns of magnetic reversals of the rocks below the floors of the oceans parallel to and on each side of the mid-oceanic ridges. This geomagnetic evidence suggests that crustal material spreads steadily from ridges across ocean floors. At ocean trenches, the spreading crust is absorbed; if there are no trenches, as in the Atlantic, the continents move apart to keep pace with the spreading. The model has been confirmed by the mechanism studies of a number of earthquakes. On the ridges earthquakes are consistent generally with tensional stresses and when offsets in ridges occur earthquakes along these scarps show mainly horizontal displacements. This theory has given new life to surmises about continental drift and provides cohesion to a great deal of geophysical and geological material which was previously a mass of independent fragments. Many geophysicists and geologists regard this recent work as a revolution in Earth science (Vine, 1966). It is, therefore, of interest to consider that computers played a fairly minor role in its working out. The magnetic and mechanical ideas were established without the need for much work on high-speed computers. What was needed was recognition of a pattern in the observational material, particularly that recently gathered. (Computers played a role, of course, in the treatment of the raw data upon which the new global scheme is based.)

A final concern regarding the long-term effect of the availability of large high-speed computers in geophysics is related to their efficiency in data processing. The concern was stated in 1964 by V. I. Keilis–Borok:

> "Changes [in geophysics] are coming so fast that utterly new conditions are cropping up. We see torrents of information which are impossible to process, if we would think about them in the habitual way; and as a result we see the information disintegrate into an infinite number of separate parts."

The evidence seems clear that, on balance, modern computing machines are enormously helpful for geophysical storage and reduction processes. On the other hand, a tendency appears now and again to amass enormous quantities of data almost for their own sake. The ease of data handling on computers makes this a relatively easy task for those who have little conception of, or interest in, geophysics itself. Such data storage is often justified in terms which reduce to essentially the old scientific method of Francis Bacon. In his philosophy, contempt for theory and the exhaustive collection of "facts" were carried to an extreme; a scientist was advised to proceed by being energetic and by handling and stroking all the objects which came within reach. Ultimately, it was held, some worthwhile property or conclusion would be discovered. Even if only for the sake of readers of geophysical journals, we should perhaps hope that this particular philosophical procedure will not find a new lease of life in geophysics through the wide availability of high-speed computers.

Acknowledgments

Dr. Cecil E. Leith and Dr. John M. Wells very kindly provided criticism of this manuscript. Some specific figures and examples were taken from recent work in the Department of Geology and Geophysics at Berkeley by Dr. John S. Derr, Dr. T. Mikumo, Dr. A. A. Nowroozi and Dr. John M. Wells.

Suggested reading

Bellamy, J. G. (1952). "Automatic Processing of Geophysical Data", *Advances in Geophysics*, ed. by H. E. Landsberg, Academic Press, 1–43.

Bullard, E. (1964). "The Language of Machines", *Endeavour*, **23**, 160–164.

Bullen, K. E. (1967). "Basic Evidence for Earth Divisions", in *The Earth's Mantle*, ed. by T. F. Gaskell, Academic Press.

Grant, F. S. and G. F. West (1965). *Interpretation Theory in Applied Geophysics*, McGraw–Hill.

Harris, D. L., N. A. Pore and R. A. Cummings (1965). "Tide and Tidal Current Prediction by High-Speed Digital Computers", *International Hydrographic Review*, **42**, 1, January.

Jeffreys, H. (1970). *The Earth* (fifth edition), Cambridge University Press.

Keilis-Borok, V.I. (1964). "Seismology and Logics", in *Research in Geophysics*, **2**, 61–79.

Kopal, Z. (1966). *An Introduction to the Study of the Moon*, in Astrophysics and Space Library, **4**, Gordon and Breach, Science Publishers.

Landsberg, H.E. (1958). "Data Processing in Geophysics", in *Contributions in Geophysics in Honor of B. Gutenberg*, Pergamon Press, 210–227.

Melchior, P. (1966). *The Earth Tides*, Pergamon Press.

Schmid, H.H. (1969). "Application of Photogrammetry to Three-Dimensional Geodesy", *Trans. Am. Geophys. Un.*, **50**, 4–12.

Panel Report (1964). "Solid-Earth Geophysics—Survey and Outlook", U.S. National Academy of Sciences—National Research Council, 1231 pp.

Various Authors (1967). *IEEE Trans. on Audio and Electroacoustics*, (Special Issue on Fast Fourier Transforms) AV–**15**.

Examples of recent specialist work in geophysics involving high-speed computers may be found in collected papers presented at successive symposia on "Geophysical Theory and Computers" under the auspices of the International Upper Mantle Committee. Most papers are seismological but terrestrial magnetism, gravity, etc. are represented.

Proceedings of the Symposia have been published as follows.

First Symposium, Moscow and Leningrad, USSR, May 15–21, 1964, *Reviews of Geophysics*, **3**, Number 1, 1965.

Second Symposium, Rehovot, Israel, June 13–23, 1965, *Geophysical Journal, Roy. Astro. Soc.*, **11**, Numbers 1–2, 1966.

Third Symposium, Cambridge, U.K., June 27–July 5, 1966. *Geophysical Journal, Roy Astro. Soc.*, **13**, Numbers 1–3, 1967.

Fourth Symposium, Trieste, Italy, Sept. 18–22, 1967, *Supplemento al Nuovo Cimento*, **6**, (First Series), 1968.

Fifth Symposium, Tokyo and Kyoto, Japan, August 1968 (Proceedings not yet published).

References

Alsop, L.E. and A.A. Nowroozi (1966). "Faster Fourier Analysis", *J. Geophys. Res.*, **71**, 5482–5483.

Blackman, R.B. and J.W. Tukey (1958). *The Measurement of Power Spectra*, Dover, New York.

Bolt, B.A. (1960). "The Revision of Earthquake Epicenters, Focal Depths and Origin-Times Using a High-Speed Computer", *Geophys. J., Roy. Astro. Soc.*, **3**, 433–440.

Bolt, B.A. and A. Marussi (1962). "Eigenvibrations of the Earth Observed at Trieste", *Geophys. J., Roy. Astro. Soc.*, **6**, 299–311.

Bolt, B.A. (1964). "Seismic Air Waves from the Great 1964 Alaskan Earthquake", *Nature*, **202**, 1095–1096.

Bolt, B.A. and J. Derr (1969). "Free Oscillations of the Terrestrial Planets", *Vistas in Astronomy*, **7**, Pergamon Press.

Bullard, E.C. and H. Gellman (1954). "Homogeneous Dynamos and Terrestrial Magnetism", *Phil. Trans., Roy. Soc. A.*, **247**, 213–278.

Butcher, J. and B.A. Bolt (1960). "Rayleigh Wave Dispersion for a Single Layer on an Elastic Half Space", *Australian Journal of Physics*, **13**, 498–504.

Cooley, J.W. and J.W. Tukey (1965). "An Algorithm for the Machine Calculation of Complex Fourier Series", *Math. of Comput.*, **19**, 297–301.

Derr, J.S. (1968). "Internal Structure of the Earth Inferred from Free Oscillations", Ph.D. Dissertation, University of California, Berkeley.

Idriss, I.M. and H.B. Seed (1968). "An Analysis of Ground Motion during the 1957 San Francisco Earthquake", *Bull. Seism. Soc. Am.*, **58**, 2013–2032.

Hoskins, L.M. (1920). "The Strain of a Gravitating Sphere of Variable Density and Elasticity", *Trans. Amer. Math. Soc.*, **21**, 1–43.

Jeffreys, H. and R.O. Vincente (1957). "The Theory of Nutations and the Variation of Latitude" (2 papers), *Mon. Not. R. Astro. Soc.*, **117**, 142–173.

Jeffreys, H. (1961). *Theory of Probability* (Third Edition), Oxford University Press.

Keilis-Borok, V.I. and T.B. Yanovskaya (1967). "Inverse Problems of Seismology (Structural Review)", *Geophys. J., Roy. Astro. Soc.*, **13**, 223–234.

Kovach, R.L. and D.L. Anderson (1967). "Study of the Energy of the Free Oscillations of the Earth", *J. Geophys. Research*, **72**, 2155–2168.

Mikumo, T. (1968). "Atmospheric Pressure Waves and Tectonic Deformation Associated with the Alaskan Earthquake of March 28, 1964", *J. Geophys. Research*, **73**, 2009–2025.

Molodensky, M.S. (1953). "Elastic Tides, Free Nutation and Some Questions Relating to the Earth's Structure", *Tr. Geofiz. in-ta Akad. Nauk. SSSR*, Sb. Statey, No. 19, 146.

Nowroozi, A.A. (1965). "Terrestrial Eigenvibrations Following the Great Alaskan Earthquake, May, 1964", Ph.D. Dissertation, University of California, Berkeley.

Pekeris, C.L. (1966). "The Internal Constitution of the Earth", *Geophys. J., Roy Astro. Soc.*, **11**, 85–132.

Press, F. (1968). "Density Distribution in Earth", *Science*, **160**, 1218–1221.

Smith, S.W. (1965). "Seismic Digital Data Acquisition System", *Reviews of Geophysics*, **3**, 151–156.

Takeuchi, H. (1950). "On the Earth Tide of the Compressible Earth of Variable Density and Elasticity", *Trans. Amer. Geophys. Un.*, **31**, 651–689.

Tobin, M. and M. Talwani (1966). "Rapid Computation of the Gravitational Attraction of Topography on a Spherical Earth", *Geophys. Prospect.*, **14**, 114–142.

Vine, F.J. (1966). "Spreading of the Ocean Floor: New Evidence", *Science*, **154**, 1405–1415.

Wells, J.M. (1969). "Spherical Harmonic Analysis of Paleomagnetic Data". Ph.D. Dissertation, University of California, Berkeley.

CHAPTER 23

Computer Simulations of the Global Circulation of the Earth's Atmosphere

AKIRA KASAHARA

1 INTRODUCTION

IN THE LAST DECADE, large, high-speed electronic computers have made a profound impact on our ability to understand the large-scale motions of the earth's atmosphere. The degree to which we understand the atmosphere can be measured by our success in simulating both day-to-day and long-term changes in the global circulation patterns.

In this article, we review the physical, mathematical, and computational aspects of the numerical simulation of large-scale motions of the earth's atmosphere. In particular, we will give a perspective view of how large, high-speed computers are being used to aid our understanding of the earth's atmosphere, not only for the sake of research, but also for social and economic benefit.

2 BASIC ATMOSPHERIC EQUATIONS

The motions of the atmosphere are governed by physical laws which are expressed in the form of the equations of hydrodynamics and thermodynamics. These equations are, in principle, well known, but some of the physical processes in the atmosphere, such as small-scale turbulence and convection, cloud formation, and precipitation mechanisms, are still difficult to express quantitatively.

Let us write the basic system of atmospheric equations as follows:

$$\frac{\partial u}{\partial t} = -\mathbf{V} \cdot \nabla u - w \frac{\partial u}{\partial z} + fv - f'w - \frac{1}{\varrho} \frac{\partial p}{\partial x} + F_x, \tag{1}$$

$$\frac{\partial v}{\partial t} = -\mathbf{V} \cdot \nabla v - w \frac{\partial v}{\partial z} - fu - \frac{1}{\varrho} \frac{\partial p}{\partial y} + F_y, \tag{2}$$

$$\frac{\partial w}{\partial t} = -\mathbf{V} \cdot \nabla w - w \frac{\partial w}{\partial z} + f'u - g - \frac{1}{\varrho} \frac{\partial p}{\partial z} + F_z, \tag{3}$$

$$\frac{\partial \varrho}{\partial t} = -\nabla \cdot (\varrho \mathbf{V}) - \frac{\partial (\varrho w)}{\partial z}, \tag{4}$$

$$\frac{\partial s}{\partial t} = -\mathbf{V} \cdot \nabla s - w \frac{\partial s}{\partial z} + \frac{Q}{T}, \tag{5}$$

$$s = C_v \ln p - C_p \ln \varrho, \tag{6}$$

$$T = \frac{p}{\varrho R}, \tag{7}$$

where

$$\mathbf{V} \cdot \nabla = u \frac{\partial}{\partial x} + v \frac{\partial}{\partial y}.$$

In these equations, x, y, and z are the three Cartesian space coordinates, which are directed eastward, northward, and upward, respectively. The time coordinate is denoted by t. The velocity components in the x, y, and z coordinates, respectively, are denoted by u, v, and w. The horizontal velocity $\mathbf{V}$ denotes $u\mathbf{i} + v\mathbf{j}$, where $\mathbf{i}$ and $\mathbf{j}$ are unit vectors in the x and y coordinates, respectively. ϱ represents density, p pressure, s specific entropy, and T temperature. $f = 2\Omega \sin \varphi$ and $f' = 2\Omega \cos \varphi$, where Ω denotes the angular speed of the earth's rotation, and φ denotes latitude. g is the acceleration due to the earth's gravity. F_x, F_y, and F_z are the three components of the frictional force per unit mass. Q is the rate of heating per unit mass. C_p and C_v are the specific heat at constant pressure and constant volume, and R is the gas constant of dry air equal to $C_p - C_v$. The first three equations are well known equations of motion under the effect of the earth's rotation. The fourth equation is that of mass continuity, and the fifth one is the thermodynamic equation. Equation (6) is the definition of specific entropy s, and Eq. (7) the equation of state. For simplicity the equations for predicting the water vapor field and describing the condensation process have been omitted.

In the system of equations (1) through (7), x, y, z, and t are called the *coordinate* variables. u, v, w, ϱ, and s are called *prognostic* variables because there are equations for the time-rate-of-change of these variables. p and T are called *diagnostic* variables because they are computed from Eqs. (6) and (7), which do not include time derivatives.

If F_x, F_y, F_z, and Q were expressed in terms of u, v, w, p, ϱ, s, and T, these equations, together with the proper boundary conditions, would be complete in the sense that there are the same number of equations as of unknowns.

The usual method of solving such a system of nonlinear partial differential equations is to replace the partial derivatives by finite differences in both space and time so that the unknowns, u, v, w, ϱ, p, s, and T, are defined only at the points of a four-dimensional lattice in space and time. If the variables u, v, w, ϱ, and s are known at $t = t_0$ at all grid points in space, the right-hand sides of Eqs. (1) through (5) may be computed by appropriate space differences. The resulting point values for the local time derivatives of u, v, w, ϱ, and s may then be multiplied by a suitable time increment, Δt, and added to the values of u, v, w, ϱ, and s at $t = t_0$ in order to define them at the time $t_0 + \Delta t$. Repetition of this process will, in principle, yield a forecast for any later time.[1-4]

It turns out, however, that a constraint on the value of Δt exists when these equations are solved by explicit difference schemes. This constraint is known as a *computational stability condition*, and was first discussed by Courant, Friedrichs, and Lewy[5] in 1928. Roughly speaking, the time increment, Δt, must satisfy the condition

$$C_m \frac{\Delta t}{\Delta s} < 1 \tag{8}$$

where Δs represents one of the space increments, Δx, Δy, or Δz, and C_m denotes the maximum magnitude of the characteristic velocity of the system. (If implicit difference schemes are used, such a constraint may not appear. However, implicit schemes are usually not economical to use.)

The maximum characteristic speed in this system is the speed of sound, 300 m/sec. If we choose 300 km as a representative horizontal grid distance, we find that Δt is less than 1000 sec, or 15 min. However, sound waves in this system propagate not only in the horizontal direction but also in the vertical. If we choose 3 km as a representative vertical grid distance, we find that Δt

must be less than 10 sec! To use such a small time step for numerical integration of a system designed to predict weather patterns for a period of ten days or so is impractical, even with modern high-speed computers. We must, therefore, consider modifications to the basic system of the atmospheric equations in order to eliminate sound waves from the system.

3 MODIFICATIONS OF BASIC ATMOSPHERIC EQUATIONS

For large to medium-scale motions in the atmosphere, terms in Eq. (3) are neglected except the fifth and sixth terms. Thus, Eq. (3) reduces to

$$\frac{\partial p}{\partial z} = -\varrho g. \tag{9}$$

Equation (9) states that the pressure difference Δp in the vertical is expressed simply by the static weight of the mass within the height increment Δz times the acceleration of gravity. This equation is called the *hydrostatic equation* and is commonly used in meteorology, oceanography, geodynamics, and astrophysics.

If we replace Eq. (3) by Eq. (9), we can show that internal sound waves are filtered out but that gravity waves undergo little change.[6] The maximum speed of gravity waves can be as large as that of sound waves, but the fact that sound waves do not propagate vertically in the hydrostatic system relaxes the severe computational stability condition mentioned earlier.

Since the equation for the time-rate-of-change of w does not exist, w is no longer a prognostic variable. Also because of the hydrostatic equation (9), one can no longer calculate $\partial\varrho/\partial t$ and $\partial s/\partial t$ (or $\partial p/\partial t$) independently from Eqs. (4) and (5) [or from Eq. (5) together with (6)]. This constraint leads to a diagnostic equation for w. In other words, w is so calculated that the computations of $\partial\varrho/\partial t$ from (4) and $\partial s/\partial t$ from (5) satisfy the hydrostatic condition (9) all the time. This important constraint was first noted in 1922 by Lewis F. Richardson[7] in his pioneering work, *Weather Prediction by Numerical Process*. Therefore, this diagnostic equation for w is now called the *Richardson equation*. The Eulerian hydrodynamic equations modified by the assumption of hydrostatic equilibrium are called, in meteorology, the *primitive equations*.

Richardson actually tried numerical integrations of primitive equations using observed meteorological data available to him at that time. As one might expect, his crude, hand-calculated forecast was hopelessly unsuccessful and even discouraged meteorologists from pursuing his objective approach to weather forecasting for the next two decades.[8] But Richardson had a dream: "Perhaps some day in the dim future it will be possible to advance the computations faster than the weather advances and at a cost less than the savings to mankind due to the information gained. But that is a dream". (Quoted from Richardson's book.[7])

In the mid-1940s, John von Neumann, of the Institute of Advanced Studies at Princeton, began to build an electronic computer, primarily for the purpose of weather prediction. At the Institute, a meteorological group was established to attack the problem of numerical weather prediction through a step-by-step investigation of models designed to approximate more and more closely the real properties of the atmosphere. In 1950, results of the first successful numerical prediction were published by Charney, Fjørtoft, and von Neumann.[9] Their atmospheric model is far simpler than Richardson's model, but is based on the so-called *quasi-geostrophic model* which will be explained later.

4 SCALES OF ATMOSPHERIC MOTION

In the days of Richardson, meteorologists did not have a clear idea about the scale of weather systems and the principal processes involved in a particular scale of motions. Perhaps this is why Richardson was reluctant to make even minor approximations and why he considered practically all the physical processes involved, including the effects of condensation nuclei, evaporation from soil and transpiration from plants, even for a one-day forecast. Recent successes in numerical weather prediction have depended upon recognition of the different scales of motion which govern weather systems, and upon simplification of the Navier–Stokes hydrodynamic equations by mathematical approximations in order to describe motions of a particular scale. Table I shows the scales of motion associated with different atmospheric phenomena.

Motions of a scale greater than a few thousand kilometers are known as large-scale motions. Figure 1 shows a pressure map of the Northern Hemisphere for a particular winter day. The lines are isobars at a height of 6 km with contour intervals of 5 mb. The symbols H and L denote the high and low pressure areas, called *cyclones* (or *troughs*) and *anticyclones* (or *ridges*), depending upon the configuration of the patterns. For this scale of motion,

Table I Scales of motion associated with typical atmospheric phenomena

Classification	Scale (km)	Typical phenomena
large scale	10,000	very long waves
	5,000	cyclones and anticyclones
medium scale	1,000	frontal cyclones
	500	tropical cyclones
mesoscale	100	local severe storms, squall lines
	50	hailstorms, thunderstorms
small scale	10	cumulonimbi
	5	cumuli
	1	tornados, waterspouts
microscale	0.5	dust devils, thermals
	0.1	
	0.05	transport of heat, momentum, and
	0.01	water vapor

the relation between the pressure field and the field of motion is given, to a first approximation, by the *geostrophic wind* law. This law states that the air moves parallel to the isobars, clockwise around a high pressure area and counterclockwise around a low pressure area in the Northern Hemisphere, and in the opposite direction in the Southern Hemisphere. The *pressure force* is, of course, directed at right angles to the isobars, from higher to lower pressure. In the mid-troposphere, friction may be neglected, and the pressure force is ordinarily balanced by the *Coriolis force* (the deflecting force which results from the earth's rotation and which acts at 90 degrees to the motion, deflecting the flow to the right in the Northern Hemisphere and to the left in the Southern Hemisphere).

Since the pressure generally decreases from the equator to the North Pole, the wind generally blows from the west to the east. It is called *westerly* when the flow is from west to east and *easterly* when the flow is from east to west. Figure 1 shows the prevailing mid-latitude westerlies meandering around the earth. The portion of the zonal flow where the intensity is maximum is called the *jet stream.*

Large-scale meandering motions, of the type seen in Fig. 1, are called *long waves.* They were discovered from the upper air analyses during World War II. The work of J. Bjerknes and C. Rossby led to the important conclusion that long waves are approximated by a state of geostrophic balance between the horizontal wind and pressure fields. The latter observation gave

Charney[10] and Eliassen[11] a basis for deriving the concept of a *quasi-geostrophic approximation* to simplify hydrodynamic equations. This simplification was later used successfully by the Princeton meteorological group to formulate a weather prediction model.

In 1955, the National Meteorological Center, U.S. Weather Bureau, began to issue numerical forecasts on an operational basis by means of an electronic computer. Until recently, their forecasting model has been of the quasi-geostrophic type mentioned earlier.[12,13]

For motions of medium scale and smaller, the deviation of wind from a geostrophic balance gradually becomes important. Figure 2 shows a sea-

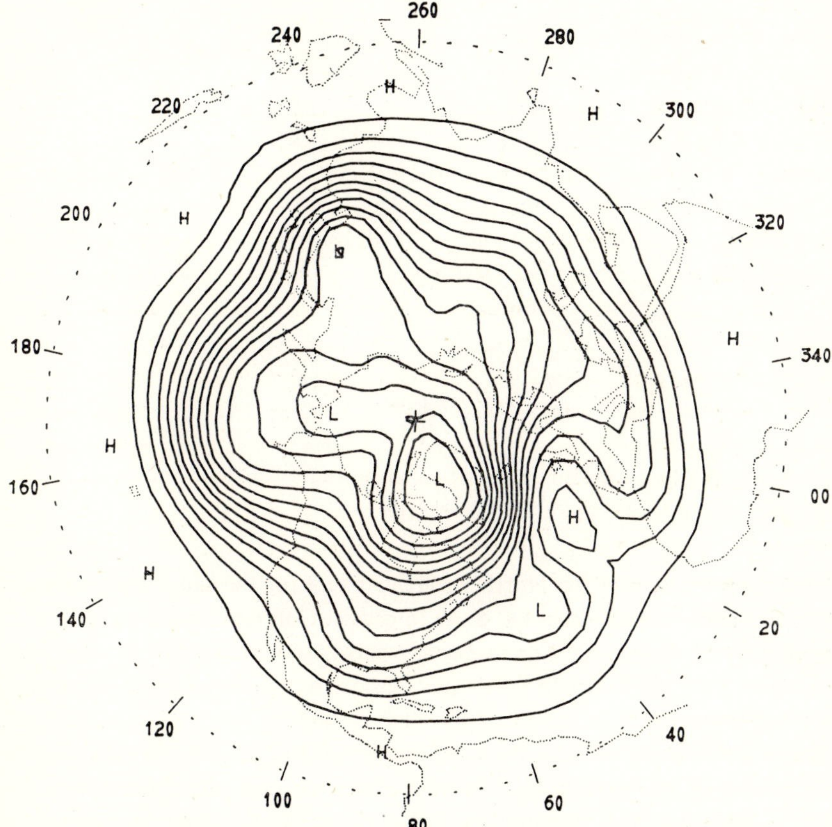

Fig. 1 Northern Hemisphere map of pressure at 6 km. Contour interval is 5 mb

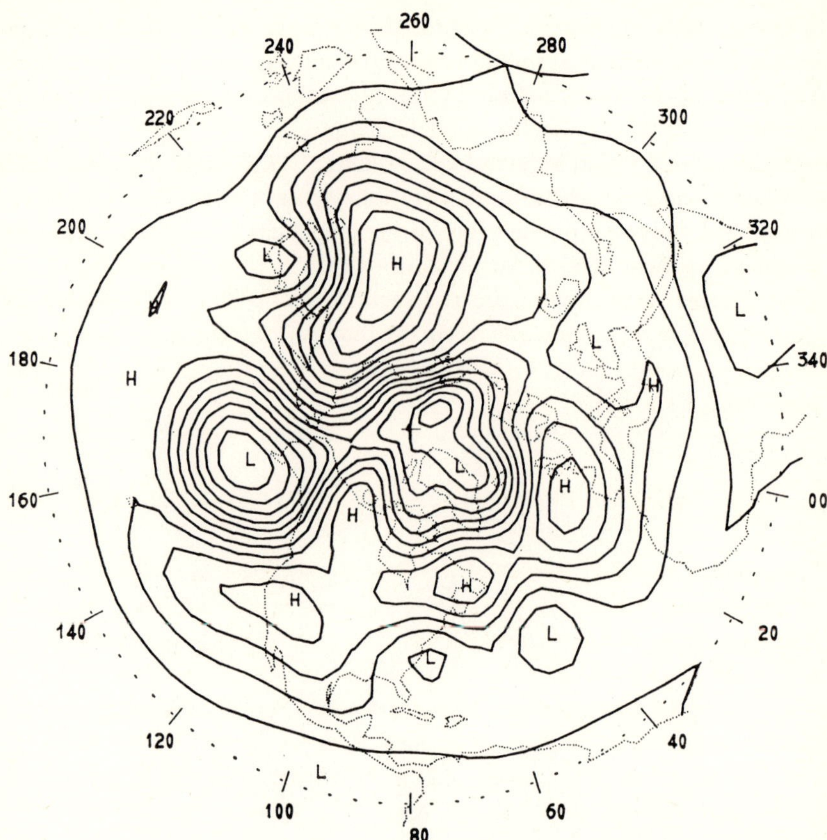

Fig. 2 Northern Hemisphere sea-level pressure map. Contour interval is 5 mb

level pressure map of the Northern Hemisphere for the same date as Fig. 1. The lines are isobars, drawn at 5 mb intervals. Compared with the flow pattern at 6 km, the sea-level pressure map shows the presence of smaller scale motions. Since these smaller scale motions are dominant in the lower part of the troposphere, they are influenced by conditions at the earth's surface. When a current of air moves over an uneven surface, the motion usually becomes irregular or turbulent. Because of the presence of surface friction, the geostrophic wind approximation is no longer a good approximation for formulating a prediction model of the flow in the lower part of the troposphere.

Ironically, it was recently found that quasi-geostrophic models are also inaccurate in describing the motions of *very long waves* on the order of 10,000 km (Table I). As mentioned earlier, motions on the order of 5000 km for cyclones and anticyclones are in geostrophic balance. This is also true in the case of very long waves, but because of the large scale, variation of the Coriolis parameter with latitude becomes very important. Therefore, quasi-geostrophic models are rather inaccurate in describing very large scale motions and medium and mesoscale motions such as *frontal cyclones* and *local severe storms* (Table I). This led to the conclusion that we can employ only the hydrostatic approximation to simplify the basic atmospheric equations of global circulation. Thus, we now have returned to the position of L. F. Richardson a half century ago.

5 PHYSICAL PROCESSES IN THE GLOBAL CIRCULATION

Let us now consider the basic physical processes involved in the mechanism of the general circulation. All energy of atmospheric motion is ultimately derived from incoming solar radiation (Fig. 3). If directly reflected radiation is ignored, the atmosphere receives from the sun 2.0 cal cm^{-2} min^{-1}, which is known as the *solar constant*. This gives a mean flux of solar energy perpendicular to the earth's surface of about 0.5 cal cm^{-2} min^{-1}. (The factor 4 is the ratio of surface area to cross-section for a sphere.) Of this, approximately

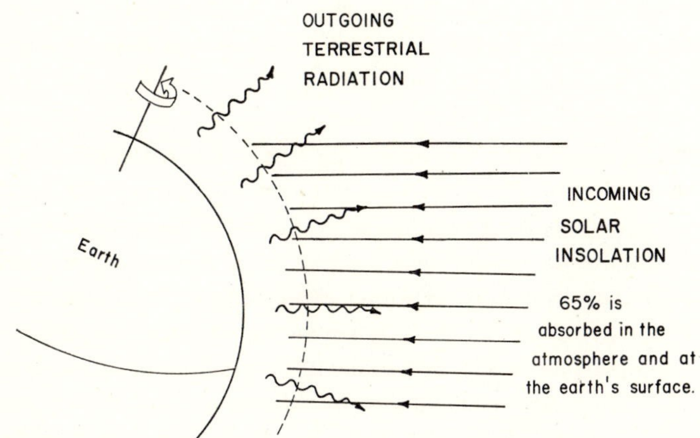

Fig. 3 Solar and terrestrial radiation

35 percent is reflected from the atmosphere (including clouds) and portions of the earth's surface. This percentage of energy reflected back to space is called the *albedo* of the earth as a whole. The remaining 65 percent is absorbed in the atmosphere and at the surface.[14]

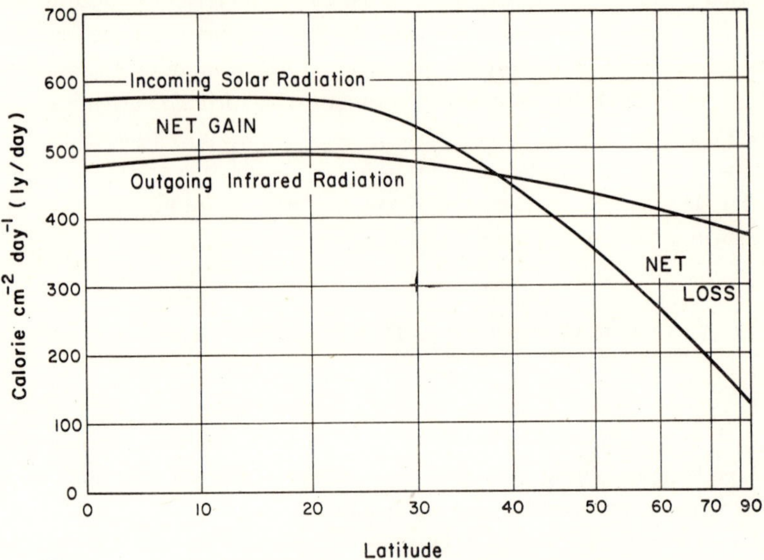

Fig. 4 Balance between incoming solar radiation and outgoing infrared radiation from the atmosphere-earth system

Since the mean temperature of the atmosphere and the earth do not change appreciably from one year to the next, the amount of energy received by the earth must be returned to outer space. This ultimate return of energy to space is in the form of low-temperature terrestrial radiation. Figure 4 shows the two curves. One represents the annual mean of solar radiation absorbed by the earth and the atmosphere, and the other the outgoing terrestrial radiation from the atmosphere, both as functions of latitude.[15,16]

At low latitudes the earth-atmosphere system gains more heat energy per unit area by the absorption of shortwave radiation from the sun than it loses to space by the emission of long-wave radiation; the reverse occurs at high latitudes. Surplus heat energy in the tropics must somehow be carried to the poles. Otherwise, tropical regions would become steadily hotter and polar regions steadily colder. About ten percent of the surplus heat is transported poleward by ocean currents. The remainder is transported by the atmospheric circulation, including both large-scale eddy motion (cyclones and anticyclones) and mean meridional circulation.

Since atmospheric motions are generated by non-uniform heating of the air, a physical prescription is needed for the heating rate, Q, in the thermodynamic equation (5). In general, the following three processes (expressed as rates) are important in the atmosphere (Fig. 5): heating from radiational sources, denoted by Q_a; the release of latent heat by condensation, denoted by Q_c; and heating/cooling due to eddy diffusion, denoted by Q_d.

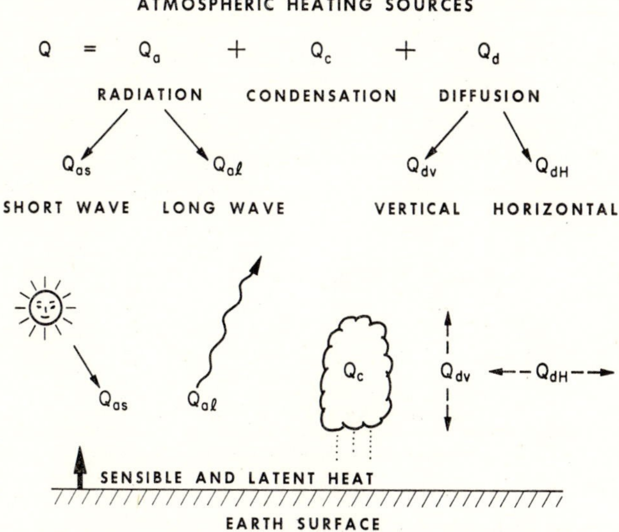

Fig. 5 Various sources of heating in the atmosphere

The rate of heating due to radiational sources, Q_a, may be divided into two parts: one is the rate of heating due to absorption of solar insolation in the atmosphere, Q_{as}, and the other is the rate of heating/cooling due to infrared radiation, Q_{al}. These terms can be evaluated by integrating the transfer equations, taking into account the distributions of water vapor, carbon dioxide, and ozone, the major absorption gases in the atmosphere.

However, the amounts of Q_{as} and Q_{al} are considerably altered by the presence of clouds in the atmosphere. The evaluation of Q_c, the rate of released latent heat of condensation, is, of course, directly dependent on the prediction of cloud formation. Thus, the prediction of clouds is one of the important aspects of a general circulation model.

The moisture field of the atmosphere is continually replenished by evaporation from the earth's surface. Since heat is required to evaporate water, and heat is released by condensation of water vapor, the evaporation process transports latent heat from the earth's surface to the atmosphere. Similarly, sensible heat is transported from the surface to the atmosphere, and vice versa, by the actions of small-scale turbulence and convection.

The direction of energy flow, in the form of latent heat and sensible heat, is determined primarily by the difference between the temperature of the earth's surface and the air temperature directly above it. Thus, it is necessary to know the temperature of the earth's surface.

The temperature of the earth's surface is determined by the balance of energy flux at the earth's surface, as shown in Fig. 6. The surface of the earth absorbs both solar radiation, S_g, and downward infrared radiation, D, from the atmosphere. A portion of the absorbed energy is transported down into the ground by heat conduction in the soils or is carried away by the motion of water in the oceans. This quantity is denoted by M.

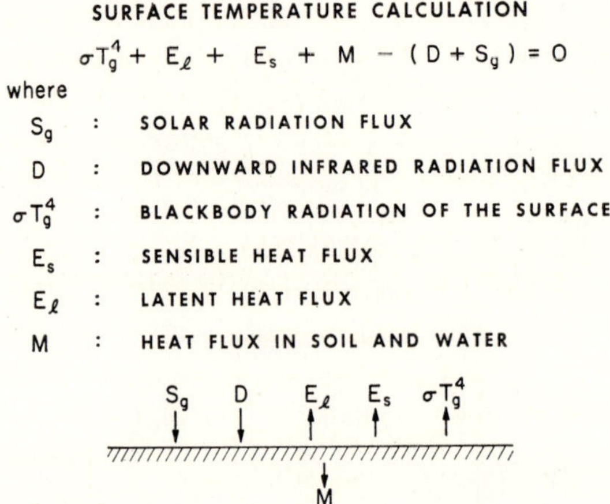

Fig. 6 Balance requirement of heat flux at the earth's surface for calculation of surface temperature

Energy is also carried away in the form of sensible heat, E_s, and of latent heat, E_l. The magnitudes of E_s and E_l are dependent on both the thermal and dynamical conditions of the atmosphere and the properties of soil and water.

The remainder of the absorbed energy is re-radiated to space as infrared radiation from the earth's surface. This amount is expressed by σT_g^4, where σ

is the Stefan–Boltzmann constant and T_g is the temperature of the earth's surface which we wish to determine.

The evaluation of M requires knowledge of the mechanism of heat and moisture transports in the *lithosphere*—and also of the circulation of water in the *hydrosphere*. Oceanic circulation is important in determining M as well as E_s. This is why there is a growing awareness of the role of the ocean's circulation in meteorology.

Perhaps equally important to meteorology is the role of soil. Knowledge of soil physics, particularly the transport processes of heat and moisture, will be essential when we try to explain, for example, the formation of deserts and swamps in connection with the long-term effects of the atmospheric circulation upon the earth's surface.

It is clear that what happens in the first hundred meters above the earth's surface is very important in accounting for the sources and sinks of energy for the atmosphere. *Micrometeorology* is a branch of meteorology dealing with various phenomena in the earth's boundary layer. Figure 7 shows a schematic structure of the lower atmosphere.[17]

In the *surface boundary layer*, which extends up to 50–100 m above the surface, shearing stress of wind and heat flux are both constant with height, and the structure of the wind is primarily determined by the characteristics of the surface and the vertical gradient of temperature. In the *planetary boundary layer*, which extends approximately 500–1000 m above the ground,

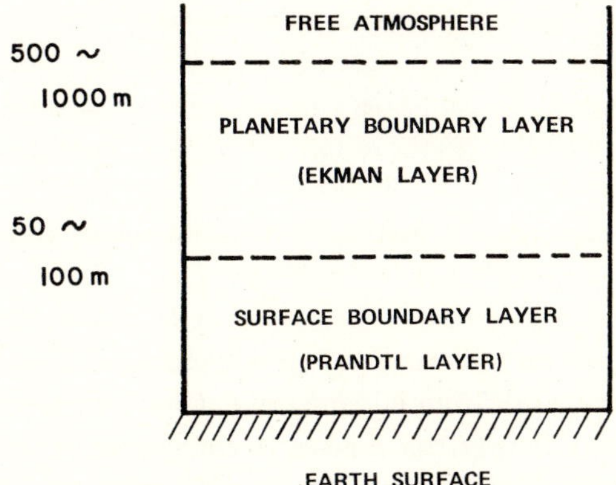

Fig. 7 Schematic structure of the lower atmosphere

the structure of the wind is influenced not only by the pressure gradient, density stratification, and surface friction, but also by the earth's rotation.

After the heating rate, Q, and the frictional terms, F_x and F_y, are suitably expressed in terms of dependent variables, the primitive equations can be integrated from prescribed initial conditions under well-posed boundary conditions.

6 GENERAL CIRCULATION MODELS

With the advent of high-speed electronic computers and the development of numerical weather prediction techniques, it became possible to perform long-term integrations of meteorological equations including complex physical processes such as those discussed in the previous section. In 1956, Phillips[18] made the first successful attempt to simulate large-scale motions through a step-by-step numerical integration of quasi-geostrophic equations. Despite many simplifications employed in setting up the model, his experiment succeeded in simulating certain gross properties of the atmospheric motion. Later Smagorinsky[19] performed a similar experiment using the primitive equations.

Following these earlier attempts, more elaborate types of general circulation experiments using the primitive equations were designed and carried out.[20–23] In 1964, when Kasahara and Washington decided to work on a numerical experiment of the general circulation of the atmosphere at the National Center for Atmospheric Research (NCAR), we made a survey of the characteristics of various general circulation models thus far reported, and we found that height as the vertical coordinate (rather than pressure, as customarily done) would be convenient in formulating a prediction model similar to Richardson's.[7] Besides the different vertical coordinate, the NCAR model includes various physical processes in the atmosphere which have been formulated somewhat differently from those incorporated in other general circulation models.[24]

Once a type of prediction model, consisting of meteorological equations, is decided upon, the next job is to select a finite-difference scheme appropriate to the relevant system of differential equations. One unique feature of the atmospheric equations is that the time-rate-of-change of the variables is about an order of magnitude less than the rest of the terms in the equations. For example, the acceleration of air in large-scale motions is about one-tenth the magnitude of the Coriolis or the horizontal pressure gradient force. This fact is in striking contrast to other hydrodynamic problems, particularly shock wave problems in which large accelerations are involved. Thus, in predicting large-scale motions, extra care must be taken to compute individual terms involving space derivatives, since the time tendency is evaluated as the sum of individual terms in the equations.

Another special requirement for the finite-difference prediction equations is that the model be run long enough to study long-term fluctuations of weather systems. To produce a 100 day forecast with a time step of 6 min requires over 20,000 iterations. This is why strongly dissipative difference schemes are not preferable for numerical weather prediction. Of course, any practical finite difference scheme must be both stable and accurate. For long-term numerical integration of the atmospheric equations, it is, therefore, necessary to design finite difference schemes which conserve certain properties of the original equations, such as momentum and energy.

Even though a finite difference equation is stable in the sense of the *von Neumann condition*,[25] there is no guarantee of stability if such a scheme is applied to nonlinear equations. As an example, consider the very simple hyperbolic nonlinear equation

$$\frac{\partial u}{\partial t} + u \frac{\partial u}{\partial x} = 0.$$

When the centered scheme is applied in order to approximate both the time and space derivatives, the finite difference analog of the above equation may become unstable, even under the usual stability condition:

$$\max |u| \frac{\Delta t}{\Delta x} < 1.$$

Since this instability typically appears in nonlinear equations, it is called *nonlinear instability*.[26] However, a similar kind of instability may occur even in linear equations if the coefficients are not constant. Phillips[26] showed that instability arises because the grid system cannot resolve wavelengths shorter than two grid intervals. When such wavelengths are formed by the nonlinear interaction of longer waves, the grid system incorrectly interprets them as long waves.

This type of instability is inherent in all meteorological equations, and many means to eliminate this difficulty have been proposed. In our general circulation experiments at NCAR, we eliminate nonlinear instability by applying a scheme which was originally proposed by Lax and Wendroff[27] and later modified by Richtmyer[28] who called it the *two-step Lax–Wendroff scheme*. The scheme consists of a combination of the first step, called "diffusing", and the second step, called "leapfrog". The diffusing step is strongly dissipative, but the leapfrog step is neutral in stable conditions. Because of the dissipative nature of the first step, the combined steps provide effective smoothing for shorter waves and suppression of nonlinear instability. We found, however, that the two-step method damps the solution too severely. In order to eliminate this difficulty, we apply the diffusing step less often, so that short waves are not over-damped. By experimenting, we have found it sufficient to apply the diffusing step only after every 135 steps of leapfrog.

7 COMPUTATIONAL ASPECTS OF THE NCAR MODEL

In order to give an idea of the magnitude of computation involved in general circulation experiments, let us discuss some computational aspects of our problem, specifically the six-layer version of the NCAR general circulation model. The model is global. Horizontal grid points are placed at 5 degree longitude and latitude intersections. However, the grid system is staggered in time (Fig. 8). The cross and dot points are at different time levels of one time increment, Δt.

Figure 9 shows a vertical cross-section of the model. The height increment,

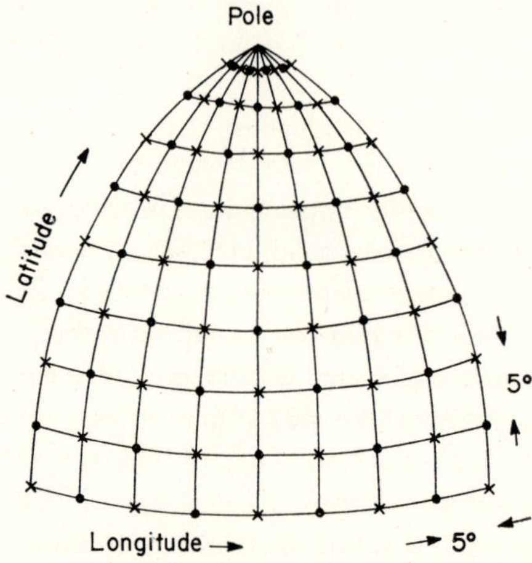

Fig. 8 Horizontal grid system of a general circulation model

Δz, is chosen to be 3 km. The top is located at $z = z_T$ ($= 6\,\Delta z$), which corresponds approximately to the 75 mb pressure level. It is assumed that vertical motion vanishes at the top in order to conserve the mass of the model atmosphere. The lower boundary of the model is a curved surface which takes into account the earth's orography, indicated by shading.

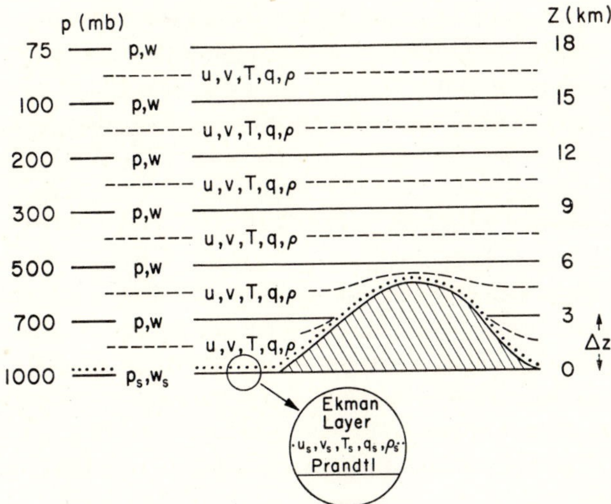

Fig. 9 Six-layer NCAR general circulation model with orography

Surface pressure, p_s, is evaluated along the mountain surface. Pressure, p, and vertical velocity, w, are placed at z = 3, 6, 9, 12, and 18 km; and wind components, u and v, temperature, T, density, ϱ, and the mixing ratio of water vapor, q, are placed at the intermediate levels.

The effects of both the surface boundary layer and the planetary boundary layer are included in the model, as shown in the inset of Fig. 9. The variables u_s, v_s, T_s, q_s, and ϱ_s, which correspond to variables in the surface boundary, are evaluated from the lower boundary conditions.

In the model, the earth's surface is divided into three parts: oceanic regions, continents, and any region where the surface temperature is below the freezing point (0°C), be it land or ocean. The third region is, therefore, considered to be covered by either ice or snow.

The ocean surface temperature is prescribed from the distribution of observed climatological mean temperatures for the appropriate season of the year. The surface temperature of the land masses is computed by the method previously described. Over ice covered regions, an upper limit of 0°C is assumed, with the implication that any excess heat flux goes into the latent form.

Let us next consider the amount of data which must be stored in the computer. In each horizontal plane, there are $72 \times 36 = 2592$ data points for a 5 degree longitude and latitude mesh. There are seven levels, including the surface, so that the total number of data points in three-dimensional space is $2592 \times 7 = 18{,}144$. Since the model is staggered in horizontal space, only half of this number (9072) is needed at one time. There are four prognostic variables, u, v, p, and q, which are needed at two consecutive time levels. There are three diagnostic variables, w, ϱ, and T. We need 15 to 30 extra fields of heating, friction, and similar terms for diagnostic purposes, in addition to such physical quantities as cloudiness and precipitation, depending on the complexity of the model. Thus, a total of more than two million pieces of information plus about 12,000 arithmetic and logic computer instructions must be stored in the computer memory. Of course, there is no need to store all of these data in the high-speed core memory all the time during a computation. Since computation is done in a sequential manner, only a portion of the data is needed at one time for the calculations.

At NCAR, we use a Control Data 6600 computer which has only 65,000 words of core memory. The grid point data are stored in six magnetic drums, each of which contains about one-half million words. Data are constantly read from the drums to the core and transferred from the core to the drums simultaneously with the arithmetic computations. About 150 floating operations are required at each mesh point, and it takes about 25 sec for one time step. Since a time step of 6 min is used, it takes about 90 min of computing time to produce a one-day forecast.

Because a large amount of data is involved, there is a serious problem in processing the output. The situation is in contrast to, for example, an eigenvalue calculation in which the output may be a single number after hours and hours of computation. Although it is possible to print out a large amount of numbers in a short time, it is practically impossible to digest the results and evaluate the experiment without resort to graphic analyses of numbers to display the patterns. At NCAR, the output data processing is aided by a cathode ray tube plotter called the Data Display 80. Using this equipment, grid point data are analyzed and graphic contour lines are then superimposed on a suitable map projection.[29]

Even though the output is analyzed in the form of weather maps, it is still difficult to comprehend the evolution of the flow patterns in a short time. We use extensively computer-generated movies not only for illustration but also for debugging the computer programs. We have found the use of color very effective in illustrating simultaneously the patterns of different variables for

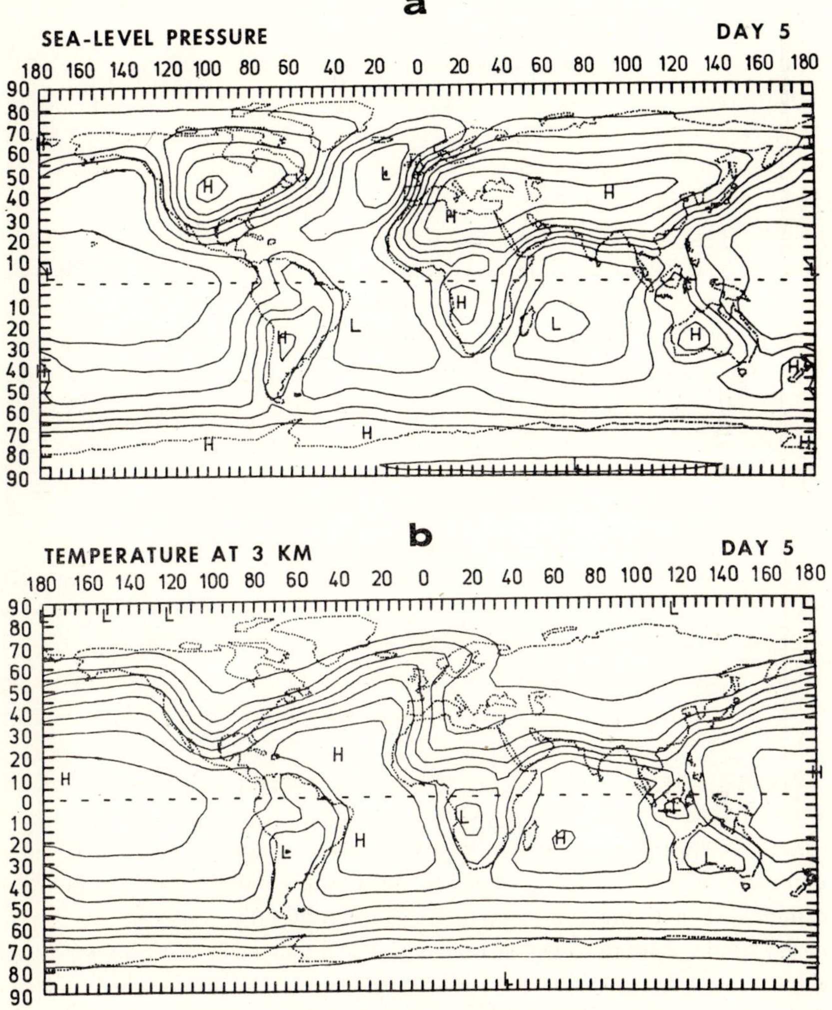

Fig. 10 Computed distributions of (a) sea-level pressure, and (b) temperature at 3 km height after 5 days. *H* denotes local maxima and *L* denotes local minima in pressure and temperature distributions

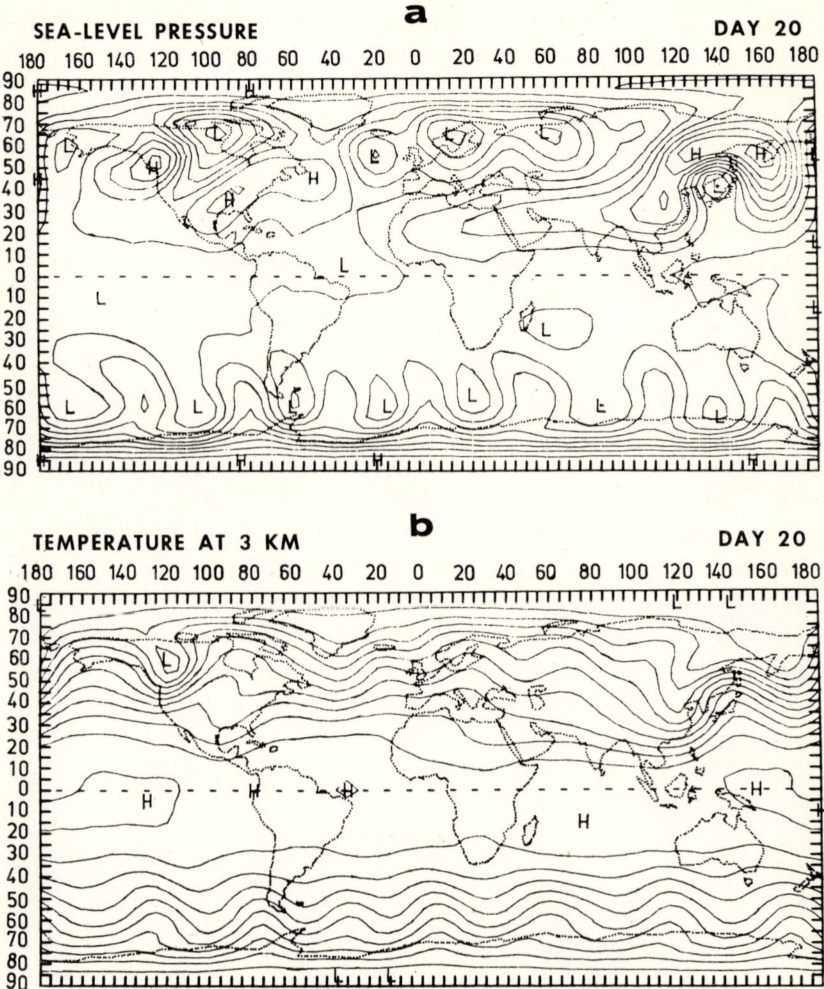

Fig. 11 Same as Figure 10, but after 20 days

comparison. Color film is processed from a black-and-white original film by a multi-exposure method using filters of different colors.

We present here the result of one experiment which shows the evolution of flow patterns starting from a motionless and isothermal atmosphere of 250 K. This particular value is chosen as the equilibrium temperature of the earth-atmosphere system, which is evaluated from the balance requirement of incoming solar radiation with outgoing infrared radiation. This experiment simulated January flow patterns. The season is determined by the declination of the sun. In addition, we prescribed a climatological mean distribution of sea surface temperature for January. As explained earlier, at low latitudes the earth-atmosphere system gains more energy by the absorption of shortwave radiation from the sun than it loses to space by the emission of long-wave radiation; the reverse occurs at high latitudes. Because of this radiational imbalance, the north-south temperature gradients develop.

Figures 10 show computed distributions of (a) sea-level pressure with contour intervals of 5 mb, and (b) temperature at 3 km height with isotherm intervals of 5 °C after five days. The circulation pattern in the early phase of the development is of large-scale monsoons. Large highs develop over the North American continent and over Siberia, and lows form over the warm North Atlantic and to some extent over the North Pacific. Meanwhile, the warming in the low latitudes slows down, but the cooling in the higher latitudes, particularly near the North Pole, continues.

These large monsoonal circulations are too inefficient to transport heat northward to make up the radiational imbalance. As a result, the north-south temperature gradients kept increasing. After approximately 15 days, the north–south temperature gradients became so large that the large-scale monsoonal circulation broke down into eddy motions. These eddies are nothing but cyclones and anticyclones, the familiar systems that appear on weather maps (Fig. 11a, b).

8 OUTLOOK

We have seen that numerical experiments help us to understand the mechanisms of the general circulation. The degree of understanding may be measured by our ability to simulate, with a model, actual atmospheric motions. A successful model, therefore, must be capable of predicting actual weather changes. Or, put in another way, we can judge how good a model is by verifying predicted patterns. For this reason, we have also been working on the use of real data in order to check the forecasting capability of our general circulation model. Incidentally, the preparation of initial data for global weather forecasting is by no means a trivial problem due to the paucity of observational data over the oceans and over the Southern Hemisphere in general. Therefore, considerable difficulty exists in judging whether forecasting errors are due to incapability of the model or incomplete real initial data. However, the forecasting skill of our model is comparable to that of the primitive equation model developed at the U.S. Weather Bureau, ESSA.[30,31]

Very recent studies by NCAR staff members and others have demonstrated that the accuracy of numerical weather simulation will increase significantly if the horizontal grid mesh is reduced to half of the 5 degrees that is presently used.[32–34] A $2\frac{1}{2}$ degree mesh will increase computing time by a factor of eight. It is obvious that to run a general circulation model which includes ocean circulation and detailed atmosphere–ground boundary layer calculations with a finer horizontal resolution will require a super-speed computer 100 to 500 times faster than the Control Data 6600.[35] A few super-computers are being developed, the best known of which is the University of Illinois' parallel-processing ILLIAC IV.

In an increasingly crowded, fast-moving industrial world, weather constantly impinges on every economic pursuit, from agriculture to super-sonic transportation. Every year in the United States alone, weather hazards, such as hurricanes, severe storms, floods, and drought, kill approximately 1300 people and cause property damage in excess of $11 billion.[36] When this problem is considered on a global basis, it is obvious that improved weather prediction can lead to immense social and economic benefits. The new technology of computers and meteorological satellites has brought within economic reach for the first time the possibility of accurate weather prediction and even of studying means of weather control.

In the last few years there has been an encouraging development in promoting a vastly improved understanding of the general circulation of the atmosphere. The program is called the Global Atmospheric Research Program (GARP), sponsored jointly by the International Council of Scientific Unions and the International Union of Geodesy and Geophysics.[37] Another important and related plan is to eliminate deficiencies in the present international system of global weather observations. The program is called the World Weather Watch (WWW). WWW is the operation-oriented partner of GARP and is sponsored by the World Meteorological Organization.

Using upper air data gathered by WWW and with the use of a super-computer, we will be able to attack the fundamental problem of an objective long-range weather forecast for two weeks and beyond. However, in order to achieve meaningful long-range forecasts, we will first have to learn more about the mechanism of long-term weather changes, an important, yet very poorly understood, problem. Results of our general circulation experiments have already indicated the importance of the condition of the earth's surface, including dynamic and thermal effects, upon atmospheric motions. We have begun to see the important effects of clouds interacting with atmospheric radiation and the equally important effects of small-scale turbulence and convection in the atmospheric boundary layer. These problems must be investigated in even greater detail before the various effects can be properly included in general circulation models.

References

1. Phillips, N.A., "Numerical weather prediction", in *Advances in Computers* (ed. F.L. Alt, Academic Press, New York, 1960), Vol. 1, 43–90.
2. Thompson, P.D., *Numerical Weather Analysis and Prediction* (The Macmillan Co., New York, 1961), 170 pp.
3. Kibel', I.A., *An Introduction to the Hydrodynamic Methods of Short Period Weather Forecasting* (translated from the original volume published in 1957, Moscow, The Macmillan Co., New York, 1963), 383 pp.
4. Marchuk, G.I., *Numerical Methods in Weather Forecasting* (in Russian, Gidromet, Izdat., Leningrad, 1967), 356 pp.
5. Courant, R., K.O. Friedrichs, and H. Lewy, "Über die partiellen Differenzengleichungen der Physik", *Math. Annalen* **100**, 32–74 (1928).
6. Monin, A.S. and A.M. Obukhov, "A note on general classification of motions in a baroclinic atmosphere", *Tellus* **11**, 159–162 (1959).
7. Richardson, L.F., *Weather Prediction by Numerical Process* (Cambridge University Press, 1922), 236 pp. (Reprinted by Dover Publications, New York, with an introduction by S. Chapman, 1965).
8. Platzman, G.W., "A retrospective view of Richardson's book on weather prediction", *Bull. Am. Meteorol. Soc.* **48**, 514–550 (1967).
9. Charney, J.G., R. Fjørtoft, and J. von Neumann, "Numerical integration of the barotropic vorticity equation", *Tellus* **2**, 237–254 (1950).
10. Charney, J.G., "On the scale of atmospheric motions", *Geofys. Publikasjoner* **17** (2), 17 pp. (1948).
11. Eliassen, A., "The quasi-static equations of motion with pressure as independent variable", *Geofys. Publikasjoner* **17** (3), 44 pp. (1949).
12. Thompson, P.D. and W.L. Gates, "A test of numerical prediction methods based on the barotropic and two-parameter baroclinic models", *J. Meteorol.* **13**, 127–141 (1956).
13. Cressman, G.P., "A three-level model suitable for daily numerical forecasting", (Tech. Memo. No. 22, National Meteorological Center, Weather Bureau, ESSA, 1963), 22 pp.
14. Goody, R.M., *Atmospheric Radiation, Vol. 1. Theoretical Basis* (Clarendon Press, Oxford, 1964), 436 pp.
15. Houghton, H.G., "On the annual heat balance of the northern hemisphere", *J. Meteorol.* **11**, 1–9 (1954).
16. London, J., *A study of the atmospheric heat balance*, Final Report, Contract AF 19(122)–165 (ASTIA No. 117227), (Department of Meteorology and Oceanography, New York University, 1957), 99 pp.

17. Sutton, O. G., *Micrometeorology* (McGraw–Hill Book Co., New York, 1953), 333 pp.
18. Phillips, N. A., "The general circulation of the atmosphere: A numerical experiment", *Quart. J. Roy. Meteorol. Soc.* **82**, 123–164 (1956).
19. Smagorinsky, J., "General circulation experiments with the primitive equations: I. The basic experiment", *Mon. Wea. Rev.* **91**, 99–164 (1963).
20. Leith, C. E., "Numerical simulation of the earth's atmosphere", in *Methods in Computational Physics*, Vol. 4 (ed. B. Alder *et al.*, Academic Press, New York, 1965), 1–28.
21. Mintz, Y., "Very long-term global integration of the primitive equations of atmospheric motion", in *Proc. WMO/IUGG Symposium on the Research and Development Aspects of Long-Range Forecasting*, Boulder, Colo., 1964 (WMO Tech. Note No. 66, 1965), 141–167.
22. Manabe, S., J. Smagorinsky, and R. F. Strickler, "Simulated climatology of a general circulation model with a hydrological cycle", *Mon. Wea. Rev.* **93**, 769–798 (1965).
23. Smagorinsky, J., S. Manabe, and J. L. Holloway, Jr., "Numerical results from a nine-level circulation model of the atmosphere", *Mon. Wea. Rev.* **93**, 727–768 (1965).
24. Kasahara, A. and W. M. Washington, "NCAR global general circulation model of the atmosphere", *Mon. Wea. Rev.* **95**, 389–402 (1967).
25. Richtmyer, R. D. and K. W. Morton, *Difference Methods for Initial-Value Problems* (Interscience Publishers, New York, 1967), 405 pp.
26. Phillips, N. A., "An example of non-linear computational instability", in *The Atmosphere and the Sea in Motion* (Rossby Memorial volume, ed. B. Bolin, Rockefeller Inst. Press, New York, 1959), 501–504.
27. Lax, P. D. and B. Wendroff, "Systems of conservation laws", *Comm. Pure Appl. Math.* **13**, 217–237 (1960).
28. Richtmyer, R. D., *A survey of difference methods for non-steady fluid dynamics* (Tech. Note 63–2, National Center for Atmospheric Research, Boulder, Colo., 1963), 25 pp.
29. Washington, W. M., B. T. O'Lear, J. Takamine, and D. Robertson, "The application of CRT contour analysis to general circulation experiments", *Bull. Am. Meteorol. Soc.* **49**, 882–888 (1968).
30. Shuman, F. G. and J. B. Hovermale, "An operational six-layer primitive equation model", *J. Appl. Meteorol.* **7**, 525–547 (1968).
31. Baumhefner, D. P., "Real-data forecasting with the NCAR general circulation model", *Proc. Tech. Exchange Conference*, Fort Monmouth, New Jersey, 118–131 (1968).
32. Gary, J. M., "A comparison of two difference schemes used for numerical weather prediction", *J. Comp. Phys.* **4**, 279–305 (1969).
33. Grammeltvedt, A., "A survey of finite difference schemes for the primitive equations for a barotropic fluid", *Mon. Wea. Rev.* **97**, 384–404 (1969).
34. Williamson, D., "Numerical integration of fluid flow over triangular grids", *Mon. Wea. Rev.* **97**, 885–895 (1969).
35. Kolsky, H. G., *Computer requirements in meteorology* (Tech. Rept. No. 38.002, IBM Scientific Center, Palo Alto, Calif., 1966), 158 pp.
36. White, R. M., "The world weather program", *Bull. Am. Meteorol. Soc.* **48**, 81–84 (1967).
37. Roberts, W. O., "The global atmospheric research program", *Bull. Am. Meteorol. Soc.* **48**, 85–88 (1967).

CHAPTER 24

The Role of the Computer in Astrophysics*

ICKO IBEN, JR.

1 GENERAL REMARKS

IN THE DEVELOPMENT of astrophysics over the past 15 years the large-scale digital computer has played an important, but by no means *uniformly* essential role. In the exploration of newly discovered phenomena whose nature is not immediately understood—the quasars, pulsars, OH sources, *X*-ray sources, and the like—pioneering questions arise that are best tackled by "back of the envelope" calculations. The discovery and exploration of each new observational phenomenon usually ushers in an era of groping and basic problem formulation during which the solution of complex or extended mathematical problems is not of overriding importance.

It is in the solution of well formulated problems that the computer has played an important and, in many instances, an essential role. There is no question but that overall progress in astrophysics, as in all of the sciences, has been accelerated by the ready availability of a powerful and fast computational tool. There are, of course, dangers inherent in the easy availability of the computer. The fact that the solution to a particular mathematical problem can be obtained rapidly, often without extensive mental strain, can have the effect of diminishing the effort expended in devising simplifying analytic approximations that contribute insight into the structure of the problem. On the other hand, the possibility of obtaining a quick answer can be of immense aid in finding which of many approximations is most suitable and can thereby

* Supported in part by the National Science Foundation (GP–8060) and in part by the National Aeronautics and Space Administration (NsG–496).

speed up the process of obtaining insight. Hopefully the latter circumstance occurs more frequently than the former since, despite the existence of an easy route to an exact solution, the achievement of insight that goes along with the recognition of the dominant factors in a problem remains one of the major goals of scientific endeavor. It is nevertheless undeniable that, as a regrettable byproduct of the computer age, there exists a much larger body of peripheral detail cluttering the literature than might have been the case if insight were always a prerequisite for solution rather than a consequence of solution.

Of primary interest here are those areas in which the computer has played a really *central* role, rather than a secondary role as a huge slide-rule whose availability has merely accelerated the solution of problems that would have been adequately solved in any case. These areas are dominated by mathematical problems of cardinal importance which either were formulated or could have been formulated long before the advent of large scale digital computers, but which were not or would not have been solved prior to the appearance of that particular generation of computers with enough speed and memory to make solution practicable.

Areas of study in which computer solution has been particularly important for growth include the internal structure and evolution of stars; the dynamics of star clusters and galaxies; the pulsational properties of Cepheids and *RR* Lyrae stars; the structure of stellar envelopes as needed for the determination of surface chemical abundances and for the understanding of other observable features of stars; and the structure of the solar wind through the analysis of satellite observations. In many of these areas, the equations to be satisfied locally in the system of interest are embarrassingly simple. The complexity arises from the fact that each system to be investigated contains many parts that are coupled to one another so that many, many arithmetical operations (each of which is simple) are necessary to obtain a solution connecting all of the parts in a self-consistent way.

2 AN EXAMPLE—QUASI-STATIC STELLAR EVOLUTION

No branch of astrophysics has benefited more from modern computer technology than has the study of stellar structure and evolution during stages when, to a good approximation, dynamic and rotational motions may be neglected. It is worthwhile to sketch what is involved in a typical evolutionary calculation, focusing on the importance of calculational speed and of a large memory in making the task practicable.

For the sake of concrete illustration here and in the following, let us define a standard computer as one with an initial 32K memory (words), a user-available memory of 28K (after system requirements are met), an average speed of 5 μsec per instruction (assuming that most instructions lead to a multiplication), and a user price of 300 dollars per hour.

1 Memory Requirements

A model star may be defined as a configuration of some 100–500 concentric shells, each containing a predetermined amount of matter. To describe the "state" of matter in each shell, three state variables—pressure, temperature, and density—must be specified at shell boundaries. To describe the location of each shell and to determine the gravitational force on it, a radius variable (distance from center to shell boundary) and a mass variable—(mass of all shells between center and shell boundary) must also be specified. Depending on the complexity or completeness desired, from three to ten (or more) chemical composition variables must also be specified at each shell boundary in order to follow stellar evolution just through hydrogen and helium burning stages. For more advanced stages, several hundred composition variables may be necessary. Finally, to complete the specification of local macroscopic properties, the energy flux across each shell boundary must be given. In summary, one must keep track of some 16 variables at each of several hundred shell boundaries in order to follow a star through relatively simple stages.

The number of shells required for sufficiently accurate calculation depends on the evolutionary phase, the number 500 representing a generous maximum during hydrogen and helium burning phases. The number of shells actually used is commonly determined by demanding that no variable change by more than, say, ten percent across a shell or by less than, say, five percent.

The coupling between variables at adjacent shell boundaries is expressed by four simple equations: (a) The pressure difference across a shell equals the gravitational force per unit area on the matter in the shell. (b) The difference in energy flux across the shell is equal to the energy released by matter in the shell. (c) The temperature difference across the shell is proportional to the rate at which matter can absorb and emit photons and is also proportional to the energy flux through the shell. (d) The mass in each shell equals the mean density in the shell times the shell volume. Although the four equations are formally quite simple, requisite quantities such as energy generation rates and average photon absorption rates (the opacity) can be rather complex functions of local variables that may be obtained only after rather formidable calculations.

One final relationship between state and composition variables must be satisfied in each shell. In its most elementary form, this relationship (the equation of state) states that the pressure is proportional to the product of the particle density and of the average particle kinetic energy. In many instances, however, the equation of state is relatively complex and can be solved only by numerical iteration.

A solution to the coupling equations will in principle give a satisfactory stellar model *at a given time*—if the distribution of composition variables is known and if the energy generation rate is known as a function of local variables.

However, since composition changes occur inevitably as a result of nuclear transmutations and since energy is continuously lost from the star, time changes in stellar structure occur that lead to the transfer of gravitational energy to and from the stellar thermal energy store. This means in practice that one must deal with *at least two* stellar models at a time; for, since the energy generation rate at any given time can only be specified if one knows how rapidly structure variables are changing with time, one requires at least two stellar models (at, say, times t and $t + \Delta t$) in order to estimate the rate of energy generation (at, say, time $t + \Delta t/2$). Thus, in order to follow the evolution of a star, we must keep track of 2×16 variables at each of 500 shell boundaries.

A numerical solution of the equations satisfied by the variables describing a model pair requires the introduction of 16 additional variables so that, finally, some $500 \times 48 \sim 25{,}000$ variables must be retained in the computer memory. This number is to be compared with a total memory capacity of about 32,000 "words" that has characterized many of the digital computers found at educational institutions during the 1960's.

Additional memory locations are required for storing the set of instructions that constitute the computer program for calculating evolution. Typically, one might need: 800 words for calculating density in terms of pressure and temperature; 800 words for calculating energy generation rates; 2000 words for composition changes; 2500 words for altering the distribution of variables preceding each evolutionary time step and estimating variables in successive trial models; 2500 words for constructing model atmospheres to serve as boundary conditions for interior models; and 1500 words for the main control and model iteration program. This means a total of about 10,000 words for the entire calculational program.

Thus, a total memory capacity of 25,000 + 10,000 = 35,000 words is required to calculate fairly completely (to the extent that the adequacy of the input physics makes this reasonable) the early stages of stellar evolution. This exceeds the available memory of 28,000 words provided by our standard computer. In order to circumvent the impasse, one may (a) reduce the number of shells or the number of composition variables and thereby reduce the accuracy and completeness of the calculation (b) use tape and/or disk storage for parts of the program (c) arrange the instructional sequence in such a way that each memory location may be used for several different variables and (d) streamline the instructional sequence. With the exception of streamlining, these procedures have the disadvantage of increasing the amount of machine time required to obtain a model pair. For example, each time a segment of the program is transmitted to or from tape, several seconds may be wasted. If, in the course of a typical calculation, transfer occurs once every ten seconds, the computer is being used at only 90 percent efficiency. Efficiency is considerably improved if disk storage is available.

It is clear that, if only because of the large number of variables involved, the calculation of stellar evolution would be a rather tedious job without a large-scale computer with a large memory and rapid access time. The importance of the computer becomes more apparent when we consider the type and number of operations involved in obtaining just one pair of models and then consider the number of model pairs required for a complete "standard" evolutionary sequence.

2 Computational times

Let us suppose that we already have one model pair that satisfies all local equations. After rezoning to ensure that no variable changes too rapidly across a shell, we extrapolate from the given model pair to a new model pair related causally to the first. After determining the rate at which the composition variables in each shell of the initial model pair are changing, we can estimate composition variables for the next model pair. Typically, the equivalent of one hundred multiplications per shell may be required. Finally, by logarithmic extrapolation, we estimate all state and position variables in the next model pair.

The next task is to determine how well or how poorly the new variables satisfy the local structure equations in each shell. This requires a calculation of the opacity and of the energy generation rate and a solution of the equation of state in each shell. Perhaps 1500 operations per shell must be executed.

In general, the structure equations are not well satisfied and one must procede to iterate in order to obtain better estimates of all variables in the new model pair. The iteration procedure requires, among other things, the calculation of derivatives of energy generation rates and of opacities in each shell and then the simultaneous solution of a linear matrix connecting the first order corrections to all variables in all shells. Perhaps a total of 4000 operations per shell must be executed for every iteration. Typically, two or three iterations are necessary before the equations relating variables in each shell are satisfied to the desired accuracy. From time to time, convergence cannot be achieved and it is necessary to begin again by extrapolating from the previous model pair with either a smaller or a larger choice of time-step size.

If all goes well, a new pair can be completed after about 10^7 instructions (mostly arithmetic operations) have been executed. With our "standard" computer that can average one mathematical operation per 5 μsec, this means a minute of computing at a cost of 5 dollars in order to obtain one model pair.

A man with a slide rule, graph paper, and a table of logarithms might be able to execute one instruction every five seconds on the average. If he were to follow the same prescriptions as those for which the computer has been programmed, he would need to work a total of two years without rest. With time out for sleep and recreation, perhaps six years might elapse before he completed a task equivalent to that accomplished by the well-programmed computer in one minute. This is, of course, an extreme example. In practice, if forced to rely on hand calculations, a moderately clever man would invent simplifying schemes, rely on graphical interpolations and analogue procedures, and reduce the complexity of the input physics, with the result that he might obtain qualitatively the same model pair after executing some 10^4–10^5 mathematical operations in a month's time (drawing a month's salary that probably exceeds the five dollars spent solving the problem more accurately on the computer).

One model pair is but a small fraction of the entire story. To follow a model star from early gravitationally contracting stages through hydrogen and helium burning phases requires that something like 500 model pairs be constructed. With our standard computer, this means about eight hours of computing time. Without the computer, a man could complete the same task only if he lived for 3000 years. After paring away all but the bare essentials and resorting to every computational trick known, the clever man could obtain roughly the same results after a lifetime of effort. Unfortunately, during his lifetime, many insufficiencies in the input physics would have been recognized and remedied and most of the equations that he had solved would have become hopelessly out of date.

The *really* clever man would form a department of stellar evolution and attract 40 graduate students, each of whom (being burdened with course work and young children) could work perhaps one tenth as rapidly as he. His department would be able to turn out one complete evolutionary track per year.

A meaningful comparison with the observations demands the availability of model sequences for stars of many different masses and of many different initial compositions. One might reasonably expect that 100 model sequences would be sufficient for a complete job. Our one department of stellar evolution could accomplish the necessary task in a century. Of course, new departments of stellar evolution would be founded across the country by graduates of the parent institution, and the task might be accomplished in several decades. On the other hand, one man with our standard computer could do the same job with 800 hours of computer time spaced out over a two year period, and at a great saving in (federal) funds.

3 Accomplishments

Prior to the advent of the first generation of digital computers, the study of stellar structure focussed primarily on the discussion of homogeneous models of uniform composition—a discussion that we now know is applicable only to initial stages of gravitational contraction, to early stages on the main sequence before an appreciable conversion of hydrogen into helium has taken place, and possibly also to final stages after a star has exhausted all of its "tappable" nuclear fuels. Although several bold efforts succeeded in estimating the effect of developing element inhomogeneities, it is fair to say that our present understanding of inhomogeneous evolution through both hydrogen and helium burning phases is in large measure a result of calculations made practicable by large scale digital computers.

Whereas before the computer-dominated era we possessed a fair qualitative understanding of only young main sequence stars and of white dwarfs, we are now able to make realistic quantitative comparisons between sophisticated models and the observable properties of evolving main sequence stars as well as of stars in much more advanced stages. We know now that Cepheids and RR Lyrae stars are burning helium at their centers and hydrogen in a shell. We know now that most red giants consist of a rapidly contracting, dense, hydrogen exhausted core capped by a hydrogen burning shell and that red supergiants may consist of a rapidly contracting, dense, helium exhausted core surrounded by both a helium-burning and a hydrogen-burning shell.

We now understand in detail how red giants cease to be red giants as a result of a violent flash that occurs when the temperature in the dense (white-dwarf-like) core of the giant becomes high enough to ignite helium explosively. We know that instabilities can arise in thin nuclear-burning shells and we suspect that such instabilities may lead to the ejection of a shell of matter and thus initiate the planetary nebula phase. We have a fair idea that the central stars of planetary nebulae contain no further nuclear fuel and are contracting to the white dwarf state. Finally, we may now understand the cause of the nova phenomena and we have found perhaps a decade of possible precursors of supernovae.

The possibility of constructing many evolving models of different initial mass and composition permits us now to interpret statistically significant regularities of large groups of stars in globular and galactic clusters in terms of age and initial composition of cluster stars. In this way, for example, we have been able to determine that the oldest stars in our galaxy are between 10 and 15 billion years old and that they each began life with a large amount of helium that was probably not synthesized by earlier generations of ordinary stars.

3 ADDITIONAL EXAMPLES

The following abbreviated set of examples should give some indication of the wide range of astrophysical problems in the solution of which the large scale digital computer has played or will play an essential role.

1 The classical many-body problem in astrophysics

This is the study of the motions of a collection of N stars held together only by the inverse square gravitational force between each pair. Systems of interest range from $N \sim 10^9$–10^{13} (as representative of galaxies), through $N \sim 10^5$

(as representative of globular clusters), to $N \sim 10^2$ (as representative of typical open or loosely bound clusters). The difficulty in handling these systems arises not only because N is large. It arises also because, paradoxically, N is small and because the systems of interest are not spatially bounded. The smallness of N (even $N \sim 10^{13}$!) and the lack of rigid retaining boundaries render inadequate those procedures that are based on the assumption of classical statistical equilibrium. It is therefore necessary to follow exactly the motion of individual stars, something that is effectively impossible to do analytically for any $N > 2$. In practice, even with a modern computer, systems with N much larger than 100 can be handled only by adopting mean field and quantization approximations. With $N \sim 100$, a direct integration through a complete period requires roughly one hour of computing time on our standard computer and to follow a system through one relaxation period (the time to reach configurations significantly different from the original) requires roughly 4 hours. The compution time required per relaxation-time varies roughly as N^2, making it clear that direct integration becomes thoroughly impracticable once $N \gtrsim 1000$.

Several features have emerged from calculations for systems with $N \sim 100$ that either were not known before computer solutions were obtained or are at variance with earlier estimates based on analytic approximations and statistical equilibrium assumptions. (a) The escape (or evaporation) rate from a small cluster is several orders of magnitude lower than that given by pre-computer estimates. (b) The escape rate depends strongly on the mass spectrum adopted (pre-computer estimates were of necessity confined to systems of equal-mass stars). (c) Equipartition in the classical sense is never reached—the distribution approaches something intermediate between equal velocity and equal energy, in the sense that more massive stars tend to have larger energies. (d) The star distribution tends towards a very high degree of central condensation—a dense core and a very extended envelope or Halo. (e) Steady state is never reached—the central concentration builds up without limit.

The analysis of systems for which $N \sim 10^{11}$ is just now getting underway, a major effort being devoted to the invention of approximation schemes to circumvent the practical impossibility of solving simultaneously, by direct integration, the classical equations of motion for all N mass points. There are indications that eventually we may be able to understand the reasons for the development of galactic spiral arms and central bars in terms of systems of gas clouds that form stars when their mean density exceeds a critical value.

2 Opacity and equation of state

Opacity relates the temperature gradient at any point in a star to the radiant flux at that point. In order to estimate the opacity for any given temperature, density, and composition, it is necessary to solve several subproblems, each of which involves many operations. First, the atomic physics must be worked out to determine the energy level structure of all atomic configurations likely to contribute to photon absorption in an environment that is in many respects more complex than that occurring in terrestial solids. Next, the probability that various levels are occupied must be worked out. Then the cross sections for scattering and absorption by each atomic configuration must be estimated as a function of photon frequency. Finally, the opacity is obtained by integrating the sum of all products of the form (cross-section x density-of-absorption-levels) over the frequency distribution of photons. In general, the level structure for any specified composition, density, and temperature is sufficiently uncertain that it is necessary to reshuffle the level ordering and redo the integration a statistically significant number of times before obtaining the most probable estimate of the opacity.

Typically, several minutes of computation on our standard computer are required to obtain the opacity for a given composition at any given temperature and density. Since opacity must be known at perhaps 200–400 temperature and density points to cover adequately the range in state variables encountered in just main sequence stars, on the order of 5–10 hours of computation are necessary to obtain just one opacity table suitable for calculating the structure of initial main sequence stars of just one composition. Since composition varies with time in a given star and since composition at birth varies from one star to another, as many as 20 opacity tables ($\sim$150–200 hours of computation) may be needed as input for a survey calculation of stellar evolution through hydrogen and helium burning stages.

It is interesting that the absorption of photons by processes that lead to transitions between bound atomic levels was not included in the calculation of opacity tables prior to the advent of fast electronic computers. In fact, several papers were published suggesting that such "bound-bound" transitions would make a minor contribution to the opacity in "all cases of interest". The first machine-aided calculations which included the contribution of bound–bound transitions in a systematic way showed that, in stellar envelope regions, the opacity is increased by factors of two to four over that obtained when bound–bound transitions are neglected.

3 Stellar pulsations

There are phases in nearly every star's life during which large-amplitude envelope pulsation is a prominent characteristic. The RR Lyrae stars with periods on the order of a few hours and the Cepheids with periods on the order of one to ten days are classical examples.

To handle pulsation phases theoretically one must include the dynamic term in the set of structure equations and limit the maximum time-step size to perhaps one twentieth of the time for one complete oscillation. The first question to answer is whether or not the pulsation artificially excited in the initial theoretical model will persist. That is, will the pulsation amplitude approach a limiting amplitude asymptotically or will the initial pulsation die away? An answer to this question requires that the pulsation be followed through many periods, the number of periods depending on the *growth* (or decay) *rate* of the amplitude. For example, it might be necessary to follow the model behavior through 10–100 periods before being convinced that pulsations will persist or fade away after millions of oscillations. Typically, 10 to 20 minutes or so of computations on the standard computer are required to answer the stability question for a single model.

Comparison with the observations on a given class of pulsating stars then demands that stability be investigated for many theoretical models characterized by different masses, power outputs, and compositions.

Finally, a more detailed comparison with the light curve and surface velocity variations of just one real star requires that several theoretical models of the appropriate period be followed until limiting amplitude is achieved. This implies perhaps 1–2 hours of computation for each of 4 or 5 models. Meaningful comparisons with representatives of just several classes of real variables can thus require hundreds of hours of computation for a given choice of input physics. As the input physics is improved additional models must be computed.

One of the many fruits of the computer-aided calculations on pulsating stars is the discovery that the oldest and most metal-poor stars in our Galaxy—the RR Lyrae stars—are characterized by a large initial helium abundance that compares favorably with the helium abundance generated in big-bang models of the early universe and with the helium abundance estimated for the oldest stars by using results of evolutionary calculations. The proximity of the three estimates of helium abundance provides strong support for the big-bang picture.

Recently, the pulsation properties of models of white dwarf stars and of neutron stars have been extensively pursued in an attempt to gauge the likelihood that pulsars might be vibrating condensed stars. By comparing model periods with pulsar periods, it can be concluded that pulsars are not vibrating white dwarfs or neutron stars, but the possibility that they could be rotating neutron stars is not excluded.

4 Implosion-explosion problems

The most spectacular stellar phenomena—supernovae, novae, planetary nebulae, and the like—are in many ways the most difficult to handle theoretically, both in principle and in practice. To all of the complexities of advanced quasistatic evolution are added dynamic motions, phase transitions, shock waves, and mass loss. Thus, memory requirements for comparable accuracy can be much larger for dynamic calculations than for quasi-static calculations.

The basic limitation on time-step size is the time for sound to cross the smallest shell in the stellar model, and small shells cannot be avoided because of the rapid variations with position of several variables in the interior. The duration of many explosive phenomena exceeds the largest permissible time-step size by many orders of magnitude. Hence many, many models are necessary to follow an explosion through its entire duration. Since the proper treatment of much of the pertinent physics is not well known, each evolutionary calculation must be repeated many times for different choices of a number of free parameters in the input physics.

The net result is that a satisfactory understanding of spectacular dynamic phases on even the qualitative level has not yet been attained, although there is some evidence that neutron stars may be the remnant of the common supernova explosion. Much of what we do know has been the result of calculations with large scale computers. Progress will be considerably accelerated with the advent of computers with speed and memory capacity that are several order of magnitude greater than those of the standard computer chosen for reference.

5 Many-element nucleosynthesis

Attempts to understand the distribution of elements in the universe and in our solar system lead to situations in which it is necessary to follow in detail nuclear transmutations involving hundreds of elements and thousands of nuclear reactions simultaneously. For example, one might wish to know the complete distribution of elements emerging from a postulated initial fireball or appearing in the wake of a shock wave passing through a supernova model.

With our standard computer it is not practicable to attempt such detailed studies at each point in a model universe or in a model star. Instead, one must estimate the variation of density and temperature with time at some hopefully typical position in a crude model that has been constructed using input physics from which most of the nuclear reaction details have been omitted. The problem then "reduces" to an examination of element transmutations at the typical position in the system of interest.

The eventual outcome of calculations of this type will be an understanding of where the elements that make up our bodies were originally synthesized—whether in an initial fireball, an exploding supernova, or in a less spectacular way in stars that regurgitate their accumulation of the heavy elements by means of quiet and steady mass loss.

4 A SUGGESTION FOR THE FUTURE

There is no need to elaborate on the desirability of developing larger and faster computers. More complex problems can be tackled and those currently requiring hundreds of hours for solution can be solved in perhaps a few hours. Areas of investigation that will benefit by an increase in both size and speed by several orders of magnitude include: (1) The study of stellar pulsation in massive stars where growth times exceed the pulsation period by factors ranging from 10^6–10^9 (2) complex nucleosynthesis in realistic models of supernovae (3) The study of rotating stars, taking large scale currents properly into account (4) The dynamics of stellar systems consisting of a collection of evolving stars that lose mass, collide, and explode.

There are finite limitations to the size and speed attainable with computers of conventional design. These are both economic and physical. For example, adding more memory banks to systems designed on the standard pattern of serial logic means an increase in mean access time since additional memory elements must be placed physically further from the central processer. The ultimate speed in communication among component parts is, of course, light volocity. The more closely such a speed is approached in practice, the more expensive the equipment. One might guess that an increase by several orders of magnitude in speed and memory capacity relative to our standard computer is the practical limit for computers of conventional design. For most astrophysical problems, an increase in *effective* speed by several more orders of magnitude could be achieved by implementing the concept of *parallel* logic. The majority of problems in astrophysics are amenable to separation into many component parts that could in principle be solved simultaneously. With the conventional computer that operates serially, only one part at a time can be tackled. The computer based on parallel logic would consist of one central computer to which perhaps 100–1000 relatively small, slow (and cheap) self-contained computers were subservient. Even though each subservient computer might be only one tenth as fast as the central processor, 100 identical operations could be completed in one tenth the time required by one computer operating serially at the speed of the central processor.

A more graphic description of the advantages of using parallel logic has been described in a previous section. It is worth reiterating. One professor with 100 graduate students (each of whom has a memory capacity equal to that of the professor, but, due to educational and family commitments, works only one-tenth as fast as the professor) can achieve the solution to a given problem in one tenth the time required by the professor working alone and can do so at less net cost. Further, thanks to the larger memory capacity of the parallel team, problems can be tackled that the professor simply could not have attempted alone.

Acknowledgements

It is a pleasure to thank Myron Lecar of the Smithsonian Astrophysical Obrervatory for aquainting me with the difficulties and the accomplishments in the solution of the many-body problem in astrophysics and to Charles McClure of the MIT Computational Center for acquainting me with the time scales and limitations of the standard computer and with the advantages of incorporating the idea of parallel logic into the design of further scientifically-oriented computers.

Subject Index

Author Index

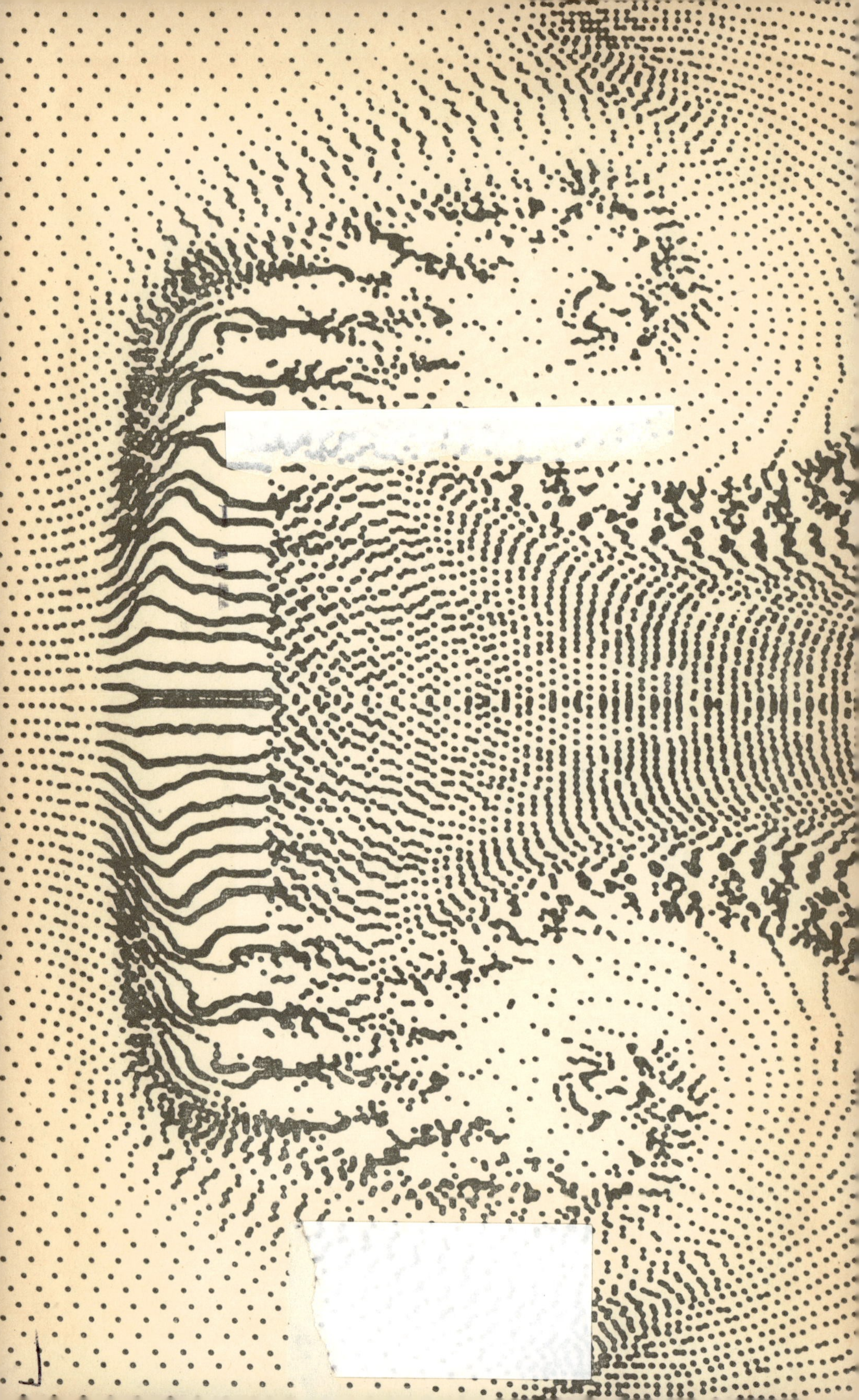